Pi: The Next Generation

David H. Bailey • Jonathan M. Borwein

Pi: The Next Generation

A Sourcebook on the Recent History of Pi and Its Computation

 Springer

David H. Bailey
Lawrence Berkeley National Laboratory
Berkeley, CA, USA

Jonathan M. Borwein
Centre for Computer Assisted Research
Mathematics and its Applications,
  School of Mathematical and Physical Sciences
University of Newcastle
Newcastle, Australia

ISBN 978-3-319-81270-0   ISBN 978-3-319-32377-0   (eBook)
DOI 10.1007/978-3-319-32377-0

Mathematics Subject Classification (2010): 01-00, 01-08, 01A05, 01A75, 11-00, 11-03, 26-03, 65-03, 68-03

© Springer International Publishing Switzerland 2016
Softcover reprint of the hardcover 1st edition 2016
This work is subject to copyright. All rights are reserved by the Publisher, whether the whole or part of the material is concerned, specifically the rights of translation, reprinting, reuse of illustrations, recitation, broadcasting, reproduction on microfilms or in any other physical way, and transmission or information storage and retrieval, electronic adaptation, computer software, or by similar or dissimilar methodology now known or hereafter developed.
The use of general descriptive names, registered names, trademarks, service marks, etc. in this publication does not imply, even in the absence of a specific statement, that such names are exempt from the relevant protective laws and regulations and therefore free for general use.
The publisher, the authors and the editors are safe to assume that the advice and information in this book are believed to be true and accurate at the date of publication. Neither the publisher nor the authors or the editors give a warranty, express or implied, with respect to the material contained herein or for any errors or omissions that may have been made.

Printed on acid-free paper

This Springer imprint is published by Springer Nature
The registered company is Springer International Publishing AG Switzerland

To Our Grandchildren

Contents

Foreword

The fundamental constant π has played an indispensable role in mathematics, science, and most world cultures for over 4000 years. One would like to calculate the circumference of a circle without actually physically measuring it. The ancient Babylonians and Hebrews recognized that it was considerably easier to determine the radius of a circle than its circumference, and so the more physically demanding task of calculating the circumference could be replaced by simply multiplying the radius by 3. However, before 2000 BCE, the Babylonians replaced this initial approximation of 3 for π by $3\frac{1}{8}$. On the Egyptian Rhind Papyrus from about 1650 BCE, we find the slightly better approximation $4(8/9)^2 = 3.16049\ldots$. We have come a long way since the heydays of the Babylonian, Hebrew, and Egyptian cultures. With the calculation of "houkouonchi" and Alexander J. Yee, we now "know" 13.3 trillion digits of π. It might be pointed out that π was not always denoted by π; William Jones first used the notation π in 1706. Euler adopted the symbol by 1737, and since that time, the notation π has been universally used.

The value π arises in many contexts in mathematics in surprising ways. One of these is the *Buffon needle problem*, which asks for the probability that a needle of length ℓ will land on one of the vertical, equally spaced parallel lines on a floor, where the distance between each two adjacent lines is equal to d. If $d = \ell$, this probability is $2/\pi$. Any calculus student will testify that the value of π often arises in the calculation of a definite integral or of an infinite series in closed form. The great Indian mathematician, Srinivasa Ramanujan, evaluated a plethora of integrals and infinite series in closed form, and he also found many beautiful series identities. Chapter 14 in his second notebook contains 87 such results, and of these series and integral evaluations in closed form and series identities, I counted 70 of them in which π appears.

Unusual or surprising formulas for π are a source of delight for many of us. When we calculate the simple continued fraction

$$\pi = 3 + \frac{1}{7} + \frac{1}{15} + \frac{1}{1} + \frac{1}{293} + \cdots,$$

of π, we are perhaps surprised by the "large" denominator 293. This indicates that if we truncate the continued fraction just prior to this large denominator, we should obtain a good approximation to π. Indeed, the associated approximation $\frac{355}{113}$ gives the first six digits of π. On the other hand, examining the simple continued fraction

$$\pi^4 = 97 + \frac{1}{2} + \frac{1}{2} + \frac{1}{3} + \frac{1}{1} + \frac{1}{16539} + \cdots,$$

for π^4, we find the "huge" denominator 16539. This leads to the approximations

$$\pi^4 \approx 97\frac{9}{22} = 97.40909090909\ldots, \quad \pi \approx 3.14159265262\ldots,$$

while

$$\pi^4 = 97.409091034002\ldots, \quad \pi = 3.14159265358\ldots.$$

Infinite series, perhaps beginning with the familiar Madhava–Gregory series for Arctan x, have played important roles in calculating the digits of π. One of the most beautiful series representations for π is given by

$$\frac{9801}{\pi\sqrt{8}} = \sum_{n=0}^{\infty} \frac{(4n)!(1103 + 26390n)}{(n!)^4(396)^{4n}}, \tag{1}$$

stated without proof along with 16 further such series for $1/\pi$ by Ramanujan in his famous paper, *Modular equations and approximations to π*. In November, 1985, R. William Gosper, Jr., employed (1) to calculate 17,526,100 digits of π, which was a world record at that time. Each term of (1) gives about 8 digits of π per term. Perhaps we should have put quotation marks about *world record*, because at that time no proof of (1) had ever been given. We now have proofs of (1), Ramanujan's 16 further series formulas for $1/\pi$, and many similar series representations for $1/\pi$, in particular, one due to David and Gregory Chudnovsky that yields 14 digits of π per term. Papers numbered 10, 11, and 19 in this volume provide readers with more details about these formulas and their use for computing digits of π.

There are many further intriguing formulas for π. One of the most popular is John Wallis's formula (1655)

$$\pi = 2\,\frac{2}{1}\,\frac{2}{3}\,\frac{4}{3}\,\frac{4}{5}\,\frac{6}{5}\,\frac{6}{7}\cdots.$$

My favorite is Lord Brouncker's (1620–1684) continued fraction for π,

$$\pi = \frac{4}{1}\,+\,\frac{1^2}{2}\,+\,\frac{3^2}{2}\,+\,\frac{5^2}{2}\,+\,\frac{7^2}{2}\,+\cdots. \tag{2}$$

Brouncker was a student of Wallis, and mathematicians have speculated for many years on how Brouncker might have derived his formula (2). It is actually a special case of a more general continued fraction

$$\frac{\Gamma(\frac{1}{4}(x+n+1))\Gamma(\frac{1}{4}(x-n+1))}{\Gamma(\frac{1}{4}(x+n+3))\Gamma(\frac{1}{4}(x-n+3))} = \frac{4}{x}\,-\,\frac{n^2-1^2}{2x}\,-\,\frac{n^2-3^2}{2x}\,-\,\frac{n^2-5^2}{2x}\,-\cdots, \tag{3}$$

which can be found as Entry 25 in Chapter 12 of Ramanujan's second notebook. If we set $x = 1$ and $n = 0$ in (3), we obtain (2).

Seemingly strange approximations to π can be obtained from *class invariants*. Let n be a positive integer, and set $q = e^{-\pi\sqrt{n}}$. The class invariant G_n is defined by

$$G_n := 2^{-1/4}q^{-1/24}(1+q)(1+q^3)(1+q^5)\cdots. \tag{4}$$

Class invariants are algebraic numbers. Thus, taking the logarithm of both sides of (4) for a certain class invariant should produce an approximate formula for π. For example, on page 300 in his second notebook, Ramanujan considered the cubic equation

$$2x^3 - 4x^2 + 6x - 1 = 0,$$

which has the real root

$$x_0 = 0.18801840607370118198183676 2895\ldots.$$

Note that the first 16 digits of

$$\frac{-24\log x_0}{\sqrt{163}} = 3.141592653589793245\ldots$$

agree with the first 16 digits of

$$\pi = 3.141592653589793238\ldots.$$

It may come as a surprise to readers that the resurgence in calculating the digits of π arose from an adaptation of the arithmetic-geometric mean of Gauss by (individually) Eugene Salamin and Richard Brent. Readers can learn about these developments in the first five papers of this volume.

Although we know a lot of digits of π, there is much that we do not know about these digits. For example, we would conjecture that the average digit should be about 4.5, but we do not know how to prove this. More precisely, it is conjectured that π is *normal*, but evidently we are far from proving this as well. This volume contains papers by Stan Wagon (no. 6); by the present authors (no. 17); by Francisco Aragon Artacho, the present authors, and Peter Borwein (no. 21); and by the present authors (no. 23) bringing readers up to date on what we know about this famous conjecture.

We have provided only a small sample of examples to illustrate the beauty and mystery of this remarkable number. Mathematicians and computer scientists continue to delve into the mysteries of π. Our goal in the present volume is to bring together some of these investigations and thoughts about π from papers published during the past half century. Many of these papers can be read by a broad audience. For readers who want to read further, most of the 25 selected papers point to more technical articles.

Bruce C. Berndt

Preface

This volume is a companion to *Pi: A Source Book* (by Lennart Berggren, Jonathan Borwein, and Peter Borwein, Springer-Verlag), which was first published in 1997, with a third edition released in 2004.

Rather than produce an even heftier fourth edition, the current authors have prepared a collection of papers written between 1975 and the present. Since a number of the collected papers contain substantial historical material, the reader can glean an accurate picture of the life of π from the current volume. That said, the focus in this book is on "π in the digital age." The reader will note that many of the papers have substantial algorithmic material and it is recommended that where possible he or she explore such material at the computer.

Each of the 25 papers comprising this volume is preceded by a brief summary of its contents, and this is accompanied by a very brief, key word, indication of some of the ways the content of the given paper relates to that of others in the collection. This information is also recorded in Table 1 below.

For the most part, however, we are happy to let the papers speak for themselves. The present authors have been fascinated by π throughout their academic lives and hope that this volume will help readers share this fascination and potentially even contribute to the ever-growing literature on the subject. The associated webpage for the collection is accessible through `http://www.springer.com/us/book/9783319323756`.

We are also delighted that Bruce Berndt agreed to add a Foreword to the volume.

Algorithms	1, 2, 4, 5, 8, 10, 11, 13, 14, 16, 17, 22, 23, 24, 25
Arithmetic-geometric mean	1, 2, 3, 4, 5, 8
Computation	1, 2, 4, 6, 7, 8, 9, 11, 14, 17, 19, 20, 22, 24, 25
Curiosities	12, 21, 22
Elliptic integrals	1, 2, 3, 8, 11
General audience	6, 10, 23, 24
Graphical representation	21
History	3, 4, 8, 10, 11, 17, 19, 20, 22, 24, 25
Approximations	2, 4, 18
Irrationality	15, 24, 25
Modular equations	10, 11
Normality	6, 7, 9, 17, 20, 21, 23
Random walks	21
Series	12, 19, 22

TABLE 1. Articles by Keyword

We wish to thank visiting Waterloo students Sally Dong and Jason Lynch for their help, Dr. Daniel Sutherland for his substantial assistance, and our Springer-Verlag editor Elizabeth Loew for her active encouragement to produce this volume.

Finally, we thank the original publishers for their permissions to republish the articles herein. In particular, we would like to thank:

- ACM, for permission to reprint paper 2
- AMS, for permission to reprint papers 1, 5, 7, 14, and 20
- ICMI, for permission to reprint paper 3
- IEEE, for permission to reprint paper 9
- MAA, for permission to reprint 8, 11, 12, 13, 15, 16, 18, 19, 23, and 25
- SIAM, for permission to reprint paper 4
- Springer-Verlag, for permission to reprint papers 6, 21, and 24
- Taylor & Francis, for permission to reprint entry 17

Berkeley, CA, USA   David H. Bailey
Newcastle, NSW, Australia   Jonathan M. Borwein
March 2016

1. Computation of π using arithmetic-geometric mean (1976)

Paper 1: Eugene Salamin, "Computation of pi using arithmetic-geometric mean," *Mathematics of Computation*, vol. 30, no. 135 (July 1976), pg. 565–570. Reprinted by permission of the American Mathematical Society.

Synopsis:

In 1976, Eugene Salamin and Richard Brent independently discovered two equivalent "quadratically convergent" algorithms for computing π, meaning that each iteration of the algorithm approximately *doubles* the number of correct digits in the result, provided each iteration is performed with a level of numeric precision that is desired for the final result. This remarkable co-discovery arguably launched the modern computer era of the computation of π.

Salamin's statement of the algorithm could hardly be more concise:

$$\pi = \frac{4\mathrm{agm}^2(1,\sqrt{2})}{1 - \sum_{j=1}^{\infty} 2^{j+1} c_j^2}, \tag{5}$$

where $\mathrm{agm}(a,b)$ is the arithmetic-geometric mean (AGM) iteration:

$$a_n = \frac{a_{n-1} + b_{n-1}}{2} \tag{6}$$

$$b_n = \sqrt{a_{n-1}b_{n-1}} \tag{7}$$

with a_0 and b_0 as the two input arguments, and $c_n^2 = a_n^2 - b_n^2$.

However, this is just a special case of his Theorem 1a, which is that

$$\pi = \frac{4\mathrm{agm}(1,k)\mathrm{agm}(1,k')}{1 - \sum_{j=1}^{\infty} 2^j (c_j^2 + c_j'^2)}, \tag{8}$$

for any k and k' satisfy $k^2 + k'^2 = 1$.

Salamin does not report any computation in his paper. He only says that some colleagues are attempting to implement this algorithm on the Illiac IV computer, one of the first parallel processors, to 33 million bit (roughly 10 million digit) precision. The present authors are not aware of any report of this calculation, if it was successful.

Keywords: Algorithms, Arithmetic-Geometric Mean, Computation, Elliptic Integrals

© Springer International Publishing Switzerland 2016
D.H. Bailey, J.M. Borwein, *Pi: The Next Generation*,
DOI 10.1007/978-3-319-32377-0_1

MATHEMATICS OF COMPUTATION, VOLUME 30, NUMBER 135
JULY 1976, PAGES 565–570

Computation of π Using Arithmetic-Geometric Mean

By Eugene Salamin

Abstract. A new formula for π is derived. It is a direct consequence of Gauss'
arithmetic-geometric mean, the traditional method for calculating elliptic integrals,
and of Legendre's relation for elliptic integrals. The error analysis shows that its
rapid convergence doubles the number of significant digits after each step. The
new formula is proposed for use in a numerical computation of π, but no actual
computational results are reported here.

1. **Introduction.** This paper announces the discovery of a new formula for π. It is based upon the arithmetic-geometric mean, a process whose rapid convergence doubles the number of significant digits at each step. The arithmetic-geometric mean, together with π as a known quantity, is the basis of Gauss' method for the calculation of elliptic integrals. But with the help of an elliptic integral relation of Legendre, Gauss' method can be turned around to express π in terms of the arithmetic-geometric mean. The resulting algorithm retains the property of doubling the number of digits at each step.

The proof of the main result (Theorem 1a) from first principles can be conducted on the elementary calculus level. The references cited here for the theorems of Landen, Gauss and Legendre have been chosen to achieve this goal, thus allowing the widest possible reader audience comprehension.

The formula presented in this paper is proposed as a method for the numerical computation of π. It has not yet been tested on a calculation of nontrivial length, although such a calculation is currently in progress [2].

2. **The Arithmetic-Geometric Mean.** Let a_0, b_0, c_0 be positive numbers satisfying $a_0^2 = b_0^2 + c_0^2$. Define a_n, the sequence of arithmetic means, and b_n, the sequence of geometric means, by

$$a_n = \tfrac{1}{2}(a_{n-1} + b_{n-1}), \qquad b_n = (a_{n-1}b_{n-1})^{1/2}.$$

Also, define a positive sequence c_n:

$$c_n^2 = a_n^2 - b_n^2.$$

Two relations easily follow from these definitions.

(1) $$c_n = \tfrac{1}{2}(a_{n-1} - b_{n-1}).$$

(2) $$c_n^2 = 4a_{n+1}c_{n+1}.$$

Received May 27, 1975; revised November 3, 1975.
AMS (MOS) subject classifications (1970). Primary 10A30, 10A40, 33A25; Secondary 41A25.
Key words and phrases. π, arithmetic-geometric mean, elliptic integral, Landen's transformation, Legendre's relation, fast Fourier transform multiplication.

The arithmetic-geometric mean is the common limit

$$\text{agm}(a_0, b_0) = \lim a_n = \lim b_n.$$

Because of the rapidity of convergence of the arithmetic-geometric mean, as exhibited by Eq. (2), the formula to be derived should be regarded as a plausible candidate for the numerical computation of π.

3. Elliptic Integrals. The complete elliptic integrals are the functions

$$K(k) = \int_0^{\pi/2} (1 - k^2 \sin^2 t)^{-\frac{1}{2}} \, dt, \qquad E(k) = \int_0^{\pi/2} (1 - k^2 \sin^2 t)^{\frac{1}{2}} \, dt.$$

Also, if $k^2 + k'^2 = 1$, then $K'(k) = K(k')$ and $E'(k) = E(k')$ are two more elliptic integrals.

There is also a symmetric form of these integrals:

$$I(a, b) = \int_0^{\pi/2} (a^2 \cos^2 t + b^2 \sin^2 t)^{-\frac{1}{2}} \, dt,$$

$$J(a, b) = \int_0^{\pi/2} (a^2 \cos^2 t + b^2 \sin^2 t)^{\frac{1}{2}} \, dt.$$

It is clear that

$$I(a, b) = a^{-1} K'(b/a), \qquad J(a, b) = a E'(b/a).$$

4. Landen's Transformation and the Computation of Elliptic Integrals. Using the notation developed in Section 2 of this paper, these transformations are [6, Section 25.15],

(3) $$I(a_n, b_n) = I(a_{n+1}, b_{n+1}),$$

(4) $$J(a_n, b_n) = 2J(a_{n+1}, b_{n+1}) - a_n b_n I(a_{n+1}, b_{n+1}).$$

From Eq. (3) it follows that

(5) $$I(a_0, b_0) = \pi/2 \, \text{agm}(a_0, b_0),$$

and, after a little work, Eq. (4) yields

(6) $$J(a_0, b_0) = \left(a_0^2 - \frac{1}{2} \sum_{j=0}^{\infty} 2^j c_j^2 \right) I(a_0, b_0).$$

For $a_0 = 1, b_0 = k'$, the integrals in Eqs. (5) and (6) are equal to $K(k)$ and $E(k)$, respectively, while for $a_0 = 1, b_0 = k$, they equal $K'(k)$ and $E'(k)$. This is the well-known method of Gauss for the numerical calculation of elliptic integrals [5, pp. 78–80], [1, Section 17.6].

5. Legendre's Relation. This relation is [4, Article 171], [1, Eq. 17.3.13],

(7) $$K(k)E'(k) + K'(k)E(k) - K(k)K'(k) = \pi/2.$$

Equivalently,

(8) $$a^2 I(a, b)J(a', b') + a'^2 I(a', b')J(a, b) - a^2 a'^2 I(a, b)I(a', b') = (\pi/2)aa',$$

where a, b, a', b' are subject to the restriction $(b/a)^2 + (b'/a')^2 = 1$.

6. New Expression for π. Take $a_0 = a_0' = 1$, $b_0 = k$, $b_0' = k'$. As in Section 2, define the sequences a_n, b_n, c_n, a_n', b_n', c_n'. In Eq. (8) eliminate the J integrals by use of (6), and then eliminate the I integrals by use of (5). Lo and behold, the resulting equation can be solved for π!

THEOREM 1a.

(9)
$$\pi = \frac{4\,\mathrm{agm}(1, k)\,\mathrm{agm}(1, k')}{1 - \sum_{j=1}^{\infty} 2^j (c_j^2 + c_j'^2)}.$$

The $j = 0$ term in the summation has been eliminated by use of $c_0^2 + c_0'^2 = k'^2 + k^2 = 1$. It is best to compute c_j from Eq. (1).

Theorem 1a is a one-dimensional continuum of formulae for π. This provides for an elegant and simple computational check. For example, π could be calculated starting with $k = k' = 2^{-1/2}$, and then checked using $k = 4/5$, $k' = 3/5$. The symmetric choice, $k = k'$, causes the two agm sequences to coincide, thus halving the computational burden.

THEOREM 1b.

$$\pi = \frac{4(\mathrm{agm}(1, 2^{-1/2}))^2}{1 - \sum_{j=1}^{\infty} 2^{j+1} c_j^2}.$$

7. Error Analysis. Although Theorem 1a is true for all complex values of k (except for a discrete set), the error analysis will assume real k and k'. Then $0 < k, k' < 1$. Let n square roots be taken in the process of computing agm $= \mathrm{agm}(1, k)$, and n' square roots in computing $\mathrm{agm}' = \mathrm{agm}(1, k')$. Then no further square roots are needed to calculate the approximation

(10)
$$\pi_{nn'} = \frac{4 a_{n+1} a_{n'+1}'}{1 - \sum_{j=1}^{n} 2^j c_j^2 - \sum_{j=1}^{n'} 2^j c_j'^2}.$$

A rough estimate shows that a_{n+1} differs from agm by c_{n+2}, and that the finite sum differs from the infinite sum by $2^{n+3} c_{n+2}$. Thus, the numerator and denominator in (10) have been truncated for compatible error contributions, and the denominator error is dominant.

To obtain rigorous error bounds, introduce the auxiliary quantity $\bar{\pi}_{nn'}$ whose denominator is taken from (10), but whose numerator is taken from (9). The first step is to establish the existence of $e_{nn'}$, $\bar{e}_{nn'}$ such that

(11)
$$0 < \pi - \bar{\pi}_{nn'} < e_{nn'},$$

(12)
$$0 < \pi_{nn'} - \bar{\pi}_{nn'} < \bar{e}_{nn'},$$
$$\bar{e}_{nn'} < e_{nn'}.$$

These three inequalities imply that $|\pi - \pi_{nn'}| < e_{nn'}$.

The left-hand inequalities in (11) and (12) are obvious. From the general inequality $(1/x) - (1/(x + y)) < y/x^2$, valid for positive x and y, it follows that

$$\pi - \bar{\pi}_{nn'} < \frac{\pi^2}{4\,\mathrm{agm}\,\mathrm{agm}'} \left(\sum_{n+1}^{\infty} 2^j c_j^2 + \sum_{n'+1}^{\infty} 2^j c_j'^2 \right).$$

This establishes (11), with error bound

$$(13) \qquad e_{nn'} = \frac{\pi^2}{2 \text{ agm agm}'} \left(\sum_{n+2}^{\infty} 2^j a_j c_j + \sum_{n'+2}^{\infty} 2^j a_j' c_j' \right).$$

Proceeding to the next inequality, we first get

$$\pi_{nn'} - \overline{\pi}_{nn'} < \frac{\pi}{\text{agm agm}'} (a_{n+1} a_{n'+1}' - \text{agm agm}').$$

Substitute $a_{n+1} = \text{agm} + s$, $a_{n'+1}' = \text{agm}' + s'$, where

$$s = \sum_{n+2}^{\infty} c_j, \qquad s' = \sum_{n'+2}^{\infty} c_j',$$

and use agm < 1, agm$' < 1$ to get

$$\pi_{nn'} - \overline{\pi}_{nn'} < \pi(s + s' + ss')/\text{agm agm}'.$$

Also, since $s < 1$, $s' < 1$, it follows that $ss' < (s + s')/2$. Thus, inequality (12) is established with error bound

$$(14) \qquad \overline{e}_{nn'} = \frac{3}{2} \frac{\pi}{\text{agm agm}'} \left(\sum_{n+2}^{\infty} c_j + \sum_{n'+2}^{\infty} c_j' \right).$$

Finally, a term-by-term comparison of (13) and (14), using $2^j a_j > 1$ and $\pi > 3$, shows that $\overline{e}_{nn'} < e_{nn'}$.

At this point, a needed inequality is derived.

$$a_j < a_j + b_j = 2a_{j+1},$$

$$2a_j c_{j+1} < 4a_{j+1} c_{j+1} = c_j^2 = (a_{j-1} - a_j)c_j,$$

$$(15) \qquad a_j(c_j + 2c_{j+1}) < a_{j-1}c_j.$$

Consider the first summation in (13), but with the upper limit ∞ replaced by finite N. Perform the following sequence of operations, each of which increases the sum. First, replace a_N by a_{N-1}. Next, repeatedly apply (15) to the pair of highest-indexed terms in the sum. At the end, we are left with the single term $2^{n+2}a_{n+1}c_{n+2} < 2^{n+2}c_{n+2}$, which is thus an upper bound for the initial summation. Since N was arbitrary, the infinite summation also has this upper bound. Therefore,

$$(16) \qquad e_{nn'} < \frac{2\pi^2}{\text{agm agm}'} (2^n c_{n+2} + 2^{n'} c_{n'+2}').$$

An upper bound for c_{n+2} is needed now. It is convenient to use the abbreviations

$$x_n = \log c_n, \qquad g_n = \log(4a_n).$$

Equation (2) gives x_n as the solution to an inhomogeneous linear difference equation.

$$x_n = 2^n \left(x_0 - \sum_{j=1}^{n} 2^{-j} g_j \right).$$

By rearrangement,

$$x_n = 2^n \left(x_0 - g_1 + \sum_{j=1}^{n-1} 2^{-j}(g_j - g_{j+1}) \right) + g_n.$$

COMPUTATION OF π USING ARITHMETIC-GEOMETRIC MEAN   **569**

Using $g_j - g_{j+1} > 0$, $g_n < \log 4$, and $x_0 - g_1 = (1/2)\log(c_1/4a_1)$, we get

$$(17) \qquad x_n < 2^{n-1}\left[\sum_{j=1}^{\infty} 2^{-j+1}\log(a_j/a_{j+1}) - \log(4a_1/c_1)\right] + \log 4.$$

For the purpose of an error analysis, the expression within brackets could be calculated numerically for any case of interest. However, it can be evaluated in closed form [7, p. 14] and is equal to $-\pi K'(k)/K(k) = -\pi\,\mathrm{agm/agm}'$. Then

$$x_n < -\pi(\mathrm{agm/agm}')2^{n-1} + \log 4.$$

Substituting this into (16) yields the final result.

THEOREM 2a.

$$|\pi - \pi_{nn'}| < \frac{8\pi^2}{\mathrm{agm\ agm}'}\left[2^n \exp\left(-\pi\frac{\mathrm{agm}}{\mathrm{agm}'}2^{n+1}\right) + 2^{n'}\exp\left(-\pi\frac{\mathrm{agm}'}{\mathrm{agm}}2^{n'+1}\right)\right].$$

In the symmetric case, with $\pi_n = \pi_{nn}$, Theorem 2a simplifies to

THEOREM 2b.

$$|\pi - \pi_n| < (\pi^2 2^{n+4}/\mathrm{agm}^2)\exp(-\pi 2^{n+1}).$$

The number of valid decimal places is then

THEOREM 2c.

$$-\log_{10}|\pi - \pi_n| > (\pi/\log 10)2^{n+1} - n\log_{10}2 - 2\log_{10}(4\pi/\mathrm{agm}).$$

8. Numerical Computation. Raphael Finkel, Leo Guibas and Charles Simonyi are currently engaged in calculating π using the method proposed in this paper [2]. The operations of multiprecision division and square root are reduced to multiplication using a Newton's method iteration. The multiplications are then performed by the Schönhage-Strassen fast Fourier transform algorithm [10], [8, p. 274]. The computation, to be run on the Illiac IV computer, is expected to yield 33 million bits of π in an estimated run time of four hours. This run time is determined by disc input-output, and the actual computation is estimated to be only a couple of minutes. Alas, they do not plan to convert to decimal.

9. Concluding Remarks. The main result of this paper, Theorem 1a, directly follows from Gauss' method for calculating elliptic integrals, Eqs. (5) and (6), which was known in 1818 [3, pp. 352, 360], and from Legendre's elliptic integral relation, Eq. (7), which was known in 1811 [9, p. 61]. It is quite surprising that such an easily derived formula for π has apparently been overlooked for 155 years. The author made his discovery in December of 1973.

The series summation which was used to simplify Eq. (17) was also discovered by Gauss [3, p. 377]. An interesting consequence of this result of Gauss is that e^{π} can be expressed as a rapidly convergent infinite product. If $a_0 = 1$, $b_0 = 2^{-\frac{1}{2}}$, then

$$e^{\pi} = 32 \prod_{j=0}^{\infty} \left(\frac{a_{j+1}}{a_j}\right)^{2^{-j+1}}.$$

Charles Stark Draper Laboratory
Cambridge, Massachusetts 02139

570   EUGENE SALAMIN

   1. M. ABRAMOWITZ & I. A. STEGUN (Editors), *Handbook of Mathematical Functions, With Formulas, Graphs and Mathematical Tables*, Nat. Bur. Standards, Appl. Math. Ser., no. 55, Superintendent of Documents, U.S. Government Printing Office, Washington, D.C., 1966. MR 34 #8607.

   2. R. FINKEL, L. GUIBAS & C. SIMONYI, private communication.

   3. K. F. GAUSS, *Werke*, Bd. 3, Gottingen, 1866, pp. 331–403.

   4. A. G. GREENHILL, *The Applications of Elliptic Functions*, Dover, New York, 1959. MR 22 #2724.

   5. H. HANCOCK, *Elliptic Integrals*, Dover, New York, 1917.

   6. H. JEFFREYS & B. S. JEFFREYS, *Methods of Mathematical Physics*, 3rd ed., Cambridge Univ. Press, London, 1962.

   7. L. V. KING, *On the Direct Numerical Calculation of Elliptic Functions and Integrals*, Cambridge Univ. Press, London, 1924.

   8. D. KNUTH, *The Art of Computer Programming*. Vol. 2: *Seminumerical Algorithms*, Addison-Wesley, Reading, Mass., 1969.   MR 44 #3531.

   9. A. M. LEGENDRE, *Exercices de calcul intégral*. Vol. 1, 1811.

   10. A. SCHÖNHAGE & V. STRASSEN, "Schnelle Multiplikation grosser Zahlen," *Computing (Arch. Elektron. Rechnen)*, v. 7, 1971, pp. 281–292.   MR 45 #1431.

2. Fast multiple-precision evaluation of elementary functions (1976)

Paper 2: Richard P. Brent, "Fast multiple-precision evaluation of elementary functions," *Journal of the Association of Computing Machinery*, vol. 23 (1976), no. 11, pg. 713–735. ©1976 Association of Computing Machinery, Inc. Reprinted by permission. http://doi.acm.org/10.1145/321941.321944

Synopsis:

As mentioned in the preface to Paper 1, Eugene Salamin and Richard Brent independently discovered two equivalent "quadratically convergent" algorithms for computing π in 1976, and this remarkable co-discovery arguably launched the modern computer era of the computation of π.

Brent's paper includes an equivalent statement of the Salamin's π algorithm, but he goes further in showing how the same mathematical structure yields algorithms for evaluating all of the "elementary" algorithms, including exp, log, arctan, sin, cosh and others.

Brent also included a detailed analysis of the computational complexity of these algorithms, showing how the functions above can be evaluated, with relative error $O(2^{-n})$, in $O(M(n)\log(n))$ operations, where $M(n)$ is the cost of a multiplication. He then adds that in the wake of the Schonhage-Strassen result, namely that a multiplication can be done with a fast Fourier transform, that this means all the elementary functions can be evaluated to $O(2^{-n})$ relative error in $O(n\log^2(n)\log\log(n))$ operations.

Keywords: Algorithms, Approximations, Arithmetic-Geometric Mean, Computation, Elliptic Integrals

© Springer International Publishing Switzerland 2016
D.H. Bailey, J.M. Borwein, *Pi: The Next Generation*,
DOI 10.1007/978-3-319-32377-0_2

Fast Multiple-Precision Evaluation of Elementary Functions

RICHARD P. BRENT

Australian National University, Canberra, Australia

ABSTRACT.   Let $f(x)$ be one of the usual elementary functions (exp, log, artan, sin, cosh, etc.), and let $M(n)$ be the number of single-precision operations required to multiply n-bit integers. It is shown that $f(x)$ can be evaluated, with relative error $O(2^{-n})$, in $O(M(n)\log (n))$ operations as $n \to \infty$, for any floating-point number x (with an n-bit fraction) in a suitable finite interval. From the Schonhage-Strassen bound on $M(n)$, it follows that an n-bit approximation to $f(x)$ may be evaluated in $O(n \log^2(n) \log \log(n))$ operations. Special cases include the evaluation of constants such as π, e, and e^π. The algorithms depend on the theory of elliptic integrals, using the arithmetic-geometric mean iteration and ascending Landen transformations.

KEY WORDS AND PHRASES: multiple-precision arithmetic, analytic complexity, arithmetic-geometric mean, computational complexity, elementary function, elliptic integral, evaluation of π, exponential, Landen transformation, logarithm, trigonometric function

CR CATEGORIES: 5.12, 5.15, 5.25

1. *Introduction*

We consider the number of operations required to evaluate the elementary functions $\exp(x)$, $\log(x)$,[1] $\operatorname{artan}(x)$, $\sin(x)$, etc., with relative error $O(2^{-n})$, for x in some interval $[a, b]$, and large n. Here, $[a, b]$ is a fixed, nontrivial interval on which the relevant elementary function is defined. The results hold for computations performed on a multi-tape Turing machine, but to simplify the exposition we assume that a standard serial computer with a random-access memory is used.

Let $M(x)$ be the number of operations required to multiply two integers in the range $[0, 2^{|x|})$. We assume the number representation is such that addition can be performed in $O(M(n))$ operations, and that $M(n)$ satisfies the weak regularity condition

$$M(\alpha n) \leq \beta M(n), \tag{1.1}$$

for some α and β in $(0, 1)$, and all sufficiently large n. Similar, but stronger, conditions are usually assumed, either explicitly [11] or implicitly [15]. Our assumptions are certainly valid if the Schonhage-Strassen method [15, 19] is used to multiply n-bit integers (in the usual binary representation) in $O(n \log(n) \log \log(n))$ operations.

The elementary function evaluations may be performed entirely in fixed point, using integer arithmetic and some implicit scaling scheme. However, it is more convenient to assume that floating-point computation is used. For example, a sign and magnitude representation could be used, with a fixed length binary exponent and an n-bit binary fraction. Our results are independent of the particular floating-point number system used, so long as the following conditions are satisfied.

Author's address: Computer Centre, Australian National University, Box 4, Canberra, ACT 2600, Australia.

[1] Log(x) denotes the natural logarithm.

Journal of the Association for Computing Machinery, Vol. 23, No. 2, April 1976, pp. 242-251.

1. Real numbers which are not too large or small can be approximated by floating-point numbers, with a relative error $O(2^{-n})$.

2. Floating-point addition and multiplication can be performed in $O(M(n))$ operations, with a relative error $O(2^{-n})$ in the result.

3. The precision n is variable, and a floating-point number with precision n may be approximated, with relative error $O(2^{-m})$ and in $O(M(n))$ operations, by a floating-point number with precision m, for any positive $m < n$.

Throughout this paper, a *floating-point number* means a number in some representation satisfying conditions 1 to 3 above, not a single-precision number. We say that an operation is performed with *precision* n if the result is obtained with a relative error $O(2^{-n})$. It is assumed that the operands and result are approximated by floating-point numbers.

The main result of this paper, established in Sections 6 and 7, is that all the usual elementary functions may be evaluated, with precision n, in $O(M(n) \log(n))$ operations. Note that $O(M(n)n)$ operations are required if the Taylor series for $\log(1 + x)$ is summed in the obvious way. Our result improves the bound $O(M(n) \log^2(n))$ given in [4], although the algorithms described there may be faster for small n.

Preliminary results are given in Sections 2 to 5. In Section 2 we give, for completeness, the known result that division and extraction of square roots to precision n require $O(M(n))$ operations. Section 3 deals briefly with methods for approximating simple zeros of nonlinear equations to precision n, and some results from the theory of elliptic integrals are summarized in Section 4. Since our algorithms for elementary functions require a knowledge of π to precision n, we show, in Section 5, how this may be obtained in $O(M(n) \log(n))$ operations. An amusing consequence of the results of Section 6 is that e^π may also be evaluated, to precision n, in $O(M(n) \log(n))$ operations.

From [4, Th. 5.1], at least $O(M(n))$ operations are required to evaluate $\exp(x)$ or $\sin(x)$ to precision n. It is plausible to conjecture that $O(M(n) \log(n))$ operations are necessary.

Most of this paper is concerned with order of magnitude results, and multiplicative constants are ignored. In Section 8, though, we give upper bounds on the constants. From these bounds it is possible to estimate how large n needs to be before our algorithms are faster than the conventional ones.

After this paper was submitted for publication, Bill Gosper drew my attention to Salamin's paper [18], where an algorithm very similar to our algorithm for evaluating π is described. A fast algorithm for evaluating $\log(x)$ was also found independently by Salamin (see [2 or 5]).

Apparently similar algorithms for evaluating elementary functions are given by Borchardt [3], Carlson [8, 9], and Thacher [23]. However, these algorithms require $O(M(n)n)$ or $O(M(n)n^{\frac{1}{2}})$ operations, so our algorithms are asymptotically faster.

We know how to evaluate certain constants and functions almost as fast as elementary functions. For example, Euler's constant $\gamma = 0.5772 \ldots$ can be evaluated with $O(M(n) \log^2 n)$ operations, using Sweeney's method [22] combined with binary splitting [4]. Similarly for $\Gamma(a)$, where a is rational (or even algebraic): see Brent [7]. Related results are given by Gosper [13] and Schroeppel [20]. It is not known whether any of these upper bounds are asymptotically the best possible.

2. *Reciprocals and Square Roots*

In this section we show that reciprocals and square roots of floating-point numbers may be evaluated, to precision n, in $O(M(n))$ operations. To simplify the statement of the following lemma, we assume that $M(x) = 0$ for all $x < 1$.

LEMMA 2.1. *If $\gamma \in (0, 1)$, then $\sum_{j=0}^{\infty} M(\gamma^j n) = O(M(n))$ as $n \to \infty$.*

PROOF. If α and β are as in (1.1), there exists k such that $\gamma^k \leq \alpha$. Thus, $\sum_{j=0}^{\infty} M(\gamma^j n) \leq k \sum_{j=0}^{\infty} M(\alpha^j n) \leq kM(n)/(1 - \beta) + O(1)$, by repeated application of (1.1). Since $M(n) \to \infty$ as $n \to \infty$, the result follows.

In the following lemma, we assume that $1/c$ is in the allowable range for floating-point numbers. Similar assumptions are implicit below.

LEMMA 2.2. *If c is a nonzero floating-point number, then $1/c$ can be evaluated, to precision n, in $O(M(n))$ operations.*

PROOF. The Newton iteration

$$x_{i+1} = x_i(2 - cx_i) \tag{2.1}$$

converges to $1/c$ with order 2. In fact, if $x_i = (1 - \epsilon_i)/c$, substitution in (2.1) gives $\epsilon_{i+1} = \epsilon_i^2$. Thus, assuming $|\epsilon_0| < \frac{1}{2}$, we have $|\epsilon_i| < 2^{-2^i}$ for all $i \geq 0$, and x_k is a sufficiently good approximation to $1/c$ if $k \geq \log_2 n$. This assumes that (2.1) is satisfied exactly, but it is easy to show that it is sufficient to use precision n at the last iteration ($i = k - 1$), precision slightly greater than $n/2$ for $i = k - 2$, etc. (Details, and more efficient methods, are given in [4, 6].) Thus the result follows from Lemma 2.1. Since $x/y = x(1/y)$, it is clear that floating-point division may also be done in $O(M(n))$ operations.

LEMMA 2.3. *If $c \geq 0$ is a floating-point number, then $c^{\frac{1}{2}}$ can be evaluated, to precision n, in $O(M(n))$ operations.*

PROOF. If $c = 0$ then $c^{\frac{1}{2}} = 0$. If $c \neq 0$, the proof is similar to that of Lemma 2.2, using the Newton iteration $x_{i+1} = (x_i + c/x_i)/2$.

LEMMA 2.4. *For any fixed $k > 0$, $M(kn) = O(M(n))$ as $n \to \infty$.*

PROOF. Since we can add integers less than 2^n in $O(M(n))$ operations, we can add integers less than 2^{kn} in $O(kM(n)) = O(M(n))$ operations. The multiplication of integers less than 2^{kn} can be split into $O(k^2)$ multiplications of integers less than 2^n, and $O(k^2)$ additions, so it can be done in $O(k^2 M(n)) = O(M(n))$ operations.

3. *Solution of Nonlinear Equations*

In Section 6 we need to solve nonlinear equations to precision n. The following lemma is sufficient for this application. Stronger results are given in [4, 6].

LEMMA 3.1. *If the equation $f(x) = c$ has a simple root $\zeta \neq 0$, f' is Lipschitz continuous near ζ, and we can evaluate $f(x)$ to precision n in $O(M(n)\phi(n))$ operations, where $\phi(n)$ is a positive, monotonic increasing function, for x near ζ, then ζ can be evaluated to precision n in $O(M(n)\phi(n))$ operations.*

PROOF. Consider the discrete Newton iteration

$$x_{i+1} = x_i - h_i(f(x_i) - c)/(f(x_i + h_i) - f(x_i)). \tag{3.1}$$

If $h_i = 2^{-n/2}$, $x_i - \zeta = O(2^{-n/2})$, and the right side of (3.1) is evaluated with precision n, then a standard analysis shows that $x_{i+1} - \zeta = O(2^{-n})$. Since a sufficiently good starting approximation x_0 may be found in $O(1)$ operations, the result follows in the same way as in the proof of Lemma 2.2, using the fact that Lemma 2.1 holds with $M(n)$ replaced by $M(n)\phi(n)$. The assumption $\zeta \neq 0$ is only necessary because we want to obtain ζ with a relative (not absolute) error $O(2^{-n})$.

Other methods, e.g. the secant method, may also be used if the precision is increased appropriately at each iteration. In our applications there is no difficulty in finding a suitable initial approximation x_0 (see Section 6).

4. *Results on Elliptic Integrals*

In this section we summarize some classical results from elliptic integral theory. Most of the results may be found in [1], so proofs are omitted. Elliptic integrals of the first and second kind are defined by

$$F(\psi, \alpha) = \int_0^\psi (1 - \sin^2\alpha \, \sin^2\theta)^{-\frac{1}{2}} d\theta \tag{4.1}$$

and

$$E(\psi, \alpha) = \int_0^\psi (1 - \sin^2\alpha \, \sin^2\theta)^{\frac{1}{2}} d\theta, \tag{4.2}$$

respectively. For our purposes we may assume that α and ψ are in $[0, \pi/2]$. The complete elliptic integrals, $F(\pi/2, \alpha)$ and $E(\pi/2, \alpha)$, are simply written as $F(\alpha)$ and $E(\alpha)$, respectively.

Legendre's Relation.   We need the identity of Legendre [17]:

$$E(\alpha)F(\pi/2 - \alpha) + E(\pi/2 - \alpha)F(\alpha) - F(\alpha)F(\pi/2 - \alpha) = \pi/2, \tag{4.3}$$

and, in particular, the special case

$$2E(\pi/4)F(\pi/4) - (F(\pi/4))^2 = \pi/2. \tag{4.4}$$

Small Angle Approximation.   From (4.1) it is clear that

$$F(\psi, \alpha) = \psi + O(\alpha^2) \tag{4.5}$$

as $\alpha \to 0$.

Large Angle Approximation.   From (4.1),

$$F(\psi, \alpha) = F(\psi, \pi/2) + O(\pi/2 - \alpha)^2, \tag{4.6}$$

uniformly for $0 \le \psi \le \psi_0 < \pi/2$, as $\alpha \to \pi/2$. Also, we note that

$$F(\psi, \pi/2) = \log \tan(\pi/4 + \psi/2). \tag{4.7}$$

Ascending Landen Transformation.   If $0 < \alpha_i < \alpha_{i+1} < \pi/2$, $0 < \psi_{i+1} < \psi_i \le \pi/2$,

$$\sin \alpha_i = \tan^2(\alpha_{i+1}/2), \tag{4.8}$$

and

$$\sin(2\psi_{i+1} - \psi_i) = \sin \alpha_i \sin \psi_i, \tag{4.9}$$

then

$$F(\psi_{i+1}, \alpha_{i+1}) = [(1 + \sin \alpha_i)/2]F(\psi_i, \alpha_i). \tag{4.10}$$

If $s_i = \sin \alpha_i$ and $v_i = \tan(\psi_i/2)$, then (4.8) gives

$$s_{i+1} = 2s_i^{\frac{1}{2}}/(1 + s_i), \tag{4.11}$$

and (4.9) gives

$$v_{i+1} = w_3/(1 + (1 + w_3^2)^{\frac{1}{2}}), \tag{4.12}$$

where

$$w_3 = \tan \psi_{i+1} = (v_i + w_2)/(1 - v_i w_2), \tag{4.13}$$

$$w_2 = \tan(\psi_{i+1} - \psi_i/2) = w_1/(1 + (1 - w_1^2)^{\frac{1}{2}}), \tag{4.14}$$

and

$$w_1 = \sin(2\psi_{i+1} - \psi_i) = 2s_i v_i/(1 + v_i^2). \tag{4.15}$$

Arithmetic-Geometric Mean Iteration.   From the ascending Landen transformation it is possible to derive the arithmetic-geometric mean iteration of Gauss [12] and Lagrange [16]: if $a_0 = 1$, $b_0 = \cos \alpha > 0$,

$$a_{i+1} = (a_i + b_i)/2, \tag{4.16}$$

and

$$b_{i+1} = (a_i b_i)^{\frac{1}{2}}, \tag{4.17}$$

then

$$\lim_{\iota \to \infty} a_{\iota} = \pi/[2F(\alpha)]. \tag{4.18}$$

Also, if $c_0 = \sin \alpha$ and

$$c_{\iota+1} = a_{\iota} - a_{\iota+1}, \tag{4.19}$$

then

$$E(\alpha)/F(\alpha) = 1 - \sum_{\iota=0}^{\infty} 2^{\iota-1}c_{\iota}^{2}. \tag{4.20}$$

An Infinite Product. Let s_{ι}, a_{ι}, and b_{ι} be as above, with $\alpha = \pi/2 - \alpha_0$, so $s_0 = b_0/a_0$. From (4.11), (4.16), and (4.17), it follows that $s_{\iota} = b_{\iota}/a_{\iota}$ for all $\iota \geq 0$. Thus, $(1 + s_{\iota})/2 = a_{\iota+1}/a_{\iota}$, and

$$\prod_{\iota=0}^{\infty} [(1 + s_{\iota})/2] = \lim_{\iota \to \infty} a_{\iota} = \pi/[2F(\pi/2 - \alpha_0)] \tag{4.21}$$

follows from (4.18). (Another connection between (4.11) and the arithmetic-geometric mean iteration is evident if $s_0 = (1 - b_0^2/a_0^2)^{\frac{1}{2}}$. Assuming (4.11) holds for $i < 0$, it follows that $s_{-\iota} = (1 - b_{\iota}^2/a_{\iota}^2)^{\frac{1}{2}}$ for all $i \geq 0$. This may be used to deduce (4.18) from (4.10).)

5. *Evaluation of* π

Let $a_0 = 1$, $b_0 = c_0 = 2^{-\frac{1}{2}}$, $A = \lim_{\iota \to \infty} a_{\iota}$, and $T = \lim_{\iota \to \infty} t_{\iota}$, where a_{ι}, b_{ι}, and c_{ι} are defined by (4.16), (4.17), and (4.19) for $i \geq 1$, and $t_{\iota} = \frac{1}{2} - \sum_{j=0}^{\iota} 2^{j-1}c_j^2$. From (4.4), (4.18), and (4.20), we have

$$\pi = A^2/T. \tag{5.1}$$

Since $a_{\iota} > b_0 > 0$ for all $\iota \geq 0$, and $c_{\iota+1} = a_{\iota} - a_{\iota+1} = a_{\iota+1} - b_{\iota}$, (4.17) gives $b_{\iota+1} = [(a_{\iota+1} + c_{\iota+1})(a_{\iota+1} - c_{\iota+1})]^{\frac{1}{2}} = a_{\iota+1} - O(c_{\iota+1}^2)$, so $c_{\iota+2} = O(c_{\iota+1}^2)$. Thus, the process converges with order at least 2, and $\log_2 n + O(1)$ iterations suffice to give an error $O(2^{-n})$ in the estimate of (5.1). A more detailed analysis shows that $a_{\iota+1}^2/t_{\iota} < \pi < a_{\iota}^2/t_{\iota}$ for all $\iota \geq 0$, and also $a_{\iota}^2/t_{\iota} - \pi \sim 8\pi \exp(-2^{\iota}\pi)$ and $\pi - a_{\iota+1}^2/t_{\iota} \sim \pi^2 2^{\iota+4} \exp(-2^{\iota+1}\pi)$ as $i \to \infty$. The speed of convergence is illustrated in Table I.

From the discussion above, it is clear that the following algorithm, given in pseudo-Algol, evaluates π to precision n.

Algorithm for π

```
A ← 1, B ← 2^{-½}; T ← ½; X ← 1;
while A − B > 2^{-n} do
    begin Y ← A; A ← (A + B)/2; B ← (BY)^½;
          T ← T − X(A − Y)²; X ← 2X
    end;
return A²/T [or, better, (A + B)²/(4T)].
```

TABLE I. CONVERGENCE OF
APPROXIMATIONS TO π

ι	$\pi - a_{\iota+1}^2/t_{\iota}$	$a_{\iota}^2/t_{\iota} - \pi$
0	2.3′−1	8.6′−1
1	1.0′−3	4.6′−2
2	7.4′−9	8 8′−5
3	1.8′−19	3.1′−10
4	5.5′−41	3.7′−21

Since $\log_2 n + O(1)$ iterations are needed, it is necessary to work with precision $n + O(\log \log(n))$, even though the algorithm is numerically stable in the conventional sense. From Lemmas 2.2–2.4, each iteration requires $O(M(n))$ operations, so π may be evaluated to precision n in $O(M(n) \log(n))$ operations. This is asymptotically faster than the usual $O(n^2)$ methods [14, 21] if a fast multiplication algorithm is used. A high-precision computation of π by a similar algorithm is described in [10]. Note that, because the arithmetic-geometric mean iteration is not self-correcting, we cannot obtain a bound $O(M(n))$ in the same way as for the evaluation of reciprocals and square roots by Newton's method.

6. *Evaluation of exp(x) and log(x)*

Suppose $\delta > 0$ fixed, and $m \in [\delta, 1 - \delta]$. If $\sin \alpha_0 = m^{\frac{1}{2}}$, we may evaluate $F(\alpha_0)$ to precision n in $O(M(n) \log(n))$ operations, using (4.18) and the arithmetic-geometric mean iteration, as for the special case $F(\pi/4)$ described in Section 5. (When using (4.18) we need π, which may be evaluated as described above.) Applying the ascending Landen transformation (4.8)–(4.10) with $\iota = 0, 1, \cdots, k - 1$ and $\psi_0 = \pi/2$ gives

$$F(\psi_k, \alpha_k) = \left\{ \sum_{\iota=0}^{k-1} [(1 + \sin\alpha_\iota)/2] \right\} F(\alpha_0). \tag{6.1}$$

Since $s_0 = \sin \alpha_0 = m^{\frac{1}{2}} \geq \delta^{\frac{1}{2}} > 0$, it follows from (4.11) that $s_\iota \to 1$ as $i \to \infty$. In fact, if $s_\iota = 1 - \epsilon_\iota$, then $\epsilon_{\iota+1} = 1 - s_{\iota+1} = 1 - 2(1 - \epsilon_\iota)^{\frac{1}{2}}/(2 - \epsilon_\iota) = \epsilon_\iota^2/8 + O(\epsilon_\iota^3)$, so $s_\iota \to 1$ with order 2. Thus, after $k \sim \log_2 n$ iterations we have $\epsilon_k = O(2^{-n})$, so $\pi/2 - \alpha_k = O(2^{-n/2})$ and, from (4.6) and (4.7),

$$F(\psi_k, \alpha_k) = \log \tan(\pi/4 + \psi_k/2) + O(2^{-n}). \tag{6.2}$$

Assuming $k > 0$, the error is uniformly $O(2^{-n})$ for all $m \in [\delta, 1 - \delta]$, since $\psi_k \leq \psi_1 < \pi/2$.

Define the functions

$$U(m) = \left\{ \sum_{\iota=0}^{\infty} [(1 + \sin\alpha_\iota)/2] \right\} F(\alpha_0) \tag{6.3}$$

and

$$T(m) = \tan(\pi/4 + \psi_\infty/2), \tag{6.4}$$

where $\psi_\infty = \lim_{\iota \to \infty} \psi_\iota$. Since $s_\iota \to 1$ with order 2, the infinite product in (6.3) is convergent, and $U(m)$ is analytic for all $m \in (0, 1)$. Taking the limit in (6.1) and (6.2) as n (and hence k) tends to ∞, we have the fundamental identity

$$U(m) = \log T(m). \tag{6.5}$$

Using (4.11)–(4.15), we can evaluate $U(m) = \{\prod_{\iota=0}^{k-1} [(1 + s_\iota)/2]\} F(\alpha_0) + O(2^{-n})$ and $T(m) = (1 + v_k)/(1 - v_k) + O(2^{-n})$, to precision n, in $O(M(n) \log(n))$ operations. The algorithms are given below in pseudo-Algol.

Algorithm for $U(m)$

```
A ← 1; B ← (1 − m)^½;
while A − B > 2^{−n/2} do
    begin C ← (A + B)/2; B ← (AB)^½; A ← C end;
A ← π/(A + B); S ← m^½;
while 1 − S > 2^{−n/2} do
    begin A ← A(1 + S)/2; S ← 2S^½/(1 + S) end;
return A(1 + S)/2.
```

Algorithm for $T(m)$

```
V ← 1; S ← m^½;
while 1 − S > 2^{−n} do
```

begin $W \leftarrow 2SV/(1 + V^2)$;
$\quad W \leftarrow W/(1 + (1 - W^2)^{\frac{1}{2}})$;
$\quad W \leftarrow (V + W)/(1 - VW)$;
$\quad V \leftarrow W/(1 + (1 + W^2)^{\frac{1}{2}})$;
$\quad S \leftarrow 2S^{\frac{1}{2}}/(1 + S)$
end;
return $(1 + V)/(1 - V)$.

Properties of $U(m)$ and $T(m)$. From (4.21) and (6.3),

$$U(m) = (\pi/2)F(\alpha_0)/F(\pi/2 - \alpha_0), \tag{6.6}$$

where $\sin \alpha_0 = m^{\frac{1}{2}}$ as before. Both $F(\alpha_0)$ and $F(\pi/2 - \alpha_0)$ may be evaluated by the arithmetic-geometric mean iteration, which leads to a slightly more efficient algorithm for $U(m)$ than the one above, because the division by $(1 + S)$ in the final "while" loop is avoided. From (6.5) and (6.6), we have the special cases $U(\frac{1}{2}) = \pi/2$ and $T(\frac{1}{2}) = e^{\pi/2}$. Also, (6.6) gives

$$U(m)U(1 - m) = \pi^2/4, \tag{6.7}$$

for all $m \in (0, 1)$.

Although we shall avoid using values of m near 0 or 1, it is interesting to obtain asymptotic expressions for $U(m)$ and $T(m)$ as $m \to 0$ or 1. From the algorithm for $T(m)$, $T(1 - \epsilon) = 4\epsilon^{-\frac{1}{2}} - \epsilon^{\frac{1}{2}} + O(\epsilon^{\frac{3}{2}})$ as $\epsilon \to 0$. Thus, from (6.5), $U(1 - \epsilon) = L(\epsilon) - \epsilon/4 + O(\epsilon^2)$, where $L(\epsilon) = \log(4/\epsilon^{\frac{1}{2}})$. Using (6.7), this gives $U(\epsilon) = \pi^2/[4L(\epsilon)] + O(\epsilon/L^2)$, and hence $T(\epsilon) = \exp(\pi^2/[4L(\epsilon)]) + O(\epsilon/L^2)$. Some values of $U(m)$ and $T(m)$ are given in Table II.

Evaluation of $\exp(x)$. To evaluate $\exp(x)$ to precision n, we first use identities such as $\exp(2x) = (\exp(x))^2$ and $\exp(-x) = 1/\exp(x)$ to reduce the argument to a suitable domain, say $1 \le x \le 2$ (see below). We then solve the nonlinear equation

$$U(m) = x, \tag{6.8}$$

obtaining m to precision n, by a method such as the one described in Section 3. From Lemma 3.1, with $\phi(n) = \log(n)$, this may be done in $O(M(n)\log(n))$ operations. Finally, we evaluate $T(m)$ to precision n, again using $O(M(n)\log(n))$ operations. From (6.5) and (6.8), $T(m) = \exp(x)$, so we have computed $\exp(x)$ to precision n. Any preliminary transformations may now be undone.

Evaluation of $\log(x)$. Since we can evaluate $\exp(x)$ to precision n in $O(M(n) \log(n))$ operations, Lemma 3.1 shows that we can also evaluate $\log(x)$ in $O(M(n) \log(n))$ operations, by solving the equation $\exp(y) = x$ to the desired accuracy. A more direct method is to solve $T(m) = x$ (after suitable domain reduction), and then evaluate $U(m)$.

Further details. If $x \in [1, 2]$ then the solution m of (6.8) lies in $(0.10, 0.75)$, and it may be verified that the secant method, applied to (6.8), converges if the starting approximations are $m_0 = 0.2$ and $m_1 = 0.7$. If desired, the discrete Newton method or some other locally convergent method may be used after a few iterations of the secant method have given a good approximation to m.

TABLE II. THE FUNCTIONS $U(m)$ AND $T(m)$

m	$U(m)$	$T(m)$	m	$U(m)$	$T(m)$
0 01	0 6693	1.9529	0.60	1.7228	5.6004
0.05	0.8593	2.3615	0.70	1.9021	6 6999
0.10	0.9824	2.6710	0.80	2.1364	8.4688
0.20	1.1549	3.1738	0.90	2.5115	12.3235
0.30	1.2972	3.6591	0.95	2.8714	17.6617
0.40	1.4322	4.1878	0.99	3.6864	39.8997
0.50	1.5708	4.8105			

Similarly, if $x \in [3, 9]$, the solution of $T(m) = x$ lies in $(0.16, 0.83)$, and the secant method converges if $m_0 = 0.2$ and $m_1 = 0.8$.

If $x = 1 + \epsilon$ where ϵ is small, and for domain reduction the relation

$$\log(x) = \log(\lambda x) - \log(\lambda) \tag{6.9}$$

is used, for some $\lambda \in (3, 9)$, then $\log(\lambda x)$ and $\log(\lambda)$ may be evaluated as above, but cancellation in (6.9) will cause some loss of precision in the computed value of $\log(x)$. If $|\epsilon| > 2^{-n}$, it is sufficient to evaluate $\log(\lambda x)$ and $\log(\lambda)$ to precision $2n$, for at most n bits are lost through cancellation in (6.9). On the other hand, there is no difficulty if $|\epsilon| \leq 2^{-n}$, for then $\log(1 + \epsilon) = \epsilon(1 + O(2^{-n}))$. When evaluating $\exp(x)$, a similar loss of precision never occurs, and it is sufficient to work with precision $n + O(\log \log(n))$, as in the evaluation of π (see Section 5). To summarize, we have proved:

THEOREM 6.1. *If* $-\infty < a < b < \infty$, *then* $O(M(n) \log(n))$ *operations suffice to evaluate* $\exp(x)$ *to precision* n, *uniformly for all floating-point numbers* $x \in [a, b]$, *as* $n \to \infty$; *and similarly for* $\log(x)$ *if* $a > 0$.

7. Evaluation of Trigonometric Functions

Suppose $\delta > 0$ fixed, and $x \in [\delta, 1]$. Let $s_0 = \sin \alpha_0 = 2^{-n/2}$ and $v_0 = \tan(\psi_0/2) = x/(1 + (1 + x^2)^{\frac{1}{2}})$, so $\tan \psi_0 = x$. Applying the ascending Landen transformation, as for (6.1), gives

$$F(\psi_k, \alpha_k) = \left\{ \prod_{i=0}^{k-1} [(1 + s_i)/2] \right\} F(\psi_0, \alpha_0). \tag{7.1}$$

Also, from (4.5) and the choice of s_0,

$$F(\psi_0, \alpha_0) = \operatorname{artan}(x) + O(2^{-n}). \tag{7.2}$$

From (4.11), $s_{i+1} \geq s_i^{\frac{1}{2}}$, so there is some $j \leq \log_2 n + O(1)$ such that $s_j \in [\frac{1}{4}, \frac{3}{4}]$. Since $s_i \to 1$ with order 2, there is some $k \leq 2 \log_2 n + O(1)$ such that $1 - s_k = O(2^{-n})$. From (4.6) and (4.7), $F(\psi_k, \alpha_k) = \log \tan(\pi/4 + \psi_k/2) + O(2^{-n})$. Thus, from (7.1) and (7.2),

$$\operatorname{artan}(x) = \left\{ \prod_{i=0}^{k-1} [2/(1 + s_i)] \right\} \log \tan(\pi/4 + \psi_k/2) + O(2^{-n}). \tag{7.3}$$

If we evaluate $\tan(\pi/4 + \psi_k/2)$ as above, and use the algorithm of Section 6 to evaluate the logarithm in (7.3), we have $\operatorname{artan}(x)$ to precision n in $O(M(n) \log(n))$ operations. The algorithm may be written as follows.

Algorithm for artan(x), $x \in [\delta, 1]$

$S \leftarrow 2^{-n/2}; \; V \leftarrow x/(1 + (1 + x^2)^{\frac{1}{2}}); \; Q \leftarrow 1;$
while $1 - S > 2^{-n}$ **do**
    begin $Q \leftarrow 2Q/(1 + S);$
        $W \leftarrow 2SV/(1 + V^2);$
        $W \leftarrow W/(1 + (1 - W^2)^{\frac{1}{2}});$
        $W \leftarrow (V + W)/(1 - VW);$
        $V \leftarrow W/(1 + (1 + W^2)^{\frac{1}{2}});$
        $S \leftarrow 2S^{\frac{1}{2}}/(1 + S)$
    end;
return $Q \log((1 + V)/(1 - V)).$

After k iterations, $Q \leq 2^k$, so at most $2 \log_2 n + O(1)$ bits of precision are lost because V is small. Thus it is sufficient to work with precision $n + O(\log(n))$, and Lemma 2.4 justifies our claim that $O(M(n) \log(n))$ operations are sufficient to obtain $\operatorname{artan}(x)$ to precision n.

If x is small, we may use the same idea as that described above for evaluating $\log(1 + \epsilon)$: work with precision $3n/2 + O(\log(n))$ if $x > 2^{-n/2}$, and use $\operatorname{artan}(x)$

250   RICHARD P. BRENT

$= x(1 + O(2^{-n}))$ if $0 \le x \le 2^{-n/2}$. (Actually, it is not necessary to increase the working precision if $\log((1 + V)/(1 - V))$ is evaluated carefully.)

Using the identity $\mathrm{artan}(x) = \pi/2 - \mathrm{artan}(1/x)$ $(x > 0)$, we can extend the domain to $[0, \infty)$. Also, since $\mathrm{artan}(-x) = -\mathrm{artan}(x)$, there is no difficulty with negative x. To summarize, we have proved the following theorem.

THEOREM 7.1. $O(M(n) \log(n))$ *operations suffice to evaluate* $\mathrm{artan}(x)$ *to precision* n, *uniformly for all floating-point numbers* x, *as* $n \to \infty$.

Suppose $\theta \in [\delta, \pi/2 - \delta]$. From Lemma 3.1 and Theorem 7.1, we can solve the equation $\mathrm{artan}(x) = \theta/2$ to precision n in $O(M(n) \log(n))$ operations, and thus evaluate $x = \tan(\theta/2)$. Now $\sin\theta = 2x/(1 + x^2)$ and $\cos\theta = (1 - x^2)/(1 + x^2)$ may easily be evaluated. For arguments outside $[\delta, \pi/2 - \delta]$, domain reduction techniques like those above may be used. Difficulties occur near certain integer multiples of $\pi/2$, but these may be overcome (at least for the usual floating-point number representations) by increasing the working precision. We state the following theorem for $\sin(x)$, but similar results hold for the other trigonometric functions (and also, of course, for the elliptic integrals and their inverse functions).

THEOREM 7.2. *If* $[a, b] \subseteq (-\pi, \pi)$, *then* $O(M(n) \log(n))$ *operations suffice to evaluate* $\sin(x)$ *to precision* n, *uniformly for all floating-point numbers* $x \in [a, b]$, *as* $n \to \infty$.

8. Asymptotic Constants

So far we have been concerned with order of magnitude results. In this section we give upper bounds on the constants K such that $w(n) \le (K + o(1))M(n) \log_2 n$, where $w(n)$ is the number of operations required to evaluate π, $\exp(x)$, etc., to precision n. The following two assumptions will be made.

1. For all $\gamma > 0$ and $\epsilon > 0$, the inequality $M(\gamma n) \le (\gamma + \epsilon)M(n)$ holds for sufficiently large n.

2. The number of operations required for floating-point addition, conversion between representations of different precision (at most n), and multiplication or division of floating-point numbers by small integers is $o(M(n))$ as $n \to \infty$.

These assumptions certainly hold if a standard floating-point representation is used and $M(n) \sim n (\log(n))^\alpha (\log\log(n))^\beta$ for some $\alpha \ge 0$, provided $\beta > 0$ if $\alpha = 0$.

The following result is proved in [4]. The algorithms used are similar to those of Section 2, but slightly more efficient.

THEOREM 8.1. *Precision-n division of floating-point numbers may be performed in* $(4 + o(1))M(n)$ *operations as* $n \to \infty$, *and square roots may be evaluated in* $(11/2 + o(1))M(n)$ *operations.*

Using Theorem 8.1 and algorithms related to those of Sections 5–7, the following result is proved in [5].

THEOREM 8.2. π *may be evaluated to precision* n *in* $(15/2 + o(1))M(n) \log_2 n$ *operations as* $n \to \infty$. *If* π *and* $\log 2$ *are precomputed, the elementary function* $f(x)$ *can be evaluated to precision* n *in* $(K + o(1))M(n) \log_2 n$ *operations, where*

$$K = \begin{cases} 13 & \text{if } f(x) = \log(x) \text{ or } \exp(x), \\ 34 & \text{if } f(x) = \mathrm{artan}(x), \sin(x), \text{ etc.,} \end{cases}$$

and x *is a floating-point number in an interval on which* $f(x)$ *is defined and bounded away from* 0 *and* ∞.

For purposes of comparison, note that evaluation of $\log(1 + x)$ or $\log((1 + x)/(1 - x))$ by the usual series expansion requires $(c + o(1))M(n)n$ operations, where c is a constant of order unity (depending on the range of x and the precise method used). Since $13 \log_2 n < n$ for $n \ge 83$, the $O(M(n) \log(n))$ method for $\log(x)$ should be faster than the $O(M(n)n)$ method for n greater than a few hundred.

ACKNOWLEDGMENTS. This paper was written while the author was visiting the Com-

Fast Multiple-Precision Evaluation of Elementary Functions   251

puter Science Department, Stanford University. The comments of R.W. Gosper, D.E. Knuth, and D. Shanks on various drafts were very useful.

REFERENCES

1   ABRAMOWITZ, M., AND STEGUN, I.A. Handbook of mathematical functions with formulas, graphs, and mathematical tables. National Bureau of Standards, Washington, D.C., 1964; Dover, 1965, Ch. 17.
2   BEELER, M, GOSPER, R.W, AND SCHROEPPEL, R   Hakmem Memo No. 239, M.I.T. Artificial Intelligence Lab., M I T., Cambridge, Mass., 1972, pp. 70–71
3. BORCHARDT, C W. *Gesammelte Werke.* Berlin, 1888, pp. 455–462
4. BRENT, R P. The complexity of multiple-precision arithmetic. Proc Seminar on Complexity of Computational Problem Solving (held at Australian National U., Dec. 1974), Queensland U. Press, Brisbane, Australia 1975, pp 126–165
5. BRENT, R P. Multiple-precision zero-finding methods and the complexity of elementary function evaluation Proc Symp. on Analytic Computational Complexity, J.F. Traub, Ed , Academic Press, New York, 1976, pp 151–176.
6. BRENT, R.P. *Computer Solution of Nonlinear Equations.* Academic Press, New York (to appear), Ch. 6.
7. BRENT, R P. A Fortran multiple-precision arithmetic package. Submitted to a technical journal.
8. CARLSON, B.C   Algorithms involving arithmetic and geometric means. *Amer. Math. Monthly* **78** (May 1971), 496–505.
9   CARLSON, B C. An algorithm for computing logarithms and arctangents. *Math. Comput. 26* (April 1972), 543–549.
10. FINKEL, R , GUIBAS, L , AND SIMONYI, C   Manuscript in preparation.
11. FISCHER, M J , AND STOCKMEYER, L J   Fast on-line integer multiplication. *J. Comput. System Scis. 9* (Dec. 1974), 317–331.
12. GAUSS, C.F   *Carl Friedrich Gauss Werke, Bd. 3* Gottingen, 1876, pp. 362–403.
13. GOSPER, R.W. Acceleration of series. Memo No 304, M.I.T. Artificial Intelligence Lab., M I T , Cambridge, Mass , 1974.
14. GUILLOUD, J., AND BOUYER, M   1,000,000 decimals de pi. Unpublished manuscript
15. KNUTH, D.E. *The Art of Computer Programming, Vol. 2* Addison-Wesley, Reading, Mass , 1969. Errata and addenda: Rep. CS 194, Computer Sci Dep., Stanford U., Stanford, Calif., 1970.
16. LAGRANGE, J.L. *Oeuvres de Lagrange, Tome 2* Gauthier-Villars, Paris, 1868, pp. 267–272.
17. LEGENDRE, A.M   *Exercices de Calcul Integral, Vol. 1.* Paris, 1811, p. 61.
18. SALAMIN, E. Computation of π using arithmetic-geometric mean. To appear in *Math. Comput.*
19 SCHONHAGE, A., AND STRASSEN, V. Schnelle Multiplikation grosser Zahlen. *Computing 7* (1971), 281–292.
20. SCHROEPPEL, R. Unpublished manuscript dated May 1975
21 SHANKS, D., AND WRENCH, J.W. Calculation of π to 100,000 decimals. *Math Comput. 16* (1962), 76–99.
22 SWEENEY, D W. On the computation of Euler's constant *Math. Comput. 17* (1963), 170–178.
23 THACHER, H.C. Iterated square root expansions for the inverse cosine and inverse hyperbolic cosine. *Math. Comput. 15* (1961), 399–403.

RECEIVED MARCH 1975; REVISED SEPTEMBER 1975

3. The arithmetic-geometric mean of Gauss (1984)

Paper 3: David A. Cox, "The arithmetic-geometric mean of Guass," *L'Enseignement Mathematique*, vol. 30 (1984), p. 275–330. Reprinted by permission.

Synopsis:

Although the arithmetic-geometric mean (AGM) is now used widely in high-precision computation, it was actually discovered nearly a century earlier, independently by Lagrange, Legendre and, in considerably more detail, by Gauss (although there is no evidence that any of them saw the connection to computing π). As David A. Cox points out in this article, Gauss did numerous computations with the AGM, such as when he numerically discovered that

$$\mathrm{agm}(\sqrt{2}, 1) = 1.19814023473355922074\ldots = \frac{\pi}{2 \int_0^1 \mathrm{d}z/\sqrt{1 - z^4}},$$

a fact that Gauss declared in his workbook "will surely open an entirely new field of analysis." Indeed it did, as Cox describes in considerable detail.

As Cox observes, Gauss' analyses of the AGM for complex arguments are particularly interesting. In this case, the square root in the defining formulas of the AGM leads to multiple-valued complex functions, and is not at all clear which of these branches one should take, nor is it clear whether any of these branches converge. But Gauss developed a rigorous theory of the complex AGM, and further saw connections to the theory of elliptic modular functions. Gauss' work in this area was almost completely unknown to others during his lifetime, but today is recognized as a major contribution to the field.

Keywords: Arithmetic-Geometric Mean, Elliptic Integrals, History

© Springer International Publishing Switzerland 2016
D.H. Bailey, J.M. Borwein, *Pi: The Next Generation*,
DOI 10.1007/978-3-319-32377-0_3

L'Enseignement Mathématique, t. 30 (1984), p. 275-330

THE ARITHMETIC-GEOMETRIC MEAN OF GAUSS

by David A. Cox

INTRODUCTION

The arithmetic-geometric mean of two numbers a and b is defined to be the common limit of the two sequences $\{a_n\}_{n=0}^{\infty}$ and $\{b_n\}_{n=0}^{\infty}$ determined by the algorithm

$$a_0 = a, \qquad b_0 = b,$$

(0.1)

$$a_{n+1} = (a_n + b_n)/2, \qquad b_{n+1} = (a_n b_n)^{1/2}, \quad n = 0, 1, 2, \dots .$$

Note that a_1 and b_1 are the respective arithmetic and geometric means of a and b, a_2 and b_2 the corresponding means of a_1 and b_1, etc. Thus the limit

$$(0.2) \qquad M(a, b) = \lim_{n \to \infty} a_n = \lim_{n \to \infty} b_n$$

really does deserve to be called the arithmetic-geometric mean of a and b. This algorithm first appeared in a paper of Lagrange, but it was Gauss who really discovered the amazing depth of this subject. Unfortunately, Gauss published little on the agM (his abbreviation for the arithmetic-geometric mean) during his lifetime. It was only with the publication of his collected works [12] between 1868 and 1927 that the full extent of his work became apparent. Immediately after the last volume appeared, several papers (see [15] and [35]) were written to bring this material to a wider mathematical audience. Since then, little has been done, and only the more elementary properties of the agM are widely known today.

In § 1 we review these elementary properties, where a and b are positive real numbers and the square root in (0.1) is also positive. The convergence of the algorithm is easy to see, though less obvious is the connection between the agM and certain elliptic integrals. As an application, we use $M(\sqrt{2}, 1)$ to determine the arc length of the lemniscate. In § 2, we allow a and b to be complex numbers, and the level of difficulty changes dramatically.

The convergence of the algorithm is no longer obvious, and as might be expected, the square root in (0.1) causes trouble. In fact, $M(a, b)$ becomes a multiple valued function, and in order to determine the relation between the various values, we will need to "uniformize" the agM using quotients of the classical Jacobian theta functions, which are modular functions for certain congruence subgroups of level four in $SL(2, \mathbf{Z})$. The amazing fact is that Gauss knew all of this! Hence in § 3 we explore some of the history of these ideas. The topics encountered will range from Bernoulli's study of elastic rods (the origin of the lemniscate) to Gauss' famous mathematical diary and his work on secular perturbations (the only article on the agM published in his lifetime).

I would like to thank my colleagues David Armacost and Robert Breusch for providing translations of numerous passages originally in Latin or German. Thanks also go to Don O'Shea for suggesting the wonderfully quick proof of (2.2) given in § 2.

1. THE ARITHMETIC-GEOMETRIC MEAN OF REAL NUMBERS

When a and b are positive real numbers, the properties of the agM $M(a, b)$ are well known (see, for example, [5] and [26]). We will still give complete proofs of these properties so that the reader can fully appreciate the difficulties we encounter in § 2.

We will assume that $a \geqslant b > 0$, and we let $\{a_n\}_{n=0}^{\infty}$ and $\{b_n\}_{n=0}^{\infty}$ be as in (0.1), where b_{n+1} is always the positive square root of $a_n b_n$. The usual inequality between the arithmetic and geometric means,

$$(a+b)/2 \geqslant (ab)^{1/2},$$

immediately implies that $a_n \geqslant b_n$ for all $n \geqslant 0$. Actually, much more is true: we have

(1.1) $a \geqslant a_1 \geqslant ... \geqslant a_n \geqslant a_{n+1} \geqslant ... \geqslant b_{n+1} \geqslant b_n \geqslant ... \geqslant b_1 \geqslant b$

(1.2) $0 \leqslant a_n - b_n \leqslant 2^{-n}(a-b).$

To prove (1.1), note that $a_n \geqslant b_n$ and $a_{n+1} \geqslant b_{n+1}$ imply

$$a_n \geqslant (a_n + b_n)/2 = a_{n+1} \geqslant b_{n+1} = (a_n b_n)^{1/2} \geqslant b_n,$$

and (1.1) follows. From $b_{n+1} \geqslant b_n$ we obtain

$$a_{n+1} - b_{n+1} \leqslant a_{n+1} - b_n = 2^{-1}(a_n - b_n),$$

and (1.2) follows by induction. From (1.1) we see immediately that $\lim\limits_{n \to \infty} a_n$ and $\lim\limits_{n \to \infty} b_n$ exist, and (1.2) implies that the limits are equal. Thus, we can use (0.2) to define the arithmetic-geometric mean $M(a, b)$ of a and b.

Let us work out two examples.

Example 1. $M(a, a) = a$.

This is obvious because $a = b$ implies $a_n = b_n = a$ for all $n \geq 0$.

Example 2. $M(\sqrt{2}, 1) = 1.1981402347355922074...$

The accuracy is to 19 decimal places. To compute this, we use the fact that $a_n \geq M(a, b) \geq b_n$ for all $n \geq 0$ and the following table (all entries are rounded off to 21 decimal places).

n	a_n	b_n
0	1.414213562373905048802	1.000000000000000000000
1	1.207106781186547524401	1.189207115002721066717
2	1.198156948094634295559	1.198123521493120122607
3	1.198140234793877209083	1.198140234677307205798
4	1.198140234735592207441	1.198140234735592207439

Such computations are not too difficult these days, though some extra programming was required since we went beyond the usual 16 digits of double-precision. The surprising fact is that these calculations were done not by computer but rather by Gauss himself. The above table is one of four examples given in the manuscript "De origine proprietatibusque generalibus numerorum mediorum arithmetico-geometricorum" which Gauss wrote in 1800 (see [12, III, pp. 361-371]). As we shall see later, this is an especially important example.

Let us note two obvious properties of the agM:

(1.3)
$$M(a, b) = M(a_1, b_1) = M(a_2, b_2) = \ldots$$

$$M(\lambda a, \lambda b) = \lambda M(a, b).$$

Both of these follow easily from the definition of $M(a, b)$.

Our next result shows that the agM is not as simple as indicated by what we have done so far. We now get our first glimpse of the depth of this subject.

THEOREM 1.1.   *If* $a \geqslant b > 0$, *then*

$$M(a, b) \cdot \int_0^{\pi/2} (a^2\cos^2\phi + b^2\sin^2\phi)^{-1/2} d\phi = \pi/2 .$$

Proof.   Let $I(a, b)$ denote the above integral, and set $\mu = M(a, b)$. Thus we need to prove $I(a, b) = (\pi/2)\mu^{-1}$. The key step is to show that

(1.4) $I(a, b) = I(a_1, b_1).$

The shortest proof of (1.4) is due to Gauss. He introduces a new variable ϕ' such that

(1.5) $\sin\phi = \dfrac{2a \sin\phi'}{a + b + (a-b)\sin^2\phi'} .$

Note that $0 \leqslant \phi' \leqslant \pi/2$ corresponds to $0 \leqslant \phi \leqslant \pi/2$. Gauss then asserts "after the development has been made correctly, it will be seen" that

(1.6) $(a^2\cos^2\phi + b^2\sin^2\phi)^{-1/2} d\phi = (a_1^2\cos^2\phi' + b_1^2\sin^2\phi')^{-1/2} d\phi'$

(see [12, III, p. 352]). Given this, (1.4) follows easily. In "Fundamenta nova theoriae functionum ellipticorum," Jacobi fills in some of the details Gauss left out (see [20, I, p. 152]). Specifically, one first proves that

$$\cos\phi = \frac{2 \cos\phi'(a_1^2\cos^2\phi' + b_1^2\sin^2\phi')^{1/2}}{a + b + (a-b)\sin^2\phi'}$$

$$(a^2\cos^2\phi + b^2\sin^2\phi)^{1/2} = a \frac{a + b - (a-b)\sin^2\phi'}{a + b + (a-b)\sin^2\phi'}$$

(these are straightforward manipulations), and then (1.6) follows from these formulas by taking the differential of (1.5).

Iterating (1.4) gives us

$$I(a, b) = I(a_1, b_1) = I(a_2, b_2) = \dots ,$$

so that $I(a, b) = \lim\limits_{n \to \infty} I(a_n, b_n) = \pi/2\mu$ since the functions

$$(a_n^2\cos^2\phi + b_n^2\sin^2\phi)^{-1/2}$$

converge uniformly to the constant function μ^{-1}.   QED

This theorem relates very nicely to the classical theory of complete elliptic integrals of the first kind, i.e., integrals of the form

$$F(k, \pi/2) = \int_0^{\pi/2} (1 - k^2 \sin^2\phi)^{-1/2}d\phi = \int_0^1 ((1-z^2)(1-k^2z^2))^{-1/2}dz.$$

To see this, we set $k = \dfrac{a-b}{a+b}$. Then one easily obtains

$$I(a, b) = a^{-1} F\left(\frac{2\sqrt{k}}{1+k}, \pi/2\right), \quad I(a_1, b_1) = a_1^{-1} F(k, \pi/2),$$

so that (1.4) is equivalent to the well-known formula

$$F\left(\frac{2\sqrt{k}}{1+k}, \pi/2\right) = (1+k) F(k, \pi/2)$$

(see [16, p. 250] or [17, p. 908]). Also, the substitution (1.5) can be written as

$$\sin\phi = \frac{(1+k)\sin\phi'}{1 + k \sin^2\phi'},$$

which is now called the Gauss transformation (see [32, p. 206]).

For someone well versed in these formulas, the derivation of (1.4) would not be difficult. In fact, a problem on the 1895 Mathematical Tripos was to prove (1.4), and the same problem appears as an exercise in Whittaker and Watson's *Modern Analysis* (see [36, p. 533]), though the agM is not mentioned. Some books on complex analysis do define $M(a, b)$ and state Theorem 1.1 (see, for example, [7, p. 417]).

There are several other ways to express Theorem 1.1. For example, if $0 \leqslant k < 1$, then one can restate the theorem as

(1.7) $$\frac{1}{M(1+k, 1-k)} = \frac{2}{\pi} \int_0^{\pi/2} (1 - k^2 \sin^2\gamma)^{-1/2}d\gamma = \frac{2}{\pi} F(k, \pi/2).$$

Furthermore, using the well-known power series expansion for $F(k, \pi/2)$ (see [16, p. 905]), we obtain

(1.8) $$\frac{1}{M(1+k, 1-k)} = \sum_{n=0}^{\infty} \left[\frac{1 \cdot 3 \cdot \ldots \cdot (2n-1)}{2^n n!}\right]^2 k^{2n}.$$

Finally, it is customary to set $k' = \sqrt{1 - k^2}$. Then, using (1.3), we can rewrite (1.7) as

(1.9) $$\frac{1}{M(1, k')} = \frac{2}{\pi} \int_0^{\pi/2} (1 - k^2 \sin^2\gamma)^{-1/2}d\gamma.$$

This last equation shows that the average value of the function $(1 - k^2 \sin^2 \gamma)^{-1/2}$ on the interval $[0, \pi/2]$ is the reciprocal of the agM of the reciprocals of the minimum and maximum values of the function, a lovely interpretation due to Gauss — see [12, III, p. 371].

One application of Theorem 1.1, in the guise of (1.7), is that the algorithm for the agM now provides a very efficient method for approximating the elliptic integral $F(k, \pi/2)$. As we will see in § 3, it was just this problem that led Lagrange to independently discover the algorithm for the agM.

Another application of Theorem 1.1 concerns the arc length of the lemniscate $r^2 = \cos 2\theta$:

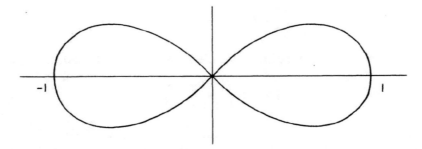

Using the formula for arc length in polar coordinates, we see that the total arc length is

$$4 \int_0^{\pi/4} (r^2 + (dr/d\theta)^2)^{1/2}\, d\theta = 4 \int_0^{\pi/4} (\cos 2\theta)^{-1/2}\, d\theta .$$

The substitution $\cos 2\theta = \cos^2 \phi$ transforms this to the integral

$$4 \int_0^{\pi/2} (1 + \cos^2 \phi)^{-1/2}\, d\phi = 4 \int_0^{\pi/2} (2 \cos^2 \phi + \sin^2 \phi)^{-1/2}\, d\phi .$$

Using Theorem 1.1 to interpret this last integral in terms of $M(\sqrt{2}, 1)$, we see that the arc length of the lemniscate $r^2 = \cos 2\theta$ is $2\pi/M(\sqrt{2}, 1)$.

From Example 2 it follows that the arc length is approximately 5.244, and much better approximations can be easily obtained. (For more on the computation of the arc length of the lemniscate, the reader should consult [33].)

On the surface, this arc length computation seems rather harmless. However, from an historical point of view, it is of fundamental importance. If we set $z = \cos\phi$, then we obtain

$$\int_0^{\pi/2} (2 \cos^2 \phi + \sin^2 \phi)^{-1/2}\, d\phi = \int_0^1 (1 - z^4)^{-1/2}\, dz .$$

The integral on the right appeared in 1691 in a paper of Jacob Bernoulli and was well known throughout the 18th century. Gauss even had a special notation for this integral, writing

$$\varpi = 2 \int_0^1 (1-z^4)^{-1/2} \, dz .$$

Then the relation between the arc length of the lemniscate and $M(\sqrt{2}, 1)$ can be written

$$M(\sqrt{2}, 1) = \frac{\pi}{\varpi} .$$

To see the significance of this equation, we turn to Gauss' mathematical diary. The 98th entry, dated May 30, 1799, reads as follows:

> We have established that the arithmetic-geometric mean between 1 and $\sqrt{2}$ is π/ϖ to the eleventh decimal place; the demonstration of this fact will surely open an entirely new field of analysis.

(See [12, X.1, p. 542].) The genesis of this entire subject lies in Gauss' observation that these two numbers are the same. It was in trying to understand the real meaning of this equality that several streams of Gauss' thought came together and produced the exceptionally rich mathematics which we will explore in § 2.

Let us first examine how Gauss actually showed that $M(\sqrt{2}, 1) = \pi/\varpi$. The proof of Theorem 1.1 given above appeared in 1818 in a paper on secular perturbations (see [12, III, pp. 331-355]), which is the only article on the agM Gauss published in his lifetime (though as we've seen, Jacobi knew this paper well). It is more difficult to tell precisely when he first proved Theorem 1.1, although his notes do reveal that he had two proofs by December 23, 1799.

Both proofs derive the power series version (1.8) of Theorem 1.1. Thus the goal is to show that $M(1 + k, 1 - k)^{-1}$ equals the function

$$(1.10) \qquad y = \sum_{n=0}^{\infty} \left(\frac{1 \cdot 3 \cdot \ldots \cdot (2n-1)}{2^n n!} \right)^2 k^{2n} .$$

The first proof, very much in the spirit of Euler, proceeds as follows. Using (1.3), Gauss derives the identity

$$(1.11) \qquad M\left(1 + \frac{2t}{1+t^2}, 1 - \frac{2t}{1+t^2}\right) = \frac{1}{1+t^2} M(1+t^2, 1-t^2) .$$

He then assumes that there is a power series expansion of the form

$$\frac{1}{M(1+k,\,1-k)} = 1 + A\,k^2 + B\,k^4 + C\,k^6 + \dots \; .$$

By letting $k = t^2$ and $2t/(1+t^2)$ in this series and using (1.11), Gauss obtains

$$1 + A\left(\frac{2t}{1+t^2}\right)^2 + B\left(\frac{2t}{1+t^2}\right)^4 + C\left(\frac{2t}{1+t^2}\right)^6 + \dots$$

$$= (1+t^2)\,(1 + At^4 + Bt^8 + Ct^{12} + \dots)\,.$$

Multiplying by $2t/(1+t^2)$, this becomes

$$\frac{2t}{1+t^2} + A\left(\frac{2t}{1+t^2}\right)^3 + B\left(\frac{2t}{1+t^2}\right)^5 + \dots = 2t(1 + At^4 + Bt^8 + \dots)\,.$$

A comparison of the coefficients of powers of t gives an infinite system of equations in $A, B, C, \dots$. Gauss showed that this system is equivalent to the equations $0 = 1 - 4A = 9A - 16B = 25B - 36C = \dots$, and (1.8) follows easily (see [12, III, pp. 367-369] for details). Gauss' second proof also uses the identity (1.11), but in a different way. Here, he first shows that the series y of (1.10) is a solution of the hypergeometric differential equation

(1.12) $(k^3 - k)y'' + (3k^2 - 1)y' + ky = 0\,.$

This enables him to show that y satisfies the identity

$$y\left(\frac{2t}{1+t^2}\right) = (1+t^2)y(t^2)\,,$$

so that by (1.11), $F(k) = M(1+k,\,1-k)y(k)$ has the property that

$$F\left(\frac{2t}{1+t^2}\right) = F(t^2)\,.$$

Gauss then asserts that $F(k)$ is clearly constant. Since $F(0) = 1$, we obtain a second proof of (1.8) (see [12, X.1, pp. 181-183]). It is interesting to note that neither proof is rigorous from the modern point of view: the first assumes without proof that $M(1+k,\,1-k)^{-1}$ has a power series expansion, and the second assumes without proof that $M(1+k,\,1-k)$ is continuous (this is needed in order to show that $F(k)$ is constant).

We can be certain that Gauss knew both of these proofs by December 23, 1799. The evidence for this is the 102nd entry in Gauss' mathematical diary. Dated as above, it states that "the arithmetic-geometric mean is itself an integral quantity" (see [12, X.1, p. 544]). However, this statement is not so easy to interpret. If we turn to Gauss' unpublished manuscript of 1800 (where we got the example $M(\sqrt{2}, 1)$), we find (1.7) and (1.8) as expected, but also the observation that a complete solution of the differential equation (1.12) is given by

$$(1.13) \qquad \frac{A}{M(1+k, 1-k)} + \frac{B}{M(1, k)}, \qquad A, B \in \mathbf{C}$$

(see [12, III, p. 370]). In eighteenth century terminology, this is the "complete integral" of (1.12) and thus may be the "integral quantity" that Gauss was referring to (see [12, X.1, pp. 544-545]). Even if this is so, the second proof must predate December 23, 1799 since it uses the same differential equation.

In §3 we will study Gauss' early work on the agM in more detail. But one thing should be already clear : none of the three proofs of Theorem 1.1 discussed so far live up to Gauss' May 30, 1799 prediction of "an entirely new field of analysis." In order to see that his claim was justified, we will need to study his work on the agM of complex numbers.

2. THE ARITHMETIC-GEOMETRIC MEAN OF COMPLEX NUMBERS

The arithmetic-geometric mean of two complex numbers a and b is not easy to define. The immediate problem is that in our algorithm

$$a_0 = a, \qquad\qquad b_0 = b,$$

(2.1)

$$a_{n+1} = (a_n + b_n)/2, \qquad b_{n+1} = (a_n b_n)^{1/2}, \qquad n = 0, 1, 2, ...,$$

there is no longer an obvious choice for b_{n+1}. In fact, since we are presented with two choices for b_{n+1} for all $n \geqslant 0$, there are uncountably many sequences $\{a_n\}_{n=0}^{\infty}$ and $\{b_n\}_{n=0}^{\infty}$ for given a and b. Nor is it clear that any of these converge!

We will see below (Proposition 2.1) that in fact all of these sequences converge, but only countably many have a non-zero limit. The limits of these particular sequences then allow us to define $M(a, b)$ as a multiple valued function of a and b. Our main result (Theorem 2.2) gives the relationship between the various values of $M(a, b)$. This theorem was discovered by Gauss in 1800, and we will follow his proof, which makes extensive use of theta functions and modular functions of level four.

We first restrict ourselves to consider only those a's and b's such that $a \neq 0$, $b \neq 0$ and $a \neq \pm b$. (If $a=0$, $b=0$ or $a=\pm b$, one easily sees that the sequences (2.1) converge to either 0 or a, and hence are not very interesting.) An easy induction argument shows that if a and b satisfy these restrictions, so do a_n and b_n for all $n \geq 0$ in (2.1).

We next give a way of distinguishing between the two possible choices for each b_{n+1}.

Definition. Let $a, b \in \mathbf{C}^*$ satisfy $a \neq \pm b$. Then a square root b_1 of ab is called the *right choice* if $|a_1 - b_1| \leq |a_1 + b_1|$ and, when $|a_1 - b_1| = |a_1 + b_1|$, we also have $\mathrm{Im}(b_1/a_1) > 0$.

To see that this definition makes sense, suppose that $\mathrm{Im}(b_1/a_1) = 0$. Then $b_1/a_1 = r \in \mathbf{R}$, and thus

$$|a_1 - b_1| = |a_1| \, |1 - r| \neq |a_1| \, |1 + r| = |a_1 + b_1|$$

since $r \neq 0$. Notice also that the right choice is unchanged if we switch a and b, and that if a and b are as in § 1, then the right choice for $(ab)^{1/2}$ is the positive one.

It thus seems natural that we should define the agM using (2.1) with b_{n+1} always the right choice for $(a_n b_n)^{1/2}$. However, this is not the only possibility: one can make some wrong choices for b_{n+1} and still get an interesting answer. For instance, in Gauss' notebooks, we find the following example:

n	a_n	b_n
0	3.0000000	1.0000000
1	2.0000000	-1.7320508
2	.1339746	1.8612098i
3	.0669873 + .9306049i	.3530969 + .3530969i
4	.2100421 + .6418509i	.2836903 + .6208239i
5	.2468676 + .6313374i	.2470649 + .6324002i
6	.2469962 + .6318688i	.2469962 + .6318685i

(see [12, III, p. 379]). Note that b_1 is the wrong choice but b_n is the right choice for $n \geq 2$. The algorithm appears to converge nicely.

Let us make this idea more precise with a definition.

THE ARITHMETIC-GEOMETRIC MEAN   285

Definition. Let $a, b \in \mathbf{C}^*$ satisfy $a \neq \pm b$. A pair of sequences $\{a_n\}_{n=0}^\infty$ and $\{b_n\}_{n=0}^\infty$ as in (2.1) is called *good* if b_{n+1} is the right choice for $(a_n b_n)^{1/2}$ for all but finitely many $n \geq 0$.

The following proposition shows the special role played by good sequences.

PROPOSITION 2.1. *If $a, b \in \mathbf{C}^*$ satisfy $a \neq \pm b$, then any pair of sequences $\{a_n\}_{n=0}^\infty$ and $\{b_n\}_{n=0}^\infty$ as in (2.1) converge to a common limit, and this common limit is non-zero if and only if $\{a_n\}_{n=0}^\infty$ and $\{b_n\}_{n=0}^\infty$ are good sequences.*

Proof. We first study the properties of the right choice b_1 of $(ab)^{1/2}$ in more detail. Let $0 \leq \text{ang}(a, b) \leq \pi$ denote the unoriented angle between a and b.

Then we have:

(2.2) $$|a_1 - b_1| \leq (1/2)|a - b|$$

(2.3) $$\text{ang}(a_1, b_1) \leq (1/2)\,\text{ang}(a, b).$$

To prove (2.2), note that

$$|a_1 - b_1|\,|a_1 + b_1| = (1/4)|a - b|^2.$$

Since $|a_1 - b_1| \leq |a_1 + b_1|$, (2.2) follows immediately. To prove (2.3), let $\theta_1 = \text{ang}(a_1, b_1)$ and $\theta = \text{ang}(a, b)$. From the law of cosines

$$|a_1 \pm b_1|^2 = |a_1|^2 + |b_1|^2 \pm 2|a_1|\,|b_1|\cos\theta_1,$$

we see that $\theta_1 \leq \pi/2$ because $|a_1 - b_1| \leq |a_1 + b_1|$. Thus

$$\text{ang}(a_1, b_1) = \theta_1 \leq \pi - \theta_1 = \text{ang}(a_1, -b_1).$$

To compare this to θ, note that one of $\pm b_1$, say b_1', satisfies $\text{ang}(a, b_1') = \text{ang}(b_1', b) = \theta/2$. Then the following picture

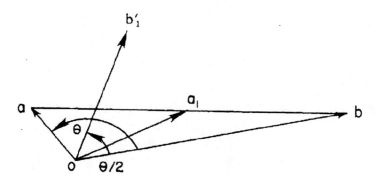

shows that $\text{ang}(a_1, b_1') \leqslant \theta/2$. Since $b_1' = \pm b_1$, the above inequalities imply that

$$\text{ang}(a_1, b_1) \leqslant \text{ang}(a_1, b_1') \leqslant (1/2)\,\text{ang}(a, b)\,,$$

proving (2.3).

Now, suppose that $\{a_n\}_{n=0}^{\infty}$ and $\{b_n\}_{n=0}^{\infty}$ are not good sequences. We set $M_n = \max\{|a_n|, |b_n|\}$, and it suffices to show that $\lim_{n\to\infty} M_n = 0$. Note that $M_{n+1} \leqslant M_n$ for $n \geqslant 0$. Suppose that for some n, b_{n+1} is not the right choice for $(a_n b_n)^{1/2}$. Then $-b_{n+1}$ is the right choice, and thus (2.2), applied to a_n and b_n, implies that

$$|a_{n+2}| = (1/2)\,|a_{n+1} - b_{n+1}| \leqslant (1/4)\,|a_n - b_n| \leqslant (1/2)M_n\,.$$

However, we also have $|b_{n+2}| \leqslant M_n$. It follows easily that

(2.4) $$M_{n+3} \leqslant (3/4)M_n\,.$$

Since $\{a_n\}_{n=0}^{\infty}$ and $\{b_n\}_{n=0}^{\infty}$ are not good sequences, (2.4) must occur infinitely often, proving that $\lim_{n\to\infty} M_n = 0$.

Next, suppose that $\{a_n\}_{n=0}^{\infty}$ and $\{b_n\}_{n=0}^{\infty}$ are good sequences. By neglecting the first N terms for N sufficiently large, we may assume that b_{n+1} is the right choice for all $n \geqslant 0$ and that $\text{ang}(a, b) < \pi$ (this is possible by (2.3)). We also set $\theta_n = \text{ang}(a_n, b_n)$. From (2.2) and (2.3) we obtain

(2.5) $$|a_n - b_n| \leqslant 2^{-n}|a - b|, \qquad \theta_n \leqslant 2^{-n}\theta_0\,.$$

Note that $a_n - a_{n+1} = (1/2)\,(a_n - b_n)$, so that by (2.5),

$$|a_n - a_{n+1}| \leqslant 2^{-(n+1)}|a - b|$$

Hence, if $m > n$, we see that

$$|a_n - a_m| \leqslant \sum_{k=n}^{m-1} |a_k - a_{k+1}| \leqslant \left(\sum_{k=n}^{m-1} 2^{-(k+1)}\right)|a - b| < 2^{-n}|a - b|\,.$$

Thus $\{a_n\}_{n=0}^{\infty}$ converges because it is a Cauchy sequence, and then (2.5) implies that $\lim_{n\to\infty} a_n = \lim_{n\to\infty} b_n$.

It remains to show that this common limit is nonzero. Let

$$m_n = \min\{|a_n|, |b_n|\}\,.$$

Clearly $|b_{n+1}| \geqslant m_n$. To relate $|a_{n+1}|$ and m_n, we use the law of cosines:

$$(2|a_{n+1}|)^2 = |a_n|^2 + |b_n|^2 + 2|a_n||b_n|\cos\theta_n$$
$$\geqslant 2m_n^2(1+\cos\theta_n) = 4m_n^2\cos^2(\theta_n/2).$$

It follows that $m_{n+1} \geqslant \cos(\theta_n/2)m_n$ since $0 \leqslant \theta_n < \pi$ (this uses (2.5) and the fact that $\theta_0 = \text{ang}(a, b) < \pi$). Using (2.5) again, we obtain

$$m_n \geqslant \left(\prod_{k=1}^{n}\cos(\theta_0/2^k)\right)m_0.$$

However, it is well known that

$$\prod_{k=1}^{\infty}\cos(\theta_0/2^k) = \frac{\sin\theta_0}{\theta_0}.$$

(See [16, p. 38]. When $\theta_0 = 0$, the right hand side is interpreted to be 1.) We thus have

$$m_n \geqslant \left(\frac{\sin\theta_0}{\theta_0}\right)m_0$$

for all $n \geqslant 1$. Since $0 \leqslant \theta_0 < \pi$, it follows that $\lim\limits_{n\to\infty} a_n = \lim\limits_{n\to\infty} b_n \neq 0$.   QED

We now define the agM of two complex numbers.

Definition. Let $a, b \in \mathbf{C}^*$ satisfy $a \neq \pm b$. A nonzero complex number μ is a value of the *arithmetic-geometric mean* $M(a, b)$ of a and b if there are good sequences $\{a_n\}_{n=0}^{\infty}$ and $\{b_n\}_{n=0}^{\infty}$ as in (2.1) such that

$$\mu = \lim_{n\to\infty} a_n = \lim_{n\to\infty} b_n.$$

Thus $M(a, b)$ is a multiple valued function of a and b and there are a countable number of values. Note, however, that there is a distinguished value of $M(a, b)$, namely the common limit of $\{a_n\}_{n=0}^{\infty}$ and $\{b_n\}_{n=0}^{\infty}$ where b_{n+1} is the right choice for $(a_n b_n)^{1/2}$ for *all* $n \geqslant 0$. We will call this the *simplest value* of $M(a, b)$. When a and b are positive real numbers, this simplest value is just the agM as defined in § 1.

We now come to the major result of this paper, which determines how the various values of $M(a, b)$ are related for fixed a and b.

THEOREM 2.2. *Fix* $a, b \in \mathbf{C}^*$ *which satisfy* $a \neq \pm b$ *and* $|a| \geqslant |b|$, *and let* μ *and* λ *denote the simplest values of* $M(a, b)$ *and* $M(a+b, a-b)$ *respectively. Then all values* μ' *of* $M(a, b)$ *are given by the formula*

$$\frac{1}{\mu'} = \frac{d}{\mu} + \frac{ic}{\lambda},$$

where   d   and   c   are arbitrary relatively prime integers satisfying $d \equiv 1 \bmod 4$ and $c \equiv 0 \bmod 4$.

Proof. Our treatment of the agM of complex numbers thus far has been fairly elementary. The proof of this theorem, however, will be quite different; we will finally discover the "entirely new field of analysis" predicted by Gauss in the diary entry quoted in § 1. In the proof we will follow Gauss' ideas and even some of his notations, though sometimes translating them to a modern setting and of course filling in the details he omitted (Gauss' notes are extremely sketchy and incomplete — see [12, III, pp. 467-468 and 477-478]).

The proof will be broken up into four steps. In order to avoid writing a treatise on modular functions, we will quote certain classical facts without proof.

Step 1.   Theta Functions

Let $\mathfrak{H} = \{\tau \in \mathbf{C}: \operatorname{Im}\tau > 0\}$ and set $q = e^{\pi i \tau}$. The Jacobi theta functions are defined as follows:

$$p(\tau) = 1 + 2 \sum_{n=1}^{\infty} q^{n^2} = \Theta_3(\tau, 0),$$

$$q(\tau) = 1 + 2 \sum_{n=1}^{\infty} (-1)^n q^{n^2} = \Theta_4(\tau, 0),$$

$$r(\tau) = 2 \sum_{n=1}^{\infty} q^{(2n-1)^2/4} = \Theta_2(\tau, 0).$$

Since $|q| < 1$ for $\tau \in \mathfrak{H}$, these are holomorphic functions of τ. The notation p, q and r is due to Gauss, though he wrote them as power series in $e^{-\pi t}$, $\operatorname{Re} t > 0$ (thus he used the right half plane rather than the upper half plane $\mathfrak{H}$ — see [12, III, pp. 383-386]). The more common notation Θ_3, Θ_4 and Θ_2 is from [36, p. 464] and [32, p. 27].

A wealth of formulas are associated with these functions, including the product expansions:

$$p(\tau) = \prod_{n=1}^{\infty} (1 - q^{2n})(1 + q^{2n-1})^2,$$

(2.6)

$$q(\tau) = \prod_{n=1}^{\infty} (1 - q^{2n})(1 - q^{2n-1})^2,$$

$$r(\tau) = 2q^{1/4} \prod_{n=1}^{\infty} (1-q^{2n})(1+q^{2n})^2 \,,$$

(which show that $p(\tau)$, $q(\tau)$ and $r(\tau)$ are nonvanishing on $\mathfrak{H}$), the transformations:

$$
\begin{array}{ll}
& p(\tau+1) = q(\tau)\,, \qquad & p(-1/\tau) = (-i\tau)^{1/2}p(\tau)\,, \\
(2.7) \quad & q(\tau+1) = p(\tau)\,, \qquad & q(-1/\tau) = (-i\tau)^{1/2}r(\tau)\,, \\
& r(\tau+1) = e^{\pi i/4}r(\tau)\,, \qquad & r(-1/\tau) = (-i\tau)^{1/2}q(\tau)\,,
\end{array}
$$

(where we assume that $\mathrm{Re}(-i\tau)^{1/2} > 0$), and finally the identities

$$
\begin{array}{l}
p(\tau)^2 + q(\tau)^2 = 2p(2\tau)^2\,, \\
(2.8) \qquad p(\tau)^2 - q(\tau)^2 = 2r(2\tau)^2\,, \\
p(\tau)q(\tau) = q(2\tau)^2\,,
\end{array}
$$

and

$$
\begin{array}{l}
p(2\tau)^2 + r(2\tau)^2 = p(\tau)^2\,, \\
(2.9) \qquad p(2\tau)^2 - r(2\tau)^2 = q(\tau)^2\,, \\
q(\tau)^4 + r(\tau)^4 = p(\tau)^4\,.
\end{array}
$$

Proofs of (2.6) and (2.7) can be found in [36, p. 469 and p. 475], while one must turn to more complete works like [32, pp. 118-119] for proofs of (2.8). (For a modern proof of (2.8), consult [34].) Finally, (2.9) follows easily from (2.8). Of course, Gauss knew all of these formulas (see [12, III, pp. 386 and 466-467]).

What do these formulas have to do with the agM? The key lies in (2.8): one sees that $p(2\tau)^2$ and $q(2\tau)^2$ are the respective arithmetic and geometric means of $p(\tau)^2$ and $q(\tau)^2$! To make the best use of this observation, we need to introduce the function $k'(\tau) = q(\tau)^2/p(\tau)^2$.

Then we have:

LEMMA 2.3. *Let* $a, b \in \mathbf{C}^*$ *satisfy* $a \neq \pm b$, *and suppose there is* $\tau \in \mathfrak{H}$ *such that* $k'(\tau) = b/a$. *Set* $\mu = a/p(\tau)^2$ *and, for* $n \geq 0$, $a_n = \mu \, p(2^n\tau)^2$ *and* $b_n = \mu \, q(2^n\tau)^2$. *Then*

(i) $\{a_n\}_{n=0}^{\infty}$ *and* $\{b_n\}_{n=0}^{\infty}$ *are good sequences satisfying (2.1),*

(ii) $\lim_{n \to \infty} a_n = \lim_{n \to \infty} b_n = \mu$.

Proof. We have $a_0 = a$ by definition, and $b_0 = b$ follows easily from $k'(\tau) = b/a$. As we observed above, the other conditions of (2.1) are clearly

satisfied. Finally, note that $\exp(\pi i 2^n \tau) \to 0$ as $n \to \infty$, so that $\lim\limits_{n \to \infty} p(2^n \tau)^2$
$= \lim\limits_{n \to \infty} q(2^n \tau)^2 = 1$, and (ii) follows. Since $\mu \neq 0$, Proposition 2.1 shows that
$\{a_n\}_{n=0}^{\infty}$ and $\{b_n\}_{n=0}^{\infty}$ are good sequences.   QED

Thus every solution τ of $k'(\tau) = b/a$ gives us a value $\mu = a/p(\tau)^2$ of $M(a, b)$. As a first step toward understanding all solutions of $k'(\tau) = b/a$, we introduce the region $F_1 \subseteq \mathfrak{H}$:

$$F_1 = \{\tau \in \mathfrak{H} : |\operatorname{Re}\tau| \leqslant 1, |\operatorname{Re}(1/\tau)| \leqslant 1\}$$

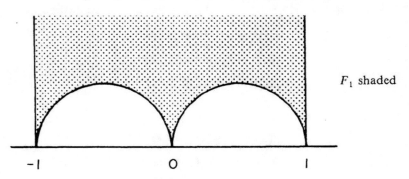

F_1 shaded

The following result is well known.

LEMMA 2.4. k'^2 *assumes every value in* $\mathbf{C} - \{0, 1\}$ *exactly once in* $F'_1 = F_1 - (\partial F_1 \cap \{\tau \in \mathfrak{H} : \operatorname{Re}\tau < 0\})$.

A proof can be found in [36, pp. 481-484]. Gauss was aware of similar results which we will discuss below. He drew F_1 as follows (see [12, III, p. 478]).

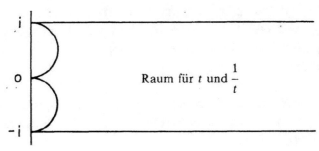

Raum für t und $\dfrac{1}{t}$

Note that our restrictions on a and b ensure that $(b/a)^2 \in \mathbf{C} - \{0, 1\}$. Thus, by Lemma 2.4, we can always solve $k'(\tau)^2 = (b/a)^2$, i.e., $k'(\tau) = \pm b/a$. We will prove below that

(2.10) $$k'\left(\frac{\tau}{2\tau+1}\right) = -k'(\tau),$$

which shows that we can always solve $k'(\tau) = b/a$. Thus, for every a and b as above, $M(a, b)$ has at least one value of the form $a/p(\tau)^2$, where $k'(\tau) = b/a$.

Three tasks now remain. We need to find *all* solutions τ of $k'(\tau) = b/a$, we need to see how the values $a/p(\tau)^2$ are related for these τ's, and we need to prove that *all* values of $M(a, b)$ arise in this way. To accomplish these goals, we must first recast the properties of $k'(\tau)$ and $p(\tau)^2$ into more modern terms.

Step 2.   Modular Forms of Weight One.

The four lemmas proved here are well known to experts, but we include their proofs in order to show how easily one can move from the classical facts of Step 1 to their modern interpretations. We will also discuss what Gauss had to say about these facts.

We will use the transformation properties (2.7) by way of the group

$$SL(2, \mathbf{Z}) = \left\{ \begin{pmatrix} a & b \\ c & d \end{pmatrix} : a, b, c, d \in \mathbf{Z},\ ad - bc = 1 \right\}$$

which acts on $\mathfrak{H}$ by linear fractional transformations as follows: if $\gamma = \begin{pmatrix} a & b \\ c & d \end{pmatrix} \in SL(2, \mathbf{Z})$ and $\tau \in \mathfrak{H}$, then $\gamma\tau = \dfrac{a\tau + b}{c\tau + d}$.

For example, if

$$S = \begin{pmatrix} 0 & -1 \\ 1 & 0 \end{pmatrix} \quad \text{and} \quad T = \begin{pmatrix} 1 & 1 \\ 0 & 1 \end{pmatrix}, \quad \text{then} \quad S\tau = \frac{-1}{\tau}, \quad T\tau = \tau + 1,$$

which are the transformations in (2.7). It can be shown that S and T generate $SL(2, \mathbf{Z})$ (see [29, Ch. VII, Thm. 2]), a fact we do not need here.

We will consider several subgroups of $SL(2, \mathbf{Z})$. The first of these is $\Gamma(2)$, the principal congruence subgroup of level 2:

$$\Gamma(2) = \left\{ \gamma \in SL(2, \mathbf{Z}): \gamma \equiv \begin{pmatrix} 1 & 0 \\ 0 & 1 \end{pmatrix} \bmod 2 \right\}.$$

Note that $-1 \in \Gamma(2)$ and that $\Gamma(2)/\{\pm 1\}$ acts on $\mathfrak{H}$.

LEMMA 2.5.

(i) $\Gamma(2)/\{\pm 1\}$ *acts freely on* $\mathfrak{H}$.

(ii) $\Gamma(2)$ *is generated by* $-1, U = \begin{pmatrix} 1 & 2 \\ 0 & 1 \end{pmatrix}$ *and* $V = \begin{pmatrix} 1 & 0 \\ 2 & 1 \end{pmatrix}$.

(iii) *Given* $\tau \in \mathfrak{H}$, *there is* $\gamma \in \Gamma(2)$ *such that* $\gamma\tau \in F_1$.

Proof. Let $\gamma = \begin{pmatrix} a & b \\ c & d \end{pmatrix}$ be an element of $\Gamma(2)$.

(i) If $\tau \in \mathfrak{H}$ and $\gamma\tau = \tau$, then we obtain $c\tau^2 + (d-a)\tau - b = 0$. If $c = 0$, then $\gamma = \pm 1$ follows immediately. If $c \neq 0$, then $(d-a)^2 + 4bc < 0$ because $\tau \in \mathfrak{H}$. Using $ad - bc = 1$, this becomes $(a+d)^2 < 4$, and thus $a + d = 0$ since a and d are odd. However, b and c are even so that

$$1 \equiv ad - bc \equiv ad \equiv -a^2 \bmod 4$$

This contradiction proves (i).

(ii) We start with a variation of the Euclidean algorithm. Given γ as above, let $r_1 = a - 2a_1c$, where $a_1 \in \mathbf{Z}$ is chosen so that $|r_1|$ is minimal. Then $|r_1| \leq |c|$, and hence $|r_1| < |c|$ since a and c have different parity. Thus

$$a = 2a_1c + r_1, \quad a_1, r_1 \in \mathbf{Z}, \quad |r_1| < |c|.$$

Note that c and r_1 also have different parity. Continuing this process, we obtain

$$c = 2a_2 r_1 + r_2, \quad |r_2| < |r_1|,$$
$$r_1 = 2a_3 r_2 + r_3, \quad |r_3| < |r_2|,$$
$$\vdots$$
$$r_{2n-1} = 2a_{2n+1} r_{2n} + r_{2n+1}, \quad r_{2n+1} = \pm 1,$$
$$r_{2n} = 2a_{2n+2} r_{2n+1} + 0,$$

since $\mathrm{GCD}(a, c) = 1$. Then one easily computes that

$$V^{-a_{2n+2}} U^{-a_{2n+1}} \ldots V^{-a_2} U^{-a_1} \gamma = \begin{pmatrix} \pm 1 & * \\ 0 & * \end{pmatrix}.$$

Since the left-hand side is in $\Gamma(2)$, the right-hand side must be of the form $\pm U^m$, and we thus obtain

$$\gamma = \pm\, U^{a_1} V^{a_2} \ldots U^{a_{2n+1}} V^{a_{2n+2}} U^m.$$

(iii) Fix $\tau \in \mathfrak{H}$. The quadratic form $|x\tau + y|^2$ is positive definite for $x, y \in \mathbf{R}$, so that for any $S \subseteq \mathbf{Z}^2$, $|x\tau + y|^2$ assumes a minimum value at some $(x, y) \in S$. In particular, $|c\tau + d|^2$, where $\gamma = \begin{pmatrix} a & b \\ c & d \end{pmatrix} \in \Gamma(2)$, assumes a minimum value at some $\gamma_0 \in \Gamma(2)$. Since $\operatorname{Im} \gamma\tau = \operatorname{Im} \tau |c\tau + d|^{-2}$, we see that $\tau' = \gamma_0 \tau$ has maximal imaginary part, i.e., $\mathrm{Im}\,\tau' \geqslant \mathrm{Im}\,\gamma\tau'$ for $\gamma \in \Gamma(2)$. Since $\mathrm{Im}\,\tau' = \mathrm{Im}\,U\tau'$, we may assume that $|\,\mathrm{Re}\,\tau'\,| \leqslant 1$. Applying the above inequality to $V^{\pm 1} \in \Gamma(2)$, we obtain

$$\mathrm{Im}\,\tau' \geqslant \mathrm{Im}\,V^{\pm 1}\tau' = \mathrm{Im}\,\tau'\,|\,2\tau' \pm 1\,|^{-2}.$$

Thus $|\,2\tau \pm 1\,| \geqslant 1$, or $|\,\tau \pm (1/2)\,| \geqslant 1/2$. This is equivalent to $|\,\mathrm{Re}\,1/\tau'\,| \leqslant 1$, and hence $\tau' \in F_1$.   QED

We next study how $p(\tau)$ and $q(\tau)$ transform under elements of $\Gamma(2)$.

LEMMA 2.6. Let $\gamma = \begin{pmatrix} a & b \\ c & d \end{pmatrix} \in \Gamma(2)$, and assume that $a \equiv d \equiv 1 \bmod 4$.
Then

(i) $p(\gamma\tau)^2 = (c\tau + d)\,p(\tau)^2$,

(ii) $q(\gamma\tau)^2 = i^c(c\tau + d)\,q(\tau)^2$.

Proof. From (2.7) and $V = \begin{pmatrix} 0 & -1 \\ 1 & 0 \end{pmatrix} U^{-1} \begin{pmatrix} 0 & 1 \\ -1 & 0 \end{pmatrix}$ we obtain

$$p(U\tau)^2 = p(\tau)^2, \qquad p(V\tau)^2 = (2\tau + 1)\,p(\tau)^2,$$

(2.11)

$$q(U\tau)^2 = q(\tau)^2, \qquad q(V\tau)^2 = -(2\tau + 1)\,q(\tau)^2.$$

Thus (i) and (ii) hold for U and V. The proof of the previous lemma shows that γ is in the subgroup of $\Gamma(2)$ generated by U and V. We now proceed by induction on the length of γ as a word in U and V.

(i) If $\gamma = \begin{pmatrix} a & b \\ c & d \end{pmatrix}$ and $p(\gamma\tau)^2 = (c\tau + d)\,p(\tau)^2$ then (2.11) implies that

$$p(U\gamma\tau)^2 = p(\gamma\tau)^2 = (c\tau + d)\,p(\tau)^2,$$
$$p(V\gamma\tau)^2 = (2\gamma\tau + 1)\,p(\gamma\tau)^2 = (2\gamma\tau + 1)\,(c\tau + d)\,p(\tau)^2$$
$$= ((2a + c)\tau + (2b + d))\,p(\tau)^2.$$

However $U\gamma = \begin{pmatrix} * & * \\ c & d \end{pmatrix}$, $V\gamma = \begin{pmatrix} * & * \\ 2a + c & 2b + d \end{pmatrix}$, so that (i) now holds for $U\gamma$ and $V\gamma$.

(ii) Using (2.11) and arguing as above, we see that if $\gamma = \begin{pmatrix} a & b \\ c & d \end{pmatrix}$
$= U^{a_1} V^{b_1} \dots U^{a_n} V^{b_n}$, then

$$q(\gamma\tau)^2 = (-1)^{\Sigma b_i}(c\tau + d)\,q(\tau)^2.$$

However, U and V commute modulo 4, so that

$$\gamma \equiv \begin{pmatrix} 1 & 2\Sigma a_i \\ 2\Sigma b_i & 1 \end{pmatrix} \text{ mod } 4 .$$

Thus $c \equiv 2\Sigma b_i \text{ mod } 4$, and (ii) follows.   QED

Note that (2.10) is an immediate consequence of Lemma 2.6.

In order to fully exploit this lemma, we introduce the following subgroups of $\Gamma(2)$:

$$\Gamma(2)_0 = \{\gamma \in \Gamma(2) : a \equiv d \equiv 1 \text{ mod } 4\} ,$$

$$\Gamma_2(4) = \{\gamma \in \Gamma(2)_0 : c \equiv 0 \text{ mod } 4\}$$

Note that $\Gamma(2) = \{\pm 1\} \cdot \Gamma(2)_0$ and that $\Gamma_2(4)$ has index 2 in $\Gamma(2)_0$. From Lemma 2.6 we obtain

$$p(\gamma\tau)^2 = (c\tau + d) \, p(\tau)^2 , \qquad \gamma \in \Gamma(2)_0 ,$$

(2.12)

$$q(\gamma\tau)^2 = (c\tau + d) \, q(\tau)^2 , \qquad \gamma \in \Gamma_2(4) .$$

Since these functions are holomorphic on $\mathfrak{H}$, one says that $p(\tau)^2$ and $q(\tau)^2$ are weak modular forms of weight one for $\Gamma(2)_0$ and $\Gamma_2(4)$ respectively. The term more commonly used is modular form, which requires that the functions be holomorphic at the cusps (see [30, pp. 28-29] for a precise definition). Because $\Gamma(2)_0$ and $\Gamma_2(4)$ are congruence subgroups of level $N = 4$, this condition reduces to proving that

(2.13) $(c\tau + d)^{-1} \, p(\gamma\tau)^2 , \qquad (c\tau + d)^{-1} \, q(\gamma\tau)^2 ,$

are holomorphic functions of $q^{1/2} = \exp(2\pi i\tau/4)$ for all $\gamma \in SL(2, \mathbf{Z})$. This will be shown later.

In general, it is well known that the square of a theta function is a modular form of weight one (see [27, Ch. I, § 9]), although the general theory only says that our functions are modular forms for the group

$$\Gamma(4) = \{\gamma \in SL(2, \mathbf{Z}) : \gamma \equiv \begin{pmatrix} 1 & 0 \\ 0 & 1 \end{pmatrix} \text{ mod } 4\}$$

(see [27, Ch. I, Prop. 9.2]). We will need the more precise information given by (2.12).

We next study the quotients of $\mathfrak{H}$ by $\Gamma(2)$ and $\Gamma_2(4)$. From Step 1, recall the region $F_1 \subseteq \mathfrak{H}$. We now define a larger region F:

$$F = \{\tau \in \mathfrak{H} : |\text{Re}\tau| \leqslant 1, \ |\tau \pm 1/4| \geqslant 1/4, \ |\tau \pm 3/4| \geqslant 1/4\} .$$

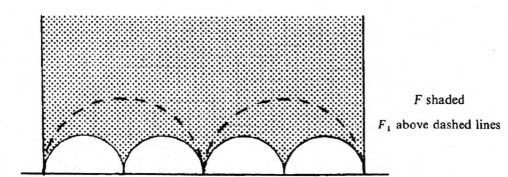

F shaded

F_1 above dashed lines

We also set

$$F'_1 = F_1 - (\partial F_1 \cap \{\tau \in \mathfrak{H}: \operatorname{Re}\tau < 0\})$$

$$F' = F - (\partial F \cap \{\tau \in \mathfrak{H}: \operatorname{Re}\tau < 0\}).$$

LEMMA 2.7. F'_1 *and* F' *are fundamental domains for* $\Gamma(2)$ *and* $\Gamma_2(4)$ *respectively, and the functions* k'^2 *and* k' *induce biholomorphic maps*

$$\overline{k'^2}: \mathfrak{H}/\Gamma(2) \xrightarrow{\sim} \mathbf{C} - \{0, 1\}$$

$$\overline{k'} : \mathfrak{H}/\Gamma_2(4) \xrightarrow{\sim} \mathbf{C} - \{0, \pm 1\}.$$

Proof. A simple modification of the proof of Lemma 2.6 shows that if $\gamma = \begin{pmatrix} a & b \\ c & d \end{pmatrix} \in \Gamma(2)$, then $p(\gamma\tau)^4 = (c\tau+d)^2 p(\tau)^4$, $q(\gamma\tau)^4 = (c\tau+d)^2 q(\tau)^4$. Thus k'^2 is invariant under $\Gamma(2)$.

Given $\tau \in \mathfrak{H}$, Lemma 2.5 shows that $\gamma\tau \in F_1$ for some $\gamma \in \Gamma(2)$. Since $U = \begin{pmatrix} 1 & 2 \\ 0 & 1 \end{pmatrix}$ maps the left vertical line in ∂F_1 to the right one and $V = \begin{pmatrix} 1 & 0 \\ 2 & 1 \end{pmatrix}$ maps the left semicircle in ∂F_1 to the right one, we may assume that $\gamma\tau \in F'_1$. If we also had $\sigma\tau \in F'_1$ for $\sigma \in \Gamma(2)$, then $k'(\sigma\tau)^2 = k'(\tau)^2 = k'(\gamma\tau)^2$, so that $\sigma\tau = \gamma\tau$ by Lemma 2.4. This shows that F'_1 is a fundamental domain for $\Gamma(2)$.

Since $\Gamma(2)_0 \simeq \Gamma(2)/\{\pm 1\}$, F'_1 is also a fundamental domain for $\Gamma(2)_0$. Since $\Gamma_2(4)$ has index 2 in $\Gamma(2)_0$ with 1 and V as coset representatives, it follows that

$$F^* = F'_1 \cup V(F'_1 \cap \{\tau \in \mathfrak{H}: \operatorname{Re}\tau \leqslant 0\}) \cup V^{-1}(F'_1 \cap \{\tau \in \mathfrak{H}: \operatorname{Re}\tau > 0\})$$

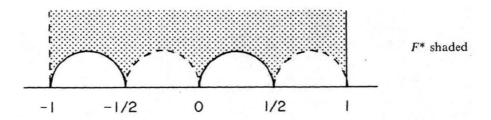

F^* shaded

is a fundamental domain for $\Gamma_2(4)$. Since $\begin{pmatrix} -3 & -2 \\ -4 & -3 \end{pmatrix} \in \Gamma_2(4)$ takes the far left semicircle in ∂F to the far right one, it follows that F' is a fundamental domain for $\Gamma_2(4)$.

It now follows easily from Lemma 2.4 that k'^2 induces a bijection $\overline{k'^2} : \mathfrak{H}/\Gamma(2) \to \mathbf{C} - \{0, 1\}$. Since $\Gamma(2)/\{\pm 1\}$ acts freely on $\mathfrak{H}$ by Lemma 2.5, $\mathfrak{H}/\Gamma(2)$ is a complex manifold and $\overline{k'^2}$ is holomorphic. A straightforward argument then shows that $\overline{k'^2}$ is biholomorphic.

Next note that k' is invariant under $\Gamma_2(4)$ by (2.12), and thus induces a map $\overline{k'} : \mathfrak{H}/\Gamma_2(4) \to \mathbf{C} - \{0, \pm 1\}$. Since $\mathfrak{H}/\Gamma(2) = \mathfrak{H}/\Gamma(2)_0$, we obtain a commutative diagram:

$$
\begin{array}{ccc}
\mathfrak{H}/\Gamma_2(4) & \overset{\overline{k'}}{\to} & \mathbf{C} - \{0, 1\} \\[2mm]
f \downarrow & & \downarrow g \\[2mm]
\mathfrak{H}/\Gamma(2)_0 & \overset{\overline{k'^2}}{\to} & \mathbf{C} - \{0, 1\}
\end{array}
$$

where f is induced by $\Gamma_2(4) \subseteq \Gamma(2)_0$ and g is just $g(z) = z^2$. Note that g is a covering space of degree 2, and the same holds for f since $[\Gamma(2)_0 : \Gamma_2(4)] = 2$ and $\Gamma(2)_0$ acts freely on $\mathfrak{H}$. We know that $\overline{k'^2}$ is a biholomorphism, and it now follows easily that $\overline{k'}$ is also.   QED

We should point out that $r(\tau)^2$ has properties similar to $p(\tau)^2$ and $q(\tau)^2$. Specifically, $r(\tau)^2$ is a modular form of weight one for the group

$$
\Gamma_2(4)^t = \left\{ \gamma \in \Gamma(2) : \gamma \equiv \begin{pmatrix} 1 & 0 \\ * & 1 \end{pmatrix} \bmod 4 \right\},
$$

which is a conjugate of $\Gamma_2(4)$. Furthermore, if we set $k(\tau) = r(\tau)^2/p(\tau)^2$, then k is invariant under $\Gamma_2(4)^t$ and induces a biholomorphism $\overline{k} : \mathfrak{H}/\Gamma_2(4)^t$

$\rightarrow \mathbf{C} - \{0, \pm 1\}$. We leave the proofs to the reader. Note also that $k(\tau)^2 + k'(\tau)^2 = 1$ by (2.9).

Our final lemma will be useful in studying the agM. Let F_2 be the region $(1/2)F_1$, pictured below. Note that $F_2 \subseteq F$.

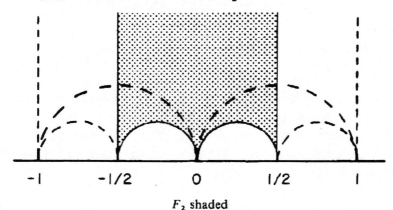

F_2 shaded

F, F_1 indicated by dashed lines

LEMMA 2.8.

$$k'(F_1) = \{z \in \mathbf{C} - \{0, \pm 1\} : \mathrm{Re}\, z \geqslant 0\},$$

$$k'(F_2) = \{z \in \mathbf{C} - \{0, \pm 1\} : |z| \leqslant 1\}.$$

Proof. We will only treat $k'(F_2)$, the proof for $k'(F_1)$ being quite similar. We first claim that $\{k'(\tau) : \mathrm{Re}\,\tau = \pm 1/2\} = S^1 - \{\pm 1\}$. To see this, note that $\mathrm{Re}\,\tau = \pm 1/2$ and the product expansions (2.6) easily imply that $\overline{k'(\tau)} = k'(\tau)^{-1}$, i.e., $|k'(\tau)| = 1$. How much of the circle is covered? It is easy to see that $k'(\pm 1/2 + it) \rightarrow 1$ as $t \rightarrow +\infty$. To study the limit as $t \rightarrow 0$, note that by (2.10) we have

$$k'(\pm 1/2 + it) = -k'\left(\pm 1/2 + \frac{i}{4t}\right).$$

As $t \rightarrow 0$, the right-hand side clearly approaches -1. Then connectivity arguments easily show that all of $S^1 - \{\pm 1\}$ is covered.

Since k' is injective on F' by Lemma 2.7, it follows that $k'(F_2) - S^1$ is connected. Since $|k'(it)| < 1$ for $t > 0$ by (2.6), we conclude that

$$k'(F_2) \subseteq \{z \in \mathbf{C} - \{0, \pm 1\} : |z| \leqslant 1\}.$$

Similar arguments show that

$$k'(F - F_2) \subseteq \{z \in \mathbf{C} : |z| > 1\}.$$

Since $k'(F) = \mathbf{C} - \{0, \pm 1\}$ by Lemma 2.7, both inclusions must be equalities.

QED

Gauss' collected works show that he was familiar with most of this material, though it's hard to tell precisely what he knew. For example, he basically has two things to say about $k'(\tau)$:

(i) $k'(\tau)$ has positive real part for $\tau \in F_1$,

(ii) the equation $k'(\tau) = A$ has one and only one solution $\tau \in F_2$.

(See [12, III, pp. 477-478].) Neither statement is correct as written. Modifications have to be made regarding boundary behavior, and Lemma 2.8 shows that we must require $|A| \leqslant 1$ in (ii). Nevertheless, these statements show that Gauss essentially knew Lemma 2.8, and it becomes clear that he would not have been greatly surprised by Lemmas 2.4 and 2.7.

Let us see what Gauss had to say about other matters we've discussed. He was quite aware of linear fractional transformations. Since he used the right half plane, he wrote

$$t' = \frac{at - bi}{cti + d}, \quad ad - bc = 1, \quad a, b, c, d \in \mathbf{Z}, \quad \mathrm{Re}t > 0$$

(see [12, III, p. 386]). To prevent confusion, we will always translate formulas into ones involving $\tau \in \mathfrak{H}$.

Gauss decomposed an element $\gamma \in SL(2, \mathbf{Z})$ into simpler ones by means of continued fractions. For example, Gauss considers those transformations $\tau^* = \gamma\tau$ which can be written as

$$\tau' = \frac{-1}{\tau} + 2a_1$$

(2.14)
$$\tau'' = \frac{-1}{\tau'} + 2a_2$$

$$\vdots$$

$$\tau^* = \tau^{(n)} = \frac{-1}{\tau^{(n-1)}} + 2a_n$$

(see [12, X.1, p. 223]). If $U = \begin{pmatrix} 1 & 2 \\ 0 & 1 \end{pmatrix}$ and $V = \begin{pmatrix} 1 & 0 \\ 2 & 1 \end{pmatrix}$, then $\tau'' = U^{a_2}V^{-a_1}\tau$, so that for n even we see a similarity to the proof of Lemma 2.5 (ii). The similarity becomes deeper once we realize that the algorithm used in the proof gives a continued fraction expansion for a/c, where $\gamma = \begin{pmatrix} a & b \\ c & d \end{pmatrix}$.

However, since n can be odd in (2.14), we are dealing with more than just elements of $\Gamma(2)$.

Gauss' real concern becomes apparent when we see him using (2.14) together with the transformation properties of $p(\tau)$. From (2.7) he obtains

$$p(\tau^*) = \sqrt{(-i\tau)(-i\tau')\cdots(-i\tau^{(n-1)})}\, p(\tau)$$

(see [12, X.1, p. 223]). The crucial thing to note is that if $\tau^* = \gamma\tau$, $\gamma = \begin{pmatrix} a & b \\ c & d \end{pmatrix}$, then $(-i\tau)\cdots(-i\tau^{(n-1)})$ is just $c\tau + d$ up to a power of i. This tells us how $p(\tau)$ transforms under those γ's described by (2.14). In general, Gauss used similar methods to determine how $p(\tau)$, $q(\tau)$ and $r(\tau)$ transform under arbitrary elements γ of $SL(2, \mathbf{Z})$. The answer depends in part on how $\gamma = \begin{pmatrix} a & b \\ c & d \end{pmatrix}$ reduces modulo 2. Gauss labeled the possible reductions as follows:

a	1	1	1	0	1	0
b	0	1	0	1	1	1
c	0	0	1	1	1	1
d	1	1	1	1	0	0
	1	2	3	4	5	6

(see [12, X.1, p. 224]). We recognize this as the isomorphism $SL(2, \mathbf{Z})/\Gamma(2) \simeq SL(2, \mathbf{F}_2)$, and note that (2.14) corresponds to cases 1 and 6. Then the transformations of $p(\tau)$, $q(\tau)$ and $r(\tau)$ under $\gamma = \begin{pmatrix} a & b \\ c & d \end{pmatrix} \in SL(2, \mathbf{Z})$ are given by

	1	2	3	4	5	6
$h^{-1} p(\gamma\tau) =$	$p(\tau)$	$q(\tau)$	$r(\tau)$	$q(\tau)$	$r(\tau)$	$p(\tau)$
$h^{-1} q(\gamma\tau) =$	$q(\tau)$	$p(\tau)$	$p(\tau)$	$r(\tau)$	$p(\tau)$	$r(\tau)$
$h^{-1} r(\gamma\tau) =$	$r(\tau)$	$r(\tau)$	$q(\tau)$	$p(\tau)$	$q(\tau)$	$q(\tau)$

(2.15)

where $h = (i^\lambda (c\tau + d))^{1/2}$ and λ is an integer depending on both γ and which one of $p(\tau)$, $q(\tau)$ or $r(\tau)$ is being transformed (see [12, X.1, p. 224]). Note that Lemma 2.6 can be regarded as giving a careful analysis of λ in case 1. An analysis of the other cases may be found in [13, pp. 117-123]. One consequence of this table is that the functions (2.13) are holomorphic functions of $q^{1/2}$, which proves that $p(\tau)^2$, $q(\tau)^2$ and $r(\tau)^2$ are modular forms, as claimed earlier.

Gauss did not make explicit use of congruence subgroups, although they appear implicitly in several places. For example, the table (2.15) shows Gauss using $\Gamma(2)$. As for $\Gamma(2)_0$, we find Gauss writing

$$k'(\gamma\tau) = i^r k'(\tau)$$

where $\gamma = \begin{pmatrix} a & -b \\ -c & d \end{pmatrix}$ and, as he carefully stipulates, "$ad - bc = 1$, $a \equiv d \equiv 1 \bmod 4$, b, c even" (see [12, III, p. 478]). Also, if we ask which of these γ's leave k' unchanged, then the above equation immediately gives us $\Gamma_2(4)$, though we should be careful not to read too much into what Gauss wrote.

More interesting is Gauss' use of the reduction theory of positive definite quadratic forms as developed in Disquisitiones Arithmeticae (see [11, § 171]). This can be used to determine fundamental domains as follows. A positive definite quadratic form $ax^2 + 2bxy + cy^2$ may be written $a \mid x - \tau y \mid^2$ where $\tau \in \mathfrak{H}$. An easy computation shows that this form is equivalent via an element γ of $SL(2, \mathbf{Z})$ to another form $a' \mid x - \tau'y \mid^2$ if and only if $\tau' = \gamma^{-1}\tau$. Then, given $\tau \in \mathfrak{H}$, Gauss applies the reduction theory mentioned above to $\mid x - \tau y \mid^2$ and obtains a $SL(2, \mathbf{Z})-$ equivalent form $A \mid x - \tau'y \mid^2 = Ax^2 + 2Bxy + Cy^2$ which is reduced, i.e.

$$2 \mid B \mid \,\leqslant A \leqslant C$$

(see [11, § 171] and [12, X.1, p. 225]). These inequalities easily imply that $\mid \mathrm{Re}\tau' \mid \,\leqslant 1/2$, $\mid \mathrm{Re}\, 1/\tau' \mid \,\leqslant 1/2$, so that τ' lies in the shaded region

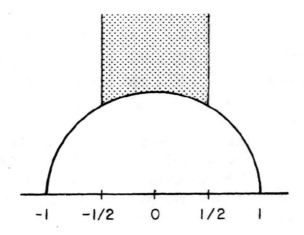

which is well known to be the fundamental domain of $SL(2, \mathbf{Z})$ acting on $\mathfrak{H}$ (see [29, Ch. VII, Thm. 1]).

This seems quite compelling, but Gauss never gave a direct connection between reduction theory and fundamental domains. Instead, he used reduction as follows: given $\tau \in \mathfrak{H}$, the reduction algorithm gives $\tau' = \gamma\tau$ as above and *at the same time* decomposes γ into a continued fraction similar to (2.14). Gauss then applies this to relate $p(\tau')$ and $p(\tau)$, etc., bringing us back to (2.15) (see [12, X.1, p. 225]). But in another place we find such continued fraction decompositions in close conjunction with geometric pictures similar to F_1 and the above (see [12, VIII, pp. 103-105]). Based on this kind of evidence, Gauss' editors decided that he did see the connection (see [12, X.2, pp. 105-106]). Much of this is still a matter of conjecture, but the fact remains that reduction theory is a powerful tool for finding fundamental domains (see [6, Ch. 12]) and that Gauss was aware of some of this power.

Having led the reader on a rather long digression, it is time for us to return to the arithmetic-geometric mean.

Step 3. The Simplest Value

Let $F^\wedge = \{\tau \in F : |\tau - 1/4| > 1/4,\ |\tau + 3/4| > 1/4\}$. We may picture $F^\wedge$ as follows.

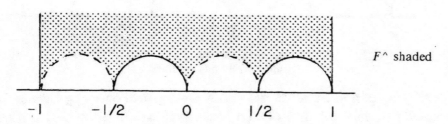

$F^\wedge$ shaded

$$-1 \qquad -1/2 \qquad 0 \qquad 1/2 \qquad 1$$

Let $a, b \in \mathbf{C}^*$ be as usual, and let $\tau \in \mathfrak{H}$ satisfy $k'(\tau) = b/a$. From Lemma 2.3 we know that $\mu = a/p(\tau)^2$ is a value of $M(a, b)$. The goal of Step 3 is to prove the following lemma.

LEMMA 2.9. *If $\tau \in F^\wedge$, then μ is the simplest value of $M(a, b)$.*

Proof. From Lemma 2.3 we know that

$$(2.16) \qquad a_n = \mu\, p(2^n\tau)^2, \qquad b_n = \mu\, q(2^n\tau)^2, \qquad n = 0, 1, 2, \ldots$$

gives us good sequences converging to μ. We need to show that b_{n+1} is the right choice for $(a_n b_n)^{1/2}$ for all $n \geqslant 0$.

The following equivalences are very easy to prove:

$$| a_{n+1} - b_{n+1} | \leqslant | a_{n+1} + b_{n+1} | \Leftrightarrow \mathrm{Re}\left(\frac{b_{n+1}}{a_{n+1}}\right) \geqslant 0$$

$$| a_{n+1} - b_{n+1} | = | a_{n+1} + b_{n+1} | \Leftrightarrow \mathrm{Re}\left(\frac{b_{n+1}}{a_{n+1}}\right) = 0.$$

Recalling the definition of the right choice, we see that we have to prove, for all $n \geqslant 0$, that $\mathrm{Re}\left(\dfrac{b_{n+1}}{a_{n+1}}\right) \geqslant 0$, and if $\mathrm{Re}\left(\dfrac{b_{n+1}}{a_{n+1}}\right) = 0$, then $\mathrm{Im}\left(\dfrac{b_{n+1}}{a_{n+1}}\right) > 0$.
From (2.16) we see that

$$\frac{b_{n+1}}{a_{n+1}} = \frac{q(2^{n+1}\tau)^2}{p(2^{n+1}\tau)^2} = k'(2^{n+1}\tau),$$

so that we are reduced to proving that if $\tau \in F^{\wedge}$, then for all $n \geqslant 0$, $\mathrm{Re}(k'(2^{n+1}\tau)) \geqslant 0$, and if $\mathrm{Re}(k'(2^{n+1}\tau)) = 0$, then $\mathrm{Im}(k'(2^{n+1}\tau)) > 0$.

Let $\tilde{F}_1$ denote the region obtained by translating F_1 by ± 2, ± 4, etc. The drawing below pictures both $\tilde{F}_1$ and F.

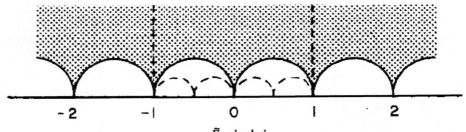

$\tilde{F}_1$ shaded

F indicated by dashed lines

Since $k'(\tau)$ has period 2 and its real part is nonnegative on F_1 by Lemma 2.8, it follows that the real part of $k'(\tau)$ is nonnegative on all of $\tilde{F}_1$. Furthermore, it is clear that on F_1, $\mathrm{Re}(k'(\tau)) = 0$ can occur only on ∂F_1. The product expansions (2.6) show that $k'(\tau)$ is real when $\mathrm{Re}\tau = \pm 1$, so that on F_1, $\mathrm{Re}(k'(\tau)) = 0$ can occur only on the boundary semicircles. From the periodicity of $k'(\tau)$ we conclude that $k'(\tau)$ has positive real part on the interior $\tilde{F}_1^0$ of $\tilde{F}_1$.

If $\tau \in F^{\wedge}$, then the above drawing makes it clear that $2^{n+1}\tau \in \tilde{F}_1$ for $n \geqslant 0$ and that $2^{n+1}\tau \in \tilde{F}_1^0$ for $n \geqslant 1$. We thus see that $\mathrm{Re}(k'(2^{n+1}\tau)) > 0$ for $n \geqslant 0$ unless $n = 0$ and $2\tau \in \partial \tilde{F}_1$. Thus the lemma will be proved once we show that $\mathrm{Im}(k'(2\tau)) > 0$ when $\tau \in F^{\wedge}$ and $2\tau \in \partial \tilde{F}_1$.

These last two conditions imply that 2τ lies on one of the semicircles A and B pictured below.

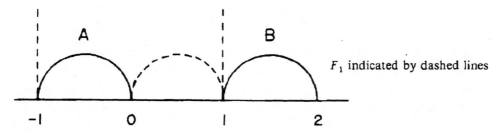

F_1 indicated by dashed lines

By periodicity, k' takes the same values on A and B. Thus it suffices to show that $\text{Im}(k'(2\tau)) > 0$ for $2\tau \in A$. Since $S = \begin{pmatrix} 0 & -1 \\ 1 & 0 \end{pmatrix}$ maps the line $\text{Re}\sigma = 1$ to A, we can write $2\tau = -1/\sigma$, where $\text{Re}\sigma = 1$. Then, using (2.7), we obtain

$$k'(2\tau) = k'(-1/\sigma) = \frac{q(-1/\sigma)^2}{p(-1/\sigma)^2} = \frac{r(\sigma)^2}{p(\sigma)^2}.$$

Since $\text{Re}\sigma = 1$, the product expansions (2.6) easily show that

$$\text{Im}(r(\sigma)^2/p(\sigma)^2) > 0,$$

which completes the proof of Lemma 2.9.   QED

Step 4. Conclusion of the Proof.

We can now prove Theorem 2.2. Recall that at the end of Step 1 we were left with three tasks: to find all solutions τ of $k'(\tau) = b/a$, to relate the values of $a/p(\tau)^2$ thus obtained, and to show that all values of $M(a, b)$ arise in this way.

We are given $a, b \in \mathbf{C}^*$ with $a \neq \pm b$ and $|a| \geqslant |b|$. We will first find $\tau_0 \in F_2 \cap F^\wedge$ such that $k'(\tau_0) = b/a$. Since $|b/a| \leqslant 1$, Lemma 2.8 gives us $\tau_0 \in F_2$ with $k'(\tau_0) = b/a$. Could τ_0 fail to lie in $F^\wedge$? From the definition of $F^\wedge$, this only happens when τ_0 lies in the semicircle B pictured below.

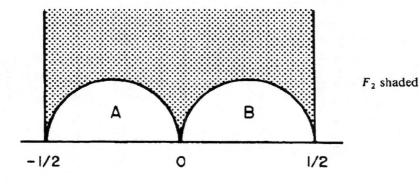

F_2 shaded

However, $\gamma = \begin{pmatrix} 1 & 0 \\ -4 & 1 \end{pmatrix} \in \Gamma_2(4)$ takes B to the semicircle A. Since k' is invariant under $\Gamma_2(4)$, we have $k'(\gamma\tau_0) = k'(\tau_0) = b/a$. Thus, replacing τ_0 by $\gamma\tau_0$, we may assume that $\tau_0 \in F_2 \cap F^\wedge$.

It is now easy to solve the first two of our tasks. Since k' induces a bijection $\mathfrak{H}/\Gamma_2(4) \cong \mathbf{C} - \{0, \pm1\}$, it follows that all solutions of $k'(\tau) = b/a$ are given by $\tau = \gamma\tau_0, \gamma \in \Gamma_2(4)$. This gives us the following set of values of $M(a, b)$:

$$\{a/p(\gamma\tau_0)^2 : \gamma \in \Gamma_2(4)\} \ .$$

Recalling the statement of Theorem 2.2, it makes sense to look at the reciprocals of these values:

$$R = \{p(\gamma\tau_0)^2/a : \gamma \in \Gamma_2(4)\}$$

By (2.12), $p(\gamma\tau_0)^2 = (c\tau_0+d)\, p(\tau_0)^2$ for $\gamma = \begin{pmatrix} a & b \\ c & d \end{pmatrix} \in \Gamma_2(4) \subseteq \Gamma(2)_0$. Setting $\mu = a/p(\tau_0)^2$, we have

$$R = \{(c\tau_0+d)\, p(\tau_0)^2/a : \gamma = \begin{pmatrix} a & b \\ c & d \end{pmatrix} \in \Gamma_2(4)\}$$

$$= \{(c\tau_0+d)/\mu : \gamma = \begin{pmatrix} a & b \\ c & d \end{pmatrix} \in \Gamma_2(4)\} \ .$$

An easy exercise in number theory shows that the bottom rows (c, d) of elements of $\Gamma_2(4)$ are precisely those pairs (c, d) satisfying $GCD(c, d) = 1$, $c \equiv 0 \bmod 4$ and $d \equiv 1 \bmod 4$. We can therefore write

$$R = \{(c\tau_0+d)/\mu : GCD(c, d) = 1, \quad c \equiv 0 \bmod 4, \quad d \equiv 1 \bmod 4\} \ .$$

Then setting $\lambda = i\mu/\tau_0$ gives us

$$(2.17) \quad R = \left\{\frac{d}{\mu} + \frac{ic}{\lambda} : GCD(c, d) = 1, \quad d \equiv 1 \bmod 4, \quad c \equiv 0 \bmod 4\right\}.$$

Finally, we will show that μ and λ are the simplest values of $M(a, b)$ and $M(a+b, a-b)$ respectively. This is easy to see for μ: since $\tau_0 \in F^\wedge$, Lemma 2.9 implies that $\mu = a/p(\tau_0)^2$ is the simplest value of $M(a, b)$. Turning to λ, recall from Lemma 2.3 that $a = \mu p(\tau_0)^2$ and $b = \mu q(\tau_0)^2$. Thus by (2.8) and (2.7),

$$a + b = \mu(p(\tau_0)^2 + q(\tau_0)^2) = 2\mu\, p(2\tau_0)^2 = 2\mu \left(\frac{i}{2\tau_0}\right) p\left(\frac{-1}{2\tau_0}\right)^2,$$

$$a - b = \mu(p(\tau_0)^2 - q(\tau_0)^2) = 2\mu\, r(2\tau_0)^2 = 2\mu\left(\frac{i}{2\tau_0}\right)q\left(\frac{-1}{2\tau_0}\right)^2,$$

which implies that

$$a + b = \lambda\, p(-1/2\tau_0)^2, \qquad a - b = \lambda\, q(-1/2\tau_0)^2.$$

Hence λ is a value of $M(a+b, a-b)$. To see that it is the simplest value, we must show that $-1/2\tau_0 \in F^\wedge$ (by Lemma 2.9). Since $\tau_0 \in F_2$, we have $2\tau_0 \in F_1$. But F_1 is stable under $S = \begin{pmatrix} 0 & -1 \\ 1 & 0 \end{pmatrix}$, so that $-1/2\tau_0 \in F_1$. The inclusion $F_1 \subseteq F^\wedge$ is obvious, and $-1/2\tau_0 \in F^\wedge$ follows. This completes our first two tasks.

Our third and final task is to show that (2.17) gives the reciprocals of *all* values of $M(a, b)$. This will finish the proof of Theorem 2.2. So let μ' be a value of $M(a, b)$, and let $\{a_n\}_{n=0}^\infty$ and $\{b_n\}_{n=0}^\infty$ be the good sequences such that $\mu' = \lim_{n\to\infty} a_n = \lim_{n\to\infty} b_n$. Then there is some m such that b_{n+1} is the right choice for $(a_n b_n)^{1/2}$ for all $n \geqslant m$; and thus μ' is the simplest value of $M(a_m, b_m)$. Since $k' : F' \to \mathbf{C} - \{0, \pm 1\}$ is surjective by Lemma 2.7, we can find $\tau \in F'$ such that $k'(\tau) = b_m/a_m$. Arguing as above, we may assume that $\tau \in F^\wedge$. Then Lemma 2.9 shows that $\mu' = a_m/p(\tau)^2$ and also that for $n \geqslant m$,

(2.18) $$a_n = \mu'\, p(2^{n-m}\tau)^2, \qquad b_n = \mu'\, q(2^{n-m}\tau)^2.$$

Let us study a_{m-1} and b_{m-1}. Their sum and product are $2\,a_m$ and b_m^2 respectively. From the quadratic formula we see that

$$\{a_{m-1}, b_{m-1}\} = \{a_m \pm (a_m^2 - b_m^2)^{1/2}\}.$$

Using (2.9), we obtain

$$a_m^2 - b_m^2 = \mu'^2(p(\tau)^4 - q(\tau)^4) = \mu'^2 r(\tau)^4,$$

so that, again using (2.9), we have

$$a_m \pm (a_m^2 - b_m^2)^{1/2} = \mu'(p(\tau)^2 \pm r(\tau)^2) = \begin{cases} \mu' p(\tau/2)^2 \\ \\ \mu' q(\tau/2)^2. \end{cases}$$

Thus, either

$$a_{m-1} = \mu'\, p(\tau/2)^2, \; b_{m-1} = \mu'\, q(\tau/2)^2 \text{ or } a_{m-1} = \mu'\, q(\tau/2)^2, \; b_m = \mu'\, p(\tau/2)^2.$$

In the former case, set $\tau_1 = \tau/2$. Then from (2.18) we easily see that for $n \geqslant m - 1$,

$$(2.19) \qquad a_n = \mu' \, p(2^{n-m+1}\tau_1)^2 \, , \qquad b_n = \mu' \, q(2^{n-m+1}\tau_1)^2 \, .$$

If the latter case holds, let $\tau_1 = \tau/2 + 1$. From (2.7) we see that $a_{m-1} = \mu' \, p(\tau_1)^2$, $b_{m-1} = \mu' \, q(\tau_1)^2$, and it also follows easily that $p(2^{n-m+1}\tau_1) = p(2^{n-m}\tau)$ and $q(2^{n-m+1}\tau_1) = q(2^{n-m}\tau)$ for all $n \geqslant m$. Thus (2.19) holds for this choice of τ_1 and $n \geqslant m - 1$.

By induction, this argument shows that there is $\tau_m \in \mathfrak{H}$ such that for all $n \geqslant 0$,

$$a_n = \mu' \, p(2^n\tau_m)^2 \, , \qquad b_n = \mu' \, q(2^n\tau_m)^2 \, .$$

In particular, $\mu' = a/p(\tau_m)^2$ and $k'(\tau_m) = b/a$. Thus $(\mu')^{-1} = p(\tau_m)^2/a$ is in the set R of (2.17). This shows that R consists of the reciprocals of all values of $M(a, b)$, and the proof of Theorem 2.2 is now complete.   QED

We should point out that the proof just given, though arrived at independently, is by no means original. The first proofs of Theorem 2.2 appeared in 1928 in [15] and [35]. Geppert's proof [15] is similar to ours in the way it uses the theory of theta functions and modular functions. The other proof [35], due to von David, is much shorter; it is a model of elegance and conciseness.

Let us discuss some consequences of the proof of Theorem 2.2. First, the formula $\lambda = i\mu/\tau_0$ obtained above is quite interesting. We say that τ_0 "uniformizes" the simplest value μ of $M(a, b)$, where

$$a = \mu \, p(\tau_0)^2 \, , \qquad b = \mu \, q(\tau_0)^2 \, .$$

Writing the above formula as $\tau_0 = i\dfrac{\mu}{\lambda}$, we see how to *explicitly compute* τ_0 in terms of the simplest values of $M(a, b)$ and $M(a+b, a-b)$. This is especially useful when $a > b > 0$. Here, if we set $c = \sqrt{a^2 - b^2}$, then, using the notation of § 1, the simplest values are $M(a, b)$ and $M(a, c)$, so that

$$(2.20) \qquad \tau_0 = i\,\frac{M(a, b)}{M(a, c)} \, .$$

A nice example is when $a = \sqrt{2}$ and $b = 1$. Then $c = 1$, which implies $\tau_0 = i$! Thus $M(\sqrt{2}, 1) = \sqrt{2}/p(i)^2 = 1/q(i)^2$. From § 1 we know $M(\sqrt{2}, 1) = \pi/\varpi$, which gives us the formulas

$$\omega/\pi = 2^{-1/2}p(i)^2 = 2^{-1/2}(1+2e^{-\pi}+2e^{-4\pi}+2e^{-9\pi}+...)^2 ,$$

(2.21)

$$\omega/\pi = q(i)^2 = (1-2e^{-\pi}+2e^{-4\pi}-2e^{-9\pi}+...)^2 .$$

We will discuss the importance of this in § 3.

Turning to another topic, note that $M(a, b)$ is clearly homogeneous of degree 1, i.e., if μ is a value of $M(a,b)$, then $c\mu$ is a value of $M(ca, cb)$ for $c \in C^*$. Thus, it suffices to study $M(1, b)$ for $b \in C - \{0, \pm 1\}$. Its values are given by $\mu = 1/p(\tau)^2$ where $k'(\tau) = b$. Since $k': \mathfrak{H} \to C - \{0, \pm 1\}$ is a local biholomorphism, it follows that $M(1, b)$ is a multiple valued holomorphic function. To make it single valued, we pull back to the universal cover via k', giving us $M(1, k'(\tau))$. We thus obtain

$$M(1, k'(\tau)) = 1/p(\tau)^2 .$$

This shows that the agM may be regarded as a meromorphic modular form of weight -1.

Another interesting multiple valued holomorphic function is the elliptic integral $\int_0^{\pi/2} (1-k^2\sin^2\phi)^{-1/2}d\phi$. This is a function of $k \in C - \{0, \pm 1\}$. If we pull back to the universal cover via $k: \mathfrak{H} \to C - \{0, \pm 1\}$ (recall from Step 2 that $k(\tau) = r(\tau)^2/p(\tau)^2$), then it is well known that

$$\frac{2}{\pi} \int_0^{\pi/2} (1-k(\tau)^2\sin^2\phi)^{-1/2}d\phi = p(\tau)^2$$

(see [36, p. 500]). Combining the above two equations, we obtain

$$\frac{1}{M(1, k'(\tau))} = p(\tau)^2 = \frac{2}{\pi} \int_0^{\pi/2} (1-k(\tau)^2\sin^2\phi)^{-1/2}d\phi ,$$

which may be viewed as a rather amazing generalization of (1.9).

Finally, let us make some remarks about the set $\mathcal{M}$ of values of $M(a, b)$, where a and b are fixed. If μ denotes the simplest value of $M(a, b)$, then it can be shown that $|\mu| \geqslant |\mu'|$ for $\mu' \in \mathcal{M}$, and $|\mu|$ is a strict maximum if $\mathrm{ang}(a, b) \neq \pi$. This may be proved directly from the definitions (see [35]). Another proof proceeds as follows. We know that any $\mu' \in \mathcal{M}$ can be written

(2.22) $$\mu' = \mu/(c\tau_0+d) ,$$

where $\tau_0 \in F_2$ and $\begin{pmatrix} a & b \\ c & d \end{pmatrix} \in \Gamma_2(4)$. Thus it suffices to prove that $|c\tau_0 + d| \geqslant 1$ whenever $\tau_0 \in F_2$ and $\begin{pmatrix} a & b \\ c & d \end{pmatrix} \in \Gamma_2(4)$. This is left as an exercise for the reader.

We can also study the accumulation points of $\mathcal{M}$. Since $|c\tau_0 + d|$ is a positive definite quadratic form in c and d, it follows from (2.22) that $0 \in \mathbb{C}$ is the only accumulation point of $\mathcal{M}$. This is very satisfying once we recall from Proposition 2.1 that $0 \in \mathbb{C}$ is the common limit of all non-good sequences $\{a_n\}_{n=0}^{\infty}$ and $\{b_n\}_{n=0}^{\infty}$ coming from (2.1).

The proof of Theorem 2.2 makes one thing very clear: we have now seen "an entirely new field of analysis." However, before we can say that Gauss' prediction of May 30, 1799 has been fulfilled, we need to show that the proof given above reflects what Gauss actually did. Since we know from Step 2 about his work with the theta functions $p(\tau)$, $q(\tau)$ and $r(\tau)$ and the modular function $k'(\tau)$, it remains to see how he applied all of this to the arithmetic-geometric mean.

The connections we seek are found in several places in Gauss' notes. For example, he states very clearly that if

(2.23) $a = \mu\, p(\tau)^2\,, \qquad b = \mu\, q(\tau)^2\,,$

then the sequences $a_n = \mu\, p(2^n\tau)^2$, $b_n = \mu\, q(2^n\tau)^2$ satisfy the agM algorithm (2.1) with μ as their common limit (see [12, III, p. 385 and pp. 467-468]). This is precisely our Lemma 2.3. In another passage, Gauss defines the "einfachste Mittel" (simplest mean) to be the limit of those sequences where $\text{Re}(b_{n+1}/a_n) > 0$ for all $n \geqslant 0$ (see [12, III, p. 477]). This is easily seen to be equivalent to our definition of simplest value when $\text{ang}(a, b) \neq \pi$. On the same page, Gauss then asserts that for $\tau \in F_2$, μ is the simplest value of $M(a, b)$ for a, b as in (2.23). This is a weak form of Lemma 2.9. Finally, consider the following quote from [12, VIII, p. 101]: "In order to solve the equation $\dfrac{q(t)}{p(t)} = A$, one sets $A^2 = n/m$ and takes the agM of m and n; let this be μ. One further takes the agM of m and $\sqrt{m^2 - n^2}$, or, what is the same, of $m + n$ and $m - n$; let this be λ. One then has $t = \mu/\lambda$. This gives only one value of t; all others are contained in the formula

$$t' = \frac{\alpha t - 2\beta i}{\delta - 2\gamma t i}\,,$$

where α, β, γ, δ signify all integers which satisfy the equation $\alpha\delta - 4\beta\gamma = 1$."
Recall that $\mathrm{Re}\,t > 0$, so that our τ is just ti. Note also that the last
assertion is not quite correct.

Unfortunately, in spite of these compelling fragments, Gauss never actually
stated Theorem 2.2. The closest he ever came is the following quote from
[12, X.1, p. 219]: "The agM changes, when one chooses the negative value for
one of n', n'', n''' etc.: however all resulting values are of the following
form:

(2.24)
$$\frac{1}{(\mu)} = \frac{1}{\mu} + \frac{4ik}{\lambda} \cdot \text{''}$$

Here, Gauss is clearly dealing with $M(m, n)$ where $m > n > 0$. The fraction
$1/\mu$ in (2.24) is correct: in fact, it can be shown that if the negative value
of $n^{(r)}$ is chosen, and all other choices are the right choice, then the corresponding value μ' of $M(m, n)$ satisfies

$$\frac{1}{\mu'} = \frac{1}{\mu} + \frac{2^{r+1}i}{\lambda}$$

(see [13, p. 140]). So (2.24) is only a very special case of Theorem 2.2.

There is one final piece of evidence to consider: the 109th entry in
Gauss' mathematical diary. It reads as follows:

> Between two given numbers there are always infinitely many means
> both arithmetic-geometric and harmonic-geometric, the observation of
> whose mutual connection has been a source of happiness for us.

(See [12, X.1, p. 550]. The harmonic-geometric mean of a and b is
$M(a^{-1}, b^{-1})^{-1}$.) What is amazing is the date of this entry: June 3, 1800,
a little more than a year after May 30, 1799. We know from § 1 that
Gauss' first proofs of Theorem 1.1 date from December 1799. So less than
six months later Gauss was aware of the multiple valued nature of $M(a, b)$
and of the relations among these values! One tantalizing question remains:
does the phrase "mutual connection" refer only to (2.24), or did Gauss have
something more like Theorem 2.2 in mind? Just how much did he know
about modular functions as of June 3, 1800? In order to answer these
questions, we need to examine the history of the whole situation more
closely.

3. HISTORICAL REMARKS

The main difficulty in writing about the history of mathematics is that so much has to be left out. The mathematics we are studying has a richness which can never be conveyed in one article. For instance, our discussion of Gauss' proofs of Theorem 1.1 in no way does justice to the complexity of his mathematical thought; several important ideas were simplified or omitted altogether. This is not entirely satisfactory, yet to rectify such gaps is beyond the scope of this paper. As a compromise, we will explore the three following topics in more detail:

A. The history of the lemniscate,

B. Gauss' work on inverting lemniscatic integrals, and

C. The chronology of Gauss' work on the agM and theta functions.

A. The lemniscate was discovered by Jacob Bernoulli in 1694. He gives the equation in the form

$$xx + yy = a\sqrt{xx - yy}$$

(in § 1 we assumed that $a = 1$), and he explains that the curve has "the form of a figure 8 on its side, as of a band folded into a knot, or of a lemniscus, or of a knot of a French ribbon" (see [2, p. 609]). "Lemniscus" is a Latin word (taken from the Greek) meaning a pendant ribbon fastened to a victor's garland.

More interesting is that the integral $\int_0^1 (1-z^4)^{-1/2}dz$, which gives one-quarter of the arc length of the lemniscate, had been discovered three years earlier in 1691! This was when Bernoulli worked out the equation of the so-called *elastic curve*. The situation is as follows: a thin elastic rod is bent until the two ends are perpendicular to a given line L.

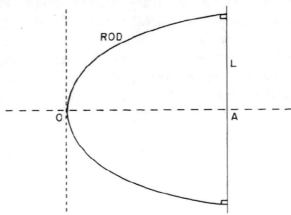

After introducing cartesian coordinates as indicated and letting a denote 0A, Bernoulli was able to show that the upper half of the curve is given by the equation

(3.1)
$$y = \int_0^x \frac{z^2 dz}{\sqrt{a^4 - z^4}},$$

where $0 \leqslant x \leqslant a$ (see [2, pp. 567-600]).

It is convenient to assume that $a = 1$. But as soon as this is done, we no longer know how long the rod is. In fact, (3.1) implies that the arc length from the origin to a point (x, y) on the rescaled elastic curve is $\int_0^x (1 - z^4)^{-1/2} dz$. Thus the length of the whole rod is $2 \int_0^1 (1 - z^4)^{-1/2} dz$, which is precisely Gauss' ϖ!

How did Bernoulli get from here to the lemniscate? He was well aware of the transcendental nature of the elastic curve, and so he used a standard seventeenth century trick to make things more manageable: he sought "an algebraic curve... whose rectification should agree with the rectification of the elastic curve" (this quote is from Euler [9, XXI, p. 276]).

Jacob actually had a very concrete reason to be interested in arc length: in 1694, just after his long paper on the elastic curve was published, he solved a problem of Leibniz concerning the "isochrona paracentrica" (see [2, pp. 601-607]). This called for a curve along which a falling weight recedes from or approaches a given point equally in equal times. Since Bernoulli's solution involved the arc length of the elastic curve, it was natural for him to seek an algebraic curve with the same arc length. Very shortly thereafter, he found the equation of the lemniscate (see [2, pp. 608-612]). So we really can say that the arc length of the lemniscate was known well before the curve itself.

But this is not the full story. In 1694 Jacob's younger brother Johann independently discovered the lemniscate! Jacob's paper on the isochrona paracentrica starts with the differential equation

$$(x dx + y dy)\sqrt{y} = (x dy - y dx)\sqrt{a},$$

which had been derived earlier by Johann, who, as Jacob rather bluntly points out, hadn't been able to solve it. Johann saw this comment for the first time when it appeared in June 1694 in Acta Eruditorum. He took up the challenge and quickly produced a paper on the isochrona paracentrica which gave the equation of the lemniscate and its relation to the elastic curve. This appeared in Acta Eruditorum in October 1694 (see [3, pp. 119-

122]), but unfortunately for Johann, Jacob's article on the lemniscate appeared in the September issue of the same journal. There followed a bitter priority dispute. Up to now relations between the brothers had been variable, sometimes good, sometimes bad, with always a strong undercurrent of competition between them. After this incident, amicable relations were never restored. (For details of this controversy, as well as a fuller discussion of Jacob's mathematical work, see [18].)

We need to mention one more thing before going on. Near the end of Jacob's paper on the lemniscate, he points out that the y-value $\int_0^x z^2(a^4 - z^4)^{-1/2}dz$ of the elastic curve can be expressed as the difference of an arc of the ellipse with semiaxes $a\sqrt{2}$ and a, and an arc of the lemniscate (see [2, pp. 611-612]). This observation is an easy consequence of the equation

$$(3.2) \qquad \int_0^x \frac{a^2 dz}{(a^4 - z^4)^{1/2}} + \int_0^x \frac{z^2 dz}{(a^4 - z^4)^{1/2}} = \int_0^x \left(\frac{a^2 + z^2}{a^2 - z^2}\right)^{1/2} dz .$$

What is especially intriguing is that the ratio $\sqrt{2} : 1$, so important in Gauss' observation of May 30, 1799, was present at the very birth of the lemniscate.

Throughout the eighteenth century the elastic curve and the lemniscate appeared in many papers. A lot of work was done on the integrals $\int_0^1 (1 - z^4)^{-1/2}dz$ and $\int_0^1 z^2(1 - z^4)^{-1/2}dz$. For example, Stirling, in a work written in 1730, gives the approximations

$$\int_0^1 \frac{dz}{\sqrt{1 - z^4}} = 1.31102877714605987$$

$$\int_0^1 \frac{z^2 dz}{\sqrt{1 - z^4}} = .59907011736779611$$

(see [31, pp. 57-58]). Note that the second number doubled is 1.19814023473559222, which agrees with $M(\sqrt{2}, 1)$ to sixteen decimal places. Stirling also comments that these two numbers add up to one half the circumference of an ellipse with $\sqrt{2}$ and 1 as axes, a special case of Bernoulli's observation (3.2).

Another notable work on the elastic curve was Euler's paper "De miris proprietatibus curvae elasticae sub equatione $y = \int \frac{xx\,dx}{\sqrt{1 - x^4}}$ contentae"

which appeared posthumously in 1786. In this paper Euler gives approximations to the above integrals (not as good as Stirling's) and, more importantly,
proves the amazing result that

(3.3)
$$\int_0^1 \frac{dz}{\sqrt{1-z^2}} \cdot \int_0^1 \frac{z^2 dz}{\sqrt{1-z^4}} = \frac{\pi}{4}$$

(see [9, XXI, pp. 91-118]). Combining this with Theorem 1.1 we see that

$$M(\sqrt{2}, 1) = 2 \int_0^1 \frac{z^2 dz}{\sqrt{1-z^4}},$$

so that the coincidence noted above has a sound basis in fact.

We have quoted these two papers on the elastic curve because, as we
will see shortly, Gauss is known to have read them. Note that each paper
has something to contribute to the equality $M(\sqrt{2}, 1) = \pi/\varpi$: from Stirling,
we get the ratio $\sqrt{2} : 1$, and from Euler we get the idea of using an
equation like (3.3).

Unlike the elastic curve, the story of the lemniscate in the eighteenth
century is well known, primarily because of the key role it played in the
development of the theory of elliptic integrals. Since this material is thoroughly
covered elsewhere (see, for example, [1, Ch. 1-3], [8, pp. 470-496], [19, § 1-§ 4]
and [21, § 19.4]), we will mention only a few highlights. One early worker
was C. G. Fagnano who, following some ideas of Johann Bernoulli, studied
the ways in which arcs of ellipses and hyperbolas can be related. One result,
known as Fagnano's Theorem, states that the sum of two appropriately
chosen arcs of an ellipse can be computed algebraically in terms of the
coordinates of the points involved. He also worked on the lemniscate,
starting with the problem of halving that portion of the arc length of the
lemniscate which lies in one quadrant. Subsequently he found methods for
dividing this arc length into n equal pieces, where $n = 2^m$, $3 \cdot 2^m$ or $5 \cdot 2^m$.
These researches of Fagnano's were published in the period 1714-1720 in an
obscure Venetian journal and were not widely known. In 1750 he had his
work republished, and he sent a copy to the Berlin Academy. It was given
to Euler for review on December 23, 1751. Less than five weeks later, on
January 27, 1752, Euler read a paper giving new derivations for Fagnano's
results on elliptic and hyperbolic arcs as well as significantly new results on
lemniscatic arcs. By 1753 he had a general addition theorem for lemniscatic
integrals, and by 1758 he had the addition theorem for elliptic integrals
(see [9, XX, pp. VII-VIII]). This material was finally published in 1761,

and for the first time there was a genuine theory of elliptic integrals. For the next twenty years Euler and Lagrange made significant contributions, paving the way for Legendre to cast the field in its classical form which we glimpsed at the end of § 1. Legendre published his definitive treatise on elliptic integrals in two volumes in 1825 and 1826. The irony is that in 1828 he had to publish a third volume describing the groundbreaking papers of Abel and Jacobi which rendered obsolete much of his own work (see [23]).

An important problem not mentioned so far is that of computing tables of elliptic integrals. Such tables were needed primarily because of the many applications of elliptic integrals to mechanics. Legendre devoted the entire second volume of his treatise to this problem. Earlier Euler had computed these integrals using power series similar to (1.8) (see also [9, XX, pp. 21-55]), but these series often converged very slowly. The real breakthrough came in Lagrange's 1785 paper "Sur une nouvelle méthode de calcul intégral" (see [22, pp. 253-312]). Among other things, Lagrange is concerned with integrals of the form

$$(3.4) \qquad \int \frac{M \, dy}{\sqrt{(1+p^2y^2)(1+q^2y^2)}} \, ,$$

where M is a rational function of y^2 and $p \geqslant q > 0$. He defines sequences $p, p', p'', ..., q, q', q'', ...$ as follows:

$$p' = p + (p^2-q^2)^{1/2} \, , q' = p - (p^2-q^2)^{1/2} \, ,$$

(3.5)

$$p'' = p' + (p'^2-q'^2)^{1/2} \, , q'' = p' - (p'^2-q'^2)^{1/2} \, ,$$

$$\vdots \qquad\qquad\qquad \vdots$$

and then, using the substitution

$$(3.6) \qquad\qquad y' = \frac{y((1+p^2y^2)(1+q^2y^2))^{1/2}}{1 + q^2y^2}$$

he shows that

$$(3.7) \quad ((1+p^2y^2)(1+q^2y^2))^{-1/2} dy = ((1+p'^2y'^2)(1+q'^2y'^2))^{-1/2} dy' \, .$$

Two methods of approximation are now given. The first starts by observing that the sequence $p, p', p'', ...$ approaches $+\infty$ while $q, q', q'', ...$ approaches 0. Thus by iterating the substitution (3.6) in the integral of (3.4), one can eventually assume that $q = 0$, which gives an easily computable integral. The second method consists of doing the first backwards: from (3.5) one easily obtains

$$p = (p' + q')/2 , \qquad q = (p'q')^{1/2} .$$

Lagrange then observes that continuing this process leads to sequences $p', p, 'p, ''p, ..., q', q, 'q, ''q, ...$ which converge to a common limit (see [22, p. 271]). Hence iterating (3.6) allows one to eventually assume $p = q$, again giving an easily computable integral.

So here we are in 1785, staring at the definition of the arithmetic-geometric mean, six years before Gauss' earliest work on the subject. By setting $py = \tan\phi$, one obtains

$$((1 + p^2 y^2)(1 + q^2 y^2))^{-1/2} dy = (p^2 \cos^2\phi + q^2 \sin^2\phi)^{-1/2} d\phi ,$$

so that (3.6) and (3.7) are precisely (1.5) and (1.6) from the proof of Theorem 1.1. Thus Lagrange not only could have defined the agM, he could have also proved Theorem 1.1 effortlessly. Unfortunately, none of this happened; Lagrange never realized the power of what he had discovered.

One question emerges from all of this: did Gauss ever see Lagrange's article? The library of the Collegium Carolinum in Brunswick had some of Lagrange's works (see [4, p. 9]) and the library at Gottingen had an extensive collection (see [12, X.2, p. 22]). On the other hand, Gauss, in the research announcement of his 1818 article containing the proof of Theorem 1.1, claims that his work is independent of that of Lagrange and Legendre (see [12, III, p. 360]). A fuller discussion of these matters is in [12, X.2, pp. 12-22]. Assuming that Gauss did discover the agM independently, we have the amusing situation of Gauss, who anticipated so much in Abel, Jacobi and others, himself anticipated by Lagrange.

The elastic curve and the lemniscate were equally well known in the eighteenth century. As we will soon see, Gauss at first associated the integral $\int (1 - z^4)^{-1/2} dz$ with the elastic curve, only later to drop it in favor of the lemniscate. Subsequent mathematicians have followed his example. Today, the elastic curve has been largely forgotten, and the lemniscate has suffered the worse fate of being relegated to the polar coordinates section of calculus books. There it sits next to the formula for arc length in polar coordinates, which can never be applied to the lemniscate since such texts know nothing of elliptic integrals.

B. Our goal in describing Gauss' work on the lemniscate is to learn more of the background to his observation of May 30, 1799. We will see that the lemniscatic functions played a key role in Gauss' development of the arithmetic-geometric mean.

Gauss began innocently enough in September 1796, using methods of Euler to find the formal power series expansion of the inverse function of first $\int (1-x^3)^{-1/2}dx$, and then more generally $\int (1-x^n)^{-1/2}dx$ (see [12, X.1, p. 502]). Things became more serious on January 8, 1797. The 51st entry in his mathematical diary, bearing this date, states that "I have begun to investigate the elastic curve depending on $\int (1-x^4)^{-1/2}dx$." Notes written at the same time show that Gauss was reading the works of Euler and Stirling on the elastic curve, as discussed earlier. Significantly, Gauss later struck out the word "elastic" and replaced it with "lemniscatic" (see [12, X.1, pp. 147 and 510]).

Gauss was strongly motivated by the analogy to the circular functions. For example, notice the similarity between $\varpi/2 = \int_0^1 (1-z^4)^{-1/2}dz$ and $\pi/2 = \int_0^1 (1-z^2)^{-1/2}dz$. (This similarity is reinforced by the fact that many eighteenth century texts used ϖ to denote π — see [12, X.2, p. 33].) Gauss then defined the lemniscatic functions as follows:

$$\text{sinlemn}\left(\int_0^x (1-z^4)^{-1/2}dz\right) = x$$

$$\text{coslemn}\left(\varpi/2 - \int_0^x (1-z^4)^{-1/2}dz\right) = x$$

(see [12, III, p. 404]). Gauss often used the abbreviations sl ϕ and cl ϕ for sinlemn ϕ and coslemn ϕ respectively, a practice we will adopt. From Euler's addition theorem one easily obtains

(3.8) $\text{sl}^2\phi + \text{cl}^2\phi + \text{sl}^2\phi\,\text{cl}^2\phi = 1$

(3.9) $\text{sl}(\phi+\phi') = \dfrac{\text{sl}\,\phi\,\text{cl}\,\phi' + \text{sl}\,\phi'\,\text{cl}\,\phi}{1 - \text{sl}\,\phi\,\text{sl}\,\phi'\,\text{cl}\,\phi\,\text{cl}\,\phi'}$

(see [12, X.1, p. 147]).

Other formulas can now be derived in analogy with the trigonometric functions (see [25, pp. 155-156] for a nice treatment), but Gauss went much, much farther. A series of four diary entries made in March 1797 reveal the amazing discoveries that he made in the first three months of 1797. We will need to describe these results in some detail.

Gauss started with Fagnano's problem of dividing the lemniscate into n equal parts. Since this involved an equation of degree n^2, Gauss realized that most of the roots were complex (see [12, X.1, p. 515]). This led him to define sl ϕ and cl ϕ for complex numbers ϕ. The first step is to show that

$$\text{sl}(iy) = i \text{ sl } y, \qquad \text{cl}(iy) = 1/\text{cl}(y),$$

(the first follows from the change of variable $z = iz'$ in $\int (1-z^4)^{-1/2} dz$, and the second follows from (3.8)). Then (3.9) implies that

$$\text{sl}(x+iy) = \frac{\text{sl } x + i \text{ sl } y \text{ cl } x \text{ cl } y}{\text{cl } y - i \text{ sl } x \text{ sl } y \text{ cl } x}$$

(see [12, X.1, p. 154]).

It follows easily that sl ϕ is doubly periodic, with periods $2\tilde{\omega}$ and $2i\tilde{\omega}$. The zeros and poles of sl ϕ are also easy to determine; they are given by $\phi = (m+in)\tilde{\omega}$ and $\phi = ((2m-1)+i(2n-1))(\tilde{\omega}/2)$, $m, n \in \mathbf{Z}$, respectively. Then Gauss shows that sl ϕ can be written as

$$\text{sl } \phi = \frac{M(\phi)}{N(\phi)}$$

where $M(\phi)$ and $N(\phi)$ are entire functions which are doubly indexed infinite products whose factors correspond to the zeros and poles respectively (see [12, X.1, pp. 153-155]). In expanding these products, Gauss writes down the first examples of Eisenstein series (see [12, X.1, pp. 515-516]). He also obtains many identities involving $M(\phi)$ and $N(\phi)$, such as

(3.10) $$N(2\phi) = M(\phi)^4 + N(\phi)^4$$

(see [12, X.1, p. 157]). Finally, Gauss notices that the numbers $N(\tilde{\omega})$ and $e^{\pi/2}$ agree to four decimal places (see [12, X.1, p. 158]). He comments that a proof of their equality would be "a most important advancement of analysis" (see 12, X.1, p. 517]).

Besides being powerful mathematics, what we have here is almost a rehearsal for what Gauss did with the arithmetic-geometric mean: the observation that two numbers are equal, the importance to analysis of proving this, and the passage from real to complex numbers in order to get at the real depth of the subject. Notice also that identities such as (3.10) are an important warm-up to the theta function identities needed in § 2.

Two other discoveries made at this time require comment. First, only a year after constructing the regular 17-gon by ruler and compass, Gauss found a ruler and compass construction for dividing the lemniscate into five equal pieces (see [12, X.1, p. 517]). This is the basis for the remarks concerning $\int (1-x^4)^{-1/2} dx$ made in Disquisitiones Arithmeticae (see [11, § 335]). Second, Gauss discovered the complex multiplication of elliptic functions when he gave formulas for $sl(1+i)\phi$, $N(1+i)\phi$, etc. (see [12, III, pp. 407 and 411]). These discoveries are linked: complex multiplication on the elliptic curve associated to the lemniscate enabled Abel to determine all n for which the lemniscate can be divided into n pieces by ruler and compass. (The answer is the same as for the circle! See [28] for an excellent modern account of Abel's theorem.)

After this burst of progress, Gauss left the lemniscatic functions to work on other things. He returned to the subject over a year later, in July 1798, and soon discovered that there was a better way to write sl ϕ as a quotient of entire functions. The key was to introduce the new variable $s = \sin\left(\dfrac{\pi}{\tilde{\omega}}\phi\right)$.

Since sl ϕ has period $2\tilde{\omega}$, it can certainly be written as a function of s. By expressing the zeros and poles of sl ϕ in terms of s, Gauss was able to prove that

$$\text{sl } \phi = \frac{P(\phi)}{Q(\phi)},$$

where

$$P(\phi) = \frac{\tilde{\omega}}{\pi} s \left(1 + \frac{4s^2}{(e^\pi - e^{-\pi})^2}\right)\left(1 + \frac{4s^2}{(e^{2\pi} - e^{-2\pi})^2}\right)\cdots$$

$$Q(\phi) = \left(1 - \frac{4s^2}{(e^{\pi/2} + e^{-\pi/2})^2}\right)\left(1 - \frac{4s^2}{(e^{3\pi/2} + e^{-3\pi/2})^2}\right)\cdots$$

(see [12, III, pp. 415-416]). Relating these to the earlier functions $M(\phi)$ and $N(\phi)$, Gauss obtains (letting $\phi = \psi\tilde{\omega}$)

$$M(\psi\tilde{\omega}) = e^{\pi\psi^2/2} P(\psi\tilde{\omega}),$$

$$N(\psi\tilde{\omega}) = e^{\pi\psi^2/2}\, Q(\psi\tilde{\omega}),$$

(see [12, III, p. 416]). Notice that $N(\tilde{\omega}) = e^{\tilde{\pi}/2}$ is an immediate consequence of the second formula.

Many other things were going on at this time. The appearance of $\pi/\tilde{\omega}$ sparked Gauss' interest in this ratio. He found several ways of expressing $\tilde{\omega}/\pi$, for example

$$(3.11) \qquad \frac{\tilde{\omega}}{\pi} = \frac{\sqrt{2}}{2}\left(1 + \left(\frac{1}{2}\right)^2\frac{1}{2} + \left(\frac{3}{8}\right)^2\frac{1}{4} + \left(\frac{5}{16}\right)^2\frac{1}{8} + \dots\right),$$

and he computed $\tilde{\omega}/\pi$ to fifteen decimal places (see [12, X.1, p. 169]). He also returned to some of his earlier notes and, where the approximation $2\int_0^1 z^2(1-z^4)^{-1/2}dz \approx 1.198$ appears, he added that this is $\pi/\tilde{\omega}$ (see [12, X.1, pp. 146 and 150]). Thus in July 1798 Gauss was intimately familiar with the right-hand side of the equation $M(\sqrt{2}, 1) = \pi/\tilde{\omega}$. Another problem he studied was the Fourier expansion of sl ϕ. Here, he first found the numerical value of the coefficients, i.e.

$$\text{sl } \psi\tilde{\omega} = .95500599 \sin \psi\pi - .04304950 \sin 3 \psi\pi + \dots,$$

and then he found a formula for the coefficients, obtaining

$$\text{sl } \psi\tilde{\omega} = \frac{4\pi}{\tilde{\omega}(e^{\pi/2}+e^{-\pi/2})} \sin \psi\pi - \frac{4\pi}{\tilde{\omega}(e^{3\pi/2}+e^{-3\pi/2})} \sin 3 \psi\pi + \dots$$

see [12, X.1, p. 168 and III, p. 417]).

The next breakthrough came in October 1798 when Gauss computed the Fourier expansions of $P(\phi)$ and $Q(\phi)$. As above, he first computed the coefficients numerically and then tried to find a general formula for them. Since he suspected that numbers like $e^{-\pi}$, $e^{-\pi/2}$, etc., would be involved, he computed several of these numbers (see [12, III, pp. 426-432]). The final formulas he found were

$$P(\psi\tilde{\omega}) =$$
$$2^{3/4}(\pi/\tilde{\omega})^{1/2}\left(e^{-\pi/4} \sin \psi\pi - e^{-9\pi/4} \sin 3 \psi\pi + e^{-25\pi/4} \sin 5\psi\pi - \dots\right)$$

$$(3.12) \qquad Q(\psi\tilde{\omega}) =$$
$$2^{-1/4}(\pi/\tilde{\omega})^{1/2}\left(1 + 2e^{-\pi} \cos 2\psi\pi + 2e^{-4\pi} \cos 4\psi\pi + 2e^{-9\pi} \cos 6\psi\pi + \dots\right)$$

(see [12, X.1, pp. 536-537]). A very brief sketch of how Gauss proved these formulas may be found in [12, X.2, pp. 38-39].

These formulas are remarkable for several reasons. First, recall the theta functions Θ_1 and Θ_3:

(3.13)

$$\Theta_1(z, q) = 2q^{1/4} \sin z - 2q^{9/4} \sin 3z + 2q^{25/4} \sin 5z - \ldots$$

$$\Theta_3(z, q) = 1 + 2q \cos 2z + 2q^4 \cos 4z + 2q^9 \cos 6z + \ldots$$

(see [36, p. 464]). Up to the constant factor $2^{-1/4}(\pi/\omega)^{1/2}$, we see that $P(\psi\omega)$ and $Q(\psi\omega)$ are precisely $\Theta_1(\psi\pi, e^{-\pi})$ and $\Theta_3(\psi\pi, e^{-\pi})$ respectively. Even though this is just a special case, one can easily discern the general form of the theta functions from (3.12). (For more on the relation between theta functions and sl ϕ, see [36, pp. 524-525]).

Several interesting formulas can be derived from (3.12) by making specific choice for ψ. For example, if we set $\psi = 1$, we obtain

$$\sqrt{\omega/\pi} = 2^{-1/4}(1 + 2e^{-\pi} + 2e^{-4\pi} + 2e^{-9\pi} + \ldots).$$

Also, if we set $\psi = 1/2$ and use the nontrivial fact that $P(\omega/2) = Q(\omega/2) = 2^{-1/4}$ (this is a consequence of the formula $Q(2\phi) = P(\phi)^4 + Q(\phi)^4$ — see (3.10)), we obtain

(3.14)

$$\sqrt{\omega/\pi} = 2(e^{-\pi/4} + e^{-9\pi/4} + e^{-25\pi/4} + \ldots)$$

$$\sqrt{\omega/\pi} = 1 - 2e^{-\pi} + 2e^{-4\pi} - 2e^{-9\pi} + \ldots.$$

Gauss wrote down these last two formulas in October 1798 (see [12, III, p. 418]). We, on the other hand, derived the first and third formulas as (2.21) in § 2, only after a very long development. Thus Gauss had some strong signposts to guide his development of modular functions.

These results, all dating from 1798, were recorded in Gauss' mathematical diary as the 91st and 92nd entries (in July) and the 95th entry (in October). The statement of the 92nd entry is especially relevant: "I have obtained most elegant results concerning the lemniscate, which surpasses all expectation—indeed, by methods which open an entirely new field to us" (see [12, X.1, p. 535]). There is a real sense of excitement here; instead of the earlier "advancement of analysis" of the 63rd entry, we have the much stronger phrase "entirely new field." Gauss knew that he had found something of importance. This feeling of excitement is confirmed by the

95th entry: "A new field of analysis is open before us, that is, the investigation of functions, etc." (see [12, X.1, p. 536]). It's as if Gauss were so enraptured he didn't even bother to finish the sentence.

More importantly, this "new field of analysis" is clearly the same "entirely new field of analysis" which we first saw in § 1 in the 98th entry. Rather than being an isolated phenomenon, it was the culmination of years of work. Imagine Gauss' excitement on May 30, 1799: this new field which he had seen grow up around the lemniscate and reveal such riches, all of a sudden expands yet again to encompass the arithmetic-geometric mean, a subject he had known since age 14. All of the powerful analytic tools he had developed for the lemniscatic functions were now ready to be applied to the agM.

C. In studying Gauss' work on the agM, it makes sense to start by asking where the observation $M(\sqrt{2}, 1) = \pi/\varpi$ came from. Using what we have learned so far, part of this question can now be answered: Gauss was very familiar with π/ϖ, and from reading Stirling he had probably seen the ratio $\sqrt{2} : 1$ associated with the lemniscate. (In fact, this ratio appears in most known methods for constructing the lemniscate—see [24, pp. 111-117].) We have also seen, in the equation $N(\varpi) = e^{\pi/2}$, that Gauss often used numerical calculations to help him discover theorems. But while these facts are enlightening, they still leave out one key ingredient, the idea of taking the agM of $\sqrt{2}$ and 1. Where did this come from? The answer is that every great mathematical discovery is kindled by some intuitive spark, and in our case, the spark came on May 30, 1799 when Gauss decided to compute $M(\sqrt{2}, 1)$.

We are still missing one piece of our picture of Gauss at this time: how much did he know about the agM? Unfortunately, this is a very difficult question to answer. Only a few scattered fragments dealing with the agM can be dated before May 30, 1799 (see [12, X.1, pp. 172-173 and 260]). As for the date 1791 of his discovery of the agM, it comes from a letter he wrote in 1816 (see [12, X.1, p. 247]), and Gauss is known to have been sometimes wrong in his recollections of dates. The only other knowledge we have about the agM in this period is an oral tradition which holds that Gauss knew the relation between theta functions and the agM in 1794 (see [12, III, 493]). We will soon see that this claim is not as outrageous as one might suspect.

It is not our intention to give a complete account of Gauss' work on the agM. This material is well covered in other places (see [10], [12, X.2, pp. 62-114], [13], [14] and [25]—the middle three references are especially complete), and furthermore it is impossible to give the full story of what happened. To explain this last statement, consider the following formulas:

$$B + (1/4)B^3 + (9/64)B^5 + ... = (2z^{1/2} + 2z^{9/2} + ...)^2 = r^2,$$

(3.15)

$$\frac{a}{M(a, b)} = 1 + (1/4)B^2 + (9/64)B^4 + ... ,$$

where $B = (1 - (b/a)^2)^{1/2}$. These come from the first surviving notes on the agM that Gauss wrote after May 30, 1799 (see [12, X.1, pp. 177-178]). If we set $a = 1$ and $b = k' = \sqrt{1 - k^2}$, then $B = k$, and we obtain

$$\frac{1}{M(1, k')} = 1 + (1/4)k^2 + (9/64)k^4 + ...$$

(3.16)

$$\frac{k}{M(1, k')} = (2z^{1/2} + 2z^{9/2} + ...)^2 = r^2.$$

The first formula is (1.8), and the second, with $z = e^{\pi i \tau/2}$, follows easily from what we learned in §2 about theta functions and the agM. Yet the formulas (3.15) appear with neither proofs nor any hint of where they came from. The discussion at the end of §1 sheds some light on the bottom formula of (3.15), but there is nothing to prepare us for the top one.

It is true that Gauss had a long-standing interest in theta functions, going back to when he first encountered Euler's wonderful formula

$$\sum_{n=-\infty}^{\infty} (-1)^n x^{(3n^2 + n)/2} = \prod_{n=1}^{\infty} (1 - x^n).$$

The right-hand side appears in a fragment dating from 1796 (see [12, X.1, p. 142]), and the 7th entry of his mathematical diary, also dated 1796, gives a continued fraction expansion for

$$1 - 2 + 8 - 64 + ...$$

Then the 58th entry, dated February 1797, generalizes this to give a continued fraction expansion for

$$1 - a + a^3 - a^6 + a^{10} - ...$$

(see [12, X.1, pp. 490 and 513]). The connection between these series and lemniscatic functions came in October 1798 with formulas such as (3.14).

This seems to have piqued his interest in the subject, for at this time he also set himself the problem of expressing

$$(3.17) \qquad\qquad 1 + x + x^3 + x^6 + x^{10} + ...$$

as an infinite product (see [12, X.1, p. 538]). Note also that the first formula of (3.14) gives r with $z = e^{-\pi/2}$.

Given these examples, we can conjecture where (3.15) came from. Gauss could easily have defined p, q and r in general and then derived identities (2.8)-(2.9) (recall the many identities obtained in 1798 for $P(\phi)$ and $Q(\phi)$—see (3.10) and [12, III, p. 410]). Then (3.15) would result from noticing that these identities formally satisfy the agM algorithm, which is the basic content of Lemma 2.3. This conjecture is consistent with the way Gauss initially treated z as a purely formal variable (the interpretation $z = e^{-\pi i t/2}$ was only to come later—see [12, X.1, pp. 262-263 and X.2, pp. 65-66]).

The lack of evidence makes it impossible to verify this or any other reasonable conjecture. But one thing is now clear: in Gauss' observation of May 30, 1799, we have not two but three distinct streams of his thought coming together. Soon after (or simultaneous with) observing that $M(\sqrt{2}, 1) = \pi/\tilde{\omega}$, Gauss knew that there were inimate connections between lemniscatic functions, the agM, and theta functions. The richness of the mathematics we have seen is in large part due to the many-sided nature of this confluence.

There remain two items of unfinished business. From § 1, we want to determine more precisely when Gauss first proved Theorem 1.1. And recall from § 2 that on June 3, 1800, Gauss discovered the "mutual connection" among the infinitely many values of $M(a, b)$. We want to see if he really knew the bulk of § 2 by this date. To answer these questions, we will briefly examine the main notebook Gauss kept between November 1799 and July 1800 (the notebook is "Scheda Ac" and appears as pp. 184-206 in [12, X.1]).

The starting date of this notebook coincides with the 100th entry of Gauss' mathematical diary, which reads "We have uncovered many new things about arithmetic-geometric means" (see [12, X.1, p. 544]). After several pages dealing with geometry, one all of a sudden finds the formula (3.11) for $\tilde{\omega}/\pi$. Since Gauss knew (3.15) at this time, we get an immediate proof of $M(\sqrt{2}, 1) = \pi/\tilde{\omega}$. Gauss must have had this in mind, for otherwise why would he so carefully recopy a formula proved in July 1798? Yet one could also ask why such a step is necessary: isn't Theorem 1.1 an immediate consequence of (3.15)? Amazingly enough, it appears that Gauss wasn't yet aware of this connection (see [12, X.1, p. 262]). Part of the problem is that he had been distracted by the power series, closely related to (3.15), which gives the arc length of the ellipse (see [12, X.1, p. 177]). This distraction was actually a bonus, for an asymptotic formula of Euler's for the arc length of the ellipse led Gauss to write

$$(3.18) \qquad M(x, 1) = \frac{(\pi/2)\,(x - \alpha x^{-1} - \beta x^{-3} - ...)}{\log(1/z)}$$

where $x = k^{-1}$, and z and k are as in (3.16) (see [12, X.1, pp. 186 and 268-270]). He was then able to show that the power series on top was $\left(k\, M(1, k')\right)^{-1}$, which implies that

$$z = \exp\left(-\frac{\pi}{2} \cdot \frac{M(1, k')}{M(1, k)}\right)$$

(see [12, X.1, pp. 187 and 190]). Letting $z = e^{\pi i \tau/2}$, we obtain formulas similar to (2.20). More importantly, we see that Gauss is now in a position to uniformize the agM; z is no longer a purely formal variable.

In the process of studying (3.18), Gauss also saw the relation between the agM and complete elliptic integrals of the first kind. The formula

$$\frac{1}{M(1, k')} = \frac{2}{\pi} \int_0^1 ((1 - x^2)\,(1 - k^2 x^2))^{-1/2} dx$$

follows easily from [12, X.1, p. 187], and this is trivially equivalent to (1.7). Furthermore, we know that this page was written on December 14, 1799 since on this date Gauss wrote in his mathematical diary that the agM was the quotient of two transcendental functions (see (3.18)), one of which was itself an integral quantity (see the 101st entry, [12, X.1, 544]). Thus Theorem 1.1 was proved on December 14, 1799, nine days earlier than our previous estimate.

Having proved this theorem, Gauss immediately notes one of its corollaries, that the "constant term" of the expression $(1 + \mu \cos^2\phi)^{-1/2}$ is $M(\sqrt{1 + \mu}, 1)^{-1}$ (see [12, X.1, p. 188]). By "constant term" Gauss means the coefficient A in the Fourier expansion

$$(1 + \mu \cos^2\phi)^{-1/2} = A + A' \cos\phi + A'' \cos 2\phi + ... \;.$$

Since A is the integral $\dfrac{2}{\pi} \displaystyle\int_0^{\pi/2} (1 + \mu \cos^2\phi)^{-1/2} d\phi$, the desired result follows from Theorem 1.1. This interpretation is important because these coefficients are useful in studying secular perturbations in astronomy (see [12, X.1, pp. 237-242]). It was in this connection that Gauss published his 1818 paper [12, III, pp. 331-355] from which we got our proof of Theorem 1.1.

What Gauss did next is unexpected: he used the agM to generalize the lemniscate functions to arbitrary elliptic functions, which for him meant inverse functions of elliptic integrals of the form

$$\int (1 + \mu^2 \sin^2 \phi)^{-1/2} d\phi = \int ((1 - x^2)(1 + \mu^2 x^2))^{-1/2} dx .$$

Note that $\mu = 1$ corresponds to the lemniscate. To start, he first set $\mu = \tan v$,

$$\varpi = \frac{\pi \cos v}{M(1, \cos v)}, \qquad \varpi' = \frac{\pi \cos v}{M(1, \sin v)}$$

and finally

(3.19) $$z = \exp\left(-\frac{\pi}{2} \cdot \frac{\varpi'}{\varpi}\right) = \exp\left(-\frac{\pi}{2} \cdot \frac{M(1, \cos v)}{M(1, \sin v)}\right).$$

Then he defined the elliptic function $S(\phi)$ by $S(\phi) = \dfrac{T(\phi)}{W(\phi)}$ where

$$T(\psi \varpi) = 2\mu^{-1/2}\sqrt{M(1, \cos v)}\,(z^{1/2}\sin\psi\pi - z^{9/2}\sin3\psi\pi + ...)$$

(3.20)

$$W(\psi \varpi) = \sqrt{M(1, \cos v)}\,(1 + 2z^2\cos2\psi\pi + 2z^8\cos4\psi\pi + ...)$$

(see [12, X.1, pp. 194-195 and 198]). In the pages that follow, we find the periods 2ϖ and $2i\varpi'$, the addition formula, and the differential equation connecting $S(\phi)$ to the above elliptic integral. Thus Gauss had a complete theory of elliptic functions.

In general, there are two basic approaches to this subject. One involves direct inversion of the elliptic integral and requires a detailed knowledge of the associated Riemann surface (see [17, Ch. VII]). The other more common approach defines elliptic functions as certain series ($\mathfrak{P}$-functions) or quotients of series (theta functions). The difficulty is proving that such functions invert all elliptic integrals. Classically, this uniformization problem is solved by studying a function such as $k(\tau)^2$ (see [36, § 20.6 and § 21.73]) or $j(\tau)$ (as in most modern texts—see [30, § 4.2]). Gauss uses the agM to solve this problem: (3.19) gives the desired uniformizing parameter! (In this connection, the reader should reconsider the from [12, VIII, p. 101] given near the end of § 2.)

For us, the most interesting aspect of what Gauss did concerns the functions T and W. Notice that (3.20) is a direct generalization of (3.12); in fact, in terms of (3.13), we have

$$T(\psi\tilde{\omega}) = \mu^{-1/2} \sqrt{M(1, \cos v)}\ \Theta_1(\psi\pi, z^2),$$

$$W(\psi\tilde{\omega}) = \sqrt{M(1, \cos v)}\ \Theta_3(\psi\pi, z^2).$$

Gauss also introduces $T(\tilde{\omega}/2 - \phi)$ and $W(\tilde{\omega}/2 - \phi)$, which are related to the theta functions Θ_2 and Θ_4 by similar formulas (see [12, X.1, pp. 196 and 275]). He then studies the squares of these functions and he obtains identities such as

$$2\Theta_3(0, z^4)\,\Theta_3(2\phi, z^4) = \Theta_3(\phi, z^2)^2 + \Theta_4(\phi, z^2)^2$$

(this, of course, is the modern formulation—see [12, X.1, pp. 196 (Eq. 14) and 275]). When $\phi =\ 0$, this reduces to the first formula

$$p(\tau)^2 + q(\tau)^2 = 2p(2\tau)^2$$

of (2.8). The other formulas of (2.8) appear similarly. Gauss also obtained product expansions for the theta functions (see [12, X.1, pp. 201-205]). In particular, one finds all the formulas of (2.6). These manipulations yielded the further result that

$$1 + z + z^3 + z^6 z^{10} + \dots - \prod_{n=1}^{\infty} (1-z)^{-1}(1-z^2),$$

solving the problem he had posed a year earlier in (3.17).

From Gauss' mathematical diary, we see that the bulk of this work was done in May 1800 (see entries 105, 106 and 108 in [12, X.1, pp. 546-549]). The last two weeks were especially intense as Gauss realized the special role played by the agM. The 108th entry, dated June 3, 1800, announces completion of a general theory of elliptic functions ("sinus lemniscatici universalissime accepti"). On the same day he recorded his discovery of the "mutual connection" among the values of the agM !

This is rather surprising. We've seen that Gauss knew the basic identities (2.6), (2.8) and (2.9), but the formulas (2.7), which tell us how theta functions behave under linear fractional transformations, are nowhere to be seen, nor do we find any hint of the fundamental domains used in § 2. Reading this notebook makes it clear that Gauss now knew the basic observation of

Lemma 2.3 that theta functions satisfy the agM algorithm, but there is no way to get from here to Theorem 2.2 without knowing (2.7). It is not until 1805 that this material appears in Gauss' notes (see [12, X.2, pp. 101-103]). Thus some authors, notably Markushevitch [25], have concluded that on June 3, 1800, Gauss had nothing approaching a proof of Theorem 2.2.

Schlesinger, the last editor of Gauss' collected works, feels otherwise. He thinks that Gauss knew (2.7) at this time, though knowledge of the fundamental domains may have not come until 1805 (see [12, X.2, p. 106]). Schlesinger often overestimates what Gauss knew about modular functions, but in this case I agree with him. As evidence, consider pp. 287-307 in [12, X.1]. These reproduce twelve consecutive pages from a notebook written in 1808 (see [12, X.1, p. 322]), and they contain the formulas (2.7), a clear statement of the basic observation of Lemma 2.3, the infinite product manipulations described above, and the equations giving the division of the agM into 3, 5 and 7 parts (in analogy with the division of the lemniscate). The last item is especially interesting because it relates to the second half of the 108th entry: "Moreover, in these same days, we have discovered the principles according to which the agM series should be interpolated, so as to enable us to exhibit by algebraic equations the terms in a given progression pertaining to any rational index" (see [12, X.1, p. 548]). There is no other record of this in 1800, yet here it is in 1808 resurfacing with other material (the infinite products) dating from 1800. Thus it is reasonable to assume that the rest of this material, including (2.7), also dates from 1800. Of course, to really check this conjecture, one would have to study the original documents in detail.

Given all of (2.6)-(2.9), it is still not clear where Gauss got the basic insight that $M(a, b)$ is a multiple valued function. One possible source of inspiration is the differential equation (1.12) whose solution (1.13) suggests linear combinations similar to those of Theorem 2.2. We get even closer to this theorem when we consiser the periods of $S(\phi)$:

$$m\bar{\omega} + in\bar{\omega}' = \pi \cos v \left(\frac{m}{M(1, \cos v)} + \frac{in}{M(1, \sin v)} \right)$$

where m, n are even integers. Gauss' struggles during May 1800 to understand the imaginary nature of these periods (see [12, X.2, pp. 70-71]) may have influenced his work on the agM. (We should point out that the above comments are related: Theorem 2.2 can be proved by analyzing the monodromy group—$\Gamma_2(4)$ in this case—of the differential equation (1.12).) On the other hand, Geppert suggests that Gauss may have taken a completely different route, involving the asymptotic formula (3.18), of arriving at Theorem 2.2 (see [14, pp. 173-175]). We will of course never really know how Gauss arrived at this theorem.

For many years, Gauss hoped to write up these results for publication. He mentions this in Disquisitiones Arithmeticae (see [11, § 335]) and in the research announcement to his 1818 article (see [12, III, p. 358]). Two manuscripts written in 1800 (one on the agM, the other on lemniscatic functions) show that Gauss made a good start on this project (see [12, III, pp. 360-371 and 413-415]). He also periodically returned to earlier work and rewrote it in more complete form (the 1808 notebook is an example of this). Aside from the many other projects Gauss had to distract him, it is clear why he never finished this one: it was simply too big. Given his predilection for completeness, the resulting work would have been enormous. Gauss finally gave up trying in 1827 when the first works of Abel and Jacobi appeared. As he wrote in 1828, "I shall most likely not soon prepare my investigations on transcendental functions which I have had for many years—since 1798—because I have many other matters which must be cleared up. Herr Abel has now, I see, anticipated me and relieved me of the burden in regard to one third of these matters, particularly since he carried out all developments with great concision and elegance" (see [12, X.1, p. 248]).

The other two thirds "of these matters" encompass Gauss' work on the agM and modular functions. The latter were studied vigorously in the nineteenth century and are still an active area of research today. The agM, on the other hand, has been relegated to the history books. This is not entirely wrong, for the history of this subject is wonderful. But at the same time the agM is also wonderful as mathematics, and this mathematics deserves to be better known.

THE ARITHMETIC-GEOMETRIC MEAN   329

REFERENCES

[1] ALLING, N. L. *Real Elliptic Curves*. North-Holland Mathematics Studies, Vol. 54, North-Holland, Amsterdam, 1981.

[2] BERNOULLI, Jacob. *Opera*, Vol. I. Geneva, 1744.

[3] BERNOULLI, Johann. *Opera omnia*, Vol. I. Lausanne, 1742.

[4] BÜHLER, W. K. *Gauss: A Biographical Study*. Springer-Verlag, Berlin-Heidelberg-New York, 1981.

[5] CARLSON, B. C. Algorithms involving Arithmetic and Geometric Means. *Amer. Math. Monthly 78* (1971), 496-505.

[6] CASSELS, J. W. S. *Rational Quadratic Forms*. Academic Press, New York, 1978.

[7] COPSON, E. T. *An Introduction to the Theory of Functions of a Complex Variable*. Oxford U. Press, London, 1935.

[8] ENNEPER, A. *Elliptische Functionen: Theorie und Geschichte*. Halle, 1876.

[9] EULER, L. *Opera Omnia*, Series Prima, Vol. XX and XXI. Teubner, Leipzig and Berlin, 1912-1913.

[10] FUCHS, W. Das arithmetisch-geometrische Mittel in den Untersuchungen von Carl Friedrich Gauss. *Gauss-Gesellschaft Göttingen, Mittelungen No. 9* (1972), 14-38.

[11] GAUSS, C. F. *Disquisitiones Arithmeticae*. Translated by A. Clark, Yale U. Press, New Haven, 1965 (see also [12, I]).

[12] —— *Werke*. Göttingen-Leipzig, 1868-1927.

[13] GEPPERT, H. *Bestimmung der Anziehung eines elliptischen Ringes*. Ostwald's Klassiker, Vol. 225, Akademische Verlag, Leipzig, 1927.

[14] —— Wie Gauss zur elliptischen Modulfunktion kam. *Deutsche Mathematik 5* (1940), 158-175.

[15] —— Zur Theorie des arithmetisch-geometrischen Mittels. *Math. Annalen 99* (1928), 162-180.

[16] GRADSHTEYN, I. S. and I. M. RYZHIK. *Table of Integrals, Series and Products*. Academic Press, New York, 1965.

[17] HANCOCK, H. *Lectures on the Theory of Elliptic Functions*. Vol. I. Wiley, New York, 1910.

[18] HOFFMAN, J. E. Über Jakob Bernoullis Beiträge zur Infinitesimalmathematik. *L'Enseignement Math. 2* (1956), 61-171.

[19] HOUZEL, C. Fonctions Elliptiques et Intégrals Abéliennes. In *Abrégé d'histoire des mathématiques 1700-1900*, Vol. II. Ed. by J. Dieudonné, Hermann, Paris, 1978, 1-112.

[20] JACOBI, C. C. J. *Gesammelte Werke*. G. Reimer, Berlin, 1881.

[21] KLINE, M. *Mathematical Thought from Ancient to Modern Times*. Oxford U. Press, New York, 1972.

[22] LAGRANGE, J. L. *Œuvres*, Vol. II. Gauthier-Villars, Paris, 1868.

[23] LEGENDRE, A. M. *Traité des Fonctions Elliptiques*. Paris, 1825-1828.

[24] LOCKWOOD, E. H. *A Book of Curves*. Cambridge U. Press, Cambridge, 1971.

[25] MARKUSHEVITCH, A. I. Die Arbeiten von C. F. Gauss über Funktionentheorie. In *C. F. Gauss Gedenkband Anlässlich des 100. Todestages am 23. Februar 1955*. Ed. by H. Reichart, Teubner, Leipzig, 1957, 151-182.

[26] MIEL, G. Of Calculations Past and Present: The Archimedean Algorithm. *Amer. Math. Monthly 90* (1983), 17-35.

[27] MUMFORD, D. *Tata Lectures on Theta I*. Progress in Mathematics Vol. 28, Birkhäuser, Boston, 1983.

330   D. A. COX

[28] ROSEN, M. Abel's Theorem on the Lemniscate. *Amer. Math. Monthly 88* (1981), 387-395.

[29] SERRE, J.-P. *Cours d'Arithmétique.* Presses U. de France, Paris, 1970.

[30] SHIMURA, G. *Introduction to the Arithmetic Theory of Automorphic Functions.* Princeton U. Press, Princeton, 1971.

[31] STIRLING, J. *Methodus Differentialis.* London, 1730.

[32] TANNERY, J. and J. MOLK. *Eléments de la Théorie des Fonctions Elliptiques,* Vol. 2. Gauthiers-Villars, Paris, 1893.

[33] TODD, J. The Lemniscate Constants. *Comm. of the ACM 18* (1975), 14-19.

[34] van der POL, B. Démonstration Elémentaire de la Relation $\Theta_3^4 = \Theta_0^4 + \Theta_2^4$ entre les Différentes Fonctions de Jacobi. *L'Enseignement Math. 1* (1955), 258-261.

[35] von DAVID, L. Arithmetisch-geometrisches Mittel und Modulfunktion. *J. für die Reine u. Ang. Math. 159* (1928), 154-170.

[36] WHITTAKER, E. T. and G. N. WATSON. *A Course of Modern Analysis,* 4th ed. Cambridge U. Press, Cambridge, 1963.

(Reçu le 21 novembre 1983)

David A. Cox

Department of Mathematics
Amherst College
Amherst, MA 01002 (USA)

4. The arithmetic-geometric mean and fast computation of elementary functions (1984)

Paper 4: J. M. Borwein and P. B. Borwein, "The arithmetic-geometric mean and fast computation of elementary functions," *SIAM Review*, vol. 26 (1984), p. 351–366. Copyright ©1984 Society for Industrial and Applied Mathematics. Reprinted with permission. All rights reserved.

Synopsis:

In this paper, brothers Jonathan and Peter Borwein present a review of the recently discovered (as of 1984) quadratically convergent formulas for π and elementary functions (including some new formulas of their own), complete with a rigorous derivation of all the requisite mathematics. The paper is thus an excellent self-contained tutorial on the theory of the arithmetic-geometric mean, quadratically convergent algorithms (including Newton's algorithm for computing square roots and roots of polynomials), and how these concepts lead to fast algorithms for π and elementary functions. They do this without needing to venture into incomplete elliptic integrals and Landen transforms, which were used to various degrees by earlier writers.

This paper has particular significance for the editors of this volume, as its appearance in *SIAM Review*, a publication mostly read in the applied mathematics and high-performance computing communities, led to Bailey becoming interested in these topics and joining with Jonathan and Peter Borwein in a multi-decade collaboration.

Keywords: Algorithms, Approximations, Arithmetic-Geometric Mean, Computation, History

© Springer International Publishing Switzerland 2016
D.H. Bailey, J.M. Borwein, *Pi: The Next Generation*,
DOI 10.1007/978-3-319-32377-0_4

SIAM REVIEW
Vol. 26, No. 3, July 1984

© 1984 Society for Industrial and Applied Mathematics
002

THE ARITHMETIC-GEOMETRIC MEAN AND FAST COMPUTATION OF ELEMENTARY FUNCTIONS*

J. M. BORWEIN† AND P. B. BORWEIN†

Abstract. We produce a self contained account of the relationship between the Gaussian arithmetic-geometric mean iteration and the fast computation of elementary functions. A particularly pleasant algorithm for π is one of the by-products.

Introduction. It is possible to calculate 2^n decimal places of π using only n iterations of a (fairly) simple three-term recursion. This remarkable fact seems to have first been explicitly noted by Salamin in 1976 [16]. Recently the Japanese workers Y. Tamura and Y. Kanada have used Salamin's algorithm to calculate π to 2^{23} decimal places in 6.8 hours. Subsequently 2^{24} places were obtained ([18] and private communication). Even more remarkable is the fact that all the elementary functions can be calculated with similar dispatch. This was proved (and implemented) by Brent in 1976 [5]. These extraordinarily rapid algorithms rely on a body of material from the theory of elliptic functions, all of which was known to Gauss. It is an interesting synthesis of classical mathematics with contemporary computational concerns that has provided us with these methods. Brent's analysis requires a number of results on elliptic functions that are no longer particularly familiar to most mathematicians. Newman in 1981 stripped this analysis to its bare essentials and derived related, though somewhat less computationally satisfactory, methods for computing π and log. This concise and attractive treatment may be found in [15].

Our intention is to provide a mathematically intermediate perspective and some bits of the history. We shall derive implementable (essentially) quadratic methods for computing π and all the elementary functions. The treatment is entirely self-contained and uses only a minimum of elliptic function theory.

1. 3.141592653589793238462643383279502884197. The calculation of π to great accuracy has had a mathematical import that goes far beyond the dictates of utility. It requires a mere 39 digits of π in order to compute the circumference of a circle of radius 2×10^{25} meters (an upper bound on the distance travelled by a particle moving at the speed of light for 20 billion years, and as such an upper bound on the radius of the universe) with an error of less than 10^{-12} meters (a lower bound for the radius of a hydrogen atom).

Such a calculation was in principle possible for Archimedes, who was the first person to develop methods capable of generating arbitrarily many digits of π. He considered circumscribed and inscribed regular n-gons in a circle of radius 1. Using $n = 96$ he obtained

$$3.1405 \cdots = \frac{6336}{2017.25} < \pi < \frac{14688}{4673.5} = 3.1428.$$

If $1/A_n$ denotes the area of an inscribed regular 2^n-gon and $1/B_n$ denotes the area of a circumscribed regular 2^n-gon about a circle of radius 1 then

(1.1) $$A_{n+1} = \sqrt{A_n B_n}, \qquad B_{n+1} = \frac{A_{n+1} + B_n}{2}.$$

---

*Received by the editors February 8, 1983, and in revised form November 21, 1983. This research was partially sponsored by the Natural Sciences and Engineering Research Council of Canada.

†Department of Mathematics, Dalhousie University, Halifax, Nova Scotia, Canada B3H 4H8.

This two-term iteration, starting with $A_2 := \frac{1}{2}$ and $B_2 := \frac{1}{4}$, can obviously be used to calculate π. (See Edwards [9, p. 34].) A_{15}^{-1}, for example, is 3.14159266 which is correct through the first seven digits. In the early sixteen hundreds Ludolph von Ceulen actually computed π to 35 places by Archimedes' method [2].

Observe that $A_n := 2^{-n} \csc(\theta/2^n)$ and $B_n := 2^{-n-1} \cot(\theta/2^{n+1})$ satisfy the above recursion. So do $A_n := 2^{-n} \operatorname{cosech}(\theta/2^n)$ and $B_n := 2^{-n-1} \coth(\theta/2^{n+1})$. Since in both cases the common limit is $1/\theta$, the iteration can be used to calculate the standard inverse trigonometric and inverse hyperbolic functions. (This is often called Borchardt's algorithm [6], [19].)

If we observe that

$$A_{n+1} - B_{n+1} = \frac{1}{2(\sqrt{A_n}/\sqrt{B_n} + 1)} (A_n - B_n)$$

we see that the error is decreased by a factor of approximately four with each iteration. This is linear convergence. To compute n decimal digits of π, or for that matter arcsin, arcsinh or log, requires $O(n)$ iterations.

We can, of course, compute π from arctan or arcsin using the Taylor expansion of these functions. John Machin (1680–1752) observed that

$$\pi = 16 \arctan\left(\frac{1}{5}\right) - 4 \arctan\left(\frac{1}{239}\right)$$

and used this to compute π to 100 places. William Shanks in 1873 used the same formula for his celebrated 707 digit calculation. A similar formula was employed by Leonhard Euler (1707–1783):

$$\pi = 20 \arctan\left(\frac{1}{7}\right) + 8 \arctan\left(\frac{3}{79}\right).$$

This, with the expansion

$$\arctan(x) = \frac{y}{x}\left(1 + \frac{2}{3} y + \frac{2.4}{3.5} y^2 + \cdots\right)$$

where $y = x^2/(1 + x^2)$, was used by Euler to compute π to 20 decimal places in an hour. (See Beckman [2] or Wrench [21] for a comprehensive discussion of these matters.) In 1844 Johann Dase (1824–1861) computed π correctly to 200 places using the formula

$$\frac{\pi}{4} = \arctan\left(\frac{1}{2}\right) + \arctan\left(\frac{1}{5}\right) + \arctan\left(\frac{1}{8}\right).$$

Dase, an "idiot savant" and a calculating prodigy, performed this "stupendous task" in "just under two months." (The quotes are from Beckman, pp. 105 and 107.)

A similar identity:

$$\pi = 24 \arctan\left(\frac{1}{8}\right) + 8 \arctan\left(\frac{1}{57}\right) + 4 \arctan\left(\frac{1}{239}\right)$$

was employed, in 1962, to compute 100,000 decimals of π. A more reliable "idiot savant", the IBM 7090, performed this calculation in a mere 8 hrs. 43 mins. [17].

There are, of course, many series, products and continued fractions for π. However, all the usual ones, even cleverly evaluated, require $O(\sqrt{n})$ operations $(+, \times, \div, \sqrt{\ })$ to arrive at n digits of π. Most of them, in fact, employ $O(n)$ operations for n digits, which is essentially linear convergence. Here we consider only full precision operations. For a time complexity analysis and a discussion of time efficient algorithms based on binary splitting see [4].

The algorithm employed in [17] requires about 1,000,000 operations to compute 1,000,000 digits of π. We shall present an algorithm that reduces this to about 200 operations. The algorithm, like Salamin's and Newman's requires some very elementary elliptic function theory. The circle of ideas surrounding the algorithm for π also provides algorithms for all the elementary functions.

2. Extraordinarily rapid algorithms for algebraic functions. We need the following two measures of speed of convergence of a sequence (a_n) with limit L. If there is a constant C_1 so that

$$|a_{n+1} - L| \leq C_1 |a_n - L|^2$$

for all n, then we say that (a_n) converges to L *quadratically*, or with *second order*. If there is a constant $C_2 > 1$ so that, for all n,

$$|a_n - L| \leq C_2^{-2^n}$$

then we say that (a_n) converges to L *exponentially*. These two notions are closely related; quadratic convergence implies exponential convergence and both types of convergence guarantee that a_n and L will "agree" through the first $O(2^n)$ digits (provided we adopt the convention that .9999...9 and 1.000...0 agree through the required number of digits).

Newton's method is perhaps the best known second order iterative method. Newton's method computes a zero of $f(x) - y$ by

$$(2.1) \qquad x_{n+1} := x_n - \frac{f(x_n) - y}{f'(x_n)}$$

and hence, can be used to compute f^{-1} quadratically from f, at least locally. For our purposes, finding suitable starting values poses little difficulty. Division can be performed by inverting $(1/x) - y$. The following iteration computes $1/y$:

$$(2.2) \qquad x_{n+1} := 2x_n - x_n^2 y.$$

Square root extraction ($\sqrt{y}$) is performed by

$$(2.3) \qquad x_{n+1} := \frac{1}{2}\left(x_n + \frac{y}{x_n}\right).$$

This ancient iteration can be traced back at least as far as the Babylonians. From (2.2) and (2.3) we can deduce that division and square root extraction are of the same order of complexity as multiplication (see [5]). Let $M(n)$ be the "amount of work" required to multiply two n digit numbers together and let $D(n)$ and $S(n)$ be, respectively, the "amount of work" required to invert an n digit number and compute its square root, to n digit accuracy. Then

$$D(n) = O(M(n))$$

and

$$S(n) = O(M(n)).$$

We are not bothering to specify precisely what we mean by work. We could for example count the number of single digit multiplications. The basic point is as follows. It requires

$O(\log n)$ iterations of Newton's method (2.2) to compute $1/y$. However, at the ith iteration, one need only work with accuracy $O(2^i)$. In this sense, Newton's method is self-correcting. Thus,

$$D(n) = O\left(\sum_{i=1}^{\log n} M(2^i)\right) = O(M(n))$$

provided $M(2^i) \geq 2M(2^{i-1})$. The constants concealed beneath the order symbol are not even particularly large. Finally, using a fast multiplication, see [12], it is possible to multiply two n digits numbers in $O(n \log (n) \log \log (n))$ single digit operations.

What we have indicated is that, for the purposes of asymptotics, it is reasonable to consider multiplication, division and root extraction as equally complicated and to consider each of these as only marginally more complicated than addition. Thus, when we refer to operations we shall be allowing addition, multiplication, division and root extraction.

Algebraic functions, that is roots of polynomials whose coefficients are rational functions, can be approximated (calculated) exponentially using Newton's method. By this we mean that the iterations converge exponentially and that each iterate is itself suitably calculable. (See [13].)

The difficult trick is to find a method to exponentially approximate just one elementary transcendental function. It will then transpire that the other elementary functions can also be exponentially calculated from it by composition, inversion and so on.

For this Newton's method cannot suffice since, if f is algebraic in (2.1) then the limit is also algebraic.

The only familiar iterative procedure that converges quadratically to a transcendental function is the arithmetic-geometric mean iteration of Gauss and Legendre for computing complete elliptic integrals. This is where we now turn. We must emphasize that it is difficult to exaggerate Gauss' mastery of this material and most of the next section is to be found in one form or another in [10].

3. The real AGM iteration. Let two positive numbers a and b with $a > b$ be given. Let $a_0 := a$, $b_0 := b$ and define

$$(3.1) \qquad a_{n+1} := \frac{1}{2}(a_n + b_n), \qquad b_{n+1} := \sqrt{a_n b_n}$$

for n in $\mathbb{N}$.

One observes, as a consequence of the arithmetic-geometric mean inequality, that $a_n \geq a_{n+1} \geq b_{n+1} \geq b_n$ for all n. It follows easily that (a_n) and (b_n) converge to a common limit L which we sometimes denote by $AG(a, b)$. Let us now set

$$(3.2) \qquad c_n := \sqrt{a_n^2 - b_n^2} \quad \text{for } n \in \mathbb{N}.$$

It is apparent that

$$(3.3) \qquad c_{n+1} = \frac{1}{2}(a_n - b_n) = \frac{c_n^2}{4a_{n+1}} \leq \frac{c_n^2}{4L},$$

which shows that (c_n) converges quadratically to zero. We also observe that

$$(3.4) \qquad a_n = a_{n+1} + c_{n+1} \quad \text{and} \quad b_n = a_{n+1} - c_{n+1}$$

which allows us to define a_n, b_n and c_n for negative n. These negative terms can also be generated by the *conjugate scale* in which one starts with $a_0' := a_0$ and $b_0' := c_0$ and defines

ARITHMETIC-GEOMETRIC MEAN AND FAST COMPUTATION   355

(a'_n) and (b'_n) by (3.1). A simple induction shows that for any integer n

(3.5) $a'_n = 2^{-n}a_{-n}, \quad b'_n = 2^{-n}c_{-n}, \quad c'_n = 2^{-n}b_{-n}.$

Thus, backward iteration can be avoided simply by altering the starting values. For future use we define the quadratic *conjugate* $k' := \sqrt{1 - k^2}$ for any k between 0 and 1.

    The limit of (a_n) can be expressed in terms of a *complete elliptic integral of the first kind*,

(3.6) $$I(a, b) := \int_0^{\pi/2} \frac{d\theta}{\sqrt{a^2 \cos^2 \theta + b^2 \sin^2 \theta}}.$$

In fact

(3.7) $$I(a, b) = \frac{1}{2} \int_{-\infty}^{\infty} \frac{dt}{\sqrt{(a^2 + t^2)(b^2 + t^2)}}$$

as the substitution $t := a \tan \theta$ shows. Now the substitution of $u := \frac{1}{2} (t - (ab/t))$ and some careful but straightforward work [15] show that

(3.8) $$I(a, b) = I\left(\left(\frac{a + b}{2}\right), \sqrt{ab}\right).$$

It follows that $I(a_n, b_n)$ is independent of n and that, on interchanging limit and integral,

$$I(a_0, b_0) = \lim_{n \to \infty} I(a_n, b_n) = I(L, L).$$

Since the last integral is a simple arctan (or directly from (3.6)) we see that

(3.9) $$I(a_0, b_0) = \frac{\pi}{2} AG(a_0, b_0).$$

Gauss, of course, had to derive rather than merely verify this remarkable formula. We note in passing that $AG(\cdot, \cdot)$ is positively homogeneous.

    We are now ready to establish the underlying limit formula.

    PROPOSITION 1.

(3.10) $$\lim_{k \to 0^+} \left[\log\left(\frac{4}{k}\right) - I(1, k)\right] = 0.$$

    Proof. Let

$$A(k) := \int_0^{\pi/2} \frac{k' \sin \theta \, d\theta}{\sqrt{k^2 + (k')^2 \cos^2 \theta}}$$

and

$$B(k) := \int_0^{\pi/2} \sqrt{\frac{1 - k' \sin \theta}{1 + k' \sin \theta}} \, d\theta.$$

Since $1 - (k' \sin \theta)^2 = \cos^2 \theta + (k \sin \theta)^2 = (k' \cos \theta)^2 + k^2$, we can check that

$$I(1, k) = A(k) + B(k).$$

Moreover, the substitution $u := k' \cos \theta$ allows one to evaluate

(3.11) $$A(k) := \int_0^{k'} \frac{du}{\sqrt{u^2 + k^2}} = \log\left(\frac{1 + k'}{k}\right).$$

Finally, a uniformity argument justifies

(3.12) $$\lim_{k \to 0^+} B(k) = B(0) = \int_0^{\pi/2} \frac{\cos \theta \, d\theta}{1 + \sin \theta} = \log 2,$$

and (3.11) and (3.12) combine to show (3.10). $\square$

It is possible to give various asymptotics in (3.10), by estimating the convergence rate in (3.12).

PROPOSITION 2. *For $k \in (0, 1]$*

(3.13) $$\left| \log \left(\frac{4}{k} \right) - I(1, k) \right| \le 4k^2 I(1, k) \le 4k^2(8 + |\log k|).$$

Proof. Let

$$\Delta(k) := \log \left(\frac{4}{k} \right) - I(1, k).$$

As in Proposition 1, for $k \in (0, 1]$,

(3.14)
$$|\Delta(k)| \le \left| \log \left(\frac{2}{k} \right) - \log \left(\frac{1 + k'}{k} \right) \right| + \left| \int_0^{\pi/2} \left[\sqrt{\frac{1 - k' \sin \theta}{1 + k' \sin \theta}} - \sqrt{\frac{1 - \sin \theta}{1 + \sin \theta}} \right] d\theta \right|.$$

We observe that, since $1 - k' = 1 - \sqrt{1 - k^2} < k^2$,

(3.15) $$\left| \log \left(\frac{2}{k} \right) - \log \left(\frac{1 + k'}{k} \right) \right| = \left| \log \left(\frac{1 + k'}{2} \right) \right| \le 1 - k' \le k^2.$$

Also, by the mean value theorem, for each θ there is a $\gamma \in [0, k]$, so that

$$0 \le \left[\sqrt{\frac{1 - k' \sin \theta}{1 + k' \sin \theta}} - \sqrt{\frac{1 - \sin \theta}{1 + \sin \theta}} \right]$$

$$\le \left[\sqrt{\frac{1 - (1 - k^2) \sin \theta}{1 + (1 - k^2) \sin \theta}} - \sqrt{\frac{1 - \sin \theta}{1 + \sin \theta}} \right]$$

$$= \left[\frac{\sqrt{1 + (1 - \gamma^2) \sin \theta}}{\sqrt{1 - (1 - \gamma^2) \sin \theta}} \cdot \frac{2\gamma \sin \theta}{(1 + (1 - \gamma^2) \sin \theta)^2} \right] k$$

$$\le \frac{2\gamma k}{\sqrt{1 - (1 - \gamma^2) \sin \theta}} \le \frac{2k^2}{\sqrt{1 - (1 - k^2) \sin \theta}}.$$

This yields

$$\left| \int_0^{\pi/2} \left[\sqrt{\frac{1 - k' \sin \theta}{1 + k' \sin \theta}} - \sqrt{\frac{1 - \sin \theta}{1 + \sin \theta}} \right] d\theta \right| \le 2k^2 \int_0^{\pi/2} \frac{d\theta}{\sqrt{1 - k' \sin \theta}} \le 2\sqrt{2} \, k^2 I(1, k)$$

which combines with (3.14) and (3.15) to show that

$$|\Delta(k)| \le (1 + 2\sqrt{2})k^2 I(1, k) \le 4k^2 I(1, k).$$

We finish by observing that

$$kI(1, k) \le \frac{\pi}{2}$$

allows us to deduce that

$$I(1, k) \leq 2\pi k + \log\left(\frac{4}{k}\right). \qquad \qquad \square$$

Similar considerations allow one to deduce that

$$(3.16) \qquad |\Delta(k) - \Delta(h)| \leq 2\pi |k - h|$$

for $0 < k, h < 1/\sqrt{2}$.

The next proposition gives all the information necessary for computing the elementary functions from the AGM.

PROPOSITION 3. *The* AGM *satisfies the following identity (for all initial values):*

$$(3.17) \qquad \lim_{n\to\infty} 2^{-n} \frac{a'_n}{a_n} \log\left(\frac{4a_n}{c_n}\right) = \frac{\pi}{2}.$$

Proof. One verifies that

$$\frac{\pi}{2} = \lim_{n\to\infty} a'_n I(a'_0, b'_0) \qquad \text{(by (3.9))}$$

$$= \lim_{n\to\infty} a'_n I(a'_{-n}, b'_{-n}) \qquad \text{(by (3.8))}$$

$$= \lim_{n\to\infty} a'_n I(2^n a_n, 2^n c_n) \qquad \text{(by 3.5)).}$$

Now the homogeneity properties of $I(\cdot, \cdot)$ show that

$$I(2^n a_n, 2^n c_n) = \frac{2^{-n}}{a_n} I\left(1, \frac{c_n}{a_n}\right).$$

Thus

$$\frac{\pi}{2} = \lim_{n\to\infty} 2^{-n} \frac{a'_n}{a_n} I\left(1, \frac{c_n}{a_n}\right),$$

and the result follows from Proposition 1. $\square$

From now on we fix $a_0 := a'_0 := 1$ and consider the iteration as a function of $b_0 := k$ and $c_0 := k'$. Let P_n and Q_n be defined by

$$(3.18) \qquad P_n(k) := \left(\frac{4a_n}{c_n}\right)^{2^{1-n}}, \qquad Q_n(k) := \frac{a_n}{a'_n},$$

and let $P(k) := \lim_{n\to\infty} P_n(k)$, $Q(k) := \lim_{n\to\infty} Q_n(k)$. Similarly let $a := a(k) := \lim_{n\to\infty} a_n$ and $a' := a'(k) := \lim_{n\to\infty} a'_n$.

THEOREM 1. *For* $0 < k < 1$ *one has ·*

$$(a) \quad P(k) = \exp(\pi Q(k)),$$

$$(3.19) \qquad (b) \quad 0 \leq P_n(k) - P(k) \leq \frac{16}{1 - k^2}\left(\frac{a_n - a}{a}\right),$$

$$(c) \quad |Q_n(k) - Q(k)| \leq \frac{a'|a - a_n| + a|a' - a'_n|}{(a')^2}.$$

Proof. (a) is an immediate rephrasing of Proposition 3, while (c) is straightforward.

To see (b) we observe that

$$(3.20) \qquad P_{n+1} = P_n \cdot \left(\frac{a_{n+1}}{a_n}\right)^{2^{1-n}}$$

because $4a_{n+1} c_{n+1} = c_n^2$, as in (3.3). Since $a_{n+1} \leq a_n$ we see that

$$(3.21) \qquad O \leq P_n - P_{n+1} \leq \left[1 - \left(\frac{a_{n+1}}{a_n}\right)^{2^{1-n}}\right] P_n \leq \left(1 - \frac{a_{n+1}}{a_n}\right) P_0,$$

$$P_n - P_{n+1} \leq \left(\frac{a_n - a_{n+1}}{a}\right) P_0$$

since a_n decreases to a. The result now follows from (3.21) on summation.   □

Thus, the theorem shows that both P and Q can be computed exponentially since (a_n) can be so calculated. In the following sections we will use this theorem to give implementable exponential algorithms for π and then for all the elementary functions.

We conclude this section by rephrasing (3.19a). By using (3.20) repeatedly we derive that

$$(3.22) \qquad P = \frac{16}{1 - k^2} \prod_{n=0}^{\infty} \left(\frac{a_{n+1}}{a_n}\right)^{2^{1-n}}.$$

Let us note that

$$\frac{a_{n+1}}{a_n} = \frac{a_n + b_n}{2a_n} = \frac{1}{2}\left(1 + \frac{b_n}{a_n}\right),$$

and $x_n := b_n/a_n$ satisfies the one-term recursion used by Legendre [14]

$$(3.23) \qquad x_{n+1} := \frac{2\sqrt{x_n}}{x_n + 1} \qquad x_0 := k.$$

Thus, also

$$(3.24) \qquad P_{n+1}(k) = \frac{16}{1 - k^2} \prod_{j=0}^{n+1} \left(\frac{1 + x_j}{2}\right)^{2^{1-j}} = \left(\frac{1 + x_n}{1 - x_n}\right)^{2^{-n}}.$$

When $k := 2^{-1/2}$, $k = k'$ and one can explicitly deduce that $P(2^{-1/2}) = e^{\pi}$. When $k = 2^{-1/2}$ (3.22) is also given in [16].

4. Some interrelationships. A centerpiece of this exposition is the formula (3.17) of Proposition 3.

$$(4.1) \qquad \lim_{n\to\infty} \frac{1}{2^n} \log\left(\frac{4a_n}{c_n}\right) = \frac{\pi}{2} \lim_{n\to\infty} \frac{a_n}{a_n'},$$

coupled with the observation that both sides converge exponentially. To approximate $\log x$ exponentially, for example, we first find a starting value for which

$$\left(\frac{4a_n}{c_n}\right)^{1/2^n} \to x.$$

This we can do to any required accuracy quadratically by Newton's method. Then we compute the right limit, also quadratically, by the AGM iteration. We can compute exp analogously and since, as we will show, (4.1) holds for complex initial values we can also get the trigonometric functions.

There are details, of course, some of which we will discuss later. An obvious detail is that we require π to desired accuracy. The next section will provide an exponentially converging algorithm for π also based only on (4.1). The principle for it is very simple. If we differentiate both sides of (4.1) we lose the logarithm but keep the π!

Formula (3.10), of Proposition 1, is of some interest. It appears in King [11, pp. 13, 38] often without the "4" in the log term. For our purposes the "4" is crucial since without it (4.1) will only converge linearly (like $(\log 4)/2^n$). King's 1924 monograph contains a wealth of material on the various iterative methods related to computing elliptic integrals. He comments [11, p. 14]:

"The limit [(4.1) without the "4"] does not appear to be generally known, although an equivalent formula is given by Legendre (*Fonctions éliptiques*, t. I, pp. 94–101)."

King adds that while Gauss did not explicitly state (4.1) he derived a closely related series expansion and that none of this "appears to have been noticed by Jacobi or by subsequent writers on elliptic functions." This series [10, p. 377] gives (4.1) almost directly.

Proposition 1 may be found in Bowman [3]. Of course, almost all the basic work is to be found in the works of Abel, Gauss and Legendre [1], [10] and [14]. (See also [7].) As was noted by both Brent and Salamin, Proposition 2 can be used to estimate log given π. We know from (3.13) that, for $0 < k \leq 10^{-3}$,

$$\left| \log\left(\frac{4}{k}\right) - I(1, k) \right| < 10k^2 |\log k|.$$

By subtraction, for $0 < x < 1$, and $n \geq 3$,

(4.2) $|\log(x) - [I(1, 10^{-n}) - I(1, 10^{-n}x)]| < n\, 10^{-2(n-1)}$

and we can compute log exponentially from the AGM approximations of the elliptic integrals in the above formula. This is in the spirit of Newman's presentation [15]. Formula (4.2) works rather well numerically but has the minor computational drawback that it requires computing the AGM for small initial values. This leads to some linear steps (roughly $\log(n)$) before quadratic convergence takes over.

We can use (3.16) or (4.2) to show directly that π is exponentially computable. With $k := 10^{-n}$ and $h := 10^{-2n} + 10^{-n}$ (3.16) yields with (3.9) that, for $n \geq 1$,

$$\left| \log(10^{-n} + 1) - \frac{\pi}{2}\left[\frac{1}{AG(1, 10^{-n})} - \frac{1}{AG(1, 10^{-n} + 10^{-2n})}\right] \right| \leq 10^{1-2n}.$$

Since $|\log(x + 1)/x - 1| \leq x/2$ for $0 < x < 1$, we derive that

(4.3) $$\left| \frac{2}{\pi} - \left[\frac{10^n}{AG(1, 10^{-n})} - \frac{10^n}{AG(1, 10^{-n} + 10^{-2n})}\right] \right| \leq 10^{1-n}.$$

Newman [15] gives (4.3) with a rougher order estimate and without proof. This analytically beautiful formula has the serious computational drawback that obtaining n digit accuracy for π demands that certain of the operations be done to twice that precision.

Both Brent's and Salamin's approaches require *Legendre's relation*: for $0 < k < 1$

(4.4) $I(1, k)J(1, k') + I(1, k')J(1, k) - I(1, k)I(1, k') = \dfrac{\pi}{2}$

where $J(a, b)$ is the *complete elliptic integral of the second kind* defined by

$$J(a, b) := \int_0^{\pi/2} \sqrt{a^2 \cos^2 \theta + b^2 \sin^2 \theta}\, d\theta.$$

360   J. M. BORWEIN AND P. B. BORWEIN

The elliptic integrals of the first and second kind are related by

(4.5) $$J(a_0, b_0) = \left(a_0^2 - \frac{1}{2} \sum_{n=0}^{\infty} 2^n c_n^2 \right) I(a_0, b_0)$$

where, as before, $c_n^2 = a_n^2 - b_n^2$ and a_n and b_n are computed from the AGM iteration.

Legendre's proof of (4.4) can be found in [3] and [8]. His elegant elementary argument is to differentiate (4.4) and show the derivative to be constant. He then evaluates the constant, essentially by Proposition 1. Strangely enough, Legendre had some difficulty in evaluating the constant since he had problems in showing that $k^2 \log (k)$ tends to zero with k [8, p. 150].

Relation (4.5) uses properties of the ascending Landen transformation and is derived by King in [11].

From (4.4) and (4.5), noting that if k equals $2^{-1/2}$ then so does k', it is immediate that

(4.6) $$\pi = \frac{[2AG(1, 2^{-1/2})]^2}{1 - \sum_{n-1}^{\infty} 2^{n+1} c_n^2}.$$

This concise and surprising exponentially converging formula for π is used by both Salamin and Brent. As Salamin points out, by 1819 Gauss was in possession of the AGM iteration for computing elliptic integrals of the first kind and also formula (4.5) for computing elliptic integrals of second kind. Legendre had derived his relation (4.4) by 1811, and as Watson puts it [20, p. 14] "in the hands of Legendre, the transformation [(3.23)] became a most powerful method for computing elliptic integrals." (See also [10], [14] and the footnotes of [11].) King [11, p. 39] derives (4.6) which he attributes, in an equivalent form, to Gauss. It is perhaps surprising that (4.6) was not suggested as a practical means of calculating π to great accuracy until recently.

It is worth emphasizing the extraordinary similarity between (1.1) which leads to linearly convergent algorithms for all the elementary functions, and (3.1) which leads to exponentially convergent algorithms.

Brent's algorithms for the elementary functions require a discussion of incomplete elliptic integrals and the Landen transform, matters we will not pursue except to mention that some of the contributions of Landen and Fagnano are entertainingly laid out in an article by G.N. Watson entitled "The Marquis [Fagnano] and the Land Agent [Landen]" [20]. We note that Proposition 1 is also central to Brent's development though he derives it somewhat tangentially. He also derives Theorem 1(a) in different variables via the Landen transform.

5. An algorithm for π. We now present the details of our exponentially converging algorithm for calculating the digits of π. Twenty iterations will provide over two million digits. Each iteration requires about ten operations. The algorithm is very stable with all the operations being performed on numbers between ½ and 7. The eighth iteration, for example, gives π correctly to 694 digits.

THEOREM 2. *Consider the three-term iteration with initial values*

$$\alpha_0 := \sqrt{2}, \quad \beta_0 := 0, \quad \pi_0 := 2 + \sqrt{2}$$

given by

(i) $\alpha_{n+1} := \dfrac{1}{2} (\alpha_n^{1/2} + \alpha_n^{-1/2}),$

ARITHMETIC-GEOMETRIC MEAN AND FAST COMPUTATION   361

(ii) $\beta_{n+1} := \alpha_n^{1/2}\left(\dfrac{\beta_n + 1}{\beta_n + \alpha_n}\right),$

(iii) $\pi_{n+1} := \pi_n\beta_{n+1}\left(\dfrac{1 + \alpha_{n+1}}{1 + \beta_{n+1}}\right).$

Then π_n converges exponentially to π and

$$|\pi_n - \pi| \leq \frac{1}{10^{2^n}}.$$

Proof. Consider the formula

(5.1)
$$\frac{1}{2^n}\log\left(4\frac{a_n}{c_n}\right) - \frac{\pi}{2}\frac{a_n}{a_n'}$$

which, as we will see later, converges exponentially at a uniform rate to zero in some (complex) neighbourhood of $1/\sqrt{2}$. (We are considering each of a_n, b_n, c_n, a_n', b_n', c_n' as being functions of a complex initial value k, i.e. $b_0 = k$, $b_0' = \sqrt{1 - k^2}$, $a_0 = a_0' = 1$.)

Differentiating (5.1) with respect to k yields

(5.2)
$$\frac{1}{2^n}\left(\frac{\dot{a}_n}{a_n} - \frac{\dot{c}_n}{c_n}\right) - \frac{\pi}{2}\frac{a_n}{a_n'}\left(\frac{\dot{a}_n}{a_n} - \frac{\dot{a}_n'}{a_n'}\right)$$

which also converges uniformly exponentially to zero in some neighbourhood of $1/\sqrt{2}$. (This general principle for exponential convergence of differentiated sequences of analytic functions is a trivial consequence of the Cauchy integral formula.) We can compute $\dot{a}_n$, $\dot{b}_n$ and $\dot{c}_n$ from the recursions

(5.3)
$$\dot{a}_{n+1} := \frac{\dot{a}_n + \dot{b}_n}{2},$$

$$\dot{b}_{n+1} := \frac{1}{2}\left(\dot{a}_n\sqrt{\frac{b_n}{a_n}} + \dot{b}_n\sqrt{\frac{a_n}{b_n}}\right),$$

$$\dot{c}_{n+1} := \frac{1}{2}(\dot{a}_n - \dot{b}_n),$$

where $\dot{a}_0 := 0$, $\dot{b}_0 := 1$, $a_0 := 1$ and $b_0 := k$.

We note that a_n and b_n map $\{z \mid Re(z) > 0\}$ into itself and that $\dot{a}_n$ and $\dot{b}_n$ (for sufficiently large n) do likewise.

It is convenient to set

(5.4)
$$\alpha_n := \frac{a_n}{b_n} \quad \text{and} \quad \beta_n := \frac{\dot{a}_n}{\dot{b}_n}$$

with

$$\alpha_0 := \frac{1}{k} \quad \text{and} \quad \beta_0 := 0.$$

We can derive the following formulae in a completely elementary fashion from the basic relationships for a_n, b_n and c_n and (5.3):

(5.5) $\dot{a}_{n+1} - \dot{b}_{n+1} = \dfrac{1}{2}(\sqrt{a_n} - \sqrt{b_n})\left(\dfrac{\dot{a}_n}{\sqrt{a_n}} - \dfrac{\dot{b}_n}{\sqrt{b_n}}\right),$

(5.6) $\qquad 1 - \dfrac{a_{n+1}\,\dot{c}_{n+1}}{\dot{a}_{n+1}\,c_{n+1}} = \dfrac{2(\alpha_n - \beta_n)}{(\alpha_n - 1)(\beta_n + 1)},$

(5.7) $\qquad \alpha_{n+1} = \dfrac{1}{2}(\alpha_n^{1/2} + \alpha_n^{-1/2}),$

(5.8) $\qquad \beta_{n+1} = \alpha_n^{1/2}\left(\dfrac{\beta_n + 1}{\beta_n + \alpha_n}\right),$

(5.9) $\qquad \alpha_{n+1} - 1 = \dfrac{1}{2\alpha_n^{1/2}}(\alpha_n^{1/2} - 1)^2,$

(5.10) $\qquad \alpha_{n+1} - \beta_{n+1} = \dfrac{\alpha_n^{1/2}}{2}\dfrac{(1 - \alpha_n)(\beta_n - \alpha_n)}{\alpha_n(\beta_n + \alpha_n)},$

(5.11) $\qquad \dfrac{\alpha_{n+1} - \beta_{n+1}}{\alpha_{n+1} - 1} = \dfrac{(1 + \alpha_n^{1/2})^2}{(\beta_n + \alpha_n)} \cdot \dfrac{(\alpha_n - \beta_n)}{(\alpha_n - 1)}.$

From (5.7) and (5.9) we deduce that $\alpha_n \to 1$ uniformly with second order in compact subsets of the open right half-plane. Likewise, we see from (5.8) and (5.10) that $\beta_n \to 1$ uniformly and exponentially. Finally, we set

(5.12) $\qquad\qquad\qquad \gamma_n := \dfrac{1}{2^n}\left(\dfrac{\alpha_n - \beta_n}{\alpha_n - 1}\right).$

We see from (5.11) that

(5.13) $\qquad\qquad\qquad \gamma_{n+1} = \dfrac{(1 + \alpha_n^{1/2})}{2(\beta_n + \alpha_n)}\gamma_n$

and also from (5.6) that

(5.14) $\qquad\qquad\qquad \dfrac{\gamma_n}{1 + \beta_n} = \dfrac{1}{2^{n+1}}\left(1 - \dfrac{a_{n+1}\,\dot{c}_{n+1}}{\dot{a}_{n+1}\,c_{n+1}}\right).$

Without any knowledge of the convergence of (5.1) one can, from the preceding relationships, easily and directly deduce the exponential convergence of (5.2), in $\{z\,|\,|z - \frac{1}{2}| \le c < \frac{1}{2}\}$. We need the information from (5.1) only to see that (5.2) converges to zero.

The algorithm for π comes from multiplying (5.2) by $a_n/\dot{a}_n$ and starting the iteration at $k := 2^{-1/2}$. For this value of k $a_n' = a_n$, $(\dot{a}_n') = -\dot{a}_n$ and

$$\dfrac{1}{2^{n+1}}\left(1 - \dfrac{a_{n+1}\,\dot{c}_{n+1}}{\dot{a}_{n+1}\,c_{n+1}}\right) \to \pi$$

which by (5.14) shows that

$$\pi_n := \dfrac{\gamma_n}{1 + \beta_n} \to \pi.$$

Some manipulation of (5.7), (5.8) and (5.13) now produces (iii). The starting values for α_n, β_n and γ_n are computed from (5.4). Other values of k will also lead to similar, but slightly more complicated, iterations for π.

To analyse the error one considers

$$\dfrac{\gamma_{n+1}}{1 + \beta_{n+1}} - \dfrac{\gamma_n}{1 + \beta_n} = \left[\dfrac{(1 + \alpha_n^{1/2})^2}{2(\beta_n + \alpha_n)(1 + \beta_{n+1})} - \dfrac{1}{(1 + \beta_n)}\right]\gamma_n$$

and notes that, from (5.9) and (5.10), for $n \geq 4$,

$$|\alpha_n - 1| \leq \frac{1}{10^{2^n+2}} \quad \text{and} \quad |\beta_n - 1| \leq \frac{1}{10^{2^n+2}}.$$

(One computes that the above holds for $n = 4$.) Hence,

$$\left| \frac{\gamma_{n+1}}{1 + \beta_{n+1}} - \frac{\gamma_n}{1 + \beta_n} \right| \leq \left| \frac{1}{10^{2^n+1}} \right| |\gamma_n|$$

and

$$\left| \frac{\gamma_n}{1 + \beta_n} - \pi \right| \leq \frac{1}{10^{2^n}}.$$ $\square$

In fact one can show that the error is of order $2^n e^{-\pi 2^{n+1}}$.

If we choose integers in $[\delta, \delta^{-1}]$, $0 < \delta < \frac{1}{2}$ and perform n operations $(+, -, \times, \div, \sqrt{\ })$ then the result is always less than or equal to δ^{2^n}. Thus, if $\gamma > \delta$, it is not possible, using the above operations and integral starting values in $[\delta, \delta^{-1}]$, for every n to compute π with an accuracy of $O(\gamma^{-2^n})$ in n steps. In particular, convergence very much faster than that provided by Theorem 2 is not possible.

The analysis in this section allows one to derive the Gauss-Salamin formula (4.6) without using Legendre's formula or second integrals. This can be done by combining our results with problems 15 and 18 in [11]. Indeed, the results of this section make quantitative sense of problems 16 and 17 in [11]. King also observes that Legendre's formula is actually equivalent to the Gauss–Salamin formula and that each may be derived from the other using only properties of the AGM which we have developed and equation (4.5).

This algorithm, like the algorithms of §4, is not self correcting in the way that Newton's method is. Thus, while a certain amount of time may be saved by observing that some of the calculations need not be performed to full precision it seems intrinsic (though not proven) that $O(\log n)$ full precision operations must be executed to calculate π to n digits. In fact, showing that π is intrinsically more complicated from a time complexity point of view than multiplication would prove that π is transcendental [5].

6. The complex AGM iteration. The AGM iteration

$$a_{n+1} := \frac{1}{2}(a_n + b_n), \qquad b_{n+1} := \sqrt{a_n b_n}$$

is well defined as a complex iteration starting with $a_0 := 1$, $b_0 := z$. Provided that z does not lie on the negative real axis, the iteration will converge (to what then must be an analytic limit). One can see this geometrically. For initial z in the right half-plane the limit is given by (3.9). It is also easy to see geometrically that a_n and b_n are always nonzero.

The iteration for $x_n := b_n/a_n$ given in the form (3.23) as $x_{n+1} := 2\sqrt{x_n}/x_{n+1}$ satisfies

(6.1)
$$(x_{n+1} - 1) = \frac{(1 - \sqrt{x_n})^2}{1 + x_n}.$$

This also converges in the cut plane $\mathbb{C} - (-\infty, 0]$. In fact, the convergence is uniformly exponential on compact subsets (see Fig. 1). With each iteration the angle θ_n between x_n and 1 is at least halved and the real parts converge uniformly to 1.

It is now apparent from (6.1) and (3.24) that

(6.2)
$$P_n(k) := \left(\frac{4a_n}{c_n} \right)^{2^{1-n}} = \left(\frac{1 + x_n}{1 - x_n} \right)^{2^{-n}}$$

364   J. M. BORWEIN AND P. B. BORWEIN

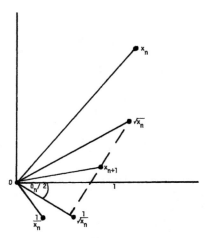

FIG. 1.

and also,

$$Q_n(k) := \frac{a_n}{a'_n}$$

converge exponentially to analytic limits on compact subsets of the complex plane that avoid

$$D := \{z \in \mathbb{C} \mid z \notin (-\infty, 0] \cup [1, \infty)\}.$$

Again we denote the limits by P and Q. By standard analytic reasoning it must be that (3.19a) still holds for k in D.

Thus one can compute the complex exponential—and so also cos and sin—exponentially using (3.19). More precisely, one uses Newton's method to approximately solve $Q(k) = z$ for k and then computes $P_n(k)$. The outcome is e^z. One can still perform the root extractions using Newton's method. Some care must be taken to extract the correct root and to determine an appropriate starting value for the Newton inversion. For example $k := 0.02876158$ yields $Q(k) = 1$ and $P_4(k) = e$ to 8 significant places. If one now uses k as an initial estimate for the Newton inversions one can compute $e^{1+i\theta}$ for $|\theta| \leq \pi/8$. Since, as we have observed, e is also exponentially computable we have produced a sufficient range of values to painlessly compute $\cos \theta + i \sin \theta$ with no recourse to any auxiliary computations (other than π and e, which can be computed once and stored). By contrast Brent's trigonometric algorithm needs to compute a different logarithm each time.

The most stable way to compute P_n is to use the fact that one may update c_n by

(6.3) $$c_{n+1} = \frac{c_n^2}{4a_{n+1}}.$$

One then computes a_n, b_n and c_n to desired accuracy and returns

$$\left(\frac{4a_n}{c_n}\right)^{1/2^n} \quad \text{or} \quad \left(\frac{2(a_n + b_n)}{c_n}\right)^{1/2^n}.$$

This provides a feasible computation of P_n, and so of exp or log.

In an entirely analogous fashion, formula (4.2) for log is valid in the cut complex plane. The given error estimate fails but the convergence is still exponential. Thus (4.2) may also be used to compute all the elementary functions.

7. Concluding remarks and numerical data. We have presented a development of the AGM and its uses for rapidly computing elementary functions which is, we hope, almost entirely self-contained and which produces workable algorithms. The algorithm for π is particularly robust and attractive. We hope that we have given something of the flavour of this beautiful collection of ideas, with its surprising mixture of the classical and the modern. An open question remains. Can one entirely divorce the central discussion from elliptic integral concerns? That is, can one derive exponential iterations for the elementary functions without recourse to some nonelementary transcendental functions? It would be particularly nice to produce a direct iteration for e of the sort we have for π which does not rely either on Newton inversions or on binary splitting.

The algorithm for π has been run in an arbitrary precision integer arithmetic. (The algorithm can be easily scaled to be integral.) The errors were as follows:

Iterate	Digits correct	Iterate	Digits correct
1	3	6	170
2	8	7	345
3	19	8	694
4	41	9	1392
5	83	10	2788

Formula (4.2) was then used to compute $2 \log (2)$ and $\log (4)$, using π estimated as above and the same integer package. Up to 500 digits were computed this way. It is worth noting that the error estimate in (4.2) is of the right order.

The iteration implicit in (3.22) was used to compute e^{π} in a double precision Fortran. Beginning with $k := 2^{-1/2}$ produced the following data:

Iterate	$P_n - e^{\pi}$	$a_n/b_n - 1$
1	1.6×10^{-1}	1.5×10^{-2}
2	2.8×10^{-9}	2.8×10^{-5}
3	1.7×10^{-20}	9.7×10^{-11}
4	$< 10^{-40}$	1.2×10^{-21}

Identical results were obtained from (6.3). In this case $y_n := 4a_n/c_n$ was computed by the two term recursion which uses x_n, given by (3.23), and

$$(7.1) \qquad y_0^2 := \frac{16}{1 - k^2}, \qquad y_{n+1} = \left(\frac{1 + x_n}{2}\right)^2 y_n^2.$$

One observes from (7.1) that the calculation of y_n is very stable.

We conclude by observing that the high precision root extraction required in the AGM [18], was actually calculated by inverting $y = 1/x^2$. This leads to the iteration

$$(7.2) \qquad x_{n+1} = \frac{3x_n - x_n^3 y}{2}$$

for computing $y^{-1/2}$. One now multiplies by y to recapture $\sqrt{y}$. This was preferred because it avoided division.

366   J. M. BORWEIN AND P. B. BORWEIN

REFERENCES

[1] N. H. ABEL, *Oeuvres complètes*, Grondahl and Son, Christiana, 1881.

[2] P. BECKMAN, *A History of Pi*, Golem Press, Ed. 4, 1977.

[3] F. BOWMAN, *Introduction to Elliptic Functions*, English Univ. Press, London, 1953.

[4] R. P. BRENT, *Multiple-precision zero-finding and the complexity of elementary function evaluation*, in Analytic Computational Complexity, J. F. Traub, ed., Academic Press, New York, 1975, pp. 151–176.

[5] ———, *Fast multiple-precision evaluation of elementary functions*, J. Assoc. Comput. Mach., 23 (1976), pp. 242–251.

[6] B. C. CARLSON, *Algorithms involving arithmetic and geometric means*, Amer. Math. Monthly, 78(1971), pp. 496–505.

[7] A. CAYLEY, *An Elementary Treatise on Elliptic Functions*, Bell and Sons, 1895, republished by Dover, New York, 1961.

[8] A. EAGLE, *The Elliptic Functions as They Should Be*, Galloway and Porter, Cambridge, 1958.

[9] C. H. EDWARDS, Jr. *The Historical Development of the Calculus*, Springer-Verlag, New York, 1979.

[10] K. F. GAUSS, *Werke*, Bd 3, Gottingen, 1866, 361–403.

[11] L. V. KING, *On the direct numerical calculation of elliptic functions and integrals*, Cambridge Univ. Press, Cambridge, 1924.

[12] D. KNUTH, *The Art of Computer Programming. Vol. 2: Seminumerical Algorithms*, Addison-Wesley, Reading, MA, 1969.

[13] H. T. KUNG AND J. F. TRAUB, *All algebraic functions can be computed fast*, J. Assoc. Comput. Mach., 25(1978), 245–260.

[14] A. M. LEGENDRE, *Exercices de calcul integral*, Vol. 1, Paris, 1811.

[15] D. J. NEUMAN, *Rational approximation versus fast computer methods*, in Lectures on Approximation and Value Distribution, Presses de l'Université de Montréal, Montreal, 1982, pp. 149–174.

[16] E. SALAMIN, *Computation of π using arithmetic-geometric mean*, Math. Comput. 135(1976), pp. 565–570.

[17] D. SHANKS AND J. W. WRENCH, Jr. *Calculation of π to 100,000 decimals*, Math. Comput., 16 (1962), pp. 76–79.

[18] Y. TAMURA AND Y. KANADA, *Calculation of π to 4,196,293 decimals based on Gauss-Legendre algorithm*, preprint.

[19] J. TODD, *Basic Numerical Mathematics*, Vol. 1, Academic Press, New York, 1979.

[20] G. N. WATSON, *The marquis and the land agent*, Math. Gazette, 17 (1933), pp. 5–17.

[21] J. W. WRENCH, Jr. *The evolution of extended decimal approximations to π*. The Mathematics Teacher, 53(1960), pp. 644–650.

5. A simplified version of the fast algorithms of Brent and Salamin (1985)

Paper 5: D. J. Newman, "A simplified version of the fast algorithms of Brent and Salamin," *Mathematics of Computation*, vol. 44 (1985), p. 207–210. Reprinted by permission of the American Mathematical Society.

Synopsis:

In this paper, Newman defines an alternate version of the arithmetic-geometric mean, which he denotes $h(a, b)$, defined as follows:

$$a_{n+1} = \sqrt{a_n b_n}, \quad b_{n+1} = 2a_n b_n/(a_n + b_n).$$

With this formulation, he is able to derive the formulas of Brent and Salamin more easily, avoiding the need for elliptic integral functions and Landen transforms, and also derives some new formulas. He shows, for instance, that $N(h(N + 1, 1) - h(N, 1))$ gives $2/\pi$ to n-bit accuracy, provided we choose $N = 2^n$, and

$$N \left(\frac{h(N + 1, 1)}{h(N, 1)} - 1 \right)$$

gives $1/\log x$ to n-bit accuracy, provided we choose $N = x^n/4$.

Keywords: Algorithms, Arithmetic-Geometric Mean

© Springer International Publishing Switzerland 2016
D.H. Bailey, J.M. Borwein, *Pi: The Next Generation*,
DOI 10.1007/978-3-319-32377-0_5

MATHEMATICS OF COMPUTATION
VOLUME 44, NUMBER 169
JANUARY 1985, PAGES 207–210

A Simplified Version of the Fast Algorithms
of Brent and Salamin

By D. J. Newman*

Abstract. We produce more elementary algorithms than those of Brent and Salamin for, respectively, evaluating e^x and π. Although the Gauss arithmetic-geometric process still plays a central role, the elliptic function theory is now unnecessary.

In their remarkable papers, Brent [1] and Salamin [3], respectively, used the theory of elliptic functions to obtain "fast" computations of the function e^x and of the number π. In both cases rather heavy use of elliptic function theory, such as the transformation law of Landen, had to be utilized. Our purpose, in this note, is to give a highly simplified version of their constructions. In our approach, for example, the incomplete elliptic integral is never used.

We begin as they did with the Gauss arithmetic-geometric process, $T(a, b) = ((a + b)/2, \sqrt{ab}\,)$ which maps couples with $a \geqslant b > 0$ into same. From the inequalities

$$\frac{(a + b)/2 - \sqrt{ab}}{(a + b)/2 + \sqrt{ab}} = \left(\frac{\sqrt{a} - \sqrt{b}}{\sqrt{a} + \sqrt{b}}\right)^2 \leqslant \left(\frac{a - b}{a + b}\right)^2,$$

and

$$\frac{(a + b)/2}{\sqrt{ab}} \leqslant \frac{a}{\sqrt{ab}} = \sqrt{\frac{a}{b}}\,,$$

we see that $T^i(a, b)$ goes to its limiting couple (m, m) ($m = m(a, b)$ the so-called arithmetic-geometric mean) with "quadratic" speed. Indeed, $m(a, b)$ is determined to n places for an i of around $\log \log a/b + \log n$. The $\log \log$ from the $\sqrt{a/b}$ inequality expressing the time till the ratio first goes below 2, and the $\log$ from the $((a - b)/(a + b))^2$ inequality expressing the time for the error squaring to do its job.

Next, we recall Gauss' beautiful formula:

$$m(a, b) = \pi \bigg/ \int_{-\infty}^{\infty} \frac{dx}{\sqrt{(x^2 + a^2)(x^2 + b^2)}}\,,$$

Received October 7, 1981; revised November 17, 1981, and March 22, 1984.
1980 *Mathematics Subject Classification*. Primary 65D15, 33A25, 41A25.
*Supported in part by NSF Grant MGS 7802171.

which follows from the fact that this (complete) elliptic integral is invariant under T. This fact, that namely

$$\int_{-\infty}^{\infty} \frac{dx}{\sqrt{(x^2 + a^2)(x^2 + b^2)}} = \int_{-\infty}^{\infty} \frac{dt}{\sqrt{(t^2 + ((a+b)/2)^2)(t^2 + ab)}},$$

is a simple consequence of the change of variables $t = (x - ab/x)/2$. Namely, we obtain

$$dt = \frac{x^2 + ab}{2x^2} dx, \qquad t^2 + \left(\frac{a+b}{2}\right)^2 = \frac{(x^2 + a^2)(x^2 + b^2)}{4x^2},$$

$$t^2 + ab = \frac{(x^2 + ab)^2}{4x^2},$$

$0 < x < \infty$, so that indeed we have

$$\int_{-\infty}^{\infty} \frac{dx}{\sqrt{(x^2 + a^2)(x^2 + b^2)}} = \int_0^{\infty} \frac{2dx}{\sqrt{(x^2 + a^2)(x^2 + b^2)}}$$

$$= \int_{-\infty}^{\infty} \frac{dt}{\sqrt{(t^2 + ((a+b)/2)^2)(t^2 + ab)}}.$$

Accordingly, a repeated use of this invariance gives

$$\int_{-\infty}^{\infty} \frac{dx}{\sqrt{(x^2 + a^2)(x^2 + b^2)}} = \cdots = \int_{-\infty}^{\infty} \frac{dx}{\sqrt{(x^2 + m^2)(x^2 + m^2)}}$$

$$= \int_{-\infty}^{\infty} \frac{dx}{x^2 + m^2} = \frac{\pi}{m},$$

and this is exactly Gauss' formula.

Actually, it is handier for us to work with what we might call the harmonic-geometric mean which can be defined by $h(a, b) = ab/m(a, b)$ or, alternatively, as the limit under repeated applications of S, rather than T, where

$$S(a, b) = (\sqrt{ab}, 2ab/(a+b)).$$

In these terms Gauss' formula reads

$$h(a, b) = \frac{1}{\pi} \int_{-\infty}^{\infty} \frac{dx}{\sqrt{(1 + x^2/a^2)(1 + x^2/b^2)}}.$$

The only place that we actually use this formula is to establish the asymptotic formula:

$$h(N, 1) = \frac{2}{\pi} \log 4N + O(1/N^2).$$

(This simple-looking formula certainly deserves an elementary proof independent of elliptic integrals, but we are unable to find one.)

So begin with

$$h(N, 1) = \frac{2}{\pi} \int_0^{\infty} \frac{dx}{\sqrt{(1 + x^2)(1 + x^2/N^2)}}$$

and observe that the map $x \to N/x$ leaves the integrand invariant. Thereby, we conclude

$$\int_0^{\sqrt{N}} \frac{dx}{\sqrt{(1 + x^2)(1 + x^2/N^2)}} = \int_{\sqrt{N}}^{\infty} \frac{dx}{\sqrt{(1 + x^2)(1 + x^2/N^2)}}$$

which gives us

$$h(N, 1) = \frac{4}{\pi} \int_0^{\sqrt{N}} \frac{dx}{\sqrt{(1 + x^2)(1 + x^2/N^2)}}$$

$$= \frac{4}{\pi} \int_0^{\sqrt{N}} \frac{1}{\sqrt{(1 + x^2)}} \left(1 - x^2/2N^2 + O(x^4/N^4)\right) dx$$

$$= \frac{4}{\pi} \int_0^{\sqrt{N}} \left(\frac{1}{\sqrt{1 + x^2}} - \frac{x}{2N^2} \right) dx + O(1/N^2)$$

$$= \frac{4}{\pi} \left(\log(\sqrt{N} + \sqrt{N + 1}) - 1/4N \right) + O(1/N^2)$$

and so, since

$$\sqrt{N} + \sqrt{N + 1} = 2\sqrt{N} \left(1 + 1 \big/ \left(2N + 2\sqrt{N(N + 1)}\right) \right),$$

we obtain

$$\log(\sqrt{N} + \sqrt{N + 1}) = \log 2\sqrt{N} + 1/4N + O\left(\frac{1}{N^2}\right),$$

which together with the previous gives

$$h(N, 1) = \frac{4}{\pi} \log 2\sqrt{N} + O\left(\frac{1}{N^2}\right) = \frac{2}{\pi} \log 4N + O\left(\frac{1}{N^2}\right),$$

as required. (This result can also be found in [2].)

Summarizing, then, we have produced a fast method for obtaining n places of $2 \log 4N/\pi$ (if N is of the size c^n). But, and here is the trick, this combination of π and the logarithm can be used to yield both of them separately, and we can thereby rederive both Salamin's and Brent's results.

To obtain π we examine the difference, $h(N + 1, 1) - h(N, 1)$, and observe that $N(h(N + 1, 1) - h(N, 1)) = 2/\pi + O(1/N)$ which gives n place accuracy for π if we choose, e.g., $N = 2^n$.

For the logarithm, on the other hand, we look to the quotient, $h(N + 1, 1)/h(N, 1)$. This time we obtain

$$N\left(\frac{h(N + 1, 1)}{h(N, 1)} - 1 \right) = N\frac{\log(1 + 1/N) + O(1/N^2)}{\log 4N + O(1/N)} = \frac{1}{\log 4N + O(1/N)}.$$

From this we will be able to evaluate $\log x$ throughout the interval $(3, 9)$, and so, of course, throughout any interval. And thereby, we will be able to obtain e^x, the inverse function, by the usual use of the (fast) Newton iteration scheme.

To obtain $\log x$, then, in the interval $(3, 9)$, we first calculate $N = \frac{1}{4}x^n$, a process that takes only $\log n$ multiplications. But then the above formula becomes, upon substitution of this value of N,

$$\frac{1}{4}nx^n\left(\frac{h(\frac{1}{4}x^n + 1, 1)}{h(\frac{1}{4}x^n, 1)} - 1 \right) = \frac{1}{\log x} + O\left(\frac{n}{x^n}\right) = \frac{1}{\log x} + O\left(\frac{n}{3^n}\right)$$

which does give the desired n place evaluation of $\log x$.

This trick of "differencing" $h(N + 1, 1)$ and $h(N, 1)$, of course, carries a price. Thus we must compute these two quantities to $2n$ places and so the running time is around twice as long as the corresponding ones of Brent and Salamin.

Mathematics Department
Temple University
Broad and Montgomery Ave.
Philadelphia, Pennsylvania 19122

1. R. BRENT, "Fast multiple evaluation of elementary functions," *J. Assoc. Comput. Mach.*, v. 23, 1976, pp. 242–251.

2. R. BRENT, *Multiple-Precision Zero-Finding Methods and the Complexity of Elementary Function Evaluation* (J. F. Traub, ed.), Academic Press, New York, 1975, pp. 151–176.

3. E. SALAMIN, "Computation of π using the arithmetic-geometric mean," *Math. Comp.*, v. 30, 1976, pp. 565–570.

6. Is pi normal? (1985)

Paper 6: Stan Wagon, "Is pi normal," *The Mathematical Intelligencer*, vol. 7 (1985), p. 65–67. With permission of Springer.

Synopsis:

As our earlier Compendium makes clear, mathematicians have been fascinated by the decimal expansion (and expansions in other bases) of π since the time of Archimedes — what sort of number is π? Questions such as whether π is rational or not, or algebraic or not, were settled in the 18th and 19th century, respectively.

But one question, originally raised by Borel in 1909, remains unanswered even today: whether or not π is *normal*. A real constant is said to be *normal base* 10, or 10-*normal*, if every m-long string of digits appears in the decimal expansion of π with limiting frequency $1/10^m$ (with a similar definition for a general base b), and is said to be *absolutely normal* if it is b-normal for all integer bases b simultaneously.

In this highly readable paper, Wagon introduces the question of the normality of π in the context of the recently discovered quadratically convergent algorithms. He also presents a statistical analysis of the digits of π provided by Yasamusa Kanada, who had, at the time, just computed π to 10,000,000-digit precision.

Keywords: Computation, General Audience, Normality

© Springer International Publishing Switzerland 2016
D.H. Bailey, J.M. Borwein, *Pi: The Next Generation*,
DOI 10.1007/978-3-319-32377-0_6

The Evidence

Is π Normal?
by Stan Wagon

The nature of the number π has intrigued mathematicians since the beginning of mathematical history. The most important properties of π are its irrationality and transcendence, which were established in 1761 and 1882, respectively. In the twentieth century the focus has been on a different sort of question, namely whether π, despite being irrational and transcendental, is normal.

The idea of normality, first introduced by E. Borel in 1909, is an attempt to formalize the notion of a real number being random. The definition is as follows: A real number x is *normal in base b* if in its representation in base b all digits occur, in an asymptotic sense, equally often. In addition, for each m, the b^m different m-strings must occur equally often. In other words, $\lim_{n \to \infty} N(s,n)/n = b^{-m}$ for each m-string s, where $N(s,n)$ is the number of occurrences of s in the first n base-b digits of x. A number that is normal in all bases is called *normal*. The apparent randomness of π's digits had been observed prior to the precise definition of normality. De Morgan, for example, pointed out that one would expect the digits to occur equally often, but yet the number of 7's in the first 608 digits is 44, much lower than expected. However, it turned out that his count was based on inaccurate data.

There are lots of normal numbers—Borel proved (see Niven or §9.12 of Hardy and Wright) that the set of non-normal numbers has measure zero—but it is difficult to provide concrete examples. While an undergraduate at Cambridge University, D. Champernowne proved that 0.12345678910111213 ... is normal in base 10, but an explicit example of a normal number is still lacking.

The question of π's normality only scratches the surface of the deeper question whether the digits of π are "random." That normality is not sufficient follows from the observation that a truly random sequence of digits ought to be normal when only digits in positions corresponding to perfect squares are examined. But if all such positions in a normal number are set to 0, the number is still normal. On the other hand, more rigorous definitions of "random" exclude π because π's decimal expansion is a recursive sequence. For an enlightening discussion of the general problem of defining randomness, see the section on "What is a Random Sequence" in volume 2 of Knuth's trilogy.

Thus deeper questions are lurking, but so little is known about π's decimal expansion that it is reasonable to focus on whether π is normal to base ten. To put our ignorance in perspective, note that it is not even known that all digits appear infinitely often: perhaps

$$\pi = 3.1415926.....01001000100001000001... \ .$$

In order to gather evidence for π's normality one would like to examine as many digits as possible. Those who have pursued the remote digits of π have often been pejoratively referred to as "digit hunters," but certain recent developments have added some glamor to the centuries-old hunt. In 1975 Brent and Salamin, independently, discovered an algorithm that dramatically lowered the time needed to compute large numbers of digits of π. Moreover, the algorithm has important connections with efficient algorithms for computing various transcendental functions (sin, arctan, exp, log, elliptic integrals) to great accuracy.

The reader is probably aware that various arctangent formulas have been central in the computation of π. Early investigators used series such as the one for arctan $1/\sqrt{3}$, but convergence was much speeded by the use of formulas such as the following, by means of which Machin computed 100 decimals of π in 1706: π = 16arctan1/5 - 4arctan1/239. Indeed, this same formula was used in the first computer calculation, the ENIAC's computation of 2037 digits in 1949. Computers have become much faster, and various opti-

Column Editor: Stan Wagon, Department of Mathematics, Smith College, Northampton, Massachusetts 01063, USA.

Year	Time	Number of digits	Computer time per digit
1949	~70 hours	2,037	~2 minutes
1958	100 minutes	10,000	0.6 second
1961	8.43 hours	100,000	1/3 second
1973	23.3 hours	1,000,000	1/12 second
1983	<30 hours	16,000,000	<1/155 second

Sophisticated algorithms, combined with ever-faster machines, have led to increasingly more efficient computations of π's digits during the computer age.

Table 1

mizing tricks help speed up computations, but even the million-digit computation in 1973 used the same sort of formula, this one due to Gauss:

$$\pi = 48\arctan 1/18 + 32\arctan 1/57 \pm 20\arctan 1/239.$$

A limiting aspect of the arctangent formulas is the number of full-precision operations that must be carried out. By an *operation* we mean one of $+$, $\times$, $\div$, $\sqrt{\ }$. Since large-precision square roots and divisions can be performed in essentially as much time as that required by a full-precision multiplication (Newton's method can be used for both; see Borwein and Borwein), this measure of complexity (number of full-precision operations) is as good as the more usual "time complexity" for comparing algorithms. Now, an examination of the rate of convergence of the arctangent series shows that the arctangent method uses $O(n)$ (i.e., at most cn, c a constant) full-precision operations to compute n decimals of π; for example, the Shanks and Wrench computation of 100,000 decimals used just under 105,000 full-precision operations. Thus there are two basic time costs involved in the pushing of a calculation from n to $10n$ digits:

(1) the number of operations increases by a factor of 10, and
(2) the time for each full-precision operation is about 10 times greater.

The Brent-Salamin algorithm requires only $O(\log n)$ full-precision operations for n digits of π. Since $\log n$ barely increases when n is replaced by $10n$, the first of the two costs just mentioned is almost entirely eliminated!

The Brent-Salamin formula uses ideas that go back to Gauss and Legendre, but prior to the 1970s no one had thought to apply these ideas to evaluate π. The formula exploits the speed of convergence of the defining sequences for the arithmetic-geometric mean of two numbers. Given positive reals $a_0 > b_0$, their *arithmetic-geometric mean*, $AG(a_0,b_0)$, is defined to be the common limit of $\{a_n\}$, $\{b_n\}$, where $a_n = (a_{n-1} + b_{n-1})/2$

and $b_n = \sqrt{a_{n-1} b_{n-1}}$. Gauss had investigated such limits and had proved that $AG(a,b)$ equals $\pi/2I$, where

$$I = \int_0^{\pi/2} \frac{dt}{\sqrt{a^2\cos^2 t + b^2\sin^2 t}},$$

a complete elliptic integral of the first kind. To get a formula for π let a_0, b_0 be 1, $1/\sqrt{2}$, respectively, and define d_n to be $a_n^2 - b_n^2$; then

$$\pi = 4(AG(1,1/\sqrt{2}))^2 / \left(1 - \sum_{j=1}^{\infty} 2^{j+1}d_j\right).$$

If π_n is defined to be $4a_{n+1}^2/(1 - \sum_{j=1}^{n} 2^{j+1}d_j)$, then π_n converges quadratically to π. This means, roughly, that the number of correct digits *doubles* from one π_n to the next.

Quadratic convergence is most familiar from Newton's method of approximating solutions to algebraic equations. Thus, from a computational perspective, π behaves like an algebraic number.

A more precise error analysis shows that π_{16} is accurate to 178,000 digits, π_{19} to over a million, π_{22} to over ten million, and π_{26} to almost 200,000,000 digits. Since the computation of π_n requires $7n$ full-precision operations, the improvement over the classical algorithm is impressive: 100,000 digits require only 112 full-precision operations!

The fact that $c \log n$ full-precision operations yield n digits of π means that the time complexity (essentially, number of bit operations) of the computation is $O(n \log^2 n \log\log n)$; this uses the fact that n-digit multiplication is of complexity $O(n \log n \log\log n)$. Thus these fast algorithms for π are just about the fastest possible: it takes n steps just to write down n digits, and there is not much room between n and $n \log^2 n \log\log n$.

For refinements of the Brent-Salamin formulas, applications to the computation of transcendental functions, and some proofs, see the paper by Borwein and Borwein, who are preparing a book on the arithmetic-geometric mean. See the paper by Cox for more on the

Type of hand	Expected number	Actual number
No two digits the same	604,800	604,976
One pair	1,008,000	1,007,151
Two pair	216,000	216,520
Three of a kind	144,000	144,375
Full house	18,000	17,891
Four of a kind	9,000	8,887
Five of a kind	200	200

Distribution of the first two million poker hands in the digits of π.

Table 2

arithmetic-geometric mean and for a historical account of Gauss's work.

There is a surprising connection between these modern algorithms and the method of Archimedes. Archimedes used inscribed and circumscribed polygons to approximate π. Now, if A_n is the reciprocal of the circumference of a 2^n-gon inscribed in a unit circle and B_n likewise for a circumscribed 2^n-gon, then A_n and B_n satisfy: $B_{n+1} = \frac{1}{2}(A_n + B_n)$ and $A_{n+1} = \sqrt{A_n B_{n+1}}$. Thus the double sequence of Archimedes obeys a recursion almost identical to the defining recursion for AG(1/2, 1/4). Archimedes' sequences converge much more slowly, however: each iteration decreases the error by, approximately, a factor of four.

The Brent-Salamin algorithm has been implemented in Japan by Kanada, Tamura, Yoshino, and Ushiro who, in 1983, used it to compute 16 million decimal places. They checked the first 10,013,395 of these using Gauss's arctangent relation. The computation of 16 million digits took less than 30 hours of CPU time, although some time (10–20%) was saved by the reuse of intermediate values from earlier computations. See Table 1 for a comparison with previous computations. The check took only 24 hours, but the two times are not comparable since the arctangent computation was performed on a much faster computer, a Japanese Hitachi supercomputer with a speed of 630 MFLOPS (million floating-point operations per second).

A forthcoming paper by Kanada contains a statistical analysis of the first ten million digits, which show no unusual deviation from expected behavior. The frequencies for each of the ten digits are: 999,440; 999,333; 1,000,306; 999,964; 1,001,093; 1,000,466; 999,337; 1,000,207; 999,814; and 1,000,040. Moreover, the speed with which the relative frequencies are approaching 1/10 agrees with theory. Consider the digit 7 for example. Its relative frequencies in the first 10^i digits ($i = 1, \ldots, 7$) are 0, .08, .095, .097, .10025, .0998, .1000207, which seem to be approaching 1/10 at the speed predicted by probability theory for random digits, namely at a speed approximately proportional to $1/\sqrt{n}$. The poker test is relevant to the question of

normality in base ten, and Table 2 contains the frequencies of poker hands from the first ten million digits; there is no significant deviation from the expected values.

Writers over the years have been fond of mentioning that 20 decimals of π suffice for any application imaginable. Moreover, the millions of digits now known shed absolutely no light on how to prove π's normality. But these criticisms miss the point. Huyghens, in using an extrapolative techinque to extend Archimedes' calculations, was the first to use an important technique that, in this century, has come to be known as the Romberg method for approximating definite integrals. And the arithmetic-geometric mean algorithms and their refinements are closely connected to the fastest known techniques for evaluating multiprecision transcendental functions. Thus digit-hunting has an importance that goes beyond the mere extension of the known decimal places of π.

References

J. M. Borwein and P. B. Borwein, The arithmetic-geometric mean and fast computation of elementary functions, *SIAM Review* **26**, (1984) 351–366.

R. P. Brent, Multiple-precision zero-findings methods and the complexity of elementary function evaluation, in *Analytic Computational Complexity*, J. F. Traub, ed., New York: Academic Press, 1976, pp. 151–176.

D. Cox, The arithmetic-geometric mean of Gauss, *Ens. Math.* **30**, (1984) 275–330.

G. H. Hardy and E. M. Wright, *An Introduction to the Theory of Numbers*, 4th edition, London: Oxford, 1975.

Y. Kanada, Y. Tamura, S. Yoshino, and Y. Ushiro, Calculation of π to 10,013,395 decimal places based on the Gauss-Legendre algorithm and Gauss arctangent relations, *Mathematics of Computation* (forthcoming).

D. E. Knuth, *The Art of Computer Programming*, vol. 2, Reading, Mass.: Addison-Wesley, 1969.

I. Niven, *Irrational Numbers*, Carus Mathematical Monographs, No. 11, The Mathematical Association of America. Distributed by Wiley, New York, 1967.

E. Salamin, Computation of π using arithmetic-geometric mean, *Mathematics of Computation* **30**, (1976) 565–570.

7. The computation of π to 29,360,000 decimal digits using Borweins' quartically convergent algorithm (1988)

Paper 7: David H. Bailey, "The computation of pi to 29,360,000 decimal digits using Borweins' quartically convergent algorithm," *Mathematics of Computation*, vol. 50 (1988), p. 283–296. Reprinted by permission of the American Mathematical Society.

Synopsis:

This paper, written by one of the present editors, describes the computation of π using a set of formulas that at the time (1988) had just been discovered by Jonathan and Peter Borwein: Let $a_0 = 6 - 4\sqrt{2}$ and $y_0 = \sqrt{2} - 1$. Iterate

$$y_{k+1} = \frac{1 - (1 - y_k^4)^{1/4}}{1 + (1 - y_k^4)^{1/4}}$$

$$a_{k+1} = a_k(1 + y_{k+1})^4 - 2^{2k+3}y_{k+1}(1 + y_{k+1} + y_{k+1}^2).$$

Then a_k converges *quartically* to $1/\pi$: each successive iteration approximately quadruples the number of correct digits in the result.

Bailey also described in detail the computational techniques required to do such a long computation, such as the observation that a fast Fourier transform (FFT) can be employed to perform high-precision multiplication, and also presented the results of detailed statistical analyses of the digits.

Interestingly, this computation, which was performed on one of the original Cray-2 supercomputers while in a test mode, disclosed at least one bug in the hardware, which was subsequently rectified. In the wake of this finding, Cray employed a similar calculation as a test code to be run on new computers to ensure hardware integrity.

Keywords: Computation, Normality

© Springer International Publishing Switzerland 2016
D.H. Bailey, J.M. Borwein, *Pi: The Next Generation*,
DOI 10.1007/978-3-319-32377-0_7

MATHEMATICS OF COMPUTATION
VOLUME 50, NUMBER 181
JANUARY 1988, PAGES 283-296

The Computation of π to 29,360,000 Decimal Digits Using Borweins' Quartically Convergent Algorithm

By David H. Bailey

Abstract. In a recent work [6], Borwein and Borwein derived a class of algorithms based on the theory of elliptic integrals that yield very rapidly convergent approximations to elementary constants. The author has implemented Borweins' quartically convergent algorithm for $1/\pi$, using a prime modulus transform multi-precision technique, to compute over 29,360,000 digits of the decimal expansion of π. The result was checked by using a different algorithm, also due to the Borweins, that converges quadratically to π. These computations were performed as a system test of the Cray-2 operated by the Numerical Aerodynamical Simulation (NAS) Program at NASA Ames Research Center. The calculations were made possible by the very large memory of the Cray-2.

Until recently, the largest computation of the decimal expansion of π was due to Kanada and Tamura [12] of the University of Tokyo. In 1983 they computed approximately 16 million digits on a Hitachi S-810 computer. Late in 1985 Gosper [9] reported computing 17 million digits using a Symbolics workstation. Since the computation described in this paper was performed, Kanada has reported extending the computation of π to over 134 million digits (January 1987).

This paper describes the algorithms and techniques used in the author's computation, both for converging to π and for performing the required multi-precision arithmetic. The results of statistical analyses of the computed decimal expansion are also included.

1. **Introduction.** The computation of the numerical value of the constant π has been pursued for centuries for a variety of reasons, both practical and theoretical. Certainly, a value of π correct to 10 decimal places is sufficient for most "practical" applications. Occasionally, there is a need for double-precision or even multi-precision computations involving π and other elementary constants and functions in order to compensate for unusually severe numerical difficulties in an extended computation. However, the author is not aware of even a single case of a "practical" scientific computation that requires the value of π to more than about 100 decimal places.

Beyond immediate practicality, the decimal expansion of π has been of interest to mathematicians, who have still not been able to resolve the question of whether the digits in the expansion of π are "random". In particular, it is widely suspected that the decimal expansions of $\pi, e, \sqrt{2}, \sqrt{2\pi}$, and a host of related mathematical constants all have the property that the limiting frequency of any digit is one tenth, and that the limiting frequency of any n-long string of digits is 10^{-n}. Such a guaranteed property could, for instance, be the basis of a reliable pseudo-random number generator. Unfortunately, this assertion has not been proven in even one instance. Thus, there is a continuing interest in performing statistical analyses on

Received January 27, 1986; revised April 27, 1987.
1980 *Mathematics Subject Classification* (1985 *Revision*). Primary 11-04, 65-04.

the decimal expansions of these numbers to see if there is any irregularity that would suggest this assertion is false.

In recent years, the computation of the expansion of π has assumed the role as a standard test of computer integrity. If even one error occurs in the computation, then the result will almost certainly be completely in error after an initial correct section. On the other hand, if the result of the computation of π to even 100,000 decimal places is correct, then the computer has performed billions of operations without error. For this reason, programs that compute the decimal expansion of π are frequently used by both manufacturers and purchasers of new computer equipment to certify system reliability.

2. History. The first serious attempt to calculate an accurate value for the constant π was made by Archimedes, who approximated π by computing the areas of equilateral polygons with increasing numbers of sides. More recently, infinite series have been used. In 1671 Gregory discovered the arctangent series

$$\tan^{-1}(x) = x - \frac{x^3}{3} + \frac{x^5}{5} - \frac{x^7}{7} + \cdots.$$

This discovery led to a number of rapidly convergent algorithms. In 1706 Machin used Gregory's series coupled with the identity

$$\pi = 16\tan^{-1}(1/5) - 4\tan^{-1}(1/239)$$

to compute 100 digits of π.

In the nearly 300 years since that time, most computations of the value of π, even those performed by computer, have employed some variation of this technique. For instance, a series based on the identity

$$\pi = 24\tan^{-1}(1/8) + 8\tan^{-1}(1/57) + 4\tan^{-1}(1/239)$$

was used in a computation of π to 100,000 decimal digits using an IBM 7090 in 1961 [15]. Readers interested in the history of the computation π are referred to Beckmann's entertaining book on the subject [2].

3. New Algorithms for π. Only very recently have algorithms been discovered that are fundamentally faster than the above techniques. In 1976 Brent [7] and Salamin [14] independently discovered an approximation algorithm based on elliptic integrals that yields quadratic convergence to π. With all of the previous techniques, the number of correct digits increases only linearly with the number of iterations performed. With this new algorithm, each additional iteration of the algorithm approximately *doubles* the number of correct digits. Kanada and Tamura employed this algorithm in 1983 to compute π to over 16 million decimal digits.

More recently, J. M. Borwein and P. B. Borwein [4] discovered another quadratically convergent algorithm for π, together with similar algorithms for fast computation of all the elementary functions. Their quadratically convergent algorithm for π can be stated as follows: Let $a_0 = \sqrt{2}$, $b_0 = 0$, $p_0 = 2 + \sqrt{2}$. Iterate

$$a_{k+1} = \frac{(\sqrt{a_k} + 1/\sqrt{a_k})}{2}, \quad b_{k+1} = \frac{\sqrt{a_k}(1 + b_k)}{a_k + b_k}, \quad p_{k+1} = \frac{p_k b_{k+1}(1 + a_{k+1})}{1 + b_{k+1}}.$$

Then p_k converges quadratically to π: Successive iterations of this algorithm yield 3, 8, 19, 40, 83, 170, 345, 694, 1392, and 2788 correct digits of the expansion of π.

However, it should be noted that this algorithm is not self-correcting for numerical errors, so that all iterations must be performed to full precision. In other words, in a computation of π to 2788 decimal digits using the above algorithm, each of the ten iterations must be performed with more than 2788 digits of precision.

Most recently, the Borweins [6] have discovered a general technique for obtaining even higher-order convergent algorithms for certain elementary constants. Their quartically convergent algorithm for $1/\pi$ can be stated as follows: Let $a_0 = 6 - 4\sqrt{2}$ and $y_0 = \sqrt{2} - 1$. Iterate

$$y_{k+1} = \frac{1 - (1 - y_k^4)^{1/4}}{1 + (1 - y_k^4)^{1/4}},$$

$$a_{k+1} = a_k(1 + y_{k+1})^4 - 2^{2k+3}y_{k+1}(1 + y_{k+1} + y_{k+1}^2).$$

Then a_k converges quartically to $1/\pi$: Each successive iteration approximately *quadruples* the number of correct digits in the result. As in the previous case, each iteration must be performed to at least the level of precision desired for the final result.

4. Multi-Precision Arithmetic Techniques. A key element of a very high precision computation of this sort is a set of high-performance routines for performing multi-precision arithmetic. A naive approach to multi-precision computation would require a prohibitive amount of processing time and would, as a result, sharply increase the probability that an undetected hardware error would occur, rendering the result invalid. In addition to employing advanced algorithms for such key operations as multi-precision multiplication, it is imperative that these algorithms be implemented in a style that is conducive for high-speed computation on the computer being used.

The computer used for these computations is the Cray-2 at the NASA Ames Research Center. This computation was performed to test the integrity of the Cray-2 hardware, as well as the Fortran compiler and the operating system. The Cray-2 is particularly well suited for this computation because of its very large main memory, which holds $2^{28} = 268,435,456$ words (one word is 64 bits of data). With this huge capacity, all data for these computations can be contained entirely within main memory, insuring ease of programming and fast execution.

No attempt was made to employ more than one of the four central processing units in the Cray-2. Thus, at the same time these calculations were being performed, the computer was executing other jobs on the other processors. However, full advantage was taken of the vector operations and vector registers of the system. Considerable care was taken in programming to insure that the multi-precision algorithms were implemented in a style that would admit vector processing. Most key loops were automatically vectorized by the Cray-2 Fortran compiler. For those few that were not automatically vectorized, compiler directives were inserted to force vectorization. As a result of this effort, virtually all arithmetic operations were performed in vector mode, which on the Cray-2 is approximately 20 times faster than scalar mode. Because of the high level of vectorization that was achieved using the Fortran compiler, it was not necessary to use assembly language, nonstandard constructs, or library subroutines.

A multi-precision number is represented in these computations as an $(n + 2)$-long array of floating-point whole numbers. The first cell contains the sign of the number, either 1, -1, or 0 (reserved for an exact zero). The second cell of the array contains the exponent (powers of the radix), and the remaining n cells contain the mantissa. The radix selected for the multi-precision numbers is 10^7. Thus the number 1.23456789 is represented by the array $1, 0, 1, 2345678, 9000000, 0, 0, \ldots, 0$.

A floating-point representation was chosen instead of an integer representation because the hardware of numerical supercomputers such as the Cray-2 is designed for floating-point computation. Indeed, the Cray-2 does not even have full-word integer multiply or divide hardware instructions. Such operations are performed by first converting the operands to floating-point form, using the floating-point unit, and converting the results back to fixed-point (integer) form. A decimal radix was chosen instead of a binary value because multiplications and divisions by powers of two are not performed any faster than normal on the Cary-2 (in vector mode). Since a decimal radix is clearly preferable to a binary radix for program troubleshooting and for input and output, a decimal radix was chosen. The value 10^7 was chosen because it is the largest power of ten that will fit in half of the mantissa of a single word. In this way two of these numbers may be multiplied to obtain the exact product using ordinary single-precision arithmetic.

Multi-precision addition and subtraction are not computationally expensive compared to multiplication, division, and square root extraction. Thus, simple algorithms suffice to perform addition and subtraction. The only part of these operations that is not immediately conducive to vector processing is releasing the carries for the final result. This is because the normal "schoolboy" approach of beginning at the last cell and working forward is a recursive operation. On a vector supercomputer this is better done by starting at the beginning and releasing the carry only one cell back for each cell processed. Unfortunately, it cannot be guaranteed that one application of this process will release all carries (consider the case of two or more consecutive 9999999's, followed by a number exceeding 10^7). Thus it is necessary to repeat this operation until all carries have been released (usually one or two additional times). In the rare cases where three applications of this vectorized process are not successful in releasing all carries, the author's program resorts to the scalar "schoolboy" method.

Provided a fast multi-precision multiplication procedure is available, multi-precision division and square root extraction may be performed economically using Newton's iteration, as follows. Let x_0 and y_0 be initial approximations to the reciprocal of a and to the reciprocal of the square root of a, respectively. Then

$$x_{k+1} = x_k(2 - ax_k), \qquad y_{k+1} = \frac{y_k(3 - ay_k^2)}{2}$$

both converge quadratically to the desired values. One additional full-precision multiplication yields the quotient and the square root, respectively. What is especially attractive about these algorithms is that the first iteration may be performed using ordinary single-precision arithmetic, and subsequent iterations may be performed using a level of precision that approximately doubles each time. Thus the total cost of computation is only about twice the cost of the final iteration, plus the one additional multiplication. As a result, a multi-precision division costs only about five times as much as a multi-precision multiplication, and a multi-precision square root costs only about seven times as much as a multi-precision multiplication.

5. Multi-Precision Multiplication. It can be seen from the above that the key component of a high-performance multi-precision arithmetic system is the multiply operation. For modest levels of precision (fewer than about 1000 digits), some variation of the usual "schoolboy" method is sufficient, although care must be taken in the implementation to insure that the operations are vectorizable. Above this level of precision, however, other more sophisticated techniques have a significant advantage. The history of the development of high-performance multiply algorithms will not be reviewed here. The interested reader is referred to Knuth [13]. It will suffice to note that all of the current state-of-the-art techniques derive from the following fact of Fourier analysis: Let $F(x)$ denote the discrete Fourier transform of the sequence $x = (x_0, x_1, x_2, \ldots, x_{N-1})$, and let $F^{-1}(x)$ denote the inverse discrete Fourier transform of x:

$$F_k(x) = \sum_{j=0}^{N-1} x_j \omega^{jk}, \qquad F_k^{-1}(x) = \frac{1}{N} \sum_{j=0}^{N-1} x_j \omega^{-jk},$$

where $\omega = e^{-2\pi i/N}$ is a primitive Nth root of unity. Let $C(x, y)$ denote the convolution of the sequences x and y:

$$C_k(x, y) = \sum_{j=0}^{N-1} x_j y_{k-j},$$

where the subscript $k - j$ is to be interpreted as $k - j + N$ if $k - j$ is negative. Then the "convolution theorem", whose proof is a straightforward exercise, states that

$$F[C(x, y)] = F(x)F(y),$$

or expressed another way,

$$C(x, y) = F^{-1}[F(x)F(y)].$$

This result is applicable to multi-precision multiplication in the following way. Let x and y be n-long representations of two multi-precision numbers (without the sign or exponent words). Extend x and y to length $2n$ by appending n zeros at the end of each. Then the multi-precision product z of x and y, except for releasing the carries, can be written as follows:

$$z_0 = x_0 y_0$$
$$z_1 = x_0 y_1 + x_1 y_0$$
$$z_2 = x_0 y_2 + x_1 y_1 + x_2 y_0$$
$$\vdots$$
$$z_{n-1} = x_0 y_{n-1} + x_1 y_{n-2} + \cdots + x_{n-1} y_0$$
$$\vdots$$
$$z_{2n-3} = x_{n-1} y_{n-2} + x_{n-2} y_{n-1}$$
$$z_{2n-2} = x_{n-1} y_{n-1}$$
$$z_{2n-1} = 0.$$

It can now be seen that this "multiplication pyramid" is precisely the convolution of the two sequences x and y, where $N = 2n$. The convolution theorem states that the multiplication pyramid can be obtained by performing two forward discrete Fourier transforms, one vector complex multiplication, and one reverse transform, each of length $N = 2n$. Once the resulting complex numbers have been rounded to the nearest integer, the final multi-precision product may be obtained by merely releasing the carries as described in the section above on addition and subtraction.

The key computational savings here is that the discrete Fourier transform may of course be economically computed using some variation of the "fast Fourier transform" (FFT) algorithm. It is most convenient to employ the radix two fast Fourier transform since there is a wealth of literature on how to efficiently implement this algorithm (see [1], [8], and [16]). Thus, it will be assumed from this point that $N = 2^m$ for some integer m.

One useful "trick" can be employed to further reduce the computational requirement for complex transforms. Note that the input data vectors x and y and the result vector z are purely real. This fact can be exploited by using a simple procedure ([8, p. 169]) for performing real-to-complex and complex-to-real transforms that obtains the result with only about half the work otherwise required.

One important item has been omitted from the above discussion. If the radix 10^7 is used, then the product of two cells will be in the neighborhood of 10^{14}, and the sum of a large number of these products cannot be represented exactly in the 48-bit mantissa of a Cray-2 floating-point word. In this case the rounding operation at the completion of the transform will not be able to recover the exact whole number result. As a result, for the complex transform method to work correctly, it is necessary to alter the above scheme slightly. The simplest solution is to use the radix 10^6 and to divide all input data into two words with only three digits each. Although this scheme greatly increases the memory space required, it does permit the complex transform method to be used for multi-precision computation up to several million digits on the Cray-2.

6. Prime Modulus Transforms. Some variation of the above method has been used in almost all high-performance multi-precision computer programs, including the program used by Kanada and Tamura. However, it appears to break down for very high-precision computation (beyond about ten million digits on the Cray-2), due to the round-off error problem mentioned above. The input data can be further divided into two digits per word or even one digit per word, but only with a substantial increase in run time and main memory. Since a principal goal in this computation was to remain totally within the Cray-2 main memory, a somewhat different method was used.

It can readily be seen that the technique of the previous section, including the usage of a fast Fourier transform algorithm, can be applied in any number field in which there exists a primitive Nth root of unity ω. This requirement holds for the field of the integers modulo p, where p is a prime of the form $p = kN + 1$ ([11, p. 85]). One significant advantage of using a prime modulus field instead of the field of complex numbers is that there is no need to worry about round-off error in the results, since all computations are exact.

However, there are some difficulties in using a prime modulus field for the transform operations above. The first is to find a prime p of the form $kN + 1$, where $N = 2^m$. The second is to find a primitive Nth root of unity modulo p. As it turns out, it is not too hard using a computer to find both of these numbers by direct search. Thirdly, one must compute the multiplicative inverse of N modulo p. This can be done using a variation of the Euclidean algorithm from elementary number theory. Note that each of these calculations needs to be performed one time only.

A more troublesome difficulty in using a prime modulus transform is the fact that the final multiplication pyramid results are only recovered modulo p. If p is greater than about 10^{24} then this is not a problem, but the usage of such a large prime would require *quadruple*-precision arithmetic operations to be performed in the inner loop of the fast Fourier transform, which would very greatly increase the run time. A simpler and faster approach to the problem is to use two primes, p_1 and p_2, each slightly greater than 10^{12}, and to perform the transform algorithm above using each prime. Then the Chinese remainder theorem may be applied to the results modulo p_1 and p_2 to obtain the results modulo the product $p_1 p_2$. Since $p_1 p_2$ is greater than 10^{24}, these results will be the exact multiplication pyramid numbers. Unfortunately, double-precision arithmetic must still be performed in the fast Fourier transform and in the Chinese remainder theorem calculation. However, the whole-number format of the input data simplifies these operations, and it is possible to program them in a vectorizable fashion.

Borodin and Munro ([3, p. 90]) have suggested using three transforms with three primes p_1, p_2 and p_3, each of which is just smaller than half of the mantissa, and using the Chinese remainder theorem to recover the results modulo $p_1 p_2 p_3$. In this way, double-precision operations are completely avoided in the inner loop of the FFT. This scheme runs quite fast, but unfortunately the largest transform that can be performed on the Cray-2 using this system is $N = 2^{19}$, which corresponds to a maximum precision of about three million digits.

Readers interested in studying about prime modulus number fields, the Euclidean algorithm, or the Chinese remainder theorem are referred to any elementary text on number theory, such as [10] or [11]. Knuth [13] and Borodin [3] also provide excellent information on using these tools for computation.

7. Computational Results. The author has implemented all three of the above techniques for multi-precision multiplication on the Cray-2. By employing special high-performance techniques [1], the complex transform can be made to run the fastest, about four times faster than the two-prime transform method. However, the memory requirement of the two-prime scheme is significantly less than either the three-prime or the complex scheme, and since the two-prime scheme permits very high-precision computation, it was selected for the computations of π.

One of the author's computations used twelve iterations of Borweins' quartic algorithm for $1/\pi$, followed by a reciprocal operation, to yield 29,360,128 digits of π. In this computation, approximately 12 trillion arithmetic operations were performed. The run took 28 hours of processing time on one of the four Cray-2 central processing units and used 138 million words of main memory. It was started on January 7, 1986 and completed January 9, 1986. The program was not running this entire time—the system was taken down for service several times, and the run was frequently interrupted by other programs. Restarting the computation after a system down was a simple matter since the two key multi-precision number arrays were saved on disk after the completion of each iteration.

This computation was checked using 24 iterations of Borweins' quadratically convergent algorithm for π. This run took 40 hours processing time and 147 million words of main memory. A comparison of these output results with the first run found no discrepancies except for the last 24 digits, a normal truncation error. Thus it can be safely assumed that at least 29,360,000 digits of the final result are correct.

It was discovered after both computations were completed that one loop in the Chinese remainder theorem computation was inadvertently performed in scalar mode instead of vector mode. As a result, both of these calculations used about 25% more run time than would otherwise have been required. This error, however, did not affect the validity of the computed decimal expansions.

8. Statistical Analysis of π. Probably the most significant mathematical motivation for the computation of π, both historically and in modern times, has been to investigate the question of the randomness of its decimal expansion. Before Lambert proved in 1766 that π is irrational, there was great interest in checking whether or not its decimal expansion eventually repeats, thus disclosing that π is rational. Since that time there has been a continuing interest in the still unanswered question of whether the expansion is statistically random. It is of course strongly suspected that the decimal expansion of π, if computed to sufficiently high precision, will pass any reasonable statistical test for randomness. The most frequently mentioned conjecture along this line is that any sequence of n digits occurs with a limiting frequency of 10^{-n}.

With 29,360,000 digits, the frequencies of n-long strings may be studied for randomness for n as high as six. Beyond that level the expected number of any one string is too low for statistical tests to be meaningful. The results of tabulated frequencies for one and two digit strings are listed in Tables 1 and 2. In the first table the Z-score numbers are computed as the deviation from the mean divided by the standard deviation, and thus these statistics should be normally distributed with mean zero and variance one.

TABLE 1

Single digit statistics

Digit	Count	Deviation	Z-score
0	2935072	-928	-0.5709
1	2936516	516	0.3174
2	2936843	843	0.5186
3	2935205	-795	-0.4891
4	2938787	2787	1.7145
5	2936197	197	0.1212
6	2935504	-496	-0.3051
7	2934083	-1917	-1.1793
8	2935698	-302	-0.1858
9	2936095	95	0.0584

THE COMPUTATION OF π TO 29,360,000 DECIMAL DIGITS   291

TABLE 2

Two digit frequency counts

00	293062	01	293970	02	293533	03	292893	04	294459
05	294189	06	292688	07	292707	08	294260	09	293311
10	294503	11	293409	12	293591	13	294285	14	294020
15	293158	16	293799	17	293020	18	293262	19	293469
20	293952	21	293226	22	293844	23	293382	24	293869
25	293721	26	293655	27	293969	28	293320	29	293905
30	293718	31	293542	32	293272	33	293422	34	293178
35	293490	36	293484	37	292694	38	294152	39	294253
40	294622	41	294793	42	293863	43	293041	44	293519
45	293998	46	294418	47	293616	48	293296	49	293621
50	292736	51	294272	52	293614	53	293215	54	293569
55	294194	56	293260	57	294152	58	293137	59	294048
60	293842	61	293105	62	294187	63	293809	64	293463
65	293544	66	293123	67	293307	68	293602	69	293522
70	292650	71	294304	72	293497	73	293761	74	293960
75	293199	76	293597	77	292745	78	293223	79	293147
80	292517	81	292986	82	293637	83	294475	84	294267
85	293600	86	293786	87	293971	88	293434	89	293025
90	293470	91	292908	92	293806	93	292922	94	294483
95	293104	96	293694	97	293902	98	294012	99	293794

The most appropriate statistical procedure for testing the hypothesis that the empirical frequencies of n-long strings of digits are random is the χ^2 test. The χ^2 statistic of the k observations $X_1, X_2, \ldots, X_k$ is defined as

$$\chi^2 = \sum_{i=1}^{k} \frac{(X_i - E_i)^2}{E_i}$$

where E_i is the expected value of the random variable X_i. In this case $k = 10^n$ and $E_i = 10^{-n}d$ for all i, where $d = 29,360,000$ denotes the number of digits. The mean of the χ^2 statistic in this case is $k - 1$ and its standard deviation is $\sqrt{2(k-1)}$. Its distribution is nearly normal for large k. The results of the χ^2 analysis are shown in Table 3.

TABLE 3

Multiple digit χ^2 statistics

Length	χ^2 value	Z-score
1	4.869696	-0.9735
2	84.52604	-1.0286
3	983.9108	-0.3376
4	10147.258	1.0484
5	100257.92	0.5790
6	1000827.7	0.5860

Another test that is frequently performed on long pseudo-random sequences is an analysis to check whether the number of n-long repeats for various n is within statistical bounds of randomness. An n-long repeat is said to occur if the n-long digit sequence beginning at two different positions is the same. The mean M and the variance V of the number of n-long repeats in d digits are (to an excellent approximation)

$$M = \frac{10^{-n}d^2}{2}, \qquad V = \frac{11 \cdot 10^{-n}d^2}{18}.$$

Tabulation of repeats in the expansion of π was performed by packing the string beginning at each position into a single Cray-2 word, sorting the resulting array, and counting equal contiguous entries in the sorted list. The results of this analysis are shown in Table 4.

TABLE 4

Long repeat statistics

10	42945	43100.	-0.677
11	4385	4310.	1.033
12	447	431.	0.697
13	48	43.1	0.675
14	6	4.31	0.736
15	1	0.43	0.784

A third test frequently performed as a test for randomness is the runs test. This test compares the observed frequency of long runs of a single digit with the number of such occurrences that would be expected at random. The mean and variance of this statistic are the same as the formulas for repeats, except that d^2 is replaced by $2d$. Table 5 lists the observed frequencies of runs for the calculated expansion of π.

The frequencies of long runs are all within acceptable limits of randomness. The only phenomenon of any note in Table 5 is the occurrence of a 9-long run of sevens. However, there is a 29% chance that a 9-long run of some digit would occur in 29,360,000 digits, so this instance by itself is not remarkable.

TABLE 5

Single-digit run counts

Digit	Length of Run				
	5	6	7	8	9
0	308	29	3	0	0
1	281	21	1	0	0
2	272	23	0	0	0
3	266	26	5	0	0
4	296	40	6	1	0
5	292	30	4	0	0
6	316	33	3	0	0
7	315	37	6	2	1
8	295	36	3	0	0
9	306	40	7	0	0

9. Conclusion. The statistical analyses that have been performed on the expansion of π to 29,360,000 decimal places have not disclosed any irregularity. The observed frequencies of n-long strings of digits for n up to 6 are entirely unremarkable. The numbers of long repeating strings and single-digit runs are completely

THE COMPUTATION OF π TO 29,360,000 DECIMAL DIGITS   **293**

acceptable. Thus, based on these tests, the decimal expansion of π appears to be completely random.

Appendix

Selected Output Listing

Initial 1000 digits:

```
3.
14159265358979323846264338327950288419716939937510
5820974944592307816406286208998628034825342117067 9
8214808651328230664709384460955058223172535940812 8
4811174502841027019385211055596446229489549303819 6
4428810975665933446128475648233786783165271201909 1
4564856692346034861045432664821339360726024914127 3
7245870066063155881748815209209628292540917153643 6
7892590360011330530548820466521384146951941511609 4
3305727036575959195309218611738193261179310511854 8
0744623799627495673518857527248912279381830119491 2
9833673362440656643086021394946395224737190702179 8
6094370277053921717629317675238467481846766940513 2
0005681271452635608277857713427577896091736371787 2
1468440901224953430146549585371050792279689258923 5
4201995611212902196086403441815981362977477130996 0
5187072113499999983729780499510597317328160963185 9
5024459455346908302642522308253344685035261931188 1
7101000313783875288658753320838142061717766914730 3
5982534904287554687311595628638823537875937519577 8
1857780532171226806613001927876611195909216420198 9
```

Digits 4,999,001 to 5,000,000:

```
4948075478455810018273193163248841280448872229695 6
7985501546485578048673653522790283699791808486723 0
6496222100452708576833503521206968480181713761632 9
9756173842516034047253710005635164034216249202717 9
6682492645893096018264502692310226657054164147534 7
2034155491377042150576445280780903524839362109303 1
0228809623848687792314524084163727118095305889004 0
6884376678143149891429989362127854526014314043904 8
4993880155633605951311673189113276577788136469070 8
4703686341119632306388650748085212568284225785252 4
0308699370325569209396081858741418123048415320404 9
2023498900273244759302032379479077644475239844551 4
6730440321096898524496196714343396489589319055233 8
4981885274684449248363146342500064216306286868588 4 8
2745331866992673473064273503636400285602221896635 0
1142918263431997416325337236879855345111125305526 2
3910408263997093450814667252138110591304721005242 8
1898862653316946933195167529620930675229159071599 9
8984617928805926200084863813881128094405648802106 0
4886585519184670236542176178350518172132076461971 5
```

DAVID H. BAILEY

Digits 9,999,001 to 10,000,000:

```
5509781824351672822784991072040028675790790446 6335
1271820297952515061772533406689498895642470326 9230
1539982090039016627522433818442480858939529365 2582
5363565858417548553674481865028924518820644785 3280
7912967550486557292908308348548393758333467101 9089
1206711453695517314092946182346647872528952997 4204
0212763523592329330577017942386522596324069402 7480
6041288030309245248103494158273593244388727310 9397
4163488960469581924539515134104343399838187465 0972
3369263522579147245424440132631296439639120960 7800
1634485119912542081973740744604589974214573104 2313
6445648650193780106352660374405656882386138937 5443
9735168129683156791161888422225114147732261233 1396
1860608037311034869266093394043841630032614344 9280
5082113157573772773982155152228650999766243258 7213
9339344590209166227290549349382717820512666902 1149
4719231138093382231122409958837224633250122232 3378
9689526902536626394126701031732786498717025714 9617
7610515549257985759204553246894468742702504639 7905
6532655319406099946978733381063171948173534895 5897
```

Digits 14,999,001 to 15,000,000:

```
7516191258272903443712327974925631151192524395 6985
4146673506919481516383722607392515188775175165 9741
0062288072644860220945693041448853988298110851 2492
3062608837596678362164975341253968308492271134 2513
9495399569362544133140173813330858481723158874 73225
6686213925193854010224947557549494715839562351 2785
6703388882449555108446230047240761216595278438 6252
8305999230222328486593456626292974843682773081 2030
1443459368987425976641551441209798413399801593 4584
3539347565062432385016043273191880512640667187 1353
7755576621467093181315116287950050971055179515 2818
0909315448105804476736412216610003242509826316 6257
4173051822048071548822461656389134404693420810 3238
3990325402988174634249658318683694748619425753 3540
3633122383822239249405627085637803305621354468 6593
0298682171495280858594941867653229106733981768 4850
7757615178505727709880627370814385794117668763 599
7581449914989031459409852596033637798998822813 8579
0395460850007618075488043395846861964109276265 3446
7964520526347339328607497932393150314117277566 9803
```

THE COMPUTATION OF π TO 29,360,000 DECIMAL DIGITS   **295**

Digits 19,999,001 to 20,000,000:

2466242165219965948681580445687019757643895160769786758526528445124126249995515004465281646092893016373961985962486271165524696863816796798989261652141998514539271654610871466425799827875023943144669024524827883001435830699295155565194378002452231513034984501651352825341097581675080414571879068219509815688966940154057556043048954713178146479692058699611799897126388736531564345333853581593559913668626084862270298656682308563913220818592052433492234189846647982105263462296862876649515069626241605624275201300452308788083860012754008114751496913646624222976304434816051167918643343026623869212978502788523588894213372112340064720173755448172632485389905485693682923700908893714354426488242078425462806740072794920355326388439531017684353590261463476307233029969045465206192626213143248919480318684240913408886185032376704408770471930796657178425684902689744570168173881678986118970643044572067493681903857815020793466156644931359073005891342758785950724478952328081911162910558013800493386345276644

Digits 24,999,001 to 25,000,000:

6462637665778840162687203583515025093238112680413224527774629670113871130617683224437149346115597163910991083622688538884847037999823966041879542473503663585952130451687270980967894865853409228442863249489360013422079559687409670921107196838565582053081604815190224085606214877412355102352998581079274189214723685203602121713995138514107079374902532543507859972884134839114349522198649483213304900746014643512125431125957394730114253118457091422408072612210306331872567179327168155609249989038137333669602575213348431548953618884362087312748886747811837398473931375007714926901146221961579804706751435050981335283641909759090614464729227662129370246470570908744501080272319698635170294417265180383673276289174186382214920853922637638290730594173963907549588865849168186491743776278287261919660505923924757388365872266493595243832978614043782282882817359631264257437061195680129735603634263779356276138037507909491563108238168922672241753290004525344607864115924597806944245511285225546774836191884322

296 DAVID H. BAILEY

Digits 29,359,001 to 29,360,000:

```
3419284178891522964336847388197769853900574621984 6
6952534757700172988654339243262184097259196825915 7
6110747629400730307400523562782978702554407540554 3
9989507153059816218961131505041969730972829060606 7
1889011613820684258998021544539575359379289882357 5
0141234748667204693563573577738064843730857329184 0
6210849633097482768941126867522297552323062395683 3
6263114891606388397766197309149915519284789410969 1
3961226532935119597872556676425646289537518090744 9
4936309292131412764088851017042258408474414931911 8
6575582572177283614497797876605228546904719759626 4
7668005536084220968951773713500861189045243301521 2
3769374570207033898894012337669396105726953527814 6
9971913630707464320185386407130799750797450988355 4
6596157578284974751264578644113084532532314940541 9
1726336489964791203287817189338731781932491238234 2
1864827176372302256172001634836858495565816511248 9
9544684872069362195779794342949464025841993908913 5
3426698523277623931436525967083202637025092477681 4
7049097142449367541433098725950780665432227288825 3
```

NAS Systems Division
NASA Ames Research Center
Mail Stop 258-5
Moffett Field, California 94035

1. D. H. BAILEY, "A high-performance fast Fourier transform algorithm for the Cray-2," *J. Supercomputing*, v. 1, 1987, pp. 43–60.

2. P. BECKMANN, *A History of Pi*, Golem Press, Boulder, CO, 1971.

3. A. BORODIN & I. MUNRO, *The Computational Complexity of Algebraic and Numeric Problems*, American Elsevier, New York, 1975.

4. J. M. BORWEIN & P. B. BORWEIN, "The arithmetic-geometric mean and fast computation of elementary functions," *SIAM Rev.*, v. 26, 1984, pp. 351–366.

5. J. M. BORWEIN & P. B. BORWEIN, "More quadratically converging algorithms for π," *Math. Comp.*, v. 46, 1986, pp. 247–253.

6. J. M. BORWEIN & P. B. BORWEIN, *Pi and the AGM—A Study in Analytic Number Theory and Computational Complexity*, Wiley, New York, 1987.

7. R. P. BRENT, "Fast multiple-precision evaluation of elementary functions," *J. Assoc. Comput. Mach.*, v. 23, 1976, pp. 242–251.

8. E. O. BRIGHAM, *The Fast Fourier Transform*, Prentice-Hall, Englewood Cliffs, N. J., 1974.

9. W. GOSPER, private communication.

10. EMIL GROSSWALD, *Topics from the Theory of Numbers*, Macmillan, New York, 1966.

11. G. H. HARDY & E. M. WRIGHT, *An Introduction to the Theory of Numbers*, 5th ed., Oxford Univ. Press, London, 1984.

12. Y. KANADA & Y. TAMURA, *Calculation of π to 10,013,395 Decimal Places Based on the Gauss-Legendre Algorithm and Gauss Arctangent Relation*, Computer Centre, University of Tokyo, 1983.

13. D. KNUTH, *The Art of Computer Programming*, Vol. 2: *Seminumerical Algorithms*, Addison-Wesley, Reading, Mass., 1981.

14. E. SALAMIN, "Computation of π using arithmetic-geometric mean," *Math. Comp.*, v. 30, 1976, pp. 565–570.

15. D. SHANKS & J. W. WRENCH, JR., "Calculation of π to 100,000 decimals," *Math. Comp.*, v. 16, 1962, pp. 76–99.

16. P. SWARZTRAUBER, "FFT algorithms for vector computers," *Parallel Comput.*, v. 1, 1984, pp. 45–64.

8. Gauss, Landen, Ramanujan, the arithmetic-geometric mean, ellipses, π, and the Ladies Diary (1988)

Paper 8: Gert Almkvist and Bruce Berndt, "Gauss, Landen, Ramanujan, the arithmetic-geometric mean, ellipses, pi, and the Ladies Diary," *American Mathematical Monthly*, vol. 95 (1988), pg. 585–608. Copyright 1988 Mathematical Association of America. All Rights Reserved.

Synopsis:

If the title of this article seems wide-ranging, it is because the article itself is, delightfully so. The authors take the reader on an entertaining but highly informative tour of the arithmetic-geometric mean (AGM), including the historical roots of Gauss's work, a mini-biography of the reclusive 19th century British mathematician John Landen, who wrote articles on mathematics (including on the AGM) for the *Ladies Diary*, a popular womens magazine that included a mathematics column, and late-20th-century developments including the new quadratically convergent algorithms of Salamin, Brent and others, and computations implementing these techniques on powerful computer systems. All of this is presented in a very rigorous style that exposes both the methods and results of the theory.

Almkvist and Bernt also discuss some extensions of this theory that were undertaken by the famed Indian mathematician Ramanujan. The authors were well-equipped to present this material, since Berndt, for instance, was in the process of editing the works of Ramanujan for a modern audience. The result is a very enlightening mathematical tour of the AGM.

Keywords: Algorithms, Arithmetic-Geometric Mean, Computation, Elliptic Integrals, History

© Springer International Publishing Switzerland 2016
D.H. Bailey, J.M. Borwein, *Pi: The Next Generation*,
DOI 10.1007/978-3-319-32377-0_8

Gauss, Landen, Ramanujan, the Arithmetic-Geometric Mean, Ellipses, π, and the *Ladies Diary*

GERT ALMKVIST, *University of Lund*

BRUCE BERNDT*, *University of Illinois*

GERT ALMKVIST received his Ph.D. at the University of California in 1966 and has been at Lund since 1967. His main interests are algebraic K-theory, invariant theory, and elliptic functions.

BRUCE BERNDT received his A.B. degree from Albion College, Albion, Michigan in 1961 and his Ph.D. from the University of Wisconsin, Madison, in 1966. Since 1977, he has devoted all of his research efforts to proving the hitherto unproven results in Ramanujan's notebooks. His book, *Ramanujan's Notebooks*, Part I (Springer-Verlag, 1985), is the first of either three or four volumes to be published on this project.

Virtue and sense, with female-softness join'd
(All that subdues and captivates mankind!)
In Britain's matchless fair resplendent shine;
They rule Love's empire by a right divine:
Justly their charms the astonished World admires,
Whom Royal Charlotte's bright example fires.

1. Introduction. The arithmetic-geometric mean was first discovered by Lagrange and rediscovered by Gauss a few years later while he was a teenager. However, Gauss's major contributions, including an elegant integral representation, were made about 7–9 years later. The first purpose of this article is, then, to explain the arithmetic-geometric mean and to describe some of its major properties, many of which are due to Gauss.

*Research partially supported by the Vaughn Foundation.

Because of its rapid convergence, the arithmetic-geometric mean has been significantly employed in the past decade in fast machine computation. A second purpose of this article is thus to delineate its role in the computation of π. We emphasize that the arithmetic-geometric mean has much broader applications, e.g., to the calculation of elementary functions such as $\log x$, e^x, $\sin x$, and $\cos x$. The interested reader should further consult the several references cited here, especially Brent's paper [14] and the Borweins' book [13].

The determination of the arithmetic-geometric mean is intimately related to the calculation of the perimeter of an ellipse. Since the days of Kepler and Euler, several approximate formulas have been devised to calculate the perimeter. The primary motivation in deriving such approximations was evidently the desire to accurately calculate the elliptical orbits of planets. A third purpose of this article is thus to describe the connections between the arithmetic-geometric mean and the perimeter of an ellipse, and to survey many of the approximate formulas that have been given in the literature. The most accurate of these is due to Ramanujan, who also found some extraordinarily unusual and exotic approximations to elliptical perimeters. The latter results are found in his notebooks and have never been published, and so we shall pay particular attention to these approximations.

Also contributing to this circle of ideas is the English mathematician John Landen. In the study of both the arithmetic-geometric mean and the determination of elliptical perimeters, there arises his most important mathematical contribution, which is now called Landen's transformation. Many very important and seemingly unrelated guises of Landen's transformation exist in the literature. Thus, a fourth purpose of this article is to delineate several formulations of Landen's transformation as well as to provide a short biography of this undeservedly, rather obscure, mathematician.

For several years, Landen published almost exclusively in the *Ladies Diary*. This is, historically, the first regularly published periodical to contain a section devoted to the posing of mathematical problems and their solutions. Because an important feature of the MONTHLY has its roots in the *Ladies Diary*, it seems then dually appropriate in this paper to provide a brief description of the *Ladies Diary*.

2. Gauss and the arithmetic-geometric mean. As we previously alluded, the arithmetic-geometric mean was first set forth in a memoir of Lagrange [30] published in 1784–85. However, in a letter, dated April 16, 1816, to a friend, H. C. Schumacher, Gauss confided that he independently discovered the arithmetic-geometric mean in 1791 at the age of 14. At about the age of 22 or 23, Gauss wrote a long paper [23] describing his many discoveries on the arithmetic-geometric mean. However, this work, like many others by Gauss, was not published until after his death. Gauss's fundamental paper thus did not appear until 1866 when E. Schering, the editor of Gauss's complete works, published the paper as part of Gauss's *Nachlass*. Gauss obviously attached considerable importance to his findings on the arithmetic-geometric mean, for several of the entries in his diary, in particular, from the years 1799 to 1800, pertain to the arithmetic-geometric mean. Some of these entries are quite vague, and we may still not know everything that Gauss discovered about the arithmetic-geometric mean. (For an English translation of Gauss's diary together with commentary, see a paper by J. J. Gray [24].)

By now, the reader is anxious to learn about the arithmetic-geometric mean and what the young Gauss discovered.

Let a and b denote positive numbers with $a > b$. Construct a sequence of arithmetic means and a sequence of geometric means as follows:

$$a_1 = \frac{1}{2}(a + b), \qquad b_1 = \sqrt{ab},$$

$$a_2 = \frac{1}{2}(a_1 + b_1), \qquad b_2 = \sqrt{a_1 b_1},$$

$$\vdots \qquad\qquad\qquad \vdots$$

$$a_{n+1} = \frac{1}{2}(a_n + b_n), \qquad b_{n+1} = \sqrt{a_n b_n},$$

$$\vdots \qquad\qquad\qquad \vdots$$

Gauss [23] gives four numerical examples, of which we reproduce one. Let $a = 1$ and $b = 0.8$. Then

$a_1 = 0.9$, $b_1 = 0.894427190999915878564$,
$a_2 = 0.897213595499957939282$, $b_2 = 0.897209268732734$,
$a_3 = 0.897211432116346$, $b_3 = 0.897211432113738$,
$a_4 = 0.897211432115042$, $b_4 = 0.897211432115042$.

(Obviously, Gauss did not shirk from numerical calculations.) It appears from this example that $\{a_n\}$ and $\{b_n\}$ converge to the same limit, and that furthermore this convergence is very rapid. This we now demonstrate.

Observe that

$$b < b_1 < a_1 < a,$$

$$b < b_1 < b_2 < a_2 < a_1 < a,$$

$$b < b_1 < b_2 < b_3 < a_3 < a_2 < a_1 < a,$$

etc. Thus, $\{b_n\}$ is increasing and bounded, and $\{a_n\}$ is decreasing and bounded. Each sequence therefore converges. Elementary algebraic manipulation now shows that

$$\frac{a_1 - b_1}{a - b} = \frac{a - b}{4(a_1 + b_1)} = \frac{a - b}{2(a + b) + 4b_1} < \frac{1}{2}.$$

Iterating this procedure, we deduce that

$$a_n - b_n < \left(\frac{1}{2}\right)^n (a - b), \qquad n \geq 1,$$

which tends to 0 as n tends to ∞. Thus, a_n and b_n converge to the same limit, which we denote by $M(a, b)$. By definition, $M(a, b)$ *is the arithmetic-geometric mean of a and b.*

To provide a more quantitative measure of the rapidity of convergence, first define

$$c_n = \sqrt{a_n^2 - b_n^2}, \qquad n \geq 0, \tag{1}$$

where $a_0 = a$ and $b_0 = b$. Observe that

$$c_{n+1} = \frac{1}{2}(a_n - b_n) = \frac{c_n^2}{4a_{n+1}} \leqslant \frac{c_n^2}{4M(a,b)}.$$

Thus, c_n tends to 0 quadratically, or the convergence is of the second order. More generally, suppose that $\{\alpha_n\}$ converges to L and assume that there exist constants $C > 0$ and $m \geqslant 1$ such that

$$|\alpha_{n+1} - L| \leqslant C|\alpha_n - L|^m, \qquad n \geqslant 1.$$

Then we say that the convergence is of the mth order.

Perhaps the most significant theorem in Gauss's paper [23] is the following representation for M for which we provide Gauss's ingenious proof.

THEOREM 1. *Let* $|x| < 1$, *and define*

$$K(x) = \int_0^{\pi/2} (1 - x^2\sin^2\varphi)^{-1/2} \, d\varphi. \tag{2}$$

Then

$$M(1 + x, 1 - x) = \frac{\pi}{2K(x)}.$$

The integral $K(x)$ is called the complete elliptic integral of the first kind. Observe that in the definition of $K(x)$, $\sin^2\varphi$ may be replaced by $\cos^2\varphi$.

Before proving Theorem 1, we give a reformulation of it. Define

$$I(a, b) = \int_0^{\pi/2} (a^2\cos^2\varphi + b^2\sin^2\varphi)^{-1/2} \, d\varphi. \tag{3}$$

It is easy to see that

$$I(a, b) = \frac{1}{a}K(x),$$

where

$$x = \frac{1}{a}\sqrt{a^2 - b^2}.$$

Since

$$M(a, b) = M(a_1, b_1) \quad \text{and} \quad M(ca, cb) = cM(a, b), \tag{4}$$

for any constant c, it follows that, with x as above,

$$M(1 + x, 1 - x) = \frac{1}{a}M(a, b).$$

The following reformulation of Theorem 1 is now immediate.

THEOREM 1'. *Let* $a > b > 0$. *Then*

$$M(a, b) = \frac{\pi}{2I(a, b)}.$$

Proof. Clearly, $M(1 + x, 1 - x)$ is an even function of x. Gauss then *assumes* that

$$\frac{1}{M(1 + x, 1 - x)} = \sum_{k=0}^{\infty} A_k x^{2k}. \tag{5}$$

Now make the substitution $x = 2t/(1 + t^2)$. From (4), it follows that

$$M(1 + x, 1 - x) = \frac{1}{1 + t^2} M\big((1 + t)^2, (1 - t)^2\big) = \frac{1}{1 + t^2} M(1 + t^2, 1 - t^2).$$

Substituting in (5), we find that

$$(1 + t^2) \sum_{k=0}^{\infty} A_k t^{4k} = \sum_{k=0}^{\infty} A_k \left(\frac{2t}{1 + t^2}\right)^{2k}.$$

Clearly, $A_0 = 1$. Expanding $(1 + t^2)^{-2k-1}$, $k \geqslant 0$, in a binomial series and equating coefficients of like powers of t on both sides, we eventually find that

$$\frac{1}{M(1 + x, 1 - x)} = 1 + \left(\frac{1}{2}\right)^2 x^2 + \left(\frac{1 \cdot 3}{2 \cdot 4}\right)^2 x^4 + \left(\frac{1 \cdot 3 \cdot 5}{2 \cdot 4 \cdot 6}\right)^2 x^6 + \cdots$$

$$= \sum_{k=0}^{\infty} \frac{\left(\frac{1}{2}\right)_k^2}{(k!)^2} x^{2k}. \tag{6}$$

Here we have introduced the notation

$$(\alpha)_k = \alpha(\alpha + 1)(\alpha + 2) \cdots (\alpha + k - 1). \tag{7}$$

Complete details for the derivation of (6) may be found in Gauss's paper [23, pp. 367–369].

We now must identify the series in (6) with $K(x)$. Expanding the integrand of $K(x)$ in a binomial series and integrating termwise, we find that

$$K(x) = \sum_{k=0}^{\infty} \frac{\left(\frac{1}{2}\right)_k}{k!} x^{2k} \int_0^{\pi/2} \sin^{2k}\varphi \, d\varphi$$

$$= \frac{\pi}{2} \sum_{k=0}^{\infty} \frac{\left(\frac{1}{2}\right)_k^2}{(k!)^2} x^{2k}. \tag{8}$$

Combining (6) and (8), we complete the proof of Gauss's theorem.

Another short, elegant proof of Theorem 1 has been given by Newman [39] and is sketched by J. M. and P. B. Borwein [10].

For a very readable, excellent account of Gauss's many contributions to the arithmetic-geometric mean, see Cox's paper [18]. We shall continue the discussion of some of Gauss's discoveries in Section 5.

3. Landen and the Ladies Diary. We next sketch another proof of Theorem 1 (or Theorem 1′) which is essentially due to the eighteenth-century English mathematician John Landen.

Second proof. Although the basic idea is due to Landen, the iterative procedure that we shall describe is apparently due to Legendre [33, pp. 79–83] some years later.

For brevity, set

$$x_n = c_n/a_n, \qquad n \geqslant 0, \tag{9}$$

where c_n is defined by (1). In the complete elliptic integral of the first kind (2), make the substitution

$$\tan \varphi_1 = \frac{\sin(2\varphi)}{x_1 + \cos(2\varphi)}. \tag{10}$$

This is called Landen's transformation. After a considerable amount of work, we find that

$$K(x) = (1 + x_1)K(x_1).$$

Upon n iterations, we deduce that

$$K(x) = (1 + x_1)(1 + x_2) \cdots (1 + x_n)K(x_n). \tag{11}$$

Since, by (1) and (9), $1 + x_k = a_{k-1}/a_k$, $k \geqslant 1$, we see that (11) reduces to

$$K(x) = \frac{a}{a_n}K(x_n).$$

We now let n tend to ∞. Since a_n tends to $M(a, b)$ and x_n tends to 0, we conclude that

$$K(x) = \frac{a}{M(a, b)} K(0) = \frac{a\pi}{2M(a, b)}.$$

Landen's transformation (10) was introduced by him in a paper [31] published in 1771 and in more developed form in his most famous paper [32] published in 1775. There exist several versions of Landen's transformation. Often Landen's transformation is expressed as an equality between two differentials in the theory of elliptic functions [17], [37]. The importance of Landen's transformation is conveyed by Mittag-Leffler who, in his very perceptive survey [37, p. 291] on the theory of elliptic functions, remarks, "Euler's addition theorem and the transformation theorem of Landen and Lagrange were the two fundamental ideas of which the theory of elliptic functions was in possession when this theory was brought up for renewed consideration by Legendre in 1786."

In Section 4, we shall prove the following theorem, which is often called Landen's transformation for complete elliptic integrals of the first kind.

THEOREM 2. *If* $0 \leqslant x < 1$, *then*

$$K\left(\frac{2\sqrt{x}}{1 + x}\right) = (1 + x)K(x).$$

In fact, Theorem 2 is the special case $\alpha = \pi$, $\beta = \pi/2$ of the following more general formula. If $x \sin \alpha = \sin(2\beta - \alpha)$, then

$$(1 + x)\int_0^\alpha (1 - x^2\sin^2\varphi)^{-1/2} \, d\varphi = 2\int_0^\beta \left(1 - \frac{4x}{(1 + x)^2} \sin^2\varphi\right)^{-1/2} d\varphi,$$

which is known as Landen's transformation for incomplete elliptic integrals of the first kind.

To describe another form of Landen's transformation, we introduce Gauss's ordinary hypergeometric series

$$F(a, b; c; x) = \sum_{k=0}^{\infty} \frac{(a)_k(b)_k}{(c)_k k!} x^k, \qquad |x| < 1, \tag{12}$$

where a, b, and c denote arbitrary complex numbers and $(\alpha)_k$ is defined by (7). Then

$$F\left(a, b; 2b; \frac{4x}{(1+x)^2}\right) = (1+x)^{2a} F\left(a, a-b+\frac{1}{2}; b+\frac{1}{2}; x^2\right) \tag{13}$$

is Landen's transformation for hypergeometric series. Theorems 1 and 2 imply the special case

$$F\left(\frac{1}{2}, \frac{1}{2}; 1; \frac{4x}{(1+x)^2}\right) = (1+x) F\left(\frac{1}{2}, \frac{1}{2}; 1; x^2\right).$$

Thus, a seemingly innocent "change of variable" (10) has many important ramifications. Indeed, Landen himself evidently never realized the importance of his idea.

Since Landen undoubtedly is not known to most readers, it seems appropriate here to give a brief biography. He was born in 1719. According to the Encyclopedia Britannica [20], "He lived a very retired life, and saw little or nothing of society; when he did mingle in it, his dogmatism and pugnacity caused him to be generally shunned." In 1762, he was appointed as the land-agent to the Earl Fitzwilliam, a post he held until two years before his death in 1790.

As a mathematician, Landen was primarily an analyst and geometer. Most of his important works were published in the latter part of his career. These include the aforementioned papers and *Mathematical Memoirs*, published in 1780 and 1789. For several years, Landen contributed many problems and solutions to the *Ladies Diary*. From 1743–1749, he posed a total of eleven problems and published thirteen solutions to problems. However, Leybourn [34] has disclosed that contributors to the *Ladies Diary* frequently employed aliases. In particular, Landen used the pseudonyms Sir Stately Stiff, Peter Walton, Waltoniensis, C. Bumpkin, and Peter Puzzlem, who, collectively, proposed ten problems and answered seventeen. Leybourn [34] has compiled in four volumes the problems and solutions from the *Ladies Diary* from 1704–1816. Especially valuable are his indices of subject classifications and contributors. (The problems and solutions from the years 1704–1760 had been previously collected by others in one volume in 1774 [50].)

First published in 1704, the annual *Ladies Diary* evidently was very popular in England with a yearly circulation of several thousand. The *Ladies Diary* is "designed principally for the amusement and instruction of the fair sex." It contains "new improvements in arts and sciences, and many entertaining particulars...for the use and diversion of the fair sex." The cover is graced by a poem dedicated to the reigning queen and which normally changed little from year to year. Our paper begins with the poem from 1776 paying eloquent homage to the beloved of King George III. Among other things, the *Ladies Diary* contains a "chronology of remarkable events," birth dates of the royal family, enigmas, and answers to enigmas from the previous year. The enigmas as well as the answers were normally set to verse.

The largest portion of the *Ladies Diary* is devoted to the solutions of mathematical problems posed in the previous issue. Despite the name of the journal, very few contributors were women. Leybourn's [34] index lists a total of 913 contributors of which 32 were women. Because proposers and solvers did occasionally employ pen names such as Plus Minus, Mathematicus, Amicus, Archimedes, Diophantoides, and the aforementioned aliases for Landen, it is possible that the number of female contributors is slightly higher. In 1747, Landen gave a solution to a problem which was "designed to improve gunnery of which there are several things wanting." Does not this have a familiar ring today? Geometrical problems were popular, and rigor was lax at times. Here is an example from 1783. Let

$$a = \sum_{k=0}^{\infty} \frac{1}{\sqrt{2k+1}} \quad \text{and} \quad b = \sum_{k=1}^{\infty} \frac{1}{\sqrt{2k}}.$$

Show that $a/b = \sqrt{2} - 1$. In 1784, Joseph French provided the following "elegant" solution. We see that

$$\sqrt{2} \sum_{k=1}^{\infty} \frac{1}{\sqrt{2k}} = \sum_{k=1}^{\infty} \frac{1}{\sqrt{k}} = a + b.$$

Thus, $b(\sqrt{2} - 1) = a$, and the result follows.

Those readers wishing to learn more about Landen's work should consult Watson's very delightful article, "The Marquis and the Land-agent" [52]. Readers desiring more knowledge of the mathematical content of the *Ladies Diary* should definitely consult Leybourn's compendium [34]. (Only a few libraries in the U.S. possess copies of the *Ladies Diary*. The University of Illinois Library has a fairly complete collection, although there are several gaps prior to 1774. T. Perl [43] has written a detailed description of the *Ladies Diary* with an emphasis on the contributions by women and an analysis of both the positive and negative sociological factors on womens' mathematical education during the years of the *Diary*. For additional historical information about the *Ladies Diary* and other obscure English journals containing mathematics, see Archibald's paper [2].)

4. Ivory and Landen's transformation. In 1796, J. Ivory [25] published a new formula for the perimeter of an ellipse. A very similar proof establishes Theorem 2, a version of Landen's transformation discussed in the previous section.

Before proving Theorem 2, we note that it implies a new version of Theorem 1.

THEOREM 1″. *If $x > 0$, then*

$$M(1 + x, 1 - x) = \frac{\pi(1 + x)}{2K\left(\dfrac{2\sqrt{x}}{1 + x}\right)}.$$

Theorem 1′ also follows from Theorem 1″; put $x = (a - b)/(a + b)$ and utilize (4).

Proof of Theorem 2. Using the definition (2) of K, employing the binomial series, and inverting the order of summation and integration below, we find that

$$
\begin{aligned}
K\left(\frac{2\sqrt{x}}{1+x}\right) &= \frac{1}{2}\int_0^\pi\left(1 - \frac{4x}{(1+x)^2}\sin^2\varphi\right)^{-1/2} d\varphi \\
&= \frac{1}{2}\int_0^\pi\left(1 - \frac{2x}{(1+x)^2}(1 - \cos(2\varphi))\right)^{-1/2} d\varphi \\
&= \frac{1}{2}(1+x)\int_0^\pi\left(1 + x^2 + 2x\cos(2\varphi)\right)^{-1/2} d\varphi \\
&= \frac{1}{2}(1+x)\int_0^\pi\left(1 + xe^{2i\varphi}\right)^{-1/2}\left(1 + xe^{-2i\varphi}\right)^{-1/2} d\varphi \\
&= \frac{1}{2}(1+x)\sum_{m=0}^\infty \frac{\left(\frac{1}{2}\right)_m(-x)^m}{m!}\sum_{n=0}^\infty \frac{\left(\frac{1}{2}\right)_n(-x)^n}{n!}\int_0^\pi e^{2i(m-n)\varphi}\,d\varphi \\
&= \frac{\pi}{2}(1+x)\sum_{n=0}^\infty \frac{\left(\frac{1}{2}\right)_n^2 x^{2n}}{(n!)^2} \\
&= (1+x)K(x),
\end{aligned}
$$

by (8). This concludes the proof.

Ivory's paper [25], establishing an analogue of Theorem 2, possesses an unusual feature in that it begins with the "cover letter" that Ivory sent to the editor John Playfair when he submitted his paper! In this letter, Ivory informs Playfair about what led him to his discovery. Evidently then, the editor deemed it fair play to publish Ivory's letter as a preamble to his paper. The letter reads as follows.

Dear Sir,

Having, as you know, bestowed a good deal of time and attention on the study of that part of physical astronomy which relates to the mutual disturbances of the planets, I have, naturally, been led to consider the various methods of resolving the formula $(a^2 + b^2 - 2ab\cos\varphi)^n$ into infinite series of the form $A + B\cos\varphi + C\cos 2\varphi + \&c.$ In the course of these investigations, a series for the rectification of the ellipsis occurred to me, remarkable for its simplicity, as well as its rapid convergency. As I believe it to be new, I send it to you, inclosed, together with some remarks on the evolution of the formula just mentioned, which if you think proper, you may submit to the consideration of the Royal Society.

I am, Dear Sir,

Yours, & c.

James Ivory

5. Calculation of π. First, we define the complete elliptic integral of the second kind,

$$
E(x) := \int_0^{\pi/2}(1 - x^2\sin^2\varphi)^{1/2}\,d\varphi,
$$

where $|x| < 1$. Two formulas relating the elliptic integrals $E(x)$ and $K(x)$ are the basis for one of the currently most efficient methods to calculate π. The first is due to Legendre [33, p. 61]. We give below a simple proof that appears not to have been, heretofore, given.

THEOREM 3. *Let* $x' = \sqrt{1 - x^2}$, *where* $0 < x < 1$. *Then*

$$K(x)E(x') + K(x')E(x) - K(x)K(x') = \frac{\pi}{2}. \tag{14}$$

Proof. Let $c = x^2$ and $c' = 1 - c$. A straightforward calculation gives

$$\frac{d}{dc}(E - K) = -\frac{d}{dc}\int_0^{\pi/2} \frac{c\sin^2\varphi}{\left(1 - c\sin^2\varphi\right)^{1/2}}\,d\varphi$$

$$= \frac{E}{2c} - \frac{1}{2c}\int_0^{\pi/2} \frac{d\varphi}{\left(1 - c\sin^2\varphi\right)^{3/2}}.$$

Since

$$\frac{d}{d\varphi}\left(\frac{\sin\varphi\cos\varphi}{\left(1 - c\sin^2\varphi\right)^{1/2}}\right) = \frac{1}{c}\left(1 - c\sin^2\varphi\right)^{1/2} - \frac{c'}{c}\left(1 - c\sin^2\varphi\right)^{-3/2},$$

we deduce that

$$\frac{d}{dc}(E - K) = \frac{E}{2c} - \frac{E}{2cc'} + \frac{1}{2c'}\int_0^{\pi/2} \frac{d}{d\varphi}\left(\frac{\sin\varphi\cos\varphi}{\left(1 - c\sin^2\varphi\right)^{1/2}}\right)d\varphi$$

$$= \frac{E}{2c}\left(1 - \frac{1}{c'}\right) = -\frac{E}{2c'}. \tag{15}$$

For brevity, put $K' = K(c')$ and $E' = E(c')$. Since $c' = 1 - c$, it follows that

$$\frac{d}{dc}(E' - K') = \frac{E'}{2c}. \tag{16}$$

Lastly, easy calculations yield

$$\frac{dE}{dc} = \frac{E - K}{2c} \quad \text{and} \quad \frac{dE'}{dc} = -\frac{E' - K'}{2c'}. \tag{17}$$

If L denotes the left side of (14), we may write L in the form

$$L = EE' - (E - K)(E' - K').$$

Employing (15)–(17), we find that

$$\frac{dL}{dc} = \frac{(E - K)E'}{2c} - \frac{E(E' - K')}{2c'} + \frac{E(E' - K')}{2c'} - \frac{(E - K)E'}{2c} = 0.$$

Hence, L is a constant, and we will find its value by letting c approach 0.
  First,

$$E - K = -c\int_0^{\pi/2} \frac{\sin^2\varphi}{\left(1 - c\sin^2\varphi\right)^{1/2}}\,d\varphi = O(c)$$

as c tends to 0. Next,

$$K' = \int_0^{\pi/2} (1 - c'\sin^2\varphi)^{-1/2}\, d\varphi \leqslant \int_0^{\pi/2} (1 - c')^{-1/2}\, d\varphi$$
$$= O(c^{-1/2}),$$

as c tends to 0. Thus,

$$\lim_{c\to 0} L = \lim_{c\to 0} \{(E - K)K' + E'K\}$$
$$= \lim_{c\to 0} \left\{ O(c^{1/2}) + 1 \cdot \frac{\pi}{2} \right\} = \frac{\pi}{2},$$

and the proof is complete.

The second key formula, given in Theorem 4 below, can be proved via an iterative process involving Landen's transformation. We forego a proof here; a proof may be found, for example, in King's book [**29**, pp. 7, 8].

THEOREM 4. *Let, for $a > b > 0$,*

$$J(a, b) = \int_0^{\pi/2} (a^2\cos^2\varphi + b^2\sin^2\varphi)^{1/2}\, d\varphi, \tag{18}$$

and recall that c_n is defined by (1). Then

$$J(a, b) = \left(a^2 - \frac{1}{2} \sum_{n=0}^{\infty} 2^n c_n^2 \right) I(a, b),$$

where $I(a, b)$ is defined by (3).

Note that

$$J(a, b) = aE(x),$$

where $x = (1/a)\sqrt{a^2 - b^2}$.

Theorems 3 and 4 now lead to a formula for π which is highly suitable for computation.

THEOREM 5. *If c_n is defined by (1), then*

$$\pi = \frac{4M^2(1, 1/\sqrt{2})}{1 - \sum_{n=1}^{\infty} 2^{n+1} c_n^2}.$$

Proof. Letting $x = x' = 1/\sqrt{2}$ in Theorem 3, we find that

$$2K\left(\frac{1}{\sqrt{2}}\right) E\left(\frac{1}{\sqrt{2}}\right) - K^2\left(\frac{1}{\sqrt{2}}\right) = \frac{\pi}{2}. \tag{19}$$

Setting $a = 1$ and $b = 1/\sqrt{2}$ in Theorem 4, we see that

$$E\left(\frac{1}{\sqrt{2}}\right) = \left(1 - \frac{1}{2} \sum_{n=0}^{\infty} 2^n c_n^2 \right) K\left(\frac{1}{\sqrt{2}}\right), \tag{20}$$

since $I(1, \sqrt{2}) = K(1/\sqrt{2})$ and $J(1, 1/\sqrt{2}) = E(1/\sqrt{2})$. Lastly, by Theorem 1',

$$M(1, 1/\sqrt{2}) = \frac{\pi}{2K(1/\sqrt{2})}. \tag{21}$$

Substituting (20) into (19), employing (21), noting that $c_0^2 = 1/2$, and solving for π, we complete the proof.

According to King [29, pp. 8, 9, 12], an equivalent form of Theorem 5 was established by Gauss. Observe that in the proof of Theorem 5, we used only the special case $x = x' = 1/\sqrt{2}$ of Legendre's identity, Theorem 3. We would like to show now that this special case is equivalent to the formula

$$\int_0^1 \frac{dx}{\sqrt{1 - x^4}} \int_0^1 \frac{x^2 \, dx}{\sqrt{1 - x^4}} = \frac{\pi}{4}, \tag{22}$$

first proved by Euler [22] in 1782. (Watson [52, p. 12] claimed that an equivalent formulation of (22) was earlier established by both Landen and Wallis, but we have been unable to verify this.) The former integral in (22) is one quarter of the arc length of the lemniscate given by $r^2 = \cos(2\varphi)$, $0 \leqslant \varphi \leqslant 2\pi$. The latter integral in (22) is intimately connected with the classical elastic curve. For a further elaboration of the connections of these two curves with the arithmetic-geometric mean, see Cox's paper [18].

In order to prove (22), make the substitution $x = \cos \varphi$. Then straightforward calculations yield

$$K\left(\frac{1}{\sqrt{2}}\right) = \sqrt{2} \int_0^1 \frac{dx}{\sqrt{1 - x^4}}$$

and

$$2E\left(\frac{1}{\sqrt{2}}\right) - K\left(\frac{1}{\sqrt{2}}\right) = \sqrt{2} \int_0^1 \frac{x^2 \, dx}{\sqrt{1 - x^4}}.$$

It is now easy to see that Legendre's relation Theorem 3 in the case $x = x' = 1/\sqrt{2}$ implies (22).

In 1976, Salamin [47] rederived the forgotten Theorem 5, from which he established a rapidly convergent algorithm for the computation of π. Recall that in Section 2 we demonstrated how rapidly the arithmetic-geometric mean converges and thus how fast c_n tends to 0. Tamura and Kanada have used this algorithm to compute π to 2^{24} (over 16 million) decimal places. An announcement about their calculation of π to 2^{23} decimal places was made in *Scientific American* [48]. Their paper [27] describes their calculation to 10,013,395 decimal places. More recently, D. H. Bailey [3] has used a quartically convergent algorithm to calculate π to 29,360,000 digits.

Newman [40] has obtained a quadratic algorithm for the computation of π that is somewhat simpler than Salamin's. His proof is quite elementary and avoids Legendre's identity. It should be remarked that Newman's estimates of some integrals are not quite correct. However, the final result (middle of p. 209) is correct.

In 1977, Brent [14] observed that the arithmetic-geometric mean could be implemented to calculate elementary functions as well. Let us briefly indicate how to calculate $\log 2$. From Whittaker and Watson's text [53, p. 522], as x tends to $1-$,

$$K(x) \sim \log \frac{4}{\sqrt{1-x^2}}.$$

From Theorem 1',

$$K(x) = \frac{\pi}{2M\left(1, \sqrt{1-x^2}\right)},$$

and so

$$\log \frac{4}{\sqrt{1-x^2}} \approx \frac{\pi}{2M\left(1, \sqrt{1-x^2}\right)}.$$

Taking $\sqrt{1-x^2} = 4 \cdot 2^{-n}$, we find that

$$\log 2 \approx \frac{\pi}{2nM\left(1, 2^{2-n}\right)},$$

for large n.

Further improvements in both the calculation of π and elementary functions have been made by J. M. and P. B. Borwein [8], [9], [10], [11], [12], [13]. In particular, in [8], [11], and [12], they have utilized elliptic integrals and modular equations to obtain algorithms of higher order convergence to approximate π. The survey article [10] by the Borwein brothers is to be especially recommended. Carlson [16] has written an earlier survey on algorithms dependent on the arithmetic-geometric mean and variants thereof.

Postscript to π. The challenge of approximating and calculating π has been with us for over 4000 years. By 1844, π was known to 200 decimal places. This stupendous feat was accomplished by a calculating prodigy named Johann Dase in less than two months. On Gauss's recommendation, the Hamburg Academy of Sciences hired Dase to compute the factors of all integers between 7,000,000 and 10,000,000. Thus, our ideas have come to a full circle. As Beckmann [5, p. 104] remarks, "It would thus appear that Carl Friedrich Gauss, who holds so many firsts in all branches of mathematics, was also the first to introduce payment for computer time." The computer time now for 29 million digits (28 hours) is considerably less than the computer time for 200 digits by Gauss's computer, Dase.

6. Approximations for the perimeter L of an ellipse. If an ellipse is given by the parametric equations $x = a \cos \varphi$ and $y = b \sin \varphi$, $0 \leqslant \varphi \leqslant 2\pi$, then from elementary calculus,

$$\begin{aligned} L = L(a, b) &= \int_0^{2\pi} \left(a^2\sin^2\varphi + b^2\cos^2\varphi\right)^{1/2} d\varphi \\ &= 4J(b, a), \end{aligned} \tag{23}$$

where $J(b, a)$ is defined by (18). Thus, we see immediately from Theorems 1' and 4 that elliptical perimeters and arithmetic-geometric means are inextricably intertwined. Ivory's letter and our concomitant comments also unmistakenly pointed to this union.

Before discussing approximations for L, we offer two exact formulas. The former is due to MacLaurin [36] in 1742, and the latter was initially found by Ivory [25] in 1796, although it is implicit in the earlier work of Landen.

THEOREM 6. *Let* $x = a \cos \varphi$ *and* $y = b \sin \varphi$, $0 \leqslant \varphi \leqslant 2\pi$. *Let* $e = (1/a)\sqrt{a^2 - b^2}$, *the eccentricity of the ellipse. Then if* F *is defined by* (12),

$$L(a, b) = 2\pi a F\left(\frac{1}{2}, -\frac{1}{2}; 1; e^2\right) \tag{24}$$

$$= \pi(a + b) F\left(-\frac{1}{2}, -\frac{1}{2}; 1; \lambda^2\right), \tag{25}$$

where

$$\lambda = \frac{a - b}{a + b}.$$

Proof. The proofs are very similar to those in Sections 2 and 4. First, using (23), expanding the integrand in a binomial series, and integrating termwise, we deduce that

$$L(a, b) = 4a \int_0^{\pi/2} (1 - e^2 \cos^2 \varphi)^{1/2} \, d\varphi$$

$$= 4a \sum_{n=0}^{\infty} \frac{\left(-\frac{1}{2}\right)_n}{n!} e^{2n} \int_0^{\pi/2} \cos^{2n} \varphi \, d\varphi \tag{26}$$

$$= 2\pi a F\left(\frac{1}{2}, -\frac{1}{2}; 1; e^2\right).$$

Thus, (24) is established.

We indicate two proofs of (25). First, in Landen's transformation (13) of hypergeometric series, set $a = -1/2$, $b = 1/2$, and $x = \lambda$. We immediately find that

$$F\left(-\frac{1}{2}, \frac{1}{2}; 1; e^2\right) = \frac{a + b}{2a} F\left(-\frac{1}{2}, -\frac{1}{2}; 1; \lambda^2\right).$$

By this formula and (24), formula (25) is demonstrated.

The second proof that we mention is that of Ivory [25]. Using (26), proceed exactly in the same fashion as in the proof of Theorem 2 in Section 4.

In fact, there exists a third early formula for $L(a, b)$. In 1773, Euler [21] proved that

$$L(a, b) = \pi \sqrt{2(a^2 + b^2)} \, F\left(-\frac{1}{4}, \frac{1}{4}; 1; \left(\frac{a^2 - b^2}{a^2 + b^2}\right)^2\right).$$

Although Euler proceeded differently, we mention that his formula may be derived from MacLaurin's via a certain quadratic transformation for hypergeometric series that is different from Landen's. Euler's formula also trivially leads to an approximation for $L(a, b)$ given in our table below.

The problem of determining $L(a, b)$ is not as venerable as that for determining π. However, some have argued (not very convincingly) that the problem goes back to the time of King Solomon, who hired a craftsman Huram to make a tank. According to 1 Kings 7 : 23, "Huram made a round tank of bronze 5 cubits deep, 10 cubits in diameter, and 30 cubits in circumference." The implication is clear that the ancient Hebrews regarded π as being equal to 3. It has been suggested, perhaps by someone who believes that "God makes no mistakes," that "round" and "depth" are to be interpreted loosely, and that the tank really was elliptical in shape, with the major axis being 10 cubits and the minor axis being about 9.53 cubits in length.

As might be expected. the primary impetus in finding methods for calculating elliptical perimeters arises from astronomy. In 1609, Kepler [28] offered perhaps the first legitimate approximations

$$L \approx \pi(a + b) \quad \text{and} \quad L \approx 2\pi\sqrt{ab},$$

although, as we shall see, his arguments were not very rigorous and $2\pi\sqrt{ab}$ was intended to be only a *lower bound* for L. Kepler [28, p. 307] first remarks that the ellipse with semiaxes a and b and the circle with radius $\sqrt{ab}$ have the same areas. Since the circle has the smaller circumference,

$$L \geqslant 2\pi\sqrt{ab}.$$

He [28, p. 368] furthermore remarks that $(1/2)(a + b) \geqslant \sqrt{ab}$, and so concludes that

$$L \approx 2\pi\frac{1}{2}(a + b).$$

Kepler appears to be using the dubious principle that quantities larger than the same number must be about equal.

Approximations of several types, depending upon the relative sizes of a and b, exist in the literature. In this section, we concentrate on estimates that are best for a close to b. Thus, we shall write all of our approximations in terms of $\lambda = (a - b)/(a + b)$ and compare them with the expansion (25). For example, Kepler's second approximation can be written in the form

$$L \approx \pi(a + b)(1 - \lambda^2)^{1/2}.$$

We now show how the formula

$$L(a, b) = 4J(a, b) = \frac{2\pi}{M(a, b)}\left(a^2 - \frac{1}{2}\sum_{n=0}^{\infty} 2^n c_n^2\right), \tag{27}$$

arising from Theorems 1′ and 4, can be used to find approximations to the perimeter of an ellipse. Replacing $M(a, b)$ by a_2 and neglecting the terms with $n \geqslant 2$, we find that

$$L(a, b) \approx \frac{2\pi}{a_2}\left(a^2 - \frac{c_0^2}{2} - c_1^2\right) = \frac{2\pi a_1^2}{a_2} = 2\pi\left(\frac{a + b}{\sqrt{a} + \sqrt{b}}\right)^2.$$

This formula was first obtained by Ekwall [19] in 1973 as a consequence of a formula by Sipos from 1792 [54].

If we replace $M(a, b)$ by a_3 in (27) and neglect all terms with $n \geqslant 3$, we find, after some calculation, that

$$L(a, b) \approx 2\pi \frac{2(a + b)^2 - (\sqrt{a} - \sqrt{b})^4}{(\sqrt{a} + \sqrt{b})^2 + 2\sqrt{2}\sqrt{a + b}\sqrt[4]{ab}}.$$

This formula is complicated enough to dissuade us from calculating further approximations by this method.

We now provide a table of approximations for $L(a, b)$ that have been given in the literature. At the left, we list the discoverer (or source) and year of discovery (if known). The approximation $A(\lambda)$ for $L(a, b)/\pi(a + b)$ is given in the second column in two forms. In the last column, the first nonzero term in the power series for

$$A(\lambda) - \frac{L(a, b)}{\pi(a + b)} = A(\lambda) - F\left(-\frac{1}{2}, -\frac{1}{2}; 1; \lambda^2\right)$$

is offered so that the accuracy of the approximating formula can be discerned. For convenience, we note that

$$F\left(-\frac{1}{2}, -\frac{1}{2}; 1; \lambda^2\right) = 1 + \frac{1}{4}\lambda^2 + \frac{1}{4^3}\lambda^4 + \frac{1}{4^4}\lambda^6 + \frac{25}{4^7}\lambda^8 + \frac{49}{4^8}\lambda^{10} + \cdots.$$

Kepler [28], 1609	$\dfrac{2\sqrt{ab}}{a + b} = (1 - \lambda^2)^{1/2}$	$-\dfrac{3}{4}\lambda^2$
Euler [21], 1773	$\dfrac{\sqrt{2(a^2 + b^2)}}{a + b} = (1 + \lambda^2)^{1/2}$	$\dfrac{1}{4}\lambda^2$
Sipos [54], 1792 Ekwall [19], 1973	$\dfrac{2(a + b)}{(\sqrt{a} + \sqrt{b})^2} = \dfrac{2}{1 + \sqrt{1 - \lambda^2}}$	$\dfrac{7}{64}\lambda^4$
Peano [42], 1889	$\dfrac{3}{2} - \dfrac{\sqrt{ab}}{a + b} = \dfrac{3}{2} - \dfrac{1}{2}(1 - \lambda^2)^{1/2}$	$\dfrac{3}{64}\lambda^4$
Muir [38], 1883	$\dfrac{2}{a + b}\left(\dfrac{a^{3/2} + b^{3/2}}{2}\right)^{2/3}$ $= \dfrac{1}{2^{2/3}}\{(1 + \lambda)^{3/2} + (1 - \lambda)^{3/2}\}^{2/3}$	$-\dfrac{1}{64}\lambda^4$
Lindner [35, p. 439], 1904–1920 Nyvoll [41], 1978	$\left\{1 + \dfrac{1}{8}\left(\dfrac{a - b}{a + b}\right)^2\right\}^2$ $= \left(1 + \dfrac{1}{8}\lambda^2\right)^2$	$-\dfrac{1}{2^8}\lambda^6$
Selmer [49], 1975	$1 + \dfrac{4(a - b)^2}{(5a + 3b)(3a + 5b)}$ $= 1 + \dfrac{1}{4}\lambda^2\dfrac{1}{1 - \dfrac{1}{16}\lambda^2}$	$-\dfrac{3}{2^{10}}\lambda^6$

Ramanujan [44], [45], 1914 Fergestad [49], 1951	$\dfrac{3 - \sqrt{(a+3b)(3a+b)}}{a+b}$ $= 3 - \sqrt{4 - \lambda^2}$	$-\dfrac{1}{2^9}\lambda^6$
Almkvist [1], 1978	$2\dfrac{2(a+b)^2 - (\sqrt{a}-\sqrt{b})^4}{(a+b)\left\{(\sqrt{a}+\sqrt{b})^2 + 2\sqrt{2}\sqrt{a+b}\sqrt[4]{ab}\right\}}$ $= 2\dfrac{\left(1+\sqrt{1-\lambda^2}\right)^2 + \lambda^2\sqrt{1-\lambda^2}}{\left(1+\sqrt{1-\lambda^2}\right)\left(1+\sqrt[4]{1-\lambda^2}\right)^2}$	$\dfrac{15}{2^{14}}\lambda^8$
Bronshtein and Semendyayev [15], 1964 Selmer [49], 1975	$\dfrac{1}{16}\dfrac{64(a+b)^4 - 3(a-b)^4}{(a+b)^2(3a+b)(a+3b)}$ $= \dfrac{64 - 3\lambda^4}{64 - 16\lambda^2}$	$-\dfrac{9}{2^{14}}\lambda^8$
Selmer [49], 1975	$\dfrac{1}{8}\left\{12 + \left(\dfrac{a-b}{a+b}\right)^2 - \dfrac{2\sqrt{2(a^2+6ab+b^2)}}{a+b}\right\}$ $= \dfrac{3}{2} + \dfrac{1}{8}\lambda^2 - \dfrac{1}{2}\sqrt{1 - \dfrac{1}{2}\lambda^2}$	$-\dfrac{5}{2^{14}}\lambda^8$
Jacobsen and Waadeland [26], 1985	$\dfrac{256 - 48\lambda^2 - 21\lambda^4}{256 - 112\lambda^2 + 3\lambda^4}$	$-\dfrac{33}{2^{18}}\lambda^{10}$
Ramanujan [44], [45], 1914	$1 + \dfrac{3\lambda^2}{10 + \sqrt{4 - 3\lambda^2}}$	$-\dfrac{3}{2^{17}}\lambda^{10}$

The two approximations by India's great mathematician, S. Ramanujan, were first stated by him in his notebooks [46, p. 217], and then later at the end of his paper [44], [45, p. 39], where he says that they were discovered empirically. Ramanujan [44], [45] also provides error approximations, but they are in a form different from that given here. Since

$$\lambda = \frac{a-b}{a+b} = \frac{1 - \sqrt{1-e^2}}{1 + \sqrt{1-e^2}} \approx \frac{e^2}{4},$$

we find that, for the first approximation,

$$\pi(a+b)\frac{\lambda^6}{2^9} \approx \pi a\left(1 + \sqrt{1-e^2}\right)\frac{(e^2/4)^6}{2^9} < 2\pi a\frac{e^{12}}{2^{21}} = \pi a\frac{e^{12}}{2^{20}},$$

which is the approximate error given by Ramanujan. Similarly, for the second approximation, Ramanujan states that the error is approximately equal to

$$3\pi a\frac{e^{20}}{2^{36}},$$

which is in agreement with our claim. The exactness of Ramanujan's second formula for eccentricities that are not too large is very good. For example, for the orbit of

Mercury ($e = 0.206$), the absolute error is about 1.5×10^{-13} meters. Note that if we set $b = 0$ in Ramanujan's second formula, we find that $\pi \approx 22/7$.

Fergestad [49] rediscovered Ramanujan's first formula several years later.

Despite Ramanujan's remark on the discovery of these two formulas, Jacobsen and Waadeland [26] have offered a very plausible explanation of Ramanujan's approximations. We confine our attention to the latter approximation, since the arguments are similar. Write

$$F\left(-\frac{1}{2}, -\frac{1}{2}; 1; \lambda^2\right) = 1 + \frac{\lambda^2}{4(1 + w)}. \tag{28}$$

Then it can be shown that w has the continued fraction expansion

$$w = \frac{1}{3}\left\{\frac{-\dfrac{3}{16}\lambda^2}{1} + \frac{-\dfrac{3}{16}\lambda^2}{1} + \frac{-\dfrac{3}{16}\lambda^2}{1} + \frac{-\dfrac{11}{48}\lambda^2}{1} + \cdots\right\}.$$

If each numerator above is replaced by $-3\lambda^2/16$, then we obtain the approximation

$$w \approx \frac{1}{12}\left(-2 + \sqrt{4 - 3\lambda^2}\right).$$

Substituting this approximation in (28) and then using (25), we are immediately led to the estimate

$$\frac{L(a, b)}{\pi(a + b)} \approx 1 + \frac{3\lambda^2}{10 + \sqrt{4 - 3\lambda^2}}.$$

Since Ramanujan's facility in representing analytic functions by continued fractions is unmatched in mathematical history, it seems likely that Ramanujan discovered his approximations in this manner.

In the next section, we examine some approximations for $L(a, b)$ of a different type given by Ramanujan in his notebooks [46].

7. Further approximations given by Ramanujan. In his notebooks [46], Ramanujan offers some very unusual formulas, expressed in sexagesimal notation, for $L(a, b)$. The first is related to his approximation $3 - \sqrt{4 - \lambda^2}$ given in Section 6.

THEOREM 7. *Put*

$$L(a, b) = \pi(a + b)\left(1 + 4\sin^2\frac{1}{2}\theta\right), \quad 0 \leqslant \theta \leqslant \pi/4, \tag{29}$$

and

$$\sin\theta = \lambda\sin\alpha, \qquad \lambda = \frac{a - b}{a + b}. \tag{30}$$

Then, when the eccentricity $e = 1$, $\alpha = 30°18'6''$, and as e tends to 0, α tends monotonically to 30°.

It is not clear how Ramanujan was led to this very unusual theorem. The variance of α over such a small interval is curious.

Proof. We shall prove Theorem 7 except for the conclusion about monotonicity. However, we shall show that $\alpha \geqslant \pi/6$ always.

For brevity, we write (25) in the form

$$\frac{L(a, b)}{\pi(a + b)} = \sum_{n=0}^{\infty} \alpha_n \lambda^{2n}, \qquad |\lambda| < 1. \tag{31}$$

It then follows from (29) and (30) that

$$3 - 2\sqrt{1 - \lambda^2 \sin^2 \alpha} = 1 + 4 \sin^2 \frac{1}{2}\theta = \sum_{n=0}^{\infty} \alpha_n \lambda^{2n}, \qquad |\lambda| < 1. \tag{32}$$

Next set

$$3 - \sqrt{4 - \lambda^2} = \sum_{n=0}^{\infty} \beta_n \lambda^{2n}, \qquad |\lambda| < 2. \tag{33}$$

As implied in Section 6, $\alpha_n = \beta_n$, $n = 0, 1, 2$. We shall further show that, for $n \geq 3$,

$$\beta_n \leq \alpha_n / 2^{n-2}. \tag{34}$$

From the definitions (31) and (33), respectively, short calculations show that, for $n \geq 1$,

$$\frac{\alpha_{n+1}}{\alpha_n} = \frac{(2n - 1)^2}{(2n + 2)^2} \quad \text{and} \quad \frac{\beta_{n+1}}{\beta_n} = \frac{2n - 1}{8(n + 1)}.$$

Thus,

$$\frac{\beta_{n+1}}{\beta_n} \bigg/ \frac{\alpha_{n+1}}{\alpha_n} = \frac{n + 1}{2(2n - 1)} \leq \frac{1}{2},$$

if $n \geq 2$. Proceeding by induction, we deduce that

$$\frac{\beta_{n+1}}{\alpha_{n+1}} \leq \frac{1}{2} \frac{\beta_n}{\alpha_n} \leq \frac{1}{2^{n-1}},$$

for $n \geq 2$, and the proof of (34) is complete.

From (32) and (34), it follows that

$$3 - \sqrt{4 - \lambda^2} \leq 3 - 2\sqrt{1 - \lambda^2 \sin^2 \alpha}.$$

Solving this inequality, we find that $\sin^2 \alpha \geq 1/4$, or $\alpha \geq \pi/6$.

Second, we calculate α when $e = 1$. Thus, $\lambda = 1$ and $\theta = \alpha$. Therefore, from (25) and (32),

$$1 + 4 \sin^2 \frac{1}{2}\alpha = F\left(-\frac{1}{2}, -\frac{1}{2}; 1; 1\right) = \frac{4}{\pi}. \tag{35}$$

This evaluation follows from a general theorem of Gauss on the evaluation of hypergeometric series at the argument 1 [4, p. 2]. Moreover, this particular series is found in Gauss's diary under the date June, 1798 [24]. Thus,

$$\sin^2 \frac{1}{2}\alpha = \frac{1}{\pi} - \frac{1}{4} = 0.0683098861.$$

It follows that $\alpha = 30°18'6''$.

Third, we calculate α when $e = 0$. From (30) and (32),

$$\lim_{\lambda \to 0} \sin^2\alpha = \lim_{\lambda \to 0} \frac{\sin^2\theta}{\lambda^2} = \lim_{\lambda \to 0} \frac{4 \sin^2 \frac{1}{2}\theta}{\lambda^2}$$

$$= \lim_{\lambda \to 0} \lambda^{-2} \sum_{n=1}^{\infty} \alpha_n \lambda^{2n} = \alpha_1 = \frac{1}{4}.$$

Thus, α tends to $\pi/6$ as e tends to 0.

Ramanujan [**46**, p. 224] offers another theorem, which we do not state, like Theorem 7 but which appears to be motivated by his second approximation for $L(a, b)$.

Ramanujan [**46**, p. 224] states two additional formulas each of which combines two approximations, one for e near 0 and the other for e close to 1. Again, we give just one of the pair. A complete proof of Theorem 8 below would be too lengthy for this paper, and so we shall just sketch the main ideas of the proof. Complete details may be found in [7].

THEOREM 8. *Set*

$$L(a, b) = \pi(a + b)\frac{\tan \theta}{\theta}, \qquad 0 \leqslant \theta < \pi/2, \tag{36}$$

and

$$\tan \theta = \lambda \cos \alpha, \qquad \lambda = \frac{a - b}{a + b}. \tag{37}$$

Then as e increases from 0 to 1, α decreases from $\pi/6$ to 0. Furthermore, α is approximately given by

$$\frac{2\sqrt{ab}}{a + b}\left\{30° + 6°18'49''\frac{(\sqrt{a} - \sqrt{b})^2}{a + b} - 1°10'55''\left(\frac{a - b}{a + b}\right)^2\right\}. \tag{38}$$

Proof. If $e = 0$, then $\lambda = 0$ and $\theta = 0$. The argument is very similar to that in the proof of Theorem 7, and we find that

$$\lim_{\lambda \to 0} \cos^2\alpha = 3\alpha_1 = 3/4.$$

Thus, $\alpha = \pi/6$ when $\lambda = 0 = e$.

We next determine α when $e = 1$. Thus, $\lambda = 1$ and $\tan \theta = \cos \alpha$ by (37). From (25), (36), and (35),

$$\frac{\tan \theta}{\theta}\bigg|_{\lambda=1} = F\left(-\frac{1}{2}, -\frac{1}{2}; 1; 1\right) = \frac{4}{\pi}.$$

Thus, $\theta = \pi/4$ and $\alpha = 0$.

It appears to be extremely difficult to show that as λ goes from 0 to 1, α *monotonically* decreases from $\pi/6$ to 0. It can be shown [7], however, that $0 \leqslant \alpha \leqslant \pi/6$, always. A proof depends upon a continued fraction for $\tan^{-1}x$.

The proof of (38) is very difficult, and we provide only a brief sketch. We observe (again) that

$$\sqrt{1 - \lambda^2} = \frac{2\sqrt{ab}}{a + b},$$

and so

$$\sqrt{1 - \lambda^2} - (1 - \lambda^2) = \frac{2\sqrt{ab}}{(a + b)^2}(\sqrt{a} - \sqrt{b})^2.$$

Thus, Ramanujan is attempting to find an approximation to α of the form

$$\sqrt{1 - \lambda^2}\left(A + B\{1 - \sqrt{1 - \lambda^2}\} + C\lambda^2\right), \tag{39}$$

which will be a good approximation both when λ is close to 0 and when λ is near 1. Our task is then to determine A, B, and C.

With a considerable amount of effort, it can be shown that [7]

$$\alpha = \frac{\pi}{6} - \frac{21\sqrt{3}}{160}\lambda^2 + O(\lambda^4) \tag{40}$$

in a neighborhood of $\lambda = 0$. The proper expansion near $\lambda = 1$ is even more difficult to obtain because $F(-\frac{1}{2}, -\frac{1}{2}; 1; \lambda^2)$ is not analytic at $\lambda = 1$. However, there does exist an asymptotic expansion for $F(-\frac{1}{2}, -\frac{1}{2}; 1; \lambda^2)$ as λ tends to $1 -$, and employing this, we can show that [7]

$$\alpha = \sqrt{\frac{4 - \pi}{2\pi - 4}} \sqrt{1 - \lambda^2} + o(\sqrt{1 - \lambda^2}), \tag{41}$$

as λ tends to $1 -$.

Having omitted the hard analysis, we now determine A, B, and C from (40) and (41) with little difficulty. When λ tends to 0, (39) tends to A. Thus, $A = \pi/6$, by (40). Next, examine $(\alpha - \pi/6)/\lambda^2$ as λ tends to 0. From (39) and (40), we find that

$$-\frac{\pi}{12} + \frac{1}{2}B + C = -\frac{21\sqrt{3}}{160}.$$

Now check $\alpha/\sqrt{1 - \lambda^2}$ as λ tends to $1 -$. From (39) and (41), we see that

$$\frac{\pi}{6} + B + C = \sqrt{\frac{4 - \pi}{2\pi - 4}}.$$

Simultaneously solving these last two equalities, we conclude that

$$B = 2\sqrt{\frac{4 - \pi}{2\pi - 4}} + \frac{21\sqrt{3}}{80} - \frac{\pi}{2} = 0.1101935$$

and

$$C = \frac{\pi}{3} - \sqrt{\frac{4 - \pi}{2\pi - 4}} - \frac{21\sqrt{3}}{80} = -0.0206291.$$

Converting A, B, and C to the sexagesimal system and substituting in (39), we complete the proof.

Although Ramanujan is well known for his approximations and asymptotic formulas in number theory, he has not been adequately recognized for his deep contributions to approximations and asymptotic series in analysis, because the vast majority of his results in the latter field have been hidden in his notebooks. These notebooks were begun in about 1903, when he was 15 or 16, and are a compilation of his mathematical discoveries without proofs. The last entries were made in 1914, when he sailed to England at the urging of G. H. Hardy. Although the editing of Ramanujan's notebooks was strongly advocated by Hardy and others immediately after Ramanujan's death in 1920, it is only recently that this has come to fruition [6].

We have not attempted to give complete proofs of some of the theorems that we have described, but we hope that the principal ideas have been made clear. We have seen that a chain of related ideas stretches back over a period exceeding two centuries and provides impetus to contemporary mathematics. Ideas and topics that appear disparate are found to have common roots and merge together. For further elaboration of these ideas, readers should consult the works cited, especially Cox's paper [18], the papers and book of J. M. and P. B. Borwein [8]–[13], a paper by Almkvist [1] written in Swedish, and Berndt's forthcoming book [7].

We are most grateful to Birger Ekwall for providing some very useful references.

REFERENCES

1. G. Almkvist, Aritmetisk-geometriska medelvärdet och ellipsens båglängd, *Nordisk Mat. Tidskr.*, 25–26 (1978) 121–130.
2. R. C. Archibald, Notes on some minor English mathematical serials, *Math. Gaz.*, 14 (1929) 379–400.
3. D. H. Bailey, The computation of π to 29,360,000 decimal digits using Borweins' quartically convergent algorithm, to appear.
4. W. N. Bailey, Generalized Hypergeometric Series, Stechert-Hafner, New York, 1964.
5. P. Beckmann, A history of π, Golem Press, Boulder, 1970.
6. B. C. Berndt, Ramanujan's Notebooks, Part I, Springer-Verlag, New York, 1985.
7. B. C. Berndt, Ramanujan's Notebooks, Part III, Springer-Verlag, New York, to appear.
8. J. Borwein, Some modular identities of Ramanujan useful in approximating π, *Proc. Amer. Math. Soc.*, 95 (1985) 365–371.
9. J. M. and P. B. Borwein, A very rapidly convergent product expansion for π, *BIT*, 23 (1983) 538–540.
10. J. M. and P. B. Borwein, The arithmetic-geometric mean and fast computation of elementary functions, *SIAM Review*, 26 (1984) 351–366.
11. J. M. and P. B. Borwein, Cubic and higher order algorithms for π, *Canad. Math. Bull.*, 27 (1984) 436–443.
12. J. M. and P. B. Borwein, Elliptic integrals and approximations to π, to appear.
13. J. M. and P. B. Borwein, Pi and the AGM—A Study in Analytic Number Theory and Computational Complexity, John Wiley, New York, 1987.
14. R. P. Brent, Fast multiple-precision evaluation of elementary functions, *J. Assoc. Comput. Mach.*, 23 (1976) 242–251.
15. I. N. Bronshtein and K. A. Semendyayev, A Guide-Book to Mathematics for Technologists and Engineers, trans. by J. Jaworowski and M. N. Bleicher, Pergamon Press, Macmillan, New York, 1964.
16. B. C. Carlson, Algorithms involving arithmetic and geometric means, *Amer. Math. Monthly*, 78 (1971) 496–505.
17. A. Cayley, An Elementary Treatise on Elliptic Functions, second ed., Dover, New York, 1961.

18. D. A. Cox, The arithmetic-geometric mean of Gauss, *L'Enseign. Math.*, 30 (1984) 275–330.
19. B. Ekwall, Approximationsformler för ellipsens omkrets, unpublished manuscript, 1973.
20. Encyclopedia Britannica, ninth ed., vol. 14, John Landen, Adam and Charles Black, Edinburgh, 1882, p. 271.
21. L. Euler, Nova series infinita maxime convergens perimetrum ellipsis exprimens, *Novi Comm. Acad. Sci. Petropolitanae* 18 (1773), 71–84; Opera Omnia, t.20, B. G. Teubner, Leipzig, 1912, pp. 357–370.
22. L. Euler, De miris proprietatibus curvae elasticae sub aequatione $y = \int xx\,dx / \sqrt{1 - x^4}$ contentae, *Acta Acad. Sci. Petropolitanae* 1782: II (1786), 34–61; Opera Omnia, t.21, B. G. Teubner, Leipzig, 1913, pp. 91–118.
23. C. F. Gauss, Nachlass. Arithmetisch geometrisches Mittel, Werke, Bd. 3, Königlichen Gesell. Wiss., Göttingen, 1876, pp. 361–403.
24. J. J. Gray, A commentary on Gauss's mathematical diary, 1796–1814, with an English translation, *Expos. Math.*, 2 (1984) 97–130.
25. J. Ivory, A new series for the rectification of the ellipsis; together with some observations on the evolution of the formula $(a^2 + b^2 - 2ab \cos \varphi)^n$, *Trans. Royal Soc. Edinburgh*, 4 (1796) 177–190.
26. L. Jacobsen and H. Waadeland, Glimt fra analytisk teori for kjedebrøker, Del II, *Nordisk Mat. Tidskr.*, 33 (1985) 168–175.
27. Y. Kanada, Y. Tamura, S. Yoshino and Y. Ushiro, Calculation of π to 10,013,395 decimal places based on the Gauss-Legendre algorithm and Gauss arctangent relation, *Math. Comp.*, to appear.
28. J. Kepler, Opera Omnia, vol. 3, Astronomia Nova, Heyder & Zimmer, Frankfurt, 1860.
29. L. V. King, On the Direct Numerical Calculation of Elliptic Functions and Integrals, Cambridge University Press, Cambridge, 1924.
30. J.-L. Lagrange, Sur une novelle méthode de calcul intégral pour les différentielles affectées d'un radical carré sous lequel la variable ne passe pas le quatrième degré, *Mem. l'Acad. Roy. Sci. Turin* 2 (1784–85); Oeuvres, t.2, Gauthier-Villars, Paris, 1868, pp. 251–312.
31. J. Landen, A disquisition concerning certain fluents, which are assignable by the arcs of the conic sections; wherein are investigated some new and useful theorems for computing such fluents, *Philos. Trans. Royal Soc. London*, 61 (1771) 298–309.
32. J. Landen, An investigation of a general theorem for finding the length of any arc of any conic hyperbola, by means of two elliptic arcs, with some other new and useful theorems deduced therefrom, *Philos. Trans. Royal Soc. London*, 65 (1775) 283–289.
33. A. M. Legendre, Traité des fonctions elliptiques, Huzard-Courcier, t.1, Paris, 1825.
34. T. Leybourn, The Mathematical Questions Posed in the Ladies Diary, 1704–1816 (4 volumes), J. Mawman, London, 1817.
35. G. Lindner, Lexikon der gesamten Technik und ihrer Hilfswissenschaften, Band 3, second ed., Deutsche Verlagsanstalt, Stuttgart, 1904.
36. C. MacLaurin, A Treatise of Fluxions in Two Books, vol. 2, T. W. and T. Ruddimans, Edinburgh, 1742.
37. G. Mittag-Leffler, An introduction to the theory of elliptic functions, *Ann. of Math.*, 24 (1923) 271–351.
38. T. Muir, On the perimeter of an ellipse, *Messenger Math.*, 12 (1883) 149–151.
39. D. J. Newman, Rational approximation versus fast computer methods, Lectures on Approximation, and Value Distribution, Presses de l'Université de Montréal, 1982, pp. 149–174.
40. D. J. Newman, A simplified version of the fast algorithms of Brent and Salamin, *Math. Comp.*, 44 (1985) 207–210.
41. M. Nyvoll, Tilnaermelseformler for ellipsebuer, *Nordisk Mat. Tidskr.*, 25–26 (1978) 70–72.
42. G. Peano, Sur une formule d'approximation pour la rectification de l'ellipse, *C. R. Acad. Sci. Paris*, 108 (1889) 960–961.
43. T. Perl, The Ladies' Diary or Woman's Almanack, 1704–1841, *Historia Math.*, 6 (1979) 36–53.
44. S. Ramanujan, Modular equations and approximations to π, *Quart. J. Math.* (Oxford), 45 (1914) 350–372.
45. S. Ramanujan, Collected Papers, Chelsea, New York, 1962.
46. S. Ramanujan, Notebooks (2 volumes), Tata Institute of Fundamental Research, Bombay, 1957.
47. E. Salamin, Computation of π using arithmetic-geometric mean, *Math. Comp.*, 30 (1976) 565–570.
48. Science and the citizen, A bigger π, *Scientific American*, 248 (1983) 66.
49. E. S. Selmer, Bemerkninger til en ellipse-beregning av en ellipses omkrets, *Nordisk Mat. Tidskr.*, 23 (1975) 55–58.

50. Society of Mathematicians, The Diarian Repository or Mathematical Register, G. Robinson, London, 1774.

51. J. O. Stubban, Fergestads formel for tilnaermet beregning av en ellipses omkrets, *Nordisk Mat. Tidskr.*, 23 (1975) 51–54.

52. G. N. Watson, The marquis and the land-agent; a tale of the eighteenth century, *Math. Gaz.*, 17 (1933) 5–17.

53. E. T. Whittaker and G. N. Watson, A Course of Modern Analysis, 4th ed., Cambridge University Press, Cambridge, 1966.

54. J. Woyciechowsky, Sipos Pál egy kézirata és a kochleoid, *Mat. Fiz. Lapok*, 41 (1934) 45–54.

9. Vectorization of multiple-precision arithmetic program and 201,326,000 decimal digits of pi calculation (1988)

Paper 9: Yasumasa Kanada, "Vectorization of multiple-precision arithmetic program and 201,326,000 decimal digits of pi calculation," ©1988 IEEE. Reprinted, with permission, from *Supercomputing 88: Vol II, Science and Applications*, 117–128.

Synopsis:

In this paper, Yasumasa Kanada describes the computation of π to over 200 million decimal digits on a Hitachi S-820 vector supercomputer in Japan. Kanada employed what he termed the Gauss-Legendre formula, which is very similar to the formulas found by Salamin and Brent, and, for a check, by using Borwein quartic algorithm (the same algorithm earlier employed by Bailey).

Kanada provides very detailed information on exactly how he implemented these formulas, how he performed high-precision multiplication using a fast Fourier transform, and, how he accelerated his code to run at the maximum possible performance on the Hitachi supercomputer. Kanada also provides an interesting statistical analysis on these 200 million digits.

Keywords: Computation, Normality

© Springer International Publishing Switzerland 2016
D.H. Bailey, J.M. Borwein, *Pi: The Next Generation*,
DOI 10.1007/978-3-319-32377-0_9

Vectorization of Multiple-Precision Arithmetic Program
and
201,326,000 Decimal Digits of π Calculation

Yasumasa Kanada

Computer Centre, University of Tokyo,
Bunkyo-ku Yayoi 2-11-16, Tokyo 113, Japan

More than 200 million decimal places of π were calculated using arithmetic-geometric mean formula independently discovered by Salamin and Brent in 1976. Correctness of the calculation were verified through Borwein's quartic convergent formula developed in 1983. The computation took CPU times of 5 hours 57 minutes for the main calculation and 7 hours 30 minutes for the verification calculation on the HITAC S-820 model 80 supercomputer with 256 Mb of main memory and 3 Gb of high speed semiconductor storage, Extended Storage, for shorten I/O time.

Computation was completed in 27th of January 1988. At that day two programs generated values up to 3×2^{26}, about 201 million. The two results agreed except for the last 21 digits. These results also agree with the 133,554,000 places calculation of π which was done by the author in January 1987. Compare to the record in 1987, 50% more decimal digits were calculated with about 1/6 of CPU time.

Computation was performed with real arithmetic based vectorized Fast Fourier Transform (FFT) multiplier and newly vectorized multiple-precision add, subtract and (single word) constant multiplication programs. Vectorizations for the later cases were realized through first order linear recurrence vector instruction on the S-820. Details of the computation and statistical tests on the first 200 million digits of π-3 are reported.

1. Introduction

Since the epoch-making calculation of π to 100,000 decimals [17], several computations have been performed as in Table 1. The development of new algorithms and programs suited to the calculation of π and new high speed computers with large memory and high speed large semiconductor disk, Extended Storage or Solid State Disk, threw more light on this fascinating number.

There are many arctangent relations for π[9]. However, all these computations until 1981 and verification for 10,000,000 decimal calculation of our previous record[21] used arctangent formulae such as:

$$\pi = 16 arctan\frac{1}{5} - 4 arctan\frac{1}{239} \ , \qquad\qquad \text{Machin}$$

$$= 24 arctan\frac{1}{8} + 8 arctan\frac{1}{57} + 4 arctan\frac{1}{239} \ , \qquad \text{Störmer}$$

$$= 48 arctan\frac{1}{18} + 32 arctan\frac{1}{57} - 20 arctan\frac{1}{239} \ , \qquad \text{Gauss}$$

$$= 32 arctan\frac{1}{10} - 4 arctan\frac{1}{239} - 16 arctan\frac{1}{515} \ . \qquad \text{Klingenstierna}$$

In 1976, an innovative quadratic convergent formula for the calculation of π was published independently by Salamin [14] and Brent [5]. Later in 1983, quadratic, cubic, quadruple and septet convergent product expansion for π, which are competitive with Salamin's and Brent's formula, were also discovered by two of Borwein[2]. These new formulae are based on the arithmetic-geometric mean, a process whose rapid convergence doubles, triples, quadruples and septates the number of significant digits at each step. The arithmetic-geometric mean is the basis of Gauss' method for the calculation of elliptic integrals. With the help of the elliptic integral relation of Legendre, π can be expressed in terms of the arithmetic-geometric mean and the resulting algorithm retains quadratic, cubic, quartic and septic convergence of the arithmetic-geometric mean process.

The author and Mr. Y. Tamura have calculated π up to more than 200 million decimal places by using the formula of Brent and Salamin and verified through Borwein's quartic convergent algorithm for π. Even for the quartic convergent algorithm, quadratic convergent algorithm of Brent and Salamin is faster through actual FORTRAN programs.

For reducing the computing time, theoretically fast multiple-precision multiplication algorithm was implemented through normal fast Fourier transform (FFA), inverse fast Fourier transform (FFS) and convolution operations[11] as before[19, 9, 21, 8]. And in order to get more speed than before, vectorization schemes to the multiple-precision add, subtract and (single word) constant multiplication were introduced for the first time.

Calculation of 200 million decimal places of π was completed in January 27, 1988 and needed 5 hours 57 minutes of CPU time on a HITAC S-820 model 80 supercomputer at Hitachi Kanagawa Works under a VOS3/HAP/ES 31 bit addressing operating system. Main memory used was about 240 Mb and 2.7 Gb of Extended Storage was also required. If the machine had more memory, CPU time could be reduced and more Extended Storage, calculation decimals could be extended with minor changes in the FORTRAN programs.

The algorithms used in the 200 million place calculation are briefly explained in section 2. Programming in FORTRAN is discussed in section 3. Results of statistical tests and some interesting figures for the 200,000,000 decimals of π appear in section 4.

2. How to Calculate π: Algorithmic Aspects

In this section we briefly explain the algorithms used in the 200 million decimal place calculations.

2.1. The Gauss-Legendre Algorithm: Main Algorithm

The theoretical basis of the Gauss-Legendre algorithm for π is explained in the references [2, 5, 14]. Here, we summarize the quadratic algorithm for π. (Refer to the references for the details.)

We first define the arithmetic-geometric mean $agm(a_0, b_0)$. Let a_0, b_0 and c_0 be positive numbers satisfying $a_0^2 = b_0^2 + c_0^2$. Define a_n, the sequence of arithmetic means, and b_n, the sequence of geometric means, by

$$a_n = \frac{(a_{n-1} + b_{n-1})}{2} \ , \ b_n = (a_{n-1} \times b_{n-1})^{1/2} \ .$$

Also, define a positive number sequence c_n:

$$c_n^2 = a_n^2 - b_n^2 \ . \tag{1}$$

Note that, two relations easily follow from these definitions.

$$c_n = \frac{(a_{n-1} - b_{n-1})}{2} = (a_{n-1} - a_n) \ , \ c_n^2 = 4 \times a_{n+1} \times c_{n+1} \ . \tag{2}$$

Then, $agm(a_0, b_0)$ is the common limit of the sequences a_n and b_n, namely
$$agm(a_0, b_0) = \lim_{n \to \infty} a_n = \lim_{n \to \infty} b_n \ .$$

Now, π can be expressed as follows:

$$\pi = \frac{4 agm(1, k) agm(1, k')}{1 - \sum_{j=1}^{\infty} 2^j (c_j^2 + c_j'^2)} \ , \tag{4}$$

where $a_0 = a'_0 = 1$, $b_0 = k$, $b'_0 = k'$ and $k^2 + k'^2 = 1$. It is easier to compute squares of c_j and c'_j by Eq. (2) than to calculate c_j^2 and $c_j'^2$ by Eq. (1).

The symmetric choice of $k = k' = 2^{-1/2}$ is recommendable for the actual calculation and it causes the two sequences of arithmetic-geometric means to coincide. Then, Eq. (3) becomes

$$\pi = \frac{4(agm(1, 2^{-1/2}))^2}{1 - \sum_{j=1}^{\infty} 2^{j+1} c_j^2} \ . \tag{4}$$

After n square root operations in computing $agm = agm(1, 2^{-1/2})$, π can be approximated by π_n:

$$\pi_n = \frac{4a_{n+1}^2}{1 - \sum_{j=1}^{n} 2^{j+1}c_j^2} = \frac{a_{n+1}^2}{0.25 - \sum_{j=1}^{n} 2^{j-1}c_j^2} \ . \tag{5}$$

Then, the absolute value of $\pi - \pi_n$ is bounded as follows (Theorem 2b in reference [14]) :

$$\left| \pi - \pi_n \right| < (\pi^2 \times 2^{n+4}/agm^2) \exp(-\pi \times 2^{n+1}) \ . \tag{6}$$

Thus, the formula has the quadratic convergence nature. We must note here that all operations and constants in Eq. (4) must be correct up to the required number of digits plus α (20 to 30 as the guard digits for 200 million decimal digits calculation).

Then the sequences of agm and agm related π are calculated by the following algorithm;

$A := 1; B := 2^{-1/2}; T := 1/4; X := 1;$

while $A - B > 2^{-n}$ **do begin**
$W := A*B; V := A; A := A+B; A := A/2;$
$V := V-A; V := V*V; V := V*X;$
$T := T-V; B := \sqrt{W}; X := 2*X$ **end;**

$A := A*B; B := 1/T; A := A*B;$
return A .

Here, A, B, T, V and W are full-precision variables and X is a double-precision variable. After twenty eight iterations of the main loop, π to 200 million decimal places is to be obtained.

2.2. Borwein's Quartic Convergent Algorithm: Verification Algorithm

Borwein's quartic convergent algorithm is explained as the following scheme:

$$a_0 = 6 - 4 \times \sqrt{2}, \ y_0 = \sqrt{2} - 1$$

$$y_{k+1} = \frac{1 - (1 - y_k^4)^{1/4}}{1 + (1 + y_k^4)^{1/4}}, \ a_{k+1} = a_k \times (1 + y_{k+1})^4 - 2^{2k+3} \times y_{k+1} \times (1 + y_{k+1} + y_{k+1}^2),$$

$$\pi \approx \frac{1}{a_k} \quad \text{for large } k.$$

Here, precisions for a_k and y_k must be more than the desired digits. This algorithm is basically the same with that used for the main run of the 29 million decimal calculation done by Dr. D.H. Bailey[1]. Dr. Bailey used the following Borwein's quadratic convergent algorithm for the verification calculation:

$$a_0 = \sqrt{2}, \ b_0 = 0, \ p_0 = 2 + \sqrt{2}$$

$$a_{k+1} = \frac{(\sqrt{a_k} + \frac{1}{\sqrt{a_k}})}{2}, \ b_{k+1} = \frac{\sqrt{a_k} \times (1 + b_k)}{a_k + b_k}, \ p_{k+1} = \frac{p_k \times b_{k+1} \times (1 + a_{k+1})}{1 + b_{k+1}},$$

$$\pi \approx p_k \quad \text{for large } k.$$

Here, precisions for a_k, b_k and p_k must be more than the desired digits. It is to be noted that the Gauss-Legendre algorithm is superior to the Borwein's algorithms explained here as in the Table 2.

2.3. Calculation of Reciprocals and Square Roots

As is easily seen from the algorithms explained above, arithmetic operations, reciprocals and square roots must be computed with high efficiency in order to reduce the computing time. Theoretical bases for fast calculation of reciprocals and square roots of multi-precision numbers appear in references [16, 12, 4, 5]. In this subsection we summarize the algorithms used in the actual calculation.

The reciprocal of C is obtained by the Newton iteration for the equation $f(x) = x^{-1} - C \equiv 0$:

$$x_{i+1} = x_i \times (2 - C \times x_i) \ . \tag{7}$$

A single-, double- or quadruple-precision approximation of $1/C$ is a reasonable selection for x_0, i.e. initial starting value for the Newton iteration.

Square roots of C should be calculated through the multiplication of C and the result obtained by the Newton iteration for the equation $f(x) = x^{-2} - C \equiv 0$:

$$x_{i+1} = \frac{x_i \times (3 - C \times x_i^2)}{2} . \tag{8}$$

In this case also, a single-, double- or quadruple-precision approximation of $1/\sqrt{C}$ is a reasonable selection for x_0, initial starting value.

Compare this to the well known Newton iteration for the equation $f(x) = x^2 - C \equiv 0$:

$$x_{i+1} = \frac{(x_i + \dfrac{C}{x_i})}{2} . \tag{9}$$

The iteration of Eq. (8) is better in both computing complexity and in actual calculation than the iteration of Eq. (9). The order of convergence for the iterations of Eq. (7) and Eq. (8) is two. This convergence speed is favorable in the actual calculation. It is to be noted that in these iterations the whole operation need not to be done with full-precision at each step. That is, if k-precision calculation is done at step j, $2k$-precision calculation is sufficient for at step $j+1$.

2.4. Multiple-precision Multiplication

Schönhage-Strassen's algorithm [15], which uses the discrete Fourier transform with modulo $2^n + 1$, could be the key multiple-precision multiplication algorithm for speeding up the π calculation. However, this algorithm is so hard to implement and needs binary to decimal radix conversion for the final result. Dr. Bailey also used discrete Fourier transform but with three prime modulo computation followed by the reconstruction through Chinese Remainder Theorem for his 29 million decimal places calculation[1].

Now, we focused the special scheme which utilizes the fact of *"the Fourier transform of a convolution product is the ordinary product of the Fourier transforms."*

Let consider the product C of two length n with radix X integers A and B. Note that the radix X need not to be a power of 2.

$$A = \sum_{i=0}^{2n-1} a_i X^i , B = \sum_{i=0}^{2n-1} b_i X^i ,$$

where $0 \le a_i < X, 0 \le b_i < X$ for $0 \le i < n-1$, $0 < a_{n-1} < X, 0 < b_{n-1} < X$ and $a_i = b_i = 0$, for $i < 0, n \le i$.

Then,

$$C \equiv A \cdot B = (\sum_{i=0}^{2n-1} a_i X^i) \cdot (\sum_{j=0}^{2n-1} b_j X^j) = \sum_{i=0}^{2n-1} X^i (\sum_{j=0}^{2n-1} a_j b_{i-j}) \equiv \sum_{i=0}^{2n-1} c_i X^i$$

Thus,

$$c_i = \sum_{j=0}^{2n-1} a_j b_{i-j}, \quad \text{for } i = 0, \cdots, 2n-2, \quad \text{and} \quad c_{2n-1} = 0.$$

If $\omega = exp(2\pi i/2n)$ is a $2n$th root of unity, the one-dimensional Fourier transform of the sequences of complex numbers $(a_0, a_1, \cdots, a_{2n-1})$, $(b_0, b_1, \cdots, b_{2n-1})$ and $(c_0, c_1, \cdots, c_{2n-1})$ are defined as the sequences of $(\hat{a}_0, \hat{a}_1, \cdots, \hat{a}_{2n-1})$, $(\hat{b}_0, \hat{b}_1, \cdots, \hat{b}_{2n-1})$ and $(\hat{c}_0, \hat{c}_1, \cdots, \hat{c}_{2n-1})$, respectively. Here,

$$\hat{a}_l = \sum_{i=0}^{2n-1} a_i \omega^{il}, \hat{b}_l = \sum_{i=0}^{2n-1} b_i \omega^{il}, \hat{c}_l = \sum_{i=0}^{2n-1} c_i \omega^{il}, \quad \text{for } 0 \le l \le 2n-1 .$$

Then, $(\hat{a}_0 \hat{b}_0, \hat{a}_1 \hat{b}_1, \cdots, \hat{a}_{2n-1} \hat{b}_{2n-1})$ is equal to $(\hat{c}_0, \hat{c}_1, \cdots, \hat{c}_{2n-1})$ as the followings:

$$\hat{a}_k \hat{b}_k = (\sum_{i=0}^{2n-1} a_i \omega^{ki}) (\sum_{j=0}^{2n-1} b_j \omega^{kj}) = \sum_{i=0}^{2n-1} \sum_{j=0}^{2n-1} a_i b_j \omega^{k(i+j)} = \sum_{l=0}^{2n-1} (\sum_{j=0}^{2n-1} a_j b_{l-j}) \omega^{kl} = \sum_{l=0}^{2n-1} c_l \omega^{kl}$$

In these discussions, radix X might be an any number. However, ω must be a complex number, namely *floating point real* arithmetic operations are needed in the process of Fourier transformation. If and only if all the *floating point real* arithmetic operations are performed exactly, this scheme should give the correct result as discussed in page 290-295 of reference[11]. Compared to the discrete Fourier transform based multiplier, *floating point real* arithmetic operations based Fourier transform seems ideal and dangerous, but attractive as for the actual multiple-precision multiplication method.

The reasons are:

1) The speed of double precision floating point operations is faster than integer operations in the available machine, especially for supercomputers. And in general, the number of bits obtainable in one double precision floating point instruction is longer than that obtainable in one single integer instruction.

2) Conversion from binary results to decimal results needs other techniques and coding. (Simple is best. Schönhage-Strassen's discrete FFT algorithm is the binary data multiplier.)

3) A qualified high-speed FFT routine was available as a library. (Qualified programming improves the reliability of the program. If we adopt Schönhage-Strassen's discrete FFT algorithm, we have to code for it.)

4) Chinese remainder based discrete FFT algorithm is not so fast as far as the Dr. Y. Ushiro's experiment in 1983[20] is concerned. (He showed the inferiority of Chinese remainder based discrete FFT to us prior to the Dr. Bailey's experiment[1]. We needed a faster multiple-precision multiplier.)

Followings are the algorithm used in our calculation. Here, let consider the multiplication of two $m \times 2^n$ bit $(= m \times (\log_{10}2) \times 2^n$ decimal digit) integers A and B through our schemes.

Step 1: Prepare two 2×2^n entry double precision floating point array.

Step 2: Convert both of $m \times 2^n$ bit integers into double precision floating point numbers. (The first half of 2×2^n entry contains information for $m \times 2^n$ bit, namely, m bit information per one double precision floating point array entry.)

Step 3: Initialize to double precision floating point zero for the second half of 2×2^n entry.

Step 4: Apply 2^{n+1} point normal Fourier transform, say FFA, operations to A and B giving A' and B', respectively.

Step 5: Do the convolution product operations between A' and B' giving new 2×2^n entry double precision array C'.

Step 6: Apply 2^{n+1} point inverse Fourier transform, say FFS, operations to C' giving C. (Now, C is the double precision floating point array of 2×2^n entry. If operations FFA, FFS and convolution product are performed in infinite precision, each entry of C should be the exact double precision floating point representation for integer with maximum value of $2^n \times (2^m-1)^2$. However, these representation are slightly deviated from exact integer in the actual operation. Because, infinite precision operations are impossible to perform.)

Step 7: Convert each entry of C (let it to be x) into integer representation. (Conversion should be done with IDNINT operation in FORTRAN. (IDNINT(x) = IDINT(x+0.5D0).) If absolute value of $(x-DFLOAT(IDNINT(x)))$ is near to 0.5D0, the multiplication is considered to be incorrect. We don't know that criterion of 0.45D0 is sufficient or not. However criterion of 0.2D0, for example, was sufficient enough in the actual multiplication.)

Step 8: Normalize C under the suitable base. The base of 2^m or $10^{m(\log_{10}2)}$ is better for binary or decimal representation. Final result is the result of multiplication between A and B.

According to the theoretical and experimental analysis of error in FFT [6,10], FFT attains rather stable error behavior. Theoretically roughly speaking, 2^l point FFT attains $l \log(l)$ bits error at the maximum (worst case). This means that FFA and the convolution product followed by FFS would acquire $2 l \log(l)$ bits error at the maximum, even if the trigonometric functions at coefficients for the butterfly operation are calculated exactly.

Available mantissa bits are 56 for double-precision floating point data of the Japanese supercomputers. (CRAY and ETA machines adopt 48 bit mantissa representation for the floating point numbers.) Then, the following equation must be satisfied for the worst case;

bits available in the calculation ≥ (*guard bits* for *FFA* and *FFS operations*) +

(*necessary bits* for *preserving the results*)

$$56 \geq 2\,(n+1)\,log\,(n+1) + log\,(2^n \times (2^m-1)^2)\ .$$

Here n is an integer. Thus, even for $m = 1$,

$$n \leq 7\ .$$

The analysis suggests a maximum n of 7. However, the maximum n depends on the method of programming, errors in trigonometric function calculations, value of m, *etc.* (There is another error analysis for the *floating point real* FFT multiplier in reference[11] which specifies the radix X in our explanation to be the power of 2.)

The error analysis for the actual calculation is hard. Then, we have checked the availability of the FFT multiplier based on the above schemes through actual programs. We monitored the deviations from integer values after the stage of conversion to integer representation as explained in Step 7 of the above explanation. The monitoring secures a maximum n of 24, for the condition of one data point, a double word, holds 3 decimals at the maximum (m is about 10) on the first half of 2×2^{24} data points and zero on the second half of 2×2^{24} data points. Then we decided that the maximum length of multiplicand to be multiplied in-core is 3×2^{24} decimal places (about 50 million decimals) in the actual multiplication.

For multiplying 3×2^{26} decimal places numbers (multiple-precision data for 200 million decimal places), we used the classical $O(n^2)$ algorithm, e.g. school boy method, for the data of $2^{26}/2^{24} = 2^2 = 4$ units with base of $2^{24} \times 1,000!$ It is possible to realize multiple-precision multiplier all through FFT, not through school boy method. In order to do so, we must write extra program for out-of-core version of FFT. We preferred to utilize the reliability of the numerical libraries.

2.5. Vectorization of Multiple-Precision Add, Subtract and (Single Word) Constant Multiplication

For the programs of multiple-precision add, subtract and (single word) constant multiplier, time consuming process is releasing the "borrow or carry."

If the machine has special instruction for vectorizing first order linear recurrence relation of the following, the process of releasing the borrow or carry could be vectorized. (Now, all of the Japanese supercomputers are equipped with such instructions.)

```
        DO 10 I = ...
   10    A(I) = B(I) * A(I-1) + C(I)
```

Here, vector A has a nature of first order linear recurrence relation. The following program which were extracted form the source codes of the actual run will explain how to rewrite the program for multiple-precision adder. Same strategies can be applied to the multiple-precision routines for subtract and (single word) constant multiplication.

```
C  integer*4 -- MA, MB, ICY(=carry), ICW(=work), IONE8(=base=10**6)
C  real*8 -- Y(=work), Z(=carry), CY(=work), CW(=work)
C  real*8 -- ONED8=DFLOAT(IONE8), ONEDM8=1.D0/ONED8, HLFDM8=.5D0*ONEDM8

      DO 10 J=NDA,1,-1                      DO 10 J=NDA,2,-1
        ICW=MA(J)+MB(J)+ICY                   Y(J)=MA(J)+MB(J)
        IF( ICW.GE.IONE8 ) THEN               Z(J-1)=Y(J)*ONEDM8+Z(J)*ONEDM8
          MA(J)=ICW-IONE8                     Z(J)=DINT(Z(J)+HLFDM8)
          ICY=1                       10   CONTINUE
        ELSE                   ==>           Z(1)=DINT(Z(J)+HLFDM8)
          MA(J)=ICW                        DO 11 J=NDA,2,-1
          ICY=0                              MA(J)=Y(J)+Z(J)-Z(J-1)*ONED8
        END IF                        11   CONTINUE
   10 CONTINUE                               CW=DFLOAT(MA(1)+MB(1))+Z(1)
                                             CY=DINT(CW*ONEDM8+HLFDM8)
                                             MA(1)=CW-CY*ONED8
```

3. How to Calculate π: Programming Aspects

The actual programs, written in FORTRAN, consisted of 3426 (main) and 3642 (verification) lines of source code with comment lines. The numbers of program units are 59 (main) and 64 (verification). In this section we briefly overview the actual programming.

About the half of the sub-programs are routines related to multiple-precision multiplication. Others are routines for addition, subtraction, reciprocal operations, square root operations, file I/O operations, *etc.*

The HITAC S-820 model 80 is a so called 2nd generation supercomputer in Japan. The machine is single processor model and has theoretical peak performance of 3Gflops. Maximum attachable main memory size is 512 Mb and Extended Storage size is 12 Gb. FFT routines, FFA and FFS, are carefully selected from the non-vectorized FORTRAN numerical library which was provided by the Hitachi and source codes were slightly modified by myself for the vectorization.

3.1. Layout of Storage

As explained in section 2, we used a *floating point real* FFT whose accuracy is secured under the conditions of

(1) 2^{24+1} point double-precision *floating point real* FFT,

(2)   maximum number at each entry point for the first half entry is non negative number and must be bounded by 1,000,

(3)   the second half entry contains zero.

These conditions allow in-core multiplication of numbers with $2^{24} \times 3$ (= 50,331,648) decimals. If we want more decimals to be calculated in-core (on main memory), we must reduce the above conditions, e.g.

(1) 2^{24+1} points -> 2^{29+1} points,

(2)   maximum number at each entry point of 1,000 -> 100.

These new conditions would allow numbers of up to $2^{29} \times 2$ (= 1,073,741,824) decimals to be manipulated in-core. (We did not check the validity of these conditions through actual run. Our examples are only for explanation!) However, currently available maximum memory size prevent such higher precision calculations in-core. In order to run with such ideal conditions, at least 8 GB of main memory should be available. Now, the cases with 2^{17+1} points or 2^{18+1} points and maximum number at each entry point of 10,000 were applied for the previous our π calculations including 133 million decimals record. The increase of in-core operable FFT points is the major factor for the speed-up to the π calculation.

It was impossible to obtain 200 million decimal places through in-core operations because of available main memory size. Therefore we introduced the user controlled virtual memory scheme for saving high precision constants of $1/\sqrt{2}$, $\sqrt{2}$, π, and several working storage with compression factor of 6 decimals / 4 bytes. (Integer representation was used for saving storages on the extended storage. For FFT, a 3 decimals / 8 bytes scheme - double precision floating point representation - was needed as explained.) These schemes needed about 240 Mb of main memory for the working storage, input and output (I/O) buffer, object codes, *etc.* As for the extended storage size, 13.5 Mb / 1 million decimal places was needed.

3.2. Optimization for Speedup

We have employed several optimization schemes:

1) Multiplication by 1 -> normal copy operation.
2) Multiplication by 2 -> normal addition operation.
3) Deletion of unnecessary FFA and FFS operations.
4) Reuse of internal iteration results.

For 1) and 2), explanation is simple enough. In the following two subsections, we explain the details of 3) and 4).

3.2.1. How to Delete FFA and FFS Operations

As explained in section 2.4, the school boy method was employed for multiple-precision multiplication (for data length longer than in-core operable length). Now as for an example of 4×4 school boy method, let consider the simple case of 2×2. That corresponds to the multiplication of two integers, both with a length of 3×2^{25} and in-core operable length of 3×2^{24}. Without optimization, each $2^{24} \times 1,000$ decimal based multiplication needs eight FFA for input and four FFS for output as follows;

$$(A_1 \times BASE + A_2) \times (B_1 \times BASE + B_2)$$

$$= (FFS \ of \ (FFA \ of \ A_1 \cdot FFA \ of \ B_1)) \times BASE^2$$

$$+((FFS \ of \ (FFA \ of \ A_1 \cdot FFA \ of \ B_2)) + (FFS \ of \ (FFA \ of \ A_2 \cdot FFA \ of \ B_1))) \times BASE$$

$$+(FFS \ of \ (FFA \ of \ A_2 \cdot FFA \ of \ B_2)) \ ,$$

where · is the convolution product for the Fourier transformed data, + is the vector-wise addition and BASE is $2^{24} \times 1,000$.

There are very many FFA and FFS operations. These operations can be reduced a lot by the following schemes:

1) First apply FFA operations to A_1, A_2, B_1 and B_2. Let the results be A_1', A_2', B_1' and B_2', respectively.

2) Now $(A_1' \cdot B_2' + A_2' \cdot B_1')$ preserve information in the sense explained in section 2.4.

3) Then, the linearity of FFS operation satisfies the following equation:

$$(A_1 \times BASE + A_2) \times (B_1 \times BASE + B_2)$$
$$= (FFS \ of \ (A_1' \cdot B_1')) \times BASE^2 + (FFS \ of \ (A_1' \cdot B_2' + A_2' \cdot B_1')) \times BASE$$
$$+ (FFS \ of \ (A_2' \cdot B_2')) \ .$$

According to the results of CPU time profile analysis for 33 million decimal places calculation, multiplication occupies 90% of CPU time. (Here, CPU time does not contain the time for input and output operations.) Thus, this optimization is rather efficient when the length of multiplicand becomes long. (In this case, FFA operations was reduced from 8 to 4 and FFS operations was reduced from 4 to 3.)

3.2.2. Reuse of Internal Iteration Results

Newton's iteration for square roots and reciprocals has second order convergence nature. This implies that internal iteration results at the calculation of the half-length precision of π can be the initial value for the iteration at the calculation of the full-length precision of π. (And this selection is the best selection for reducing CPU time.) According to measurements of CPU time, this scheme reduced the CPU time 10-20% for the 16 or 33 million decimal places calculation.

If we utilize this fact, we can reduce the computation time probably by 10-20%. To do so, however, we had to prepare the permanent storage of about 1.5 Gb. This was completely impossible at the time of actual run. For the history of π calculation, we had saved the internal iteration results from the calculation of 16 million decimal places (data size is around 100 Mb). That data helped for the calculation of 32 million decimal places of π substantially. And also for the calculation of 16 million decimal places of π, we had utilized the internal iteration results from the calculation of 8 million decimal places.

4. Statistical Analysis of 200,000,000 Decimals of π

In order to analyze the statistics of 200,000,000 decimals of π, we used the same statistical tests as Pathria[13], except that the number of digits expanded from 100,000 to 200,000,000. In this section we briefly show the statistical data and some interesting figures from the 200,000,000 decimals of π. Now, the 200,000,000-th decimal number of π -- 3 is 9.

4.1. Results of Statistical tests

Five kinds of statistical tests were performed. These are the frequency test, the serial tests, the Poker hand test, the gap test and the five-digit sum test. We have reproduced only the results of frequency test in the Table 3. The other results are to be published through the reference[7].

4.2. Some Interesting Figures

Analysis of digit sequences for 200,000,000 decimal places of $\pi - 3$ gives some interesting figures;

1) A longest descending sequence of 2109876543 appears (from 26,160,634) only once. The next longest descending sequence is 876543210 (from 2,747,956) only. The next longest descending sequence of length 8 appears 9 times.

2) The longest ascending sequence is 901234567 which appears from 197,090,144. The next longest ascending sequence is 23456789 (from 995,998), 89012345 (from 33,064,267, 39,202,678, 62,632,993 and 78,340,559), 90123456 (from 35,105,378, 44,994,887, 98,647,533 and 127,883,114), 56789012 (from 100,800,812 and 139,825,562), 67890123 (from 102,197,548, 135,721,079 and 178,278,161), 01234567 (from 112,099,767), 78901234 (from

119,172,322, 122,016,838, 182,288,028 and 195,692,744), 12345678 (from 186,557,266) and 45678901 (from 194,981,709). The next longest ascending sequence of length 7 appears 170 times.

3) A sequence of maximum multiplicity (of 9) appears 3 times. These are 7 (from 24,658,601), 6 (from 45,681,781) and 8 (from 46,663,520). The next longest sequence of multiplicity (of 8) appears 16 times.

4) The longest sequence of 27182818 appears from 73,154,827, 143,361,474 and 183,026,622. The next longest sequence of 2718281 appears 22 times.

5) The longest sequence of 14142135 appears from 52,638, 10,505,872 and 143,965,527. The next longest sequence of 1414213 appears from 13,816,189, 40,122,589, 72,670,122, 87,067,359, 104,717,213, 115,301,872, 145,035,762, 147,685,125, 155,299,021, 165,871,476, 166,005,277, 166,491,213 and 191,208,533. The next longest sequence of 141421 appears 169 times.

6) The longest sequence of 31415926 appears 2 times. These are from 50,366,472 and 157,060,182. The next longest sequence of 3141592 appears 7 times.

5. Conclusion

Details of 201,326,000 decimal digits of π calculation and some results of statistical tests on 200,000,000 decimals of π are presented. The original program is written in FORTRAN 77 and heavily utilizes floating point operations. Multiple-precision add, subtract and constant multiplication programs were also vectorized through linear recurrence special instruction. Programs do not depend on the scheme of round-off and cut-off to the results for the floating point operations. Then, not only round-off machine, e.g. supercomputers of CRAY Inc. and ETA systems, but also cut-off machine, e.g. Japanese supercomputers, can generate correct π digits. The program calls few system specific subroutines, e.g. CLOCK, TIME, but adaptation to the new machine is easy. In order to get speed, available main memory size is crucial and shorten elapsed time, availability of high speed I/O devices is also crucial. Thus, the π calculation program based on the scheme explained here can be a good benchmark program for the supercomputer.

We have programmed the fast multiple-precision multiplication through a *floating point real* FFT package, which was available as one of the program libraries. This means that half a hundred lines of code is sufficient for the fast multiple-precision multiplication. (A few lines of array declaration, one line for calling the FFA routine, a few lines of convolution products, one line for calling the FFS routine and a few lines of normalization under a suitable base.) These schemes would be of benefit to other high-precision constant and function calculations. And another scheme, high utilization of floating point operations in the processing of integers, is also favorable for the integer calculation, especially to number crunchers.

As the Table 1 shows, it took 12 years for extending the length of known π value from 100,000 to 1,000,000, 10 years from 1,000,000 to 10,000,000 and 4 years from 10,000,000 to the order of 100,000,000. When can we unveil the digits after 1,000,000,000, how and by whom?

6. Acknowledgments

The author would like to express my thanks to Mr. Yoshiaki Tamura for his collaboration with me and to Dr. Kazunori Miyoshi for utilizing his statistical analysis program for π. Thanks are also due to Mr. Akihiko Shibata for his information concerning the history of π calculation. I also thank to Hitachi Co. for giving me a chance to extend the length of π value.

Calculated by	Machine used	Date	Precision			Time	Formula
			(calculated)	declared	correct	(check)	(check)
Reitwiesner et al.	ENIAC	1949	(2040)		2037	~70h (~70h)	M(M)
Nicholson & Jeenel	NORC	1954	(3093)		3092	13m (13m)	M(M)
Felton	Pegasus	1957	(10021)		7480	33h (33h)	K(G)
Genuys	IBM 704	1958	(10000)		10000	1h 40m (1h 40m)	M(M)
Felton	Pegasus	"	(10021)		10020	33h (33h)	K(G)
Guilloud	IBM 704	1959	(16167)		16167	4.3h (4.3h)	M(M)
Shanks & Wrench	IBM 7090	1961	(100265)		100265	8h 43m (4h 22m)	S(G)
Guilloud & Filliatre	IBM 7030	1966	(250000)		250000	41h 55m (24h 35m)	G(S)
Guilloud & Dichampt	CDC 6600	1967	(500000)		500000	28h 10m (16h 35m)	G(S)
Guilloud & Bouyer	CDC 7600	1973	(1001250)		1001250	23h 18m (13h 40m)	G(S)
Miyoshi & Kanada	FACOM M-200	1981	(2000040)	2000000	2000036	137.3h (143.3h)	K(M)
Guilloud	x	1981-82	(x)	2000050	2000050	x (x)	x
Tamura	MELCOM 900II	1982	(2097152)		2097144	7h 14m (2h 21m)	L(L)
Tamura & Kanada	HITAC M-280H	"	(4194304)		4194288	2h 21m (6h 52m)	L(L)
Tamura & Kanada	"	"	(8388608)		8388576	6h 52m (< 30h)	L(L)
Kanada, Yoshino & Tamura	"	1983	(16777216)		16777206	< 30h (6h 36m)	L(L)
Ushiro & Kanada	HITAC S-810/20	1983 Oct.	(10013400)	10000000	10013395	< 24h (< 30h)	G(L)
Gosper	Symbolics 3670	1985 Oct.	(>=17526200)		17526200	x (28h)	R(B4)
Bailey	CRAY-2	1986 Jan.	(29360128)	29360000	29360111	28h (40h)	B4(B2)
Kanada & Tamura	HITAC S-810/20	1986 Sep.	(33554432)	33554000	33554414	6h 36m (23h)	L(L)
Kanada & Tamura	HITAC S-810/20	1986 Oct.	(67108864)		67108839	23h (35h 15m)	L(L)
Kanada, Tamura, Kubo etc.	NEC SX-2	1987 Jan.	(134214728)	133554000	134214700	35h 15m (48h 2m)	L(B4)
Kanada & Tamura	HITAC S-820/80	1988 Jan.	(201326572)	201326000	201326551	5h 57m (7h 30m)	L(B4)

Table 1. Historical records of the calculation of π performed on electronic computers. M, K, G, S, L, R, B4 and B2 are the formulae of Machin, Klingenstierna, Gauss, Störmer, and Gauss-Legendre, formula explained in the reference[3], Borwein's quartic convergent, Borwein's quadratic convergent, respectively. Symbol 'x' means 'unknown'. Check time means the additional time for the calculated value checking. This information was basically obtained from Mr. Shibata[18].

Type of operation	Brent et. al.	Borwein's quartic	Borwein's quadratic
$X^{1/2}$ or $X^{1/4}$	n	n	n
$\times$	2n+2	5n	6n
reciprocal	1	n+1	2n
+,−	3n+1	6n	3n

Table 2. Comparison of basic, time consuming, operations in the three historically important algorithms for π calculation. Here n is a number of iterations. In order to compare with fare, both side columns numbers should be doubled.

Digit	0	1	2	3	4	5	6	7	8	9	χ^2
100	8	8	12	11	10	8	9	8	12	14	4.20
200	19	20	24	19	22	20	16	12	25	23	6.80
500	45	59	54	50	53	50	48	36	53	52	6.88
1k	93	116	103	102	93	97	94	95	101	106	4.74
2k	182	212	207	188	195	205	200	197	202	212	4.34
5k	466	532	496	459	508	525	513	488	492	521	10.77
10K	968	1026	1021	974	1012	1046	1021	970	948	1014	9.32
20K	1954	1997	1986	1986	2043	2082	2017	1953	1962	2020	7.72
50K	5033	5055	4867	4947	5011	5052	5018	4977	5030	5010	5.86
100K	9999	10137	9908	10025	9971	10026	10029	10025	9978	9902	4.09
200K	20104	20063	19892	20010	19874	20199	19898	20163	19956	19841	7.31
500K	49915	49984	49753	50000	50357	50235	49824	50230	49911	49791	7.73
1M	99959	99758	100026	100229	100230	100359	99548	99800	99985	100106	5.51
2M	199792	199535	200077	200141	200083	200521	199403	200310	199447	200691	9.00
4M	399419	399463	399822	399913	400792	400032	399032	400650	400183	400694	7.92
5M	499620	499898	499508	499933	500544	500025	498758	500880	499880	500954	7.88
8M	799111	800110	799788	800234	800202	800154	798885	800560	800638	800318	3.79
10M	999440	999333	1000306	999964	1001093	1000466	999337	1000207	999814	1000040	2.78
15M	1500081	1499675	1501044	1499917	1501166	1500417	1498447	1499584	1500435	1499234	4.07
20M	2001162	1999832	2001409	1999343	2001106	2000125	1999269	1998404	1999720	1999630	4.17
25M	2500496	2499915	2500707	2499313	2502826	2500139	2499603	2498290	2499189	2499522	5.28
30M	2999157	3000554	3000969	2999222	3002593	2999997	2999548	2998175	2999592	3000193	4.34
50M	4999632	5002220	5000573	4998630	5004009	4999797	4998017	4998895	4998494	4999733	6.17
80M	7998807	8002788	8001828	7997656	8003525	7996500	7998165	7999389	8000308	8001034	5.95
100M	9999922	10002475	10001092	9998442	10003863	9993478	9999417	9999610	10002180	9999521	7.27
150M	14998689	15001880	15001586	14999130	15003829	14993562	14998434	14999462	15001416	15002012	4.90
200M	19997437	20003774	20002185	20001410	19999846	19993031	19999161	20000287	20002307	20000562	4.13

Table 3. Summary of frequency for the first 200,000,000 digits of π - 3 and corresponding χ^2 values.

References

1. Bailey, D.H., "The Computation of πi to 29,360,000 Decimal Digits Using Borwein's Quartically Convergent Algorithm," *Math. Comp.*, vol. 50, no. 181, pp. 66-88, Jan. 1988.

2. Borwein, J.M. and Borwein, P.B., *Pi and the AGM - A study in Analytic Number Theory and Computational Complexity*, A Wiley-Interscience Publication, New York, 1987.

3. Borwein, J.M. and Borwein, P.B., "Ramanujan and Pi," *Scientific American*, vol. 258, pp. 66-73, Feb. 1988.

4. Brent, R.P., "Fast Multiple-Precision Evaluation of Elementary Functions," *J. ACM.*, vol. 23, no. 2, pp. 242-251, April 1976.

5. Brent, R.P., "Multiple-Precision Zero-Finding Methods and the Complexity of Elementary Function Evaluation," in *Analytic Computational Complexity*, ed. Traub, J.F., pp. 151-176, Academic Press, New York, 1976.

6. Gentleman, W.M. and Sande, G., "Fast Fourier Transforms - For Fun and Profit," *AFIPS Conf. Proc. FJCC*, vol. 29, pp. 563-578, Spartan Books, San Francisco, Calif., November 1966.

7. Kanada, Y., "Statistical Analysis of π up to 200,000,000 Decimal Places," CCUT-TR-88-02, in preparation, Computer Centre, University of Tokyo, Bunkyo-ku, Yayoi 2-11-16, Tokyo 113, Japan, Nov. 1988.

8. Tamura, Y., Yoshino, S., Ushiro, Y. and Kanada, Y., "Circular Constant - It's High Speed Calculation Method and Statistics -," in *Proceedings Programming Symposium, in Japanese*, Information Processing Society of Japan, Jan. 1984.

9. Tamura, Y. and Kanada, Y., "Calculation of π to 4,194,293 Decimals Based on the Gauss-Legendre Algorithm," CCUT-TR-83-01, Computer Centre, University of Tokyo, Bunkyo-ku, Yayoi 2-11-16, Tokyo 113, Japan, May 1983.

10. Kaneko, T. and Liu, B., "Accumulation of Round-Off Error in Fast Fourier Transforms," *J. ACM.*, vol. 17, no. 4, pp. 637-654, Oct. 1970.

11. Knuth, D.E., *The Art of Computer Programming, Vol. 2: Seminumerical Algorithms*, Addison-Wesley, Reading, Mass., 1981.

12. Laasonen, P., "On the Iterative Solution of the Matrix Equation A X^2 - I = 0," *Math. Tables and Other Aids to Comp.*, vol. 12, pp. 109-116, 1958.

13. Pathria, R.K., "A Statistical Study of Randomness Among the First 10,000 Digits of π," *Math. Comp.*, vol. 16, pp. 188-197, 1962.

14. Salamin, E., "Computation of π Using Arithmetic-Geometric Mean," *Math. Comp.*, vol. 30, no. 135, pp. 565-570, July 1976.

15. Schönhage, A. and Strassen, V., "Schnelle Multiplikation Grosser Zahlen," *Computing*, vol. 7, pp. 281-292, 1971.

16. Schutz, G., "Iterative Berechnung der Reziproken Matrix," *Z. Angew. Math. Mech.*, vol. 13, pp. 57-59, 1933.

17. Shanks, D. and Wrench, Jr. W., "Calculation of π to 100,000 Decimals," *Math. Comp.*, vol. 16, pp. 76-79, 1962.

18. Shibata, A., 1982-1983. Private communications

19. Tamura, Y., "Calculation of π to 2 Million Decimals Using Legendre-Gauss' Relation," Proceeding of the International Latitude Observatory of Mizusawa No.22 (in Japanese) in press, Internal Latitude Observatory of Mizusawa, Hoshigaoka-cho 2-12, Mizusawa City, 023 Iwate Prefecture, Japan, 1983.

20. Ushiro, Y., "Calculation of Multiple-Precision Numbers through Vectorizing Processor," in *Numerical Calculation Workshop, in Japanese*, Information Processing Society of Japan, 26-28 May. 1983.

21. Kanada, Y., Tamura, Y., Yoshino, S. and Ushiro, Y., "Calculation of π to 10,013,395 Decimal Places Based on the Gauss-Legendre Algorithm and Gauss Arctangent Relation," CCUT-TR-84-01, Computer Centre, University of Tokyo, Bunkyo-ku, Yayoi 2-11-16, Tokyo 113, Japan, Dec 1983.

10. Ramanujan and pi (1988)

Paper 10: Jonathan M. Borwein and Peter B. Borwein, "Ramanujan and Pi"
Scientific American, vol. 256 (February 1988), 112–117.

Synopsis:

In this paper, Jonathan and Peter Borwein wrote about the recent developments in the computation of π, and their roots in the work of Archimedes, Gauss, Ramanujan and others, for a broad, semi-popular audience — *Scientific American*, which is read by several hundred thousand readers worldwide.

One of the more interesting items in this article is its presentation of a full set of formulas enabling one to compute one billion digits of π on a "pocket calculator": namely to iterate the Borwein quartically convergent algorithm 15 times. As they jokingly point out in a footnote, though, you would need a pretty big calculator or else the computation would be pretty uninteresting after the second iteration.

Keywords: Algorithms, General Audience, History, Modular Equations

© Springer International Publishing Switzerland 2016
D.H. Bailey, J.M. Borwein, *Pi: The Next Generation*,
DOI 10.1007/978-3-319-32377-0_10

Ramanujan and Pi

Some 75 years ago an Indian mathematical genius developed ways of calculating pi with extraordinary efficiency. His approach is now incorporated in computer algorithms yielding millions of digits of pi

by Jonathan M. Borwein and Peter B. Borwein

Pi, the ratio of any circle's circumference to its diameter, was computed in 1987 to an unprecedented level of accuracy: more than 100 million decimal places. Last year also marked the centenary of the birth of Srinivasa Ramanujan, an enigmatic Indian mathematical genius who spent much of his short life in isolation and poor health. The two events are in fact closely linked, because the basic approach underlying the most recent computations of pi was anticipated by Ramanujan, although its implementation had to await the formulation of efficient algorithms (by various workers including us), modern supercomputers and new ways to multiply numbers.

Aside from providing an arena in which to set records of a kind, the quest to calculate the number to millions of decimal places may seem rather pointless. Thirty-nine places of pi suffice for computing the circumference of a circle girdling the known universe with an error no greater than the radius of a hydrogen atom. It is hard to imagine physical situations requiring more digits. Why are mathematicians and computer scientists not satisfied with, say, the first 50 digits of pi?

Several answers can be given. One is that the calculation of pi has become something of a benchmark computation: it serves as a measure of the sophistication and reliability of the computers that carry it out. In addition, the pursuit of ever more accurate values of pi leads mathematicians to intriguing and unexpected niches of number theory. Another and more ingenuous motivation is simply "because it's there." In fact, pi has been a fixture of mathematical culture for more than two and a half millenniums.

Furthermore, there is always the chance that such computations will shed light on some of the riddles surrounding pi, a universal constant that is not particularly well understood, in spite of its relatively elementary nature. For example, although it has been proved that pi cannot ever be exactly evaluated by subjecting positive integers to any combination of adding, subtracting, multiplying, dividing or extracting roots, no one has succeeded in proving that the digits of pi follow a random distribution (such that each number from 0 to 9 appears with equal frequency). It is possible, albeit highly unlikely, that after a while all the remaining digits of pi are 0's and 1's or exhibit some other regularity. Moreover, pi turns up in all kinds of unexpected places that have nothing to do with circles. If a number is picked at random from the set of integers, for instance, the probability that it will have no repeated prime divisors is six divided by the square of pi. No different from other eminent mathematicians, Ramanujan was prey to the fascinations of the number.

The ingredients of the recent approaches to calculating pi are among the mathematical treasures unearthed by renewed interest in Ramanujan's work. Much of what he did, however, is still inaccessible to investigators. The body of his work is contained in his "Notebooks," which are personal records written in his own nomenclature. To make matters more frustrating for mathematicians who have studied the "Notebooks," Ramanujan generally did not include formal proofs for his theorems. The task of deciphering and editing the "Notebooks" is only now nearing completion, by Bruce C. Berndt of the University of Illinois at Urbana-Champaign.

To our knowledge no mathematical redaction of this scope or difficulty has ever been attempted. The effort is certainly worthwhile. Ramanujan's legacy in the "Notebooks" promises not only to enrich pure mathematics but also to find application in various fields of mathematical physics. Rodney J. Baxter of the Australian National University, for example, acknowledges that Ramanujan's findings helped him to solve such problems in statistical mechanics as the so-called hard-hexagon model, which considers the behavior of a system of interacting particles laid out on a honeycomblike grid. Similarly, Carlos J. Moreno of the City University of New York and Freeman J. Dyson of the Institute for Advanced Study have pointed out that Ramanujan's work is beginning to be applied by physicists in superstring theory.

Ramanujan's stature as a mathematician is all the more astonishing when one considers his limited formal education. He was born on December 22, 1887, into a somewhat impoverished family of the Brahmin caste in the town of Erode in southern India and grew up in Kumbakonam, where his father was an accountant to a clothier. His mathematical precocity was recognized early, and at the age of seven he was given a scholarship to the Kumbakonam Town High School. He is said to have recited mathematical formulas to his schoolmates—including the value of pi to many places.

When he was 12, Ramanujan mastered the contents of S. L. Loney's rather comprehensive *Plane Trigonometry,* including its discussion of the sum and products of infinite sequences, which later were to figure prominently in his work. (An infinite sequence is an unending string of terms, often generated by a simple formula. In this context the interesting sequences are those whose terms can be added or multiplied to yield

ll 2

an identifiable, finite value. If the terms are added, the resulting expression is called a series; if they are multiplied, it is called a product.) Three years later he borrowed the *Synopsis of Elementary Results in Pure Mathematics,* a listing of some 6,000 theorems (most of them given without proof) compiled by G. S. Carr, a tutor at the University of Cambridge. Those two books were the basis of Ramanujan's mathematical training.

In 1903 Ramanujan was admitted to a local government college. Yet total absorption in his own mathematical diversions at the expense of everything else caused him to fail his examinations, a pattern repeated four years later at another college in Madras. Ramanujan did set his avocation aside—if only temporarily—to look for a job after his marriage in 1909. Fortunately in 1910 R. Ramachandra Rao, a well-to-do patron of mathematics, gave him a monthly stipend largely on the strength of favorable recommendations from various sympathetic Indian mathematicians and the findings he already had jotted down in the "Notebooks."

In 1912, wanting more conventional work, he took a clerical position in the Madras Port Trust, where the chairman was a British engineer, Sir Francis Spring, and the manager was V. Ramaswami Aiyar, the founder of the Indian Mathematical Society. They encouraged Ramanujan to communicate his results to three prominent British mathematicians. Two apparently did not respond; the one who did was G. H. Hardy of Cambridge, now regarded as the foremost British mathematician of the period.

Hardy, accustomed to receiving crank mail, was inclined to disregard Ramanujan's letter at first glance the day it arrived, January 16, 1913. But after dinner that night Hardy and a close colleague, John E. Littlewood, sat down to puzzle through a list of 120 formulas and theorems Ramanujan had appended to his letter. Some hours later they had reached a verdict: they were seeing the work of a genius and not a crackpot. (According to his own "pure-talent scale" of mathematicians, Hardy was later to rate Ramanujan a 100, Littlewood a 30 and himself a 25. The German mathematician David Hilbert, the most influential figure of the time, merited only an 80.) Hardy described the revelation and its consequences as the one romantic incident in his life. He wrote that some of Ramanujan's formulas defeated him

completely, and yet "they must be true, because if they were not true, no one would have had the imagination to invent them."

Hardy immediately invited Ramanujan to come to Cambridge. In spite of his mother's strong objections as well as his own reservations, Ramanujan set out for England in March of 1914. During the next five years Hardy and Ramanujan worked together at Trinity College. The blend of Hardy's technical expertise and Ramanujan's raw brilliance produced an unequaled collaboration. They published a series of seminal papers on the properties of various arithmetic functions, laying the groundwork for the answer to such questions as: How many prime divisors is a given number likely to have? How many ways can one express a number as a sum of smaller positive integers?

In 1917 Ramanujan was made a Fellow of the Royal Society of London and a Fellow of Trinity College—the first Indian to be awarded either honor. Yet as his prominence grew his health deteriorated sharply, a decline perhaps accelerated by the difficulty of maintaining a strict vegetarian diet in war-rationed England. Although Ramanujan was in and out of sanatoriums, he continued to pour forth new results. In 1919, when peace made travel abroad safe again, Ramanujan returned to India. Already an icon for young Indian intellectuals, the 32-year-old Ramanujan died on April 26, 1920, of what was then diagnosed as tuberculosis but now is thought to have been a severe vitamin deficiency. True to mathematics until the end, Ramanujan did not slow down during his last, painracked months, producing the re-

SRINIVASA RAMANUJAN, born in 1887 in India, managed in spite of limited formal education to reconstruct almost single-handedly much of the edifice of number theory and to go on to derive original theorems and formulas. Like many illustrious mathematicians before him, Ramanujan was fascinated by pi: the ratio of any circle's circumference to its diameter. Based on his investigation of modular equations (*see box on page 114*), he formulated exact expressions for pi and derived from them approximate values. As a result of the work of various investigators (including the authors), Ramanujan's methods are now better understood and have been implemented as algorithms.

112A

markable work recorded in his so-called "Lost Notebook."

Ramanujan's work on pi grew in large part out of his investigation of modular equations, perhaps the most thoroughly treated subject in the "Notebooks." Roughly speaking, a modular equation is an algebraic relation between a function expressed in terms of a variable x—in mathematical notation, $f(x)$—and the same function expressed in terms of x raised to an integral power, for example $f(x^2)$, $f(x^3)$ or $f(x^4)$. The "order" of the modular equation is given by the integral power. The simplest modular equation is the second-order one: $f(x) = 2\sqrt{f(x^2)}/[1 + f(x^2)]$. Of course, not every function will satisfy a modular equation, but there is a class of functions, called modular functions, that do. These functions have various surprising symmetries that give them a special place in mathematics.

Ramanujan was unparalleled in his ability to come up with solutions to modular equations that also satisfy other conditions. Such solutions are called singular values. It turns out that solving for singular values in certain cases yields numbers whose natural logarithms coincide with pi (times a constant) to a surprising number of places [see box on page 114]. Applying this general approach with extraordinary virtuosity, Ramanujan produced many remarkable infinite series as well as single-term approximations for pi. Some of them are given in Ramanujan's one formal paper on the subject, *Modular Equations and Approximations to* π, published in 1914.

Ramanujan's attempts to approximate pi are part of a venerable tradition. The earliest Indo-European civilizations were aware that the area of a circle is proportional to the square of its radius and that the circumference of a circle is directly proportional to its diameter. Less clear, however, is when it was first realized that the ratio of any circle's circumference to its diameter and the ratio of any circle's area to the square of its radius are in fact the same constant, which today is designated by the symbol π. (The symbol, which gives the constant its name, is a latecomer in the history of mathematics, having been introduced in 1706 by the English mathematical writer William Jones and popularized by the Swiss mathematician Leonhard Euler in the 18th century.)

Archimedes of Syracuse, the greatest mathematician of antiquity, rigorously established the equivalence of the two ratios in his treatise *Measurement of a Circle*. He also calculated a value for pi based on mathematical principles rather than on direct measurement of a circle's circumference, area and diameter. What Archimedes did was to inscribe and circumscribe regular polygons (polygons whose sides are all the same length) on a circle assumed to have a diameter of one unit and to consider

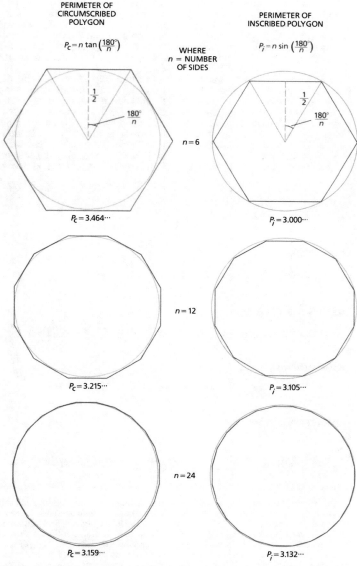

PERIMETER OF CIRCUMSCRIBED POLYGON

$$P_C = n\tan\left(\frac{180°}{n}\right)$$

WHERE
n = NUMBER OF SIDES

PERIMETER OF INSCRIBED POLYGON

$$P_i = n\sin\left(\frac{180°}{n}\right)$$

$\frac{1}{2}$ $\frac{180°}{n}$ $n=6$ $\frac{1}{2}$ $\frac{180°}{n}$

$P_C = 3.464\cdots$ $P_i = 3.000\cdots$

$n=12$

$P_C = 3.215\cdots$ $P_i = 3.105\cdots$

$n=24$

$P_C = 3.159\cdots$ $P_i = 3.132\cdots$

ARCHIMEDES' METHOD for estimating pi relied on inscribed and circumscribed regular polygons (polygons with sides of equal length) on a circle having a diameter of one unit (or a radius of half a unit). The perimeters of the inscribed and circumscribed polygons served respectively as lower and upper bounds for the value of pi. The sine and tangent functions can be used to calculate the polygons' perimeters, as is shown here, but Archimedes had to develop equivalent relations based on geometric constructions. Using 96-sided polygons, he determined that pi is greater than $3^{10}/_{71}$ and less than $3^{1}/_{7}$.

the polygons' respective perimeters as lower and upper bounds for possible values of the circumference of the circle, which is numerically equal to pi [*see illustration on opposite page*].

This method of approaching a value for pi was not novel: inscribing polygons of ever more sides in a circle had been proposed earlier by Antiphon, and Antiphon's contemporary, Bryson of Heraclea, had added circumscribed polygons to the procedure. What was novel was Archimedes' correct determination of the effect of doubling the number of sides on both the circumscribed and the inscribed polygons. He thereby developed a procedure that, when repeated enough times, enables one in principle to calculate pi to any number of digits. (It should be pointed out that the perimeter of a regular polygon can be readily calculated by means of simple trigonometric functions: the sine, cosine and tangent functions. But in Archimedes' time, the third century B.C., such functions were only partly understood. Archimedes therefore had to rely mainly on geometric constructions, which made the calculations considerably more demanding than they might appear today.)

Archimedes began with inscribed and circumscribed hexagons, which yield the inequality $3 < \pi < 2\sqrt{3}$. By doubling the number of sides four times, to 96, he narrowed the range of pi to between $3\frac{10}{71}$ and $3\frac{1}{7}$, obtaining the estimate $\pi \approx 3.14$. There is some evidence that the extant text of *Measurement of a Circle* is only a fragment of a larger work in which Archimedes described how, starting with decagons and doubling them six times, he got a five-digit estimate: $\pi \approx 3.1416$.

Archimedes' method is conceptually simple, but in the absence of a ready way to calculate trigonometric functions it requires the extraction of roots, which is rather time-consuming when done by hand. Moreover, the estimates converge slowly to pi: their error decreases by about a factor of four per iteration. Nevertheless, all European attempts to calculate pi before the mid-17th century relied in one way or another on the method. The 16th-century Dutch mathematician Ludolph van Ceulen dedicated much of his career to a computation of pi. Near the end of his life he obtained a 32-digit estimate by calculating the perimeter of inscribed and circumscribed polygons having 2^{62} (some 10^{18}) sides. His value for pi, called the Ludolphian num-

ber in parts of Europe, is said to have served as his epitaph.

The development of calculus, largely by Isaac Newton and Gottfried Wilhelm Leibniz, made it possible to calculate pi much more expeditiously. Calculus provides efficient techniques for computing a function's derivative (the rate of change in the function's value as its variables change) and its integral (the sum of the function's-values over a range of variables). Applying the techniques, one can demonstrate that inverse trigonometric functions are given by integrals of quadratic functions that describe the curve of a circle. (The inverse of a trigonometric function gives the angle that corresponds to a particular value of the function. For example, the inverse tangent of 1 is 45 degrees or, equivalently, $\pi/4$ radians.)

(The underlying connection between trigonometric functions and algebraic expressions can be appreciated by considering a circle that has a radius of one unit and its center at the origin of a Cartesian x-y plane. The equation for the circle—whose area is numerically equal to pi—is $x^2 + y^2 = 1$, which is a restatement of the Pythagorean theorem for a right triangle with a hypotenuse equal to 1. Moreover, the sine and cosine of the angle between the positive x axis and any point on the circle are equal respectively to the point's coordinates, y and x; the angle's tangent is simply y/x.)

Of more importance for the purposes of calculating pi, however, is the fact that an inverse trigonometric function can be "expanded" as a series, the terms of which are computable from the derivatives of the function. Newton himself calculated pi to 15 places by adding the first few terms of a series that can be derived

WALLIS' PRODUCT (1665)

$$\frac{\pi}{2} = \frac{2 \times 2}{1 \times 3} \times \frac{4 \times 4}{3 \times 5} \times \frac{6 \times 6}{5 \times 7} \times \frac{8 \times 8}{7 \times 9} \times \cdots = \prod_{n=1}^{\infty} \frac{4n^2}{4n^2 - 1}$$

GREGORY'S SERIES (1671)

$$\frac{\pi}{4} = 1 - \frac{1}{3} + \frac{1}{5} - \frac{1}{7} + \cdots = \sum_{n=0}^{\infty} \frac{(-1)^n}{2n+1}$$

MACHIN'S FORMULA (1706)

$$\frac{\pi}{4} = 4 \arctan(1/5) - \arctan(1/239), \quad \text{where } \arctan X = X - \frac{X^3}{3} + \frac{X^5}{5} - \frac{X^7}{7} + \cdots = \sum_{n=0}^{\infty} (-1)^n \frac{X^{(2n+1)}}{2n+1}$$

RAMANUJAN (1914)

$$\frac{1}{\pi} = \frac{\sqrt{8}}{9,801} \sum_{n=0}^{\infty} \frac{(4n)! [1,103 + 26,390n]}{(n!)^4 \, 396^{4n}}, \quad \text{where } n! = n \times (n-1) \times (n-2) \times \cdots \times 1 \text{ and } 0! = 1$$

BORWEIN AND BORWEIN (1987)

$$\frac{1}{\pi} = $$

$$12 \sum_{n=0}^{\infty} \frac{(-1)^n (6n)! [212,175,710,912\sqrt{61} + 1,657,145,277,365 + n(13,773,980,892,672\sqrt{61} + 107,578,229,802,750)]}{(n)!^3 (3n)! [5,280(236,674 + 30,303\sqrt{61})]^{(3n - 3/2)}}$$

TERMS OF MATHEMATICAL SEQUENCES can be summed or multiplied to yield values for pi (divided by a constant) or its reciprocal. The first two sequences, discovered respectively by the mathematicians John Wallis and James Gregory, are probably among the best-known, but they are practically useless for computational purposes. Not even 100 years of computing on a supercomputer programmed to add or multiply the terms of either sequence would yield 100 digits of pi. The formula discovered by John Machin made the calculation of pi feasible, since calculus allows the inverse tangent (arc tangent) of a number, x, to be expressed in terms of a sequence whose sum converges more rapidly to the value of the arc tangent the smaller x is. Virtually all calculations for pi from the beginning of the 18th century until the early 1970's have relied on variations of Machin's formula. The sum of Ramanujan's sequence converges to the true value of $1/\pi$ much faster: each successive term in the sequence adds roughly eight more correct digits. The last sequence, formulated by the authors, adds about 25 digits per term; the first term (for which n is 0) yields a number that agrees with pi to 24 digits.

113

as an expression for the inverse of the sine function. He later confessed to a colleague: "I am ashamed to tell you to how many figures I carried these calculations, having no other business at the time."

In 1674 Leibniz derived the formula $1 - 1/3 + 1/5 - 1/7\ldots = \pi/4$, which is the inverse tangent of 1. (The general inverse-tangent series was originally discovered in 1671 by the Scottish mathematician James Gregory. Indeed, similar expressions appear to have been developed independently several centuries earlier in India.) The error of the approximation, defined as the difference between the sum of n terms and the exact value of $\pi/4$, is roughly equal to the $n+1$th term in the series. Since the denominator of each successive term increases by only 2, one must add approximately 50 terms to get two-digit accuracy, 500 terms for three-digit accuracy and so on. Summing the terms of the series to calculate a value for pi more than a few digits long is clearly prohibitive.

An observation made by John Ma-

chin, however, made it practicable to calculate pi by means of a series expansion for the inverse-tangent function. He noted that pi divided by 4 is equal to 4 times the inverse tangent of 1/5 minus the inverse tangent of 1/239. Because the inverse-tangent series for a given value converges more quickly the smaller the value is, Machin's formula greatly simplified the calculation. Coupling his formula with the series expansion for the inverse tangent, Machin computed 100 digits of pi in 1706. Indeed, his technique proved to be so powerful that all extended calculations of pi from the beginning of the 18th century until recently relied on variants of the method.

Two 19th-century calculations deserve special mention. In 1844 Johann Dase computed 205 digits of pi in a matter of months by calculating the values of three inverse tangents in a Machin-like formula. Dase was a calculating prodigy who could multiply 100-digit numbers entirely in his head—a feat that took him rough-

ly eight hours. (He was perhaps the closest precursor of the modern supercomputer, at least in terms of memory capacity.) In 1853 William Shanks outdid Dase by publishing his computation of pi to 607 places, although the digits that followed the 527th place were wrong. Shank's task took years and was a rather routine, albeit laborious, application of Machin's formula. (In what must itself be some kind of record, 92 years passed before Shank's error was detected, in a comparison between his value and a 530-place approximation produced by D. F. Ferguson with the aid of a mechanical calculator.)

The advent of the digital computer saw a renewal of efforts to calculate ever more digits of pi, since the machine was ideally suited for lengthy, repetitive "number crunching." ENIAC, one of the first digital computers, was applied to the task in June, 1949, by John von Neumann and his colleagues. ENIAC produced 2,037 digits in 70 hours. In 1957 G. E. Felton attempted to compute 10,000 digits of pi, but owing to a machine error only the first 7,480 digits were correct. The 10,000-digit goal was reached by F. Genuys the following year on an IBM 704 computer. In 1961 Daniel Shanks and John W. Wrench, Jr., calculated 100,000 digits of pi in less than nine hours on an IBM 7090. The million-digit mark was passed in 1973 by Jean Guilloud and M. Bouyer, a feat that took just under a day of computation on a CDC 7600. (The computations done by Shanks and Wrench and by Guilloud and Bouyer were in fact carried out twice using different inverse-tangent identities for pi. Given the history of both human and machine error in these calculations, it is only after such verification that modern "digit hunters" consider a record officially set.)

Although an increase in the speed of computers was a major reason ever more accurate calculations for pi could be performed, it soon became clear that there were inescapable limits. Doubling the number of digits lengthens computing time by at least a factor of four, if one applies the traditional methods of performing arithmetic in computers. Hence even allowing for a hundredfold increase in computational speed, Guilloud and Bouyer's program would have required at least a quarter century to produce a billion-digit value for pi. From the perspective of the early 1970's such a computation did not seem realistically practicable.

Yet the task is now feasible, thanks

MODULAR FUNCTIONS AND APPROXIMATIONS TO PI

A modular function is a function, $\lambda(q)$, that can be related through an algebraic expression called a modular equation to the same function expressed in terms of the same variable, q, raised to an integral power: $\lambda(q^p)$. The integral power, p, determines the "order" of the modular equation. An example of a modular function is

$$\lambda(q) = 16q \prod_{n=1}^{\infty} \left(\frac{1+q^{2n}}{1+q^{2n-1}}\right)^8.$$

Its associated seventh-order modular equation, which relates $\lambda(q)$ to $\lambda(q^7)$, is given by

$$\sqrt[8]{\lambda(q)\lambda(q^7)} + \sqrt[8]{[1-\lambda(q)][1-\lambda(q^7)]} = 1.$$

Singular values are solutions of modular equations that must also satisfy additional conditions. One class of singular values corresponds to computing a sequence of values, k_p, where

$$k_p = \sqrt{\lambda(e^{-\pi\sqrt{p}})}$$

and p takes integer values. These values have the curious property that the logarithmic expression

$$\frac{-2}{\sqrt{p}} \log\left(\frac{k_p}{4}\right)$$

coincides with many of the first digits of pi. The number of digits the expression has in common with pi increases with larger values of p.

Ramanujan was unparalleled in his ability to calculate these singular values. One of his most famous is the value when p equals 210, which was included in his original letter to G. H. Hardy. It is

$$k_{210} = (\sqrt{2}-1)^2(2-\sqrt{3})(\sqrt{7}-\sqrt{6})^2(8-3\sqrt{7})(\sqrt{10}-3)^2(\sqrt{15}-\sqrt{14})(4-\sqrt{15})^2(6-\sqrt{35}).$$

This number, when plugged into the logarithmic expression, agrees with pi through the first 20 decimal places. In comparison, k_{240} yields a number that agrees with pi through more than one million digits.

Applying this general approach, Ramanujan constructed a number of remarkable series for pi, including the one shown in the illustration on the preceding page. The general approach also underlies the two-step, iterative algorithms in the top illustration on the opposite page. In each iteration the first step (calculating y_n) corresponds to computing one of a sequence of singular values by solving a modular equation of the appropriate order; the second step (calculating α_n) is tantamount to taking the logarithm of the singular value.

114

not only to faster computers but also to new, efficient methods for multiplying large numbers in computers. A third development was also crucial: the advent of iterative algorithms that quickly converge to pi. (An iterative algorithm can be expressed as a computer program that repeatedly performs the same arithmetic operations, taking the output of one cycle as the input for the next.) These algorithms, some of which we constructed, were in many respects anticipated by Ramanujan, although he knew nothing of computer programming. Indeed, computers not only have made it possible to apply Ramanujan's work but also have helped to unravel it. Sophisticated algebraic-manipulation software has allowed further exploration of the road Ramanujan traveled alone and unaided 75 years ago.

One of the interesting lessons of theoretical computer science is that many familiar algorithms, such as the way children are taught to multiply in grade school, are far from optimal. Computer scientists gauge the efficiency of an algorithm by determining its bit complexity: the number of times individual digits are added or multiplied in carrying out an algorithm. By this measure, adding two n-digit numbers in the normal way has a bit complexity that increases in step with n; multiplying two n-digit numbers in the normal way has a bit complexity that increases as n^2. By traditional methods, multiplication is much "harder" than addition in that it is much more time-consuming.

Yet, as was shown in 1971 by A. Schönhage and V. Strassen, the multiplication of two numbers can in theory have a bit complexity only a little greater than addition. One way to achieve this potential reduction in bit complexity is to implement so-called fast Fourier transforms (FFT's). FFT-based multiplication of two large numbers allows the intermediary computations among individual digits to be carefully orchestrated so that redundancy is avoided. Because division and root extraction can be reduced to a sequence of multiplications, they too can have a bit complexity just slightly greater than that of addition. The result is a tremendous saving in bit complexity and hence in computation time. For this reason all recent efforts to calculate pi rely on some variation of the FFT technique for multiplication.

Yet for hundreds of millions of dig-

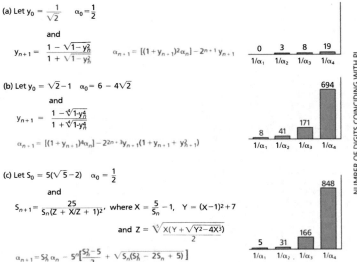

(a) Let $y_0 = \dfrac{1}{\sqrt{2}}$ $\alpha_0 = \dfrac{1}{2}$

and

$y_{n+1} = \dfrac{1 - \sqrt{1 - y_n^2}}{1 + \sqrt{1 - y_n^2}}$ $\alpha_{n+1} = [(1 + y_{n+1})^2 \alpha_n] - 2^{n+1} y_{n+1}$

(b) Let $y_0 = \sqrt{2} - 1$ $\alpha_0 = 6 - 4\sqrt{2}$

and

$y_{n+1} = \dfrac{1 - \sqrt[4]{1 - y_n^4}}{1 + \sqrt[4]{1 - y_n^4}}$

$\alpha_{n+1} = [(1 + y_{n+1})^4 \alpha_n] - 2^{2n+3} y_{n+1}(1 + y_{n+1} + y_{n+1}^2)$

(c) Let $S_0 = 5(\sqrt{5} - 2)$ $\alpha_0 = \dfrac{1}{2}$

and

$S_{n+1} = \dfrac{25}{S_n(Z + X/Z + 1)^2}$, where $X = \dfrac{5}{S_n} - 1$, $Y = (X-1)^2 + 7$

and $Z = \sqrt[5]{\dfrac{X(Y + \sqrt{Y^2 - 4X^3})}{2}}$

$\alpha_{n+1} = S_n^2 \alpha_n - 5^n\left[\dfrac{S_n^2 - 5}{2} + \sqrt{S_n(S_n^2 - 2S_n + 5)}\right]$

ITERATIVE ALGORITHMS that yield extremely accurate values of pi were developed by the authors. (An iterative algorithm is a sequence of operations repeated in such a way that the ouput of one cycle is taken as the input for the next.) Algorithm a converges to $1/\pi$ quadratically: the number of correct digits given by α_n more than doubles each time n is increased by 1. Algorithm b converges quartically and algorithm c converges quintically, so that the number of coinciding digits given by each iteration increases respectively by more than a factor of four and by more than a factor of five. Algorithm b is possibly the most efficient known algorithm for calculating pi; it was run on supercomputers in the last three record-setting calculations. As the authors worked on the algorithms it became clear to them that Ramanujan had pursued similar methods in coming up with his approximations for pi. In fact, the computation of s_n in algorithm c rests on a remarkable fifth-order modular equation discovered by Ramanujan.

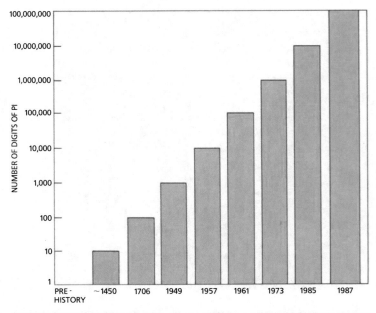

NUMBER OF KNOWN DIGITS of pi has increased by two orders of magnitude (factors of 10) in the past decade as a result of the development of iterative algorithms that can be run on supercomputers equipped with new, efficient methods of multiplication.

its of pi to be calculated practically a beautiful formula known a century and a half earlier to Carl Friedrich Gauss had to be rediscovered. In the mid-1970's Richard P. Brent and Eugene Salamin independently noted that the formula produced an algorithm for pi that converged quadratically, that is, the number of digits doubled with each iteration. Between 1983 and the present Yasumasa Kanada and his colleagues at the University of Tokyo have employed this algorithm to set several world records for the number of digits of pi.

We wondered what underlies the remarkably fast convergence to pi of the Gauss-Brent-Salamin algorithm, and in studying it we developed general techniques for the construction of similar algorithms that rapidly converge to pi as well as to other quantities. Building on a theory outlined by the German mathematician

Karl Gustav Jacob Jacobi in 1829, we realized we could in principle arrive at a value for pi by evaluating integrals of a class called elliptic integrals, which can serve to calculate the perimeter of an ellipse. (A circle, the geometric setting of previous efforts to approximate pi, is simply an ellipse with axes of equal length.)

Elliptic integrals cannot generally be evaluated as integrals, but they can be easily approximated through iterative procedures that rely on modular equations. We found that the Gauss-Brent-Salamin algorithm is actually a specific case of our more general technique relying on a second-order modular equation. Quicker convergence to the value of the integral, and thus a faster algorithm for pi, is possible if higher-order modular equations are used, and so we have also constructed various algorithms based on modular equations

of third, fourth and higher orders.

In January, 1986, David H. Bailey of the National Aeronautics and Space Administration's Ames Research Center produced 29,360,000 decimal places of pi by iterating one of our algorithms 12 times on a Cray-2 supercomputer. Because the algorithm is based on a fourth-order modular equation, it converges on pi quartically, more than quadrupling the number of digits with each iteration. A year later Kanada and his colleagues carried out one more iteration to attain 134,217,000 places on an NEC SX-2 supercomputer and thereby verified a similar computation they had done earlier using the Gauss-Brent-Salamin algorithm. (Iterating our algorithm twice more—a feat entirely feasible if one could somehow monopolize a supercomputer for a few weeks—would yield more than two billion digits of pi.)

Iterative methods are best suited for calculating pi on a computer, and so it is not surprising that Ramanujan never bothered to pursue them. Yet the basic ingredients of the iterative algorithms for pi—modular equations in particular—are to be found in Ramanujan's work. Parts of his original derivation of infinite series and approximations for pi more than three-quarters of a century ago must have paralleled our own efforts to come up with algorithms for pi. Indeed, the formulas he lists in his paper on pi and in the "Notebooks" helped us greatly in the construction of some of our algorithms. For example, although we were able to prove that an 11th-order algorithm exists and knew its general formulation, it was not until we stumbled on Ramanujan's modular equations of the same order that we discovered its unexpectedly simple form.

Conversely, we were also able to derive all Ramanujan's series from the general formulas we had developed. The derivation of one, which converged to pi faster than any other series we knew at the time, came about with a little help from an unexpected source. We had justified all the quantities in the expression for the series except one: the coefficient 1,103, which appears in the numerator of the expression [see illustration on page 113]. We were convinced—as Ramanujan must have been—that 1,103 had to be correct. To prove it we had either to simplify a daunting equation containing variables raised to powers of several thousand or

---

HOW TO GET TWO BILLION DIGITS OF PI WITH A CALCULATOR*

Let

$y_0 = \sqrt{2} - 1$

$\alpha_0 = 6 - 4\sqrt{2}$

$y_1 = [1 - \sqrt[4]{1 - y_0{}^4}]/[1 + \sqrt[4]{1 - y_0{}^4}]$

$\alpha_1 = (1 + y_1)^4 \alpha_0 - 2^3 y_1(1 + y_1 + y_1{}^2)$

$y_2 = [1 - \sqrt[4]{1 - y_1{}^4}]/[1 + \sqrt[4]{1 - y_1{}^4}]$

$\alpha_2 = (1 + y_2)^4 \alpha_1 - 2^5 y_2(1 + y_2 + y_2{}^2)$

$y_3 = [1 - \sqrt[4]{1 - y_2{}^4}]/[1 + \sqrt[4]{1 - y_2{}^4}]$

$\alpha_3 = (1 + y_3)^4 \alpha_2 - 2^7 y_3(1 + y_3 + y_3{}^2)$

$y_4 = [1 - \sqrt[4]{1 - y_3{}^4}]/[1 + \sqrt[4]{1 - y_3{}^4}]$

$\alpha_4 = (1 + y_4)^4 \alpha_3 - 2^9 y_4(1 + y_4 + y_4{}^2)$

$y_5 = [1 - \sqrt[4]{1 - y_4{}^4}]/[1 + \sqrt[4]{1 - y_4{}^4}]$

$\alpha_5 = (1 + y_5)^4 \alpha_4 - 2^{11} y_5(1 + y_5 + y_5{}^2)$

$y_6 = [1 - \sqrt[4]{1 - y_5{}^4}]/[1 + \sqrt[4]{1 - y_5{}^4}]$

$\alpha_6 = (1 + y_6)^4 \alpha_5 - 2^{13} y_6(1 + y_6 + y_6{}^2)$

$y_7 = [1 - \sqrt[4]{1 - y_6{}^4}]/[1 + \sqrt[4]{1 - y_6{}^4}]$

$\alpha_7 = (1 + y_7)^4 \alpha_6 - 2^{15} y_7(1 + y_7 + y_7{}^2)$

$y_8 = [1 - \sqrt[4]{1 - y_7{}^4}]/[1 + \sqrt[4]{1 - y_7{}^4}]$

$\alpha_8 = (1 + y_8)^4 \alpha_7 - 2^{17} y_8(1 + y_8 + y_8{}^2)$

$y_9 = [1 - \sqrt[4]{1 - y_8{}^4}]/[1 + \sqrt[4]{1 - y_8{}^4}]$

$\alpha_9 = (1 + y_9)^4 \alpha_8 - 2^{19} y_9(1 + y_9 + y_9{}^2)$

$y_{10} = [1 - \sqrt[4]{1 - y_9{}^4}]/[1 + \sqrt[4]{1 - y_9{}^4}]$

$\alpha_{10} = (1 + y_{10})^4 \alpha_9 - 2^{21} y_{10}(1 + y_{10} + y_{10}{}^2)$

$y_{11} = [1 - \sqrt[4]{1 - y_{10}{}^4}]/[1 + \sqrt[4]{1 - y_{10}{}^4}]$

$\alpha_{11} = (1 + y_{11})^4 \alpha_{10} - 2^{23} y_{11}(1 + y_{11} + y_{11}{}^2)$

$y_{12} = [1 - \sqrt[4]{1 - y_{11}{}^4}]/[1 + \sqrt[4]{1 - y_{11}{}^4}]$

$\alpha_{12} = (1 + y_{12})^4 \alpha_{11} - 2^{25} y_{12}(1 + y_{12} + y_{12}{}^2)$

$y_{13} = [1 - \sqrt[4]{1 - y_{12}{}^4}]/[1 + \sqrt[4]{1 - y_{12}{}^4}]$

$\alpha_{13} = (1 + y_{13})^4 \alpha_{12} - 2^{27} y_{13}(1 + y_{13} + y_{13}{}^2)$

$y_{14} = [1 - \sqrt[4]{1 - y_{13}{}^4}]/[1 + \sqrt[4]{1 - y_{13}{}^4}]$

$\alpha_{14} = (1 + y_{14})^4 \alpha_{13} - 2^{29} y_{14}(1 + y_{14} + y_{14}{}^2)$

$y_{15} = [1 - \sqrt[4]{1 - y_{14}{}^4}]/[1 + \sqrt[4]{1 - y_{14}{}^4}]$

$\alpha_{15} = (1 + y_{15})^4 \alpha_{14} - 2^{31} y_{15}(1 + y_{15} + y_{15}{}^2)$

$1/\alpha_{15}$ agress with π for more than two billion decimal digits

*Of course, the calculator needs to have a two-billion-digit display; on a pocket calculator the computation would not be very interesting after the second iteration.

EXPLICIT INSTRUCTIONS for executing algorithm b in the top illustration on the preceding page makes it possible in principle to compute the first two billion digits of pi in a matter of minutes. All one needs is a calculator that has two memory registers and the usual capacity to add, subtract, multiply, divide and extract roots. Unfortunately most calculators come with only an eight-digit display, which makes the computation moot.

to delve considerably further into somewhat arcane number theory.

By coincidence R. William Gosper, Jr., of Symbolics, Inc., had decided in 1985 to exploit the same series of Ramanujan's for an extended-accuracy value for pi. When he carried out the calculation to more than 17 million digits (a record at the time), there was to his knowledge no proof that the sum of the series actually converged to pi. Of course, he knew that millions of digits of his value coincided with an earlier Gauss-Brent-Salamin calculation done by Kanada. Hence the possibility of error was vanishingly small.

As soon as Gosper had finished his calculation and verified it against Kanada's, however, we had what we needed to prove that 1,103 was the number needed to make the series true to within one part in $10^{10,000,000}$. In much the same way that a pair of integers differing by less than 1 must be equal, his result sufficed to specify the number: it is precisely 1,103. In effect, Gosper's computation became part of our proof. We knew that the series (and its associated algorithm) is so sensitive to slight inaccuracies that if Gosper had used any other value for the coefficient or, for that matter, if the computer had introduced a single-digit error during the calculation, he would have ended up with numerical nonsense instead of a value for pi.

Ramanujan-type algorithms for approximating pi can be shown to be very close to the best possible. If all the operations involved in the execution of the algorithms are totaled (assuming that the best techniques known for addition, multiplication and root extraction are applied), the bit complexity of computing n digits of pi is only marginally greater than that of multiplying two n-digit numbers. But multiplying two n-digit numbers by means of an FFT-based technique is only marginally more complicated than summing two n-digit numbers, which is the simplest of the arithmetic operations possible on a computer.

Mathematics has probably not yet felt the full impact of Ramanujan's genius. There are many other wonderful formulas contained in the "Notebooks" that revolve around integrals, infinite series and continued fractions (a number plus a fraction, whose denominator can be expressed as a number plus a fraction, whose denominator can be ex-

RAMANUJAN'S "NOTEBOOKS" were personal records in which he jotted down many of his formulas. The page shown contains various third-order modular equations—all in Ramanujan's nonstandard notation. Unfortunately Ramanujan did not bother to include formal proofs for the equations; others have had to compile, edit and prove them. The formulas in the "Notebooks" embody subtle relations among numbers and functions that can be applied in other fields of mathematics or even in theoretical physics.

pressed as a number plus a fraction, and so on). Unfortunately they are listed with little—if any—indication of the method by which Ramanujan proved them. Littlewood wrote: "If a significant piece of reasoning occurred somewhere, and the total mixture of evidence and intuition gave him certainty, he looked no further."

The herculean task of editing the "Notebooks," initiated 60 years ago by the British analysts G. N. Watson and B. N. Wilson and now being completed by Bruce Berndt, requires providing a proof, a source or an occasional correction for each of many thousands of asserted theorems and identities. A single line in the "Notebooks" can easily elicit many pages

of commentary. The task is made all the more difficult by the nonstandard mathematical notation in which the formulas are written. Hence a great deal of Ramanujan's work will not become accessible to the mathematical community until Berndt's project is finished.

Ramanujan's unique capacity for working intuitively with complicated formulas enabled him to plant seeds in a mathematical garden (to borrow a metaphor from Freeman Dyson) that is only now coming into bloom. Along with many other mathematicians, we look forward to seeing which of the seeds will germinate in future years and further beautify the garden.

117

11. Ramanujan, modular equations, and approximations to pi or how to compute one billion digits of pi (1989)

Paper 11: Jonathan M. Borwein, Peter B. Borwein, and David H. Bailey, "Ramanujan, modular equations, and approximations to pi, or how to compute one billion digits of pi," *American Mathematical Monthly*, vol. 96 (1989), p. 201–219. Copyright 1989 Mathematical Association of America. All Rights Reserved.

Synopsis:

Here the authors reprise the history of π, describe the recently discovered quadratically convergent algorithms, describe (briefly) the computational techniques required to compute, say, one billion digits of π, and then explore in considerable detail the connections between these mathematical developments and the recently uncovered writings of Ramanujan, the hundredth anniversary of whose birth was celebrated in 1987. As the article observes, if Ramanujan had accomplished so much with paper, pencil and slateboard, how much further could he have seen if he had access to a modern symbolic computing environment such as *Maple* or *Mathematica*?

Interestingly, although not as asymptotically fast as the quadratically convergent formulas, Ramanujan discovered some very interesting formulas, e.g.

$$\frac{1}{\pi} = \frac{\sqrt{8}}{9801} \sum_{n=0}^{\infty} \frac{(4n)!(1103 + 26390n)}{(n!)^4 396^{4n}},$$

a variation of which was used by David and Gregory Chudnovsky to compute up to two billion digits (when accompanied with some very clever reorganization of the computation) in the early 1990s. Each term of this formula adds roughly eight correct digits.

Keywords: Algorithms, Computation, Elliptic Integrals, History, Modular Equations

© Springer International Publishing Switzerland 2016
D.H. Bailey, J.M. Borwein, *Pi: The Next Generation*,
DOI 10.1007/978-3-319-32377-0_11

Ramanujan, Modular Equations, and Approximations to Pi or How to Compute One Billion Digits of Pi

J. M. BORWEIN AND P. B. BORWEIN

Mathematics Department, Dalhousie University, Halifax, N.S. B3H 3J5 Canada

and

D. H. BAILEY

NASA Ames Research Center, Moffett Field, CA 94035

Preface. The year 1987 was the centenary of Ramanujan's birth. He died in 1920 Had he not died so young, his presence in modern mathematics might be more immediately felt. Had he lived to have access to powerful algebraic manipulation software, such as MACSYMA, who knows how much more spectacular his already astonishing career might have been.

This article will follow up one small thread of Ramanujan's work which has found a modern computational context, namely, one of his approaches to approximating pi. Our experience has been that as we have come to understand these pieces of Ramanujan's work, as they have become mathematically demystified, and as we have come to realize the intrinsic complexity of these results, we have come to realize how truly singular his abilities were. This article attempts to present a considerable amount of material and, of necessity, little is presented in detail. We have, however, given much more detail than Ramanujan provided. Our intention is that the circle of ideas will become apparent and that the finer points may be pursued through the indicated references.

1. Introduction. There is a close and beautiful connection between the transformation theory for elliptic integrals and the very rapid approximation of pi. This connection was first made explicit by Ramanujan in his 1914 paper "Modular Equations and Approximations to π" [26]. We might emphasize that Algorithms 1 and 2 are not to be found in Ramanujan's work, indeed no recursive approximation of π is considered, but as we shall see they are intimately related to his analysis. Three central examples are:

Sum 1. (Ramanujan)

$$\frac{1}{\pi} = \frac{\sqrt{8}}{9801} \sum_{n=0}^{\infty} \frac{(4n)!}{(n!)^4} \frac{[1103 + 26390n]}{396^{4n}}.$$

Algorithm 1. Let $\alpha_0 := 6 - 4\sqrt{2}$ and $y_0 := \sqrt{2} - 1$.
Let

$$y_{n+1} := \frac{1 - \left(1 - y_n^4\right)^{1/4}}{1 + \left(1 - y_n^4\right)^{1/4}}$$

and

$$\alpha_{n+1} := (1 + y_{n+1})^4 \alpha_n - 2^{2n+3} y_{n+1}\left(1 + y_{n+1} + y_{n+1}^2\right).$$

Then

$$0 < \alpha_n - 1/\pi < 16 \cdot 4^n e^{-2 \cdot 4^n \pi}$$

and α_n converges to $1/\pi$ *quartically* (that is, with order four).

Algorithm 2. Let $s_0 := 5(\sqrt{5} - 2)$ and $\alpha_0 := 1/2$.
Let

$$s_{n+1} := \frac{25}{(z + x/z + 1)^2 s_n},$$

where

$$x := 5/s_n - 1 \qquad y := (x - 1)^2 + 7$$

and

$$z := \left[\frac{1}{2} x \left(y + \sqrt{y^2 - 4x^3} \right) \right]^{1/5}.$$

Let

$$\alpha_{n+1} := s_n^2 \alpha_n - 5^n \left\{ \frac{s_n^2 - 5}{2} + \sqrt{s_n \left(s_n^2 - 2s_n + 5 \right)} \right\}.$$

Then

$$0 < \alpha_n - \frac{1}{\pi} < 16 \cdot 5^n e^{-5^n \pi}$$

and α_n converges to $1/\pi$ *quintically* (that is, with order five).

Each additional term in Sum 1 adds roughly eight digits, each additional iteration of Algorithm 1 quadruples the number of correct digits, while each additional iteration of Algorithm 2 quintuples the number of correct digits. Thus a mere thirteen iterations of Algorithm 2 provide in excess of one billion decimal digits of pi. In general, for us, pth-order convergence of a sequence $\{\alpha_n\}$ to α means that α_n tends to α and that

$$|\alpha_{n+1} - \alpha| \le C |\alpha_n - \alpha|^p$$

for some constant $C > 0$. Algorithm 1 is arguably the most efficient algorithm currently known for the extended precision calculation of pi. While the rates of convergence are impressive, it is the subtle and thoroughly nontransparent nature of these results and the beauty of the underlying mathematics that intrigue us most.

Watson [37], commenting on certain formulae of Ramanujan, talks of

a thrill which is indistinguishable from the thrill which I feel when I enter the Sagrestia Nuovo of the Capella Medici and see before me the austere beauty of the four statues representing "Day," "Night," "Evening," and "Dawn" which Michelangelo has set over the tomb of Giuliano de'Medici and Lorenzo de'Medici.

Sum 1 is directly due to Ramanujan and appears in [26]. It rests on a modular identity of order 58 and, like much of Ramanujan's work, appears without proof and with only scanty motivation. The first complete derivation we know of appears in [11]. Algorithms 1 and 2 are based on modular identities of orders 4 and 5, respectively. The underlying quintic modular identity in Algorithm 2 (the relation for s_n) is also due to Ramanujan, though the first proof is due to Berndt and will appear in [7].

One intention in writing this article is to explain the genesis of Sum 1 and of Algorithms 1 and 2. It is not possible to give a short self-contained account without assuming an unusual degree of familiarity with modular function theory. Also, parts of the derivation involve considerable algebraic calculation and may most easily be done with the aid of a symbol manipulation package (MACSYMA, MAPLE, REDUCE, etc.). We hope however to give a taste of methods involved. The full details are available in [11].

A second intention is very briefly to describe the role of these and related approximations in the recent extended precision calculations of pi. In part this entails a short discussion of the complexity and implementation of such calculations. This centers on a discussion of multiplication by fast Fourier transform methods. Of considerable related interest is the fact that these algorithms for π are provably close to the theoretical optimum.

2. The State of Our Current Ignorance. Pi is almost certainly the most natural of the transcendental numbers, arising as the circumference of a circle of unit diameter. Thus, it is not surprising that its properties have been studied for some twenty-five hundred years. What is surprising is how little we actually know.

We know that π is irrational, and have known this since Lambert's proof of 1771 (see [5]). We have known that π is transcendental since Lindemann's proof of 1882 [23]. We also know that π is not a Liouville number. Mahler proved this in 1953. An irrational number β is *Liouville* if, for any n, there exist integers p and q so that

$$0 < \left| \beta - \frac{p}{q} \right| < \frac{1}{q^n}.$$

Liouville showed these numbers are all transcendental. In fact we know that

$$\left| \pi - \frac{p}{q} \right| > \frac{1}{q^{14.65}} \tag{2.1}$$

for p, q integral with q sufficiently large. This *irrationality estimate*, due to Chudnovsky and Chudnovsky [16] is certainly not best possible. It is likely that 14.65 should be replaced by $2 + \varepsilon$ for any $\varepsilon > 0$. Almost all transcendental numbers satisfy such an inequality. We know a few related results for the rate of algebraic approximation. The results may be pursued in [4] and [11].

We know that e^π is transcendental. This follows by noting that $e^\pi = (-1)^{-i}$ and applying the Gelfond-Schneider theorem [4]. We know that $\pi + \log 2 + \sqrt{2} \log 3$ is transcendental. This result is a consequence of the work that won Baker a Fields Medal in 1970. And we know a few more than the first two hundred million digits of the decimal expansion for π (Kanada, see Section 3).

The state of our ignorance is more profound. We do not know whether such basic constants as $\pi + e$, π/e, or $\log \pi$ are irrational, let alone transcendental. The best we can say about these three particular constants is that they cannot satisfy any polynomial of degree eight or less with integer coefficients of average size less than 10^9 [3]. This is a consequence of some recent computations employing the

Ferguson-Forcade algorithm [17]. We don't know anything of consequence about the single continued fraction of pi, except (numerically) the first 17 million terms, which Gosper computed in 1985 using Sum 1. Likewise, apart from listing the first many millions of digits of π, we know virtually nothing about the decimal expansion of π. It is possible, albeit not a good bet, that all but finitely many of the decimal digits of pi are in fact 0's and 1's. Carl Sagan's recent novel *Contact* rests on a similar possibility. Questions concerning the normality of or the distribution of digits of particular transcendentals such as π appear completely beyond the scope of current mathematical techniques. The evidence from analysis of the first thirty million digits is that they are very uniformly distributed [2]. The next one hundred and seventy million digits apparently contain no surprises.

In part we perhaps settle for computing digits of π because there is little else we can currently do. We would be amiss, however, if we did not emphasize that the extended precision calculation of pi has substantial application as a test of the "global integrity" of a supercomputer. The extended precision calculations described in Section 3 uncovered hardware errors which had to be corrected before those calculations could be successfully run. Such calculations, implemented as in Section 4, are apparently now used routinely to check supercomputers before they leave the factory. A large-scale calculation of pi is entirely unforgiving; it soaks into all parts of the machine and a single bit awry leaves detectable consequences.

3. Matters Computational

I am ashamed to tell you to how many figures I carried these calculations, having no other business at the time.

                                                                    Isaac Newton

Newton's embarrassment at having computed 15 digits, which he did using the arcsinlike formula

$$\pi = \frac{3\sqrt{3}}{4} + 24\left(\frac{1}{12} - \frac{1}{5 \cdot 2^5} - \frac{1}{28 \cdot 2^7} - \frac{1}{72 \cdot 2^9} - \cdots \right)$$

$$= \frac{3\sqrt{3}}{4} + 24\int_0^{\frac{1}{4}} \sqrt{x - x^2}\, dx,$$

is indicative both of the spirit in which people calculate digits and the fact that a surprising number of people have succumbed to the temptation [5].

The history of efforts to determine an accurate value for the constant we now know as π is almost as long as the history of civilization itself. By 2000 B.C. both the Babylonians and the Egyptians knew π to nearly two decimal places. The Babylonians used, among others, the value 3 1/8 and the Egyptians used 3 13/81. Not all ancient societies were as accurate, however—nearly 1500 years later the Hebrews were perhaps still content to use the value 3, as the following quote suggests.

Also, he made a molten sea of ten cubits from brim to brim, round in compass, and five cubits the height thereof; and a line of thirty cubits did compass it round about.

                                                        Old Testament, 1 Kings 7:23

Despite the long pedigree of the problem, all nonempirical calculations have employed, up to minor variations, only three techniques.

i) The first technique due to Archimedes of Syracuse (287–212 B.C.) is, recursively, to calculate the length of circumscribed and inscribed regular $6 \cdot 2^n$-gons about a circle of diameter 1. Call these quantities a_n and b_n, respectively. Then $a_0 := 2\sqrt{3}$, $b_0 := 3$ and, as Gauss's teacher Pfaff discovered in 1800,

$$a_{n+1} := \frac{2a_n b_n}{a_n + b_n} \quad \text{and} \quad b_{n+1} := \sqrt{a_{n+1} b_n}.$$

Archimedes, with $n = 4$, obtained

$$3\tfrac{10}{71} < \pi < 3\tfrac{1}{7}.$$

While hardly better than estimates one could get with a ruler, this is the first method that can be used to generate an arbitrary number of digits, and to a nonnumerical mathematician perhaps the problem ends here. Variations on this theme provided the basis for virtually all calculations of π for the next 1800 years, culminating with a 34 digit calculation due to Ludolph van Ceulen (1540–1610). This demands polygons with about 2^{60} sides and so is extraordinarily time consuming.

ii) Calculus provides the basis for the second technique. The underlying method relies on Gregory's series of 1671

$$\arctan x = \int_0^x \frac{dt}{1 + t^2} = x - \frac{x^3}{3} + \frac{x^5}{5} - \cdots \qquad |x| \le 1$$

coupled with a formula which allows small x to be used, like

$$\frac{\pi}{4} = 4 \arctan\left(\frac{1}{5}\right) - \arctan\left(\frac{1}{239}\right).$$

This particular formula is due to Machin and was employed by him to compute 100 digits of π in 1706. Variations on this second theme are the basis of all the calculations done until the 1970's including William Shanks' monumental hand-calculation of 527 digits. In the introduction to his book [32], which presents this calculation, Shanks writes:

> Towards the close of the year 1850 the Author first formed the design of rectifying the circle to upwards of 300 places of decimals. He was fully aware at that time, that the accomplishment of his purpose would add little or nothing to his fame as a Mathematician though it might as a Computer; nor would it be productive of anything in the shape of pecuniary recompense.

Shanks actually attempted to hand-calculate 707 digits but a mistake crept in at the 527th digit. This went unnoticed until 1945, when D. Ferguson, in one of the last "nondigital" calculations, computed 530 digits. Even with machine calculations mistakes occur, so most record-setting calculations are done twice—by sufficiently different methods.

The advent of computers has greatly increased the scope and decreased the toil of such calculations. Metropolis, Reitwieser, and von Neumann computed and analyzed 2037 digits using Machin's formula on ENIAC in 1949. In 1961, Dan Shanks and Wrench calculated 100,000 digits on an IBM 7090 [31]. By 1973, still using Machin-like arctan expansions, the million digit mark was passed by Guillard and Bouyer on a CDC 7600.

iii) The third technique, based on the transformation theory of elliptic integrals, provides the algorithms for the most recent set of computations. The most recent records are due separately to Gosper, Bailey, and Kanada. Gosper in 1985 calculated over 17 million digits (in fact over 17 million terms of the continued fraction) using a carefully orchestrated evaluation of Sum 1.

Bailey in January 1986 computed over 29 million digits using Algorithm 1 on a Cray 2 [2]. Kanada, using a related quadratic algorithm (due in basis to Gauss and made explicit by Brent [12] and Salamin [27]) and using Algorithm 1 for a check, verified 33,554,000 digits. This employed a HITACHI S-810/20, took roughly eight hours, and was completed in September of 1986. In January 1987 Kanada extended his computation to 2^{27} decimal places of π and the hundred million digit mark had been passed. The calculation took roughly a day and a half on a NEC SX2 machine. Kanada's most recent feat (Jan. 1988) was to compute 201,326,000 digits, which required only six hours on a new Hitachi S-820 supercomputer. Within the next few years many hundreds of millions of digits will no doubt have been similarly computed. Further discussion of the history of the computation of pi may be found in [5] and [9].

4. Complexity Concerns. One of the interesting morals from theoretical computer science is that many familiar algorithms are far from optimal. In order to be more precise we introduce the notion of *bit complexity*. Bit complexity counts the number of single operations required to complete an algorithm. The single-digit operations we count are $+, -, \times$. (We could, if we wished, introduce storage and logical comparison into the count. This, however, doesn't affect the order of growth of the algorithms in which we are interested.) This is a good measure of time on a serial machine. Thus, addition of two n-digit integers by the usual method has bit complexity $O(n)$, and straightforward uniqueness considerations show this to be asymptotically best possible.

Multiplication is a different story. Usual multiplication of two n-digit integers has bit complexity $O(n^2)$ and no better. However, it is possible to multiply two n-digit integers with complexity $O(n(\log n)(\log \log n))$. This result is due to Schönhage and Strassen and dates from 1971 [29]. It provides the best bound known for multiplication. No multiplication can have speed better than $O(n)$. Unhappily, more exact results aren't available.

The original observation that a faster than $O(n^2)$ multiplication is possible was due to Karatsuba in 1962. Observe that

$$(a + b10^n)(c + d10^n) = ac + [(a - b)(c - d) - ac - bd]10^n + bd10^{2n},$$

and thus multiplication of two $2n$-digit integers can be reduced to three multiplications of n-digit integers and a few extra additions. (Of course multiplication by 10^n is just a shift of the decimal point.) If one now proceeds recursively one produces a multiplication with bit complexity

$$O(n^{\log_2 3}).$$

Note that $\log_2 3 = 1.58 \ldots < 2$.

We denote by $M(n)$ the bit complexity of multiplying two n-digit integers together by any method that is at least as fast as usual multiplication.

The trick to implementing high precision arithmetic is to get the multiplication right. Division and root extraction piggyback off multiplication using Newton's method. One may use the iteration

$$x_{k+1} = 2x_k - x_k^2 y$$

to compute $1/y$ and the iteration

$$x_{k+1} = \frac{1}{2}\left(x_k + \frac{y}{x_k}\right)$$

to compute $\sqrt{y}$. One may also compute $1/\sqrt{y}$ from

$$x_{k+1} = \frac{x_k(3 - yx_k^2)}{2}$$

and so avoid divisions in the computation of $\sqrt{y}$. Not only do these iterations converge quadratically but, because Newton's method is self-correcting (a slight perturbation in x_k does not change the limit), it is possible at the kth stage to work only to precision 2^k. If division and root extraction are so implemented, they both have bit complexity $O(M(n))$, in the sense that n-digit input produces n-digit accuracy in a time bounded by a constant times the speed of multiplication. This extends in the obvious way to the solution of any algebraic equation, with the startling conclusion that every algebraic number can be computed (to n-digit accuracy) with bit complexity $O(M(n))$. Writing down n-digits of $\sqrt{2}$ or $3\sqrt{7}$ is (up to a constant) no more complicated than multiplication.

The Schönhage-Strassen multiplication is hard to implement. However, a multiplication with complexity $O((\log n)^{2+\varepsilon}n)$ based on an ordinary complex (floating point) fast Fourier transform is reasonably straightforward. This is Kanada's approach, and the recent records all rely critically on some variations of this technique.

To see how the fast Fourier transform may be used to accelerate multiplication, let $x := (x_0, x_1, x_2, \ldots, x_{n-1})$ and $y := (y_0, y_1, y_2, \ldots, y_{n-1})$ be the representations of two high-precision numbers in some radix b. The radix b is usually selected to be some power of 2 or 10 whose square is less than the largest integer exactly representable as an ordinary floating-point number on the computer being used. Then, except for releasing each "carry," the product $z := (z_0, z_1, z_2, \ldots, z_{2n-1})$ of x and y may be written as

$$z_0 = x_0 y_0$$
$$z_1 = x_0 y_1 + x_1 y_0$$
$$z_2 = x_0 y_2 + x_1 y_1 + x_2 y_0$$
$$\vdots$$
$$z_{n-1} = x_0 y_{n-1} + x_1 y_{n-2} + \cdots + x_{n-1} y_0$$
$$\vdots$$
$$z_{2n-3} = x_{n-1} y_{n-2} + x_{n-2} y_{n-1}$$
$$z_{2n-2} = x_{n-1} y_{n-1}$$
$$z_{2n-1} = 0.$$

Now consider x and y to have n zeros appended, so that x, y, and z all have length $N = 2n$. Then a key observation may be made: the product sequence z is precisely the discrete convolution $C(x, y)$:

$$z_k = C_k(x, y) = \sum_{j=0}^{N-1} x_j y_{k-j},$$

where the subscript $k - j$ is to be interpreted as $k - j + N$ if $k - j$ is negative.

Now a well-known result of Fourier analysis may be applied. Let $F(x)$ denote the *discrete Fourier transform* of the sequence x, and let $F^{-1}(x)$ denote the inverse discrete Fourier transform of x:

$$F_k(x) := \sum_{j=0}^{N-1} x_j e^{-2\pi i jk/N}$$

$$F_k^{-1}(x) := \frac{1}{N} \sum_{j=0}^{N-1} x_j e^{2\pi i jk/N}.$$

Then the "convolution theorem," whose proof is a straightforward exercise, states that

$$F[C(x, y)] = F(x)F(y)$$

or, expressed another way,

$$C(x, y) = F^{-1}[F(x)F(y)].$$

Thus the entire multiplication pyramid z can be obtained by performing two forward discrete Fourier transforms, one vector complex multiplication and one inverse transform, each of length $N = 2n$. Once the real parts of the resulting complex numbers have been rounded to the nearest integer, the final multiprecision product may be obtained by merely releasing the carries modulo b. This may be done by starting at the end of the z vector and working backward, as in elementary school arithmetic, or by applying other schemes suitable for vector processing on more sophisticated computers.

A straightforward implementation of the above procedure would not result in any computational savings—in fact, it would be several times more costly than the usual "schoolboy" scheme. The reason this scheme is used is that the discrete Fourier transform may be computed much more rapidly using some variation of the well-known "fast Fourier transform" (FFT) algorithm [13]. In particular, if $N = 2^m$, then the discrete Fourier transform may be evaluated in only $5m2^m$ arithmetic operations using an FFT. Direct application of the definition of the discrete Fourier transform would require 2^{2m+3} floating-point arithmetic operations, even if it is assumed that all powers of $e^{-2\pi i/N}$ have been precalculated.

This is the basic scheme for high-speed multiprecision multiplication. Many details of efficient implementations have been omitted. For example, it is possible to take advantage of the fact that the input sequences x and y and the output sequence z are all purely real numbers, and thereby sharply reduce the operation count. Also, it is possible to dispense with complex numbers altogether in favor of performing computations in fields of integers modulo large prime numbers. Interested readers are referred to [2], [8], [13], and [22].

When the costs of all the constituent operations, using the best known techniques, are totalled both Algorithms 1 and 2 compute n digits of π with bit complexity $O(M(n)\log n)$, and use $O(\log n)$ full precision operations.

The bit complexity for Sum 1, or for π using any of the arctan expansions, is between $O((\log n)^2 M(n))$ and $O(nM(n))$ depending on implementation. In each case, one is required to sum $O(n)$ terms of the appropriate series. Done naively, one obtains the latter bound. If the calculation is carefully orchestrated so that the terms are grouped to grow evenly in size (as rational numbers) then one can achieve the former bound, but with no corresponding reduction in the number of operations.

The Archimedean iteration of section 2 converges like $1/4^n$ so in excess of n iterations are needed for n-digit accuracy, and the bit complexity is $O(nM(n))$.

Almost any familiar transcendental number such as e, γ, $\zeta(3)$, or Catalan's constant (presuming the last three to be nonalgebraic) can be computed with bit complexity $O((\log n)M(n))$ or $O((\log n)^2 M(n))$. None of these numbers is known to be computable essentially any faster than this. In light of the previous observation that algebraic numbers are all computable with bit complexity $O(M(n))$, a proof that π cannot be computed with this speed would imply the transcendence of π. It would, in fact, imply more, as there are transcendental numbers which have complexity $O(M(n))$. An example is $0.10100100001\ldots$.

It is also reasonable to speculate that computing the nth digit of π is not very much easier than computing all the first n digits. We think it very probable that computing the nth digit of π cannot be $O(n)$.

5. The Miracle of Theta Functions

When I was a student, abelian functions were, as an effect of the Jacobian tradition, considered the uncontested summit of mathematics, and each of us was ambitious to make progress in this field. And now? The younger generation hardly knows abelian functions.

Felix Klein [21]

Felix Klein's lament from a hundred years ago has an uncomfortable timelessness to it. Sadly, it is now possible never to see what Bochner referred to as "the miracle of the theta functions" in an entire university mathematics program. A small piece of this miracle is required here [6], [11], [28]. First some standard notations. The *complete elliptic integrals of the first and second kind*, respectively,

$$K(k) := \int_0^{\pi k} \frac{dt}{\sqrt{1 - k^2 \sin^2 t}} \tag{5.1}$$

and

$$E(k) := \int_0^{\pi/2} \sqrt{1 - k^2 \sin^2 t}\, dt. \tag{5.2}$$

The second integral arises in the rectification of the ellipse, hence the name elliptic integrals. The *complementary modulus* is

$$k' := \sqrt{1 - k^2}$$

and the *complementary integrals* K' and E' are defined by

$$K'(k) := K(k') \quad \text{and} \quad E'(k) := E(k').$$

The first remarkable identity is *Legendre's relation* namely

$$E(k)K'(k) + E'(k)K(k) - K(k)K'(k) = \frac{\pi}{2} \tag{5.3}$$

(for $0 < k < 1$), which is pivotal in relating these quantities to pi. We also need to define two *Jacobian theta functions*

$$\Theta_2(q) := \sum_{n=-\infty}^{\infty} q^{(n+1/2)^2} \tag{5.4}$$

and

$$\Theta_3(q) := \sum_{n=-\infty}^{\infty} q^{n^2}. \tag{5.5}$$

These are in fact specializations with ($t = 0$) of the general theta functions. More generally

$$\Theta_3(t, q) := \sum_{n=-\infty}^{\infty} q^{n^2} e^{\mathrm{im}\, t} \qquad (\mathrm{im}\, t > 0)$$

with similar extensions of Θ_2. In Jacobi's approach these general theta functions provide the basic building blocks for elliptic functions, as functions of t (see [11], [39]).

The complete elliptic integrals and the special theta functions are related as follows. For $|q| < 1$

$$K(k) = \frac{\pi}{2} \Theta_3^2(q) \tag{5.6}$$

and

$$E(k) = (k')^2 \left[K(k) + k \frac{dK(k)}{dk} \right], \tag{5.7}$$

where

$$k := k(q) = \frac{\Theta_2^2(q)}{\Theta_3^2(q)}, \qquad k' := k'(q) = \frac{\Theta_3^2(-q)}{\Theta_3^2(q)} \tag{5.8}$$

and

$$q = e^{-\pi K'(k)/K(k)}. \tag{5.9}$$

The *modular function* λ is defined by

$$\lambda(t) := \lambda(q) := k^2(q) := \left[\frac{\Theta_2(q)}{\Theta_3(q)} \right]^4, \tag{5.10}$$

where

$$q := e^{i\pi t}.$$

We wish to make a few comments about modular functions in general before restricting our attention to the particular modular function λ. *Modular functions* are functions which are meromorphic in H, the upper half of the complex plane, and which are invariant under a group of linear fractional transformations, G, in the sense that

$$f(g(z)) = f(z) \qquad \forall g \in G.$$

[Additional growth conditions on f at certain points of the associated fundamental region (see below) are also demanded.] We restrict G to be a subgroup of the *modular group* Γ where Γ is the set of all transformations w of the form

$$w(t) = \frac{at + b}{ct + d},$$

with a, b, c, d integers and $ad - bc = 1$. Observe that Γ is a group under composition. A *fundamental region* F_G is a set in H with the property that any element in H is uniquely the image of some element in F_G under the action of G. Thus the behaviour of a modular function is uniquely determined by its behaviour on a fundamental region.

Modular functions are, in a sense, an extension of elliptic (or doubly periodic) functions—functions such as sn which are invariant under linear transformations and which arise naturally in the inversion of elliptic integrals.

The definitions we have given above are not complete. We will be more precise in our discussion of λ. One might bear in mind that much of the theory for λ holds in considerably greater generality.

The *fundamental region* F we associate with λ is the set of complex numbers

$$F := \{\operatorname{im} t \geq 0\} \cap [\{|\!\operatorname{re} t| < 1 \quad \text{and}$$

$$|2t \pm 1| > 1\} \cup \{\operatorname{re} t = -1\} \cup \{|2t + 1| = 1\}].$$

The λ-*group* (or theta-subgroup) is the set of linear fractional transformations w satisfying

$$w(t) := \frac{at + b}{ct + d},$$

where a, b, c, d are integers and $ad - bc = 1$, while in addition a and d are odd and b and c are even. Thus the corresponding matrices are unimodular. What makes λ a λ-modular function is the fact that λ is meromorphic in $\{\operatorname{im} t > 0\}$ and that

$$\lambda(w(t)) := \lambda(t)$$

for all w in the λ-group, plus the fact that λ tends to a definite limit (possibly infinite) as t tends to a vertex of the fundamental region (one of the three points $(0, -1), (0, 0), (i, \infty)$). Here we only allow convergence from within the fundamental region.

Now some of the miracle of modular functions can be described. Largely because every point in the upper half plane is the image of a point in F under an element of the λ-group, one can deduce that any λ-modular function that is bounded on F is constant. Slightly further into the theory, but relying on the above, is the result that any two modular functions are algebraically related, and resting on this, but further again into the field, is the following remarkable result. Recall that q is given by (5.9).

THEOREM 1. *Let z be a primitive pth root of unity for p an odd prime. Consider the pth order modular equation for λ as defined by*

$$W_p(x, \lambda) := (x - \lambda_0)(x - \lambda_1) \cdots (x - \lambda_p), \tag{5.11}$$

where

$$\lambda_i := \lambda\left(z^i q^{1/p}\right) \qquad i < p$$

and

$$\lambda_p := \lambda(q^p).$$

*Then the function W_p is a polynomial in x and λ (**independent** of q), which has integer coefficients and is of degree $p + 1$ in both x and λ.*

The modular equation for λ usually has a simpler form in the associated variables $u := x^{1/8}$ and $v := \lambda^{1/8}$. In this form the 5th-order modular equation is given by

$$\Omega_5(u, v) := u^6 - v^6 + 5u^2 v^2(u^2 - v^2) + 4uv(1 - u^4 v^4). \qquad (5.12)$$

In particular

$$\frac{\Theta_2(q^p)}{\Theta_3(q^p)} = v^2 \quad \text{and} \quad \frac{\Theta_2(q)}{\Theta_3(q)} = u^2$$

are related by an algebraic equation of degree $p + 1$.

The miracle is not over. The *pth-order multiplier* (for λ) is defined by

$$M_p(k(q), k(q^p)) := \frac{K(k(q^p))}{K(k(q))} = \left[\frac{\Theta_3(q^p)}{\Theta_3(q)}\right]^2 \qquad (5.13)$$

and turns out to be a rational function of $k(q^p)$ and $k(q)$.

One is now in possession of a pth-order algorithm for K/π, namely: Let $k_i := k(q^{p^i})$. Then

$$\frac{2K(k_0)}{\pi} = M_p^{-1}(k_0, k_1) M_p^{-1}(k_1, k_2) M_p^{-1}(k_2, k_3) \cdots.$$

This is an entirely algebraic algorithm. One needs to know the pth-order modular equation for λ to compute k_{i+1} from k_i and one needs to know the rational multiplier M_p. The speed of convergence ($O(c^{p^i})$, for some $c < 1$) is easily deduced from (5.13) and (5.9).

The function $\lambda(t)$ is 1-1 on F and has a well-defined inverse, λ^{-1}, with branch points only at 0, 1 and ∞. This can be used to provide a one line proof of the "big" Picard theorem that a nonconstant entire function misses at most one value (as does exp). Indeed, suppose g is an entire function and that it is never zero or one; then $\exp(\lambda^{-1}(g(z)))$ is a bounded entire function and is hence constant.

Littlewood suggested that, at the right point in history, the above would have been a strong candidate for a 'one line doctoral thesis'.

6. Ramanujan's Solvable Modular Equations. Hardy [19] commenting on Ramanujan's work on elliptic and modular functions says

It is here that both the profundity and limitations of Ramanujan's knowledge stand out most sharply.

We present only one of Ramanujan's modular equations.

THEOREM 2.

$$\frac{5\Theta_3(q^{25})}{\Theta_3(q)} = 1 + r_1^{1/5} + r_2^{1/5}, \tag{6.1}$$

where for $i = 1$ and 2

$$r_1 := \tfrac{1}{2}x\left(y \pm \sqrt{y^2 - 4x^3}\right)$$

with

$$x := \frac{5\Theta_3(q^5)}{\Theta_3(q)} - 1 \quad \text{and} \quad y := (x - 1)^2 + 7.$$

This is a slightly rewritten form of entry 12(iii) of Chapter 19 of Ramanujan's *Second Notebook* (see [7], where Berndt's proofs may be studied). One can think of Ramanujan's quintic modular equation as an equation in the multiplier M_p of (5.13). The initial surprise is that it is solvable. The quintic modular relation for λ, W_5, and the related equation for $\lambda^{1/8}$, Ω_5, are both nonsolvable. The Galois group of the sixth-degree equation Ω_5 (see (5.12)) over $\mathbf{Q}(v)$ is A_5 and is nonsolvable. Indeed both Hermite and Kronecker showed, in the middle of the last century, that the solution of a general quintic may be effected in terms of the solution of the 5th-order modular equation (5.12) and the roots may thus be given in terms of the theta functions.

In fact, in general, the Galois group for W_p of (5.11) has order $p(p + 1)(p - 1)$ and is never solvable for $p \geq 5$. The group is quite easy to compute, it is generated by two permutations. If

$$q := e^{i\pi t}, \quad \text{then} \quad \tau \to \tau + 2 \quad \text{and} \quad \tau \to \frac{\tau}{(2\tau + 1)}$$

are both elements of the λ-group and induce permutations on the λ_i of Theorem 1. For any fixed p, one can use the q-expansion of (5.10) to compute the effect of these transformations on the λ_i, and can thus easily write down the Galois group.

While W_p is not solvable over $\mathbf{Q}(\lambda)$, it is solvable over $\mathbf{Q}(\lambda, \lambda_0)$. Note that λ_0 is a root of W_p. It is of degree $p + 1$ because W_p is irreducible. Thus the Galois group for W_p over $\mathbf{Q}(\lambda, \lambda_0)$ has order $p(p - 1)$. For $p = 5$, 7, and 11 this gives groups of order 20, 42, and 110, respectively, which are obviously solvable and, in fact, for general primes, the construction always produces a solvable group.

From (5.8) and (5.10) one sees that Ramanujan's modular equation can be rewritten to give λ_5 solvable in terms of λ_0 and λ. Thus, we can hope to find an explicit solvable relation for λ_p in terms of λ and λ_0. For $p = 3$, W_p is of degree 4 and is, of course, solvable. For $p = 7$, Ramanujan again helps us out, by providing a solvable seventh-order modular identity for the closely related *eta function* defined by

$$\eta(q) := q^{\frac{1}{12}} \prod_{n=1}^{\infty} (1 - q^{2n}).$$

The first interesting prime for which an explicit solvable form is not known is the "endecadic" ($p = 11$) case. We consider only prime values because for nonprime values the modular equation factors.

This leads to the interesting problem of mechanically constructing these equations. In principle, and to some extent in practice, this is a purely computational problem. Modular equations can be computed fairly easily from (5.11) and even more easily in the associated variables u and v. Because one knows a priori bounds on the size of the (integer) coefficients of the equations one can perform these calculations exactly. The coefficients of the equation, in the variables u and v, grow at most like 2^n. (See [11].) Computing the solvable forms and the associated computational problems are a little more intricate—though still in principle entirely mechanical. A word of caution however: in the variables u and v the endecadic modular equation has largest coefficient 165, a three digit integer. The endecadic modular equation for the intimately related function J (Klein's *absolute invariant*) has coefficients as large as

$$270909647855313899315632002810352263119290522273303 \times 2^{92}3^{19}5^{20}11^{2}53.$$

It is, therefore, one thing to solve these equations, it is entirely another matter to present them with the economy of Ramanujan.

The paucity of Ramanujan's background in complex analysis and group theory leaves open to speculation Ramanujan's methods. The proofs given by Berndt are difficult. In the seventh-order case, Berndt was aided by MACSYMA—a sophisticated algebraic manipulation package. Berndt comments after giving the proof of various seventh-order modular identities:

Of course, the proof that we have given is quite unsatisfactory because it is a verification that could not have been achieved without knowledge of the result. Ramanujan obviously possessed a more natural, transparent, and ingenious proof.

7. Modular Equations and Pi. We wish to connect the modular equations of Theorem 1 to pi. This we contrive via the function *alpha* defined by:

$$\alpha(r) := \frac{E'(k)}{K(k)} - \frac{\pi}{(2K(k))^2}, \tag{7.1}$$

where

$$k := k(q) \quad \text{and} \quad q := e^{-\pi\sqrt{r}}.$$

This allows one to rewrite Legendre's equation (5.3) in a one-sided form without the conjugate variable as

$$\frac{\pi}{4} = K\left[\sqrt{r}E - (\sqrt{r} - \alpha(r))K\right]. \tag{7.2}$$

We have suppressed, and will continue to suppress, the k variable. With (5.6) and (5.7) at hand we can write a q-expansion for α, namely,

$$\alpha(r) = \frac{\dfrac{1}{\pi} - \sqrt{r}\,4\dfrac{\sum\limits_{n=-\infty}^{\infty} n^2(-q)^{n^2}}{\sum\limits_{n=-\infty}^{\infty} (-q)^{n^2}}}{\left[\sum\limits_{n=-\infty}^{\infty} q^{n^2}\right]^4}, \tag{7.3}$$

and we can see that as r tends to infinity $q = e^{-\pi\sqrt{r}}$ tends to zero and $\alpha(r)$ tends to $1/\pi$. In fact

$$\alpha(r) - \frac{1}{\pi} \approx 8\left(\sqrt{r} - \frac{1}{\pi}\right)e^{-\pi\sqrt{r}}. \tag{7.4}$$

The key now is iteratively to calculate α. This is the content of the next theorem.

Theorem 3. *Let* $k_0 := k(q), k_1 := k(q^p)$ *and* $M_p := M_p(k_0, k_1)$ *as in* (5.13). *Then*

$$\alpha(p^2 r) = \frac{\alpha(r)}{M_p^2} - \sqrt{r}\left[\frac{k_0^2}{M_p^2} - pk_1^2 + \frac{pk_1'^2 k_1 \dot{M}_p}{M_p}\right],$$

where represents the full derivative of M_p *with respect to* k_0. *In particular,* α *is algebraic for rational arguments.*

We know that $K(k_1)$ is related via M_p to $K(k)$ and we know that $E(k)$ is related via differentiation to K. (See (5.7) and (5.13).) Note that $q \to q^p$ corresponds to $r \to p^2 r$. Thus from (7.2) some relation like that of the above theorem must exist. The actual derivation requires some careful algebraic manipulation. (See [11], where it has also been made entirely explicit for $p := 2, 3, 4, 5,$ and 7, and where numerous algebraic values are determined for $\alpha(r)$.) In the case $p := 5$ we can specialize with some considerable knowledge of quintic modular equations to get:

Theorem 4. *Let* $s := 1/M_5(k_0, k_1)$. *Then*

$$\alpha(25r) = s^2\alpha(r) - \sqrt{r}\left[\frac{(s^2 - 5)}{2} + \sqrt{s(s^2 - 2s + 5)}\right].$$

This couples with Ramanujan's quintic modular equation to provide a derivation of Algorithm 2.

Algorithm 1 results from specializing Theorem 3 with $p := 4$ and coupling it with a quartic modular equation. The quartic equation in question is just two steps of the corresponding quadratic equation which is Legendre's form of the *arithmetic geometric mean iteration*, namely:

$$k_1 = \frac{2\sqrt{k}}{1 + k}.$$

An algebraic pth-order algorithm for π is derived from coupling Theorem 3 with a pth-order modular equation. The substantial details which are skirted here are available in [11].

8. Ramanujan's sum. This amazing sum,

$$\frac{1}{\pi} = \frac{\sqrt{8}}{9801} \sum_{n=0}^{\infty} \frac{(4n)! \, [1103 + 26390n]}{(n!)^4 \quad 396^{4n}}$$

is a specialization ($N = 58$) of the following result, which gives reciprocal series for π in terms of our function alpha and related modular quantities.

THEOREM 5.

$$\frac{1}{\pi} = \sum_{n=0}^{\infty} \frac{\left(\frac{1}{4}\right)_n \left(\frac{1}{2}\right)_n \left(\frac{3}{4}\right)_n d_n(N)}{(n!)^3} x_N^{2n+1}, \tag{8.1}$$

where,

$$x_N := \frac{4k_N(k_N')^2}{\left(1 + k_N^2\right)^2} := \left(\frac{g_N^{12} + g_N^{-12}}{2}\right)^{-1},$$

with

$$d_n(N) = \left[\frac{\alpha(N)x_N^{-1}}{1 + k_N^2} - \frac{\sqrt{N}}{4}g_N^{-12}\right] + n\sqrt{N}\left(\frac{g_N^{12} - g_N^{-12}}{2}\right)$$

and

$$k_N := k\left(e^{-\pi\sqrt{N}}\right), \qquad g_N^{12} = (k_N')^2/(2k_N).$$

Here $(c)_n$ *is the rising factorial:* $(c)_n := c(c + 1)(c + 2) \cdots (c + n - 1).$

Some of the ingredients for the proof of Theorem 5, which are detailed in [11], are the following. Our first step is to write (7.2) as a sum after replacing the E by K and dK/dk using (5.7). One then uses an identity of Clausen's which allows one to write the square of a hypergeometric function $_2F_1$ in terms of a generalized hypergeometric $_3F_2$, namely, for all k one has

$$(1 + k^2)\left[\frac{2K(k)}{\pi}\right]^2 = {}_3F_2\left(\frac{1}{4}, \frac{3}{4}, \frac{1}{2}, 1, 1 : \left(\frac{2}{g^{12} + g^{-12}}\right)^2\right)$$

$$= \sum_{n=0}^{\infty} \frac{\left(\frac{1}{4}\right)_n \left(\frac{3}{4}\right)_n \left(\frac{1}{2}\right)_n}{(1)_n (1)_n} \frac{\left(\frac{2}{g^{12} + g^{-12}}\right)^{2n}}{n!}.$$

Here g is related to k by

$$\frac{4k(k')^2}{(1 + k^2)^2} = \left(\frac{g^{12} + g^{-12}}{2}\right)^{-1}$$

as required in Theorem 5. We have actually done more than just use Clausen's identity, we have also transformed it once using a standard hypergeometric substitution due to Kummer. Incidentally, Clausen was a nineteenth-century mathematician who, among other things, computed 250 digits of π in 1847 using Machin's formula. The desired formula (8.1) is obtained on combining these pieces.

Even with Theorem 5, our work is not complete. We still have to compute

$$k_{58} := k\left(e^{-\pi\sqrt{58}}\right) \quad \text{and} \quad \alpha_{58} := \alpha(58).$$

In fact

$$g_{58}^2 = \left(\frac{\sqrt{29} + 5}{2}\right)$$

is a well-known *invariant* related to the fundamental solution to Pell's equation for 29 and it turns out that

$$\alpha_{58} = \left(\frac{\sqrt{29} + 5}{2} \right)^6 (99\sqrt{29} - 444)(99\sqrt{2} - 70 - 13\sqrt{29}).$$

One can, in principle, and for $N := 58$, probably in practice, solve for k_N by directly solving the Nth-order equation

$$W_N\left(k_N^2, 1 - k_N^2 \right) = 0.$$

For $N = 58$, given that Ramanujan [26] and Weber [38] have calculated g_{58} for us, verification by this method is somewhat easier though it still requires a tractable form of W_{58}. Actually, more sophisticated number-theoretic techniques exist for computing k_N (these numbers are called *singular moduli*). A description of such techniques, including a reconstruction of how Ramanujan might have computed the various singular moduli he presents in [26]; is presented by Watson in a long series of papers commencing with [36]; and some more recent derivations are given in [11] and [30]. An inspection of Theorem 5 shows that all the constants in Series 1 are determined from g_{58}. Knowing α is equivalent to determining that the number 1103 is correct.

It is less clear how one explicitly calculates α_{58} in algebraic form, except by brute force, and a considerable amount of brute force is required; but a numerical calculation to any reasonable accuracy is easily obtained from (7.3) and 1103 appears! The reader is encouraged to try this to, say, 16 digits. This presumably is what Ramanujan observed. Ironically, when Gosper computed 17 million digits of π using Sum 1, he had no mathematical proof that Sum 1 actually converged to $1/\pi$. He compared ten million digits of the calculation to a previous calculation of Kanada et al. This verification that Sum 1 is correct to ten million places also provided the first complete proof that α_{58} is as advertised above. A nice touch—that the calculation of the sum should prove itself as it goes.

Roughly this works as follows. One knows enough about the exact algebraic nature of the components of $d_n(N)$ and x_N to know that if the purported sum (of positive terms) were incorrect, that before one reached 3 million digits, this sum must have ceased to agree with $1/\pi$. Notice that the components of Sum 1 are related to the solution of an equation of degree 58, but virtually no irrationality remains in the final packaging. Once again, there are very good number-theoretic reasons, presumably unknown to Ramanujan, why this must be so (58 is at least a good candidate number for such a reduction). Ramanujan's insight into this marvellous simplification remains obscure.

Ramanujan [26] gives 14 other series for $1/\pi$, some others almost as spectacular as Sum 1—and one can indeed derive some even more spectacular related series.* He gives almost no explanation as to their genesis, saying only that there are "corresponding theories" to the standard theory (as sketched in section 5) from which they follow. Hardy, quoting Mordell, observed that "it is unfortunate that Ramanujan has not developed the corresponding theories." By methods analogous

---

*(Added in proof) Many related series due to Borwein and Borwein and to Chudnovsky and Chudnovsky appear in papers in *Ramanujan Revisited*, Academic Press, 1988.

to those used above, all his series can be derived from the classical theory [11]. Again it is unclear what passage Ramanujan took to them, but it must in some part have diverged from ours.

We conclude by writing down another extraordinary series of Ramanujan's, which also derives from the same general body of theory,

$$\frac{1}{\pi} = \sum_{n=0}^{\infty} \binom{2n}{n}^3 \frac{42n + 5}{2^{12n+4}}.$$

This series is composed of fractions whose numerators grow like 2^{6n} and whose denominators are exactly $16 \cdot 2^{12n}$. In particular this can be used to calculate the second block of n binary digits of π without calculating the first n binary digits. This beautiful observation, due to Holloway, results, disappointingly, in no intrinsic reduction in complexity.

9. Sources. References [7], [11], [19], [26], [36], and [37] relate directly to Ramanujan's work. References [2], [8], [9], [10], [12], [22], [24], [27], [29], and [31] discuss the computational concerns of the paper.

Material on modular functions and special functions may be pursued in [1], [6], [9], [14], [15], [18], [20], [28], [34], [38], and [39]. Some of the number-theoretic concerns are touched on in [3], [6], [9], [11], [16], [23], and [35].

Finally, details of all derivations are given in [11].

REFERENCES

1. M. Abramowitz and I. Stegun, Handbook of Mathematical Functions, Dover, New York, 1964.
2. D. H. Bailey, The Computation of π to 29,360,000 decimal digits using Borweins' quartically convergent algorithm, Math. Comput., 50 (1988) 283–96.
3. _____, Numerical results on the transcendence of constants involving π, e, and Euler's constant, Math. Comput., 50 (1988) 275–81.
4. A. Baker, Transcendental Number Theory, Cambridge Univ. Press, London, 1975.
5. P. Beckmann, A History of Pi, 4th ed., Golem Press, Boulder, CO, 1977.
6. R. Bellman, A Brief Introduction to Theta Functions, Holt, Reinhart and Winston, New York, 1961.
7. B. C. Berndt, Modular Equations of Degrees 3, 5, and 7 and Associated Theta Functions Identities, chapter 19 of Ramanujan's Second Notebook, Springer—to be published.
8. A. Borodin and I. Munro, The Computational Complexity of Algebraic and Numeric Problems, American Elsevier, New York, 1975.
9. J. M. Borwein and P. B. Borwein, The arithmetic-geometric mean and fast computation of elementary functions, SIAM Rev., 26 (1984), 351–365.
10. _____, An explicit cubic iteration for pi, BIT, 26 (1986) 123–126.
11. _____, Pi and the AGM—A Study in Analytic Number Theory and Computational Complexity, Wiley, N.Y., 1987.
12. R. P. Brent, Fast multiple-precision evaluation of elementary functions, J. ACM, 23 (1976) 242–251.
13. E. O. Brigham, The Fast Fourier Transform, Prentice-Hall, Englewood Cliffs, N.J., 1974.
14. A. Cayley, An Elementary Treatise on Elliptic Functions, Bell and Sons, 1885; reprint Dover, 1961.
15. A. Cayley, A memoir on the transformation of elliptic functions, Phil. Trans. T., 164 (1874) 397–456.
16. D. V. Chudnovsky and G. V. Chudnovsky, Padé and Rational Approximation to Systems of Functions and Their Arithmetic Applications, Lecture Notes in Mathematics 1052, Springer, Berlin, 1984.
17. H. R. P. Ferguson and R. W. Forcade, Generalization of the Euclidean algorithm for real numbers to all dimensions higher than two, Bull. AMS, 1 (1979) 912–914.
18. C. F. Gauss, Werke, Göttingen 1866–1933, Bd 3, pp. 361–403.

19. G. H. Hardy, Ramanujan, Cambridge Univ. Press, London, 1940.
20. L. V. King, On The Direct Numerical Calculation of Elliptic Functions and Integrals, Cambridge Univ. Press, 1924.
21. F. Klein, Development of Mathematics in the 19th Century, 1928, Trans Math Sci. Press, R. Hermann ed., Brookline, MA, 1979.
22. D. Knuth, The Art of Computer Programming, vol. 2: Seminumerical Algorithms, Addison-Wesley, Reading, MA, 1981.
23. F. Lindemann, Über die Zahl π, Math. Ann., 20 (1882) 213–225.
24. G. Miel, On calculations past and present: the Archimedean algorithm, Amer. Math. Monthly, 90 (1983) 17–35.
25. D. J. Newman, Rational Approximation Versus Fast Computer Methods, in Lectures on Approximation and Value Distribution, Presses de l'Université de Montreal, 1982, pp. 149–174.
26. S. Ramanujan, Modular equations and approximations to π, Quart. J. Math, 45 (1914) 350–72.
27. E. Salamin, Computation of π using arithmetic-geometric mean, Math. Comput., 30 (1976) 565–570.
28. B. Schoenberg, Elliptic Modular Functions, Springer, Berlin, 1976.
29. A. Schönhage and V. Strassen, Schnelle Multiplikation Grosser Zahlen, Computing, 7 (1971) 281–292.
30. D. Shanks, Dihedral quartic approximations and series for π, J. Number Theory, 14 (1982) 397–423.
31. D. Shanks and J. W. Wrench, Calculation of π to 100,000 decimals, Math. Comput., 16 (1962) 76–79.
32. W. Shanks, Contributions to Mathematics Comprising Chiefly of the Rectification of the Circle to 607 Places of Decimals, G. Bell, London, 1853.
33. Y. Tamura and Y. Kanada, Calculation of π to 4,196,393 decimals based on Gauss-Legendre algorithm, preprint (1983).
34. J. Tannery and J. Molk, Fonctions Elliptiques, vols. 1 and 2, 1893; reprint Chelsea, New York, 1972.
35. S. Wagon, Is π normal?, The Math Intelligencer, 7 (1985) 65–67.
36. G. N. Watson, Some singular moduli (1), Quart. J. Math., 3 (1932) 81–98.
37. _____, The final problem: an account of the mock theta functions, J. London Math. Soc., 11 (1936) 55–80.
38. H. Weber, Lehrbuch der Algebra, Vol. 3, 1908; reprint Chelsea, New York, 1980.
39. E. T. Whittaker and G. N. Watson, A Course of Modern Analysis, 4th ed, Cambridge Univ. Press, London, 1927.

12. Pi, Euler numbers, and asymptotic expansions (1989)

Paper 12: Jonathan M. Borwein, Peter B. Borwein and Karl Dilcher, "Pi, Euler numbers and asymptotic expansions," *American Mathematical Monthly*, vol. 96 (October 1989), p. 681–687. Copyright 1989 Mathematical Association of America. All Rights Reserved.

Synopsis:

In this paper, Jonathan Borwein, Peter Borwein and Karl Dilcher described a remarkable phenomenon originally brought to their attention by R. D. North of Colorado Springs, Colorado. North found that if one uses Gregory's series for π, namely $\pi = 4 \sum_{k=1}^{\infty} (-1)^{k-1}/(2k-1)$, to compute decimal digits of π, but truncates the sum to 500,000 terms, one finds, not surprisingly, that the sum is incorrect in the sixth digit. But surprisingly, the next ten digits are correct, and a similar on-again, off-again patterns continues much further.

Borwein, Borwein and Dilcher were above to explain this curious phenomenon — first by looking up the sequence of discrepancy values in Sloane's *Handbook of Integer Sequences* (this was before the online version was available), identifying these numbers as Euler numbers, applying the Boole summation formula, and extending the result to some other series. The resulting study not only explains a very interesting feature of the decimal expansion of π, but it is also a classic of experimental mathematics in action.

Keywords: Curiosities, Series

© Springer International Publishing Switzerland 2016
D.H. Bailey, J.M. Borwein, *Pi: The Next Generation*,
DOI 10.1007/978-3-319-32377-0_12

Pi, Euler Numbers, and Asymptotic Expansions

J. M. BORWEIN, P. B. BORWEIN, and K. DILCHER[1], *Dalhousie University, Halifax, Canada*

JONATHAN M. BORWEIN was an Ontario Rhodes Scholar (1971) at Jesus College, Oxford, where he completed a D. Phil. (1974) with Michael Dempster. Since 1974 he has worked at Dalhousie University where he is professor of mathematics. He has also been on faculty at Carnegie-Mellon University (1980–82). He was the 1987 Coxeter-James lecturer of the Canadian Mathematical Society and was awarded the Atlantic Provinces Council on the Sciences 1988 Gold Medal for Research. His research interests include functional analysis, classical analysis, and optimization theory.

PETER B. BORWEIN obtained a Ph.D. (1979) from the University of British Columbia, under the supervision of David Boyd. He spent 1979–80 as a NATO research fellow in Oxford. Since then he has been on faculty at Dalhousie (except for a sabbatical year at the University of Toronto) and is now Associate Professor of Mathematics. His research interests include approximation theory, classical analysis, and complexity theory.

KARL DILCHER received his undergraduate education and a Dipl. Math. degree at Technische Universität Clausthal in West Germany. He completed his Ph.D (1983) at Queen's University in Kingston, Ontario with Paulo Ribenboim. Since 1984 he has been teaching at Dalhousie, where he is now Assistant Professor. His research interests include Bernoulli numbers and polynomials, and classical complex analysis.

1. Introduction. Gregory's series for π, truncated at 500,000 terms, gives to forty places

$$4 \sum_{k=1}^{500,000} \frac{(-1)^{k-1}}{2k-1} = 3.14159\underline{0}653589793\underline{2}404626433832\underline{6}9502884197.$$

The number on the right is not π to forty places. As one would expect, the 6th digit after the decimal point is wrong. The surprise is that the next 10 digits are correct. In fact, only the 4 underlined digits aren't correct. This intriguing observation was sent to us by R. D. North [10] of Colorado Springs with a request for an explanation. The point of this article is to provide that explanation. Two related

[1] Research of the authors supported in part by NSERC of Canada.

examples, to fifty digits, are

$$\frac{\pi}{2} \doteq 2 \sum_{k=1}^{50,000} \frac{(-1)^{k-1}}{2k-1}$$

$$= 1.5707\underline{8}63267948\underline{9}7619231321\underline{1}916397520520985833147388$$

$$1 \qquad -1 \qquad 5 \qquad -61$$

and

$$\log 2 \doteq \sum_{k=1}^{50,000} \frac{(-1)^{k+1}}{k}$$

$$= .6931\underline{3}71806\underline{5}9945309397232121474176568048300013446572,$$

$$1 \qquad -1 \qquad 2 \qquad -16 \qquad 272$$

where all but the underlined digits are correct. The numbers under the underlined digits are the numbers that must be added to correct these. The numbers 1, -1, 5, -61 are the first four *Euler numbers* while 1, -1, 2, -16, 272 are the first five *tangent numbers*. Our process of discovery consisted of generating these sequences and then identifying them with the aid of Sloane's *Handbook of Integer Sequences* [11]. What one is observing, in each case, is an asymptotic expansion of the error in Euler summation. The amusing detail is that the coefficients of the expansion are integers. All of this is explained by Theorem 1.

The standard facts we need about the *Euler numbers* $\{E_i\}$, the *tangent numbers* $\{T_i\}$, and the *Bernoulli numbers* $\{B_i\}$, may all be found in [1] or in [6]. The numbers are defined as the coefficients of the power series

$$\sec z = \sum_{n=0}^{\infty} (-1)^n \frac{E_{2n} z^{2n}}{(2n)!}, \tag{1.1}$$

$$\tan z = \sum_{n=0}^{\infty} (-1)^{n+1} \frac{T_{2n+1} z^{2n+1}}{(2n+1)!} \qquad \text{and } T_0 = 1, \tag{1.2}$$

$$\frac{z}{e^z - 1} = \sum_{n=0}^{\infty} \frac{B_n z^n}{n!}. \tag{1.3}$$

They satisfy the relations

$$\sum_{k=0}^{n} \binom{2n}{2k} E_{2k} = 0, \qquad E_{2n+1} = 0, \tag{1.4}$$

$$B_n = \frac{-nT_{n-1}}{2^n(2^n - 1)} \qquad n \geqslant 1, \tag{1.5}$$

and

$$\sum_{k=0}^{n} \binom{n+1}{k} B_k = 0. \tag{1.6}$$

These three identities allow for the easy generation of $\{E_n\}$, $\{T_n\}$, and $\{B_n\}$. The first few values are recorded below.

n	0	1	2	3	4	5	6	7	8
E_n	1	0	-1	0	5	0	-61	0	1365
T_n	1	-1	0	2	0	-16	0	272	0
B_n	1	$\dfrac{-1}{2}$	$\dfrac{1}{6}$	0	$\dfrac{-1}{30}$	0	$\dfrac{1}{42}$	0	$\dfrac{-1}{30}$

It is clear from (1.4) that the Euler numbers are integral. From (1.5) and (1.6) it follows that the tangent numbers are integers. Also,

$$|E_{2n}| \sim \frac{4^{n+1}(2n)!}{\pi^{2n+1}} \quad \text{and} \quad |B_{2n}| \sim \frac{2(2n)!}{(2\pi)^{2n}}$$

as follows from (5.1) and (5.2) below. The main content of this note is the following theorem. The simple proof we offer relies on the Boole Summation Formula, which is a pretty but less well-known analogue of Euler summation. The details are contained in Sections 2 and 3 (except for c] which is a straightforward application of Euler summation). More complicated developments can be based directly on Euler summation or on results in [9].

THEOREM 1. *The following asymptotic expansions hold*:

a]
$$\frac{\pi}{2} - 2 \sum_{k=1}^{N/2} \frac{(-1)^{k-1}}{2k-1} \sim \sum_{m=0}^{\infty} \frac{E_{2m}}{N^{2m+1}}$$
$$= \frac{1}{N} - \frac{1}{N^3} + \frac{5}{N^5} - \frac{61}{N^7} + \cdots$$

b]
$$\log 2 - \sum_{k=1}^{N/2} \frac{(-1)^{k-1}}{k} \sim \frac{1}{N} + \sum_{m=1}^{\infty} \frac{T_{2m-1}}{N^{2m}}$$
$$= \frac{1}{N} - \frac{1}{N^2} + \frac{2}{N^4} - \frac{16}{N^6} + \frac{272}{N^8} - \cdots$$

and

c]
$$\frac{\pi^2}{6} - \sum_{k=1}^{N-1} \frac{1}{k^2} \sim \frac{1}{2N^2} + \sum_{m=0}^{\infty} \frac{B_{2m}}{N^{2m+1}}$$
$$= \frac{1}{N} + \frac{1}{2N^2} + \frac{1}{6N^3} - \frac{1}{30N^5} + \frac{1}{42N^7} \cdots.$$

From the asymptotics of $\{E_n\}$ and $\{B_n\}$ and (1.5) we see that each of the above infinite series is everywhere divergent; the correct interpretation of their asymptotics is

a']
$$\sum_{m=1}^{\infty} \frac{E_{2m}}{N^{2m}} = \sum_{m=1}^{K} \frac{E_{2m}}{N^{2m}} + 0\left(\frac{(2K+1)!}{(\pi N)^{2K+1}} \right)$$

b']
$$\sum_{m=1}^{\infty} \frac{T_{2m-1}}{N^{2m}} = \sum_{m=1}^{K} \frac{T_{2m-1}}{N^{2m}} + 0\left(\frac{(2K+1)!}{(\pi N)^{2K-1}} \right)$$

c']
$$\sum_{m=1}^{\infty} \frac{B_{2m}}{N^{2m+1}} = \sum_{m=1}^{K} \frac{B_{2m}}{N^{2m+1}} + 0\left(\frac{(2K+1)!}{(2\pi N)^{2K+1}} \right),$$

where in each case the constant concealed by the order symbol is independent of N and K. In fact, the constant 10 works in all cases.

2. The Boole Summation Formula. The Euler polynomials $E_n(x)$ can be defined by the generating function

$$\frac{2e^{tx}}{e^t + 1} = \sum_{n=0}^{\infty} E_n(x)\frac{t^n}{n!} \qquad (|t| < \pi); \tag{2.1}$$

(see [1, p. 804]). Each $E_n(x)$ is a polynomial of degree n with leading coefficient 1. We also define the periodic Euler function $\bar{E}_n(x)$ by

$$\bar{E}_n(x + 1) = -\bar{E}_n(x)$$

for all x, and

$$\bar{E}_n(x) = E_n(x) \quad \text{for} \quad 0 \leqslant x < 1.$$

It can be shown that $\bar{E}_n(x)$ has continuous derivatives up to the $(n-1)$st order.

The following is known as **Boole's summation formula** (see, for example, [9, p. 34]).

LEMMA 1. *Let $f(t)$ be a function with m continuous derivatives, defined on the interval $x \leqslant t \leqslant x + \omega$. Then for $0 \leqslant h \leqslant 1$*

$$f(x + h\omega) = \sum_{k=0}^{m-1} \frac{\omega^k}{k!} E_k(h) \cdot \frac{1}{2}\left(f^{(k)}(x + \omega) + f^{(k)}(x)\right) + R_m,$$

where

$$R_m = \frac{1}{2}\omega^m \int_0^1 \frac{\bar{E}_{m-1}(h - t)}{(m-1)!} f^{(m)}(x + \omega t)\,dt.$$

This summation formula is easy to establish by repeated integration by parts of the above integral. It is remarked in [9, p. 26] that this formula was known to Euler, for polynomial f and without the remainder term. Also note that Lemma 1 turns into Taylor's formula with Lagrange's remainder term if we replace h by h/ω and let ω approach zero.

To derive a convenient version of Lemma 1 for the applications we have in mind, we set $\omega = 1$ and impose further restrictions on f.

LEMMA 2. *Let f be a function with m continuous derivatives, defined on $t \geqslant x$. Suppose that $f^{(k)}(t) \to 0$ as $t \to \infty$ for all $k = 0, 1, ..., m$. Then for $0 \leqslant h \leqslant 1$*

$$\sum_{v=0}^{\infty} (-1)^v f(x + h + v) = \sum_{k=0}^{m-1} \frac{E_k(h)}{2k!} f^{(k)}(x) + R_m,$$

where

$$R_m = \frac{1}{2} \int_0^{\infty} \frac{\bar{E}_{m-1}(h - t)}{(m-1)!} f^{(m)}(x + t)\,dt.$$

3. The Remainder for Gregory's Series. The Euler numbers E_n may also be defined by the generating function

$$\frac{2}{e^t + e^{-t}} = \sum_{n=0}^{\infty} E_n \frac{t^n}{n!}.$$ (3.1)

Comparing (3.1) with (2.1), we see that

$$E_n = 2^n E_n\left(\frac{1}{2}\right).$$ (3.2)

The phenomenon mentioned in the introduction is entirely explained by the next proposition—if we set $n = 500,000$. It is also clear that we will get similar patterns for $n = 10^m/2$ with any positive integer m.

PROPOSITION 1. *For positive integers n and M we have*

$$4 \sum_{k=n}^{\infty} \frac{(-1)^k}{2k+1} = (-1)^n \sum_{k=0}^{M} \frac{2E_{2k}}{(2n)^{2k+1}} + R_1(M),$$ (3.3)

where

$$|R_1(M)| \leq \frac{2|E_{2M}|}{(2n)^{2M+1}}.$$

Proof. Apply Lemma 2 with $f(x) = 1/x$; then set $x = n$ and $h = 1/2$. We get

$$\sum_{v=0}^{\infty} \frac{(-1)^v}{n+v+1/2} = \sum_{k=0}^{m-1} \frac{E_k(1/2)}{2k!} \frac{(-1)^k k!}{n^{k+1}} + R_m,$$ (3.4)

with

$$R_m = \frac{1}{2} \int_0^{\infty} \frac{\overline{E}_{m-1}(h-t)}{(m-1)!} \frac{(-1)^m m!}{(x+t)^{m+1}} dt.$$

We multiply both sides of (3.4) by $2(-1)^n$. Then the left-hand side is seen to be identical with the left-hand side of (3.3). After replacing m by $2M + 1$ and taking into account (3.2) and the fact that odd-index Euler numbers vanish, we see that the first terms on the right-hand sides of (3.3) and (3.4) agree. To estimate the error term, we use the following inequality,

$$|E_{2M}(x)| \leq 2^{-2M}|E_{2M}| \quad \text{for } 0 \leq x \leq 1$$

(see, e.g., [1, p. 805]). Carrying out the integration now leads to the error estimate given in Proposition 1.   □

4. An Analogue For log 2. Lemma 2 can also be used to derive a result similar to Proposition 1, concerning truncations of the series

$$\log 2 = \sum_{k=1}^{\infty} \frac{(-1)^{k+1}}{k}.$$ (4.1)

In this case the tangent numbers T_n will play the role of the E_n in Proposition 1. It follows from the identity

$$\tan z = \frac{1}{i}\left(\frac{2e^{2iz}}{e^{2iz}+1}-1\right)$$

together with (1.2) and (2.1) that

$$T_n = (-1)^n 2^n E_n(1) \tag{4.2}$$

as in [9, p. 28]. The T_n can be computed using the recurrence relation $T_0 = 1$ and

$$\sum_{k=0}^{n}\binom{n}{k}2^k T_{n-k} + T_n = 0 \quad \text{for } n \geqslant 1.$$

Other properties can be found, e.g., in [8] or [9, Ch. 2].

PROPOSITION 2. *For positive integers n and M we have*

$$\sum_{k=n+1}^{\infty}\frac{(-1)^{k+1}}{k} = (-1)^{n+1}\left\{\frac{1}{2n} + \sum_{k=1}^{M}\frac{T_{2k-1}}{(2n)^{2k}}\right\} + R_2(M), \tag{4.3}$$

where

$$|R_2(M)| \leqslant \frac{|E_{2M}|}{(2n)^{2M+1}}.$$

Proof. We proceed as in the proof of Proposition 1. Here we take $x = n$ and $h = 1$. Using (4.2) and the fact that $T_0 = 1$ and $T_{2k} = 0$ for $k \geqslant 1$, we get the summation on the right-hand side of (4.3). The remainder term is estimated as in the proof of Proposition 1. $\qquad\square$

Using Proposition 2 with $n = 10^m/2$ one again gets many more correct digits of $\log 2$ than is suggested by the error term of the Taylor series.

5. Generalizations. Proposition 1 and 2 can be extended easily in two different directions.

i). The well-known infinite series (see, e.g., [1, p. 807])

$$\sum_{k=0}^{\infty}\frac{(-1)^k}{(2k+1)^{2n+1}} = \frac{|E_{2n}|}{2^{2n+2}(2n)!}\pi^{2n+1} \quad (n = 0, 1, \ldots), \tag{5.1}$$

and

$$\sum_{k=1}^{\infty}\frac{(-1)^{k-1}}{k^{2n}} = (1 - 2^{1-2n})\zeta(2n)$$

$$= (2^{2n-1} - 1)\frac{|B_{2n}|}{(2n)!}\pi^{2n} \quad (n = 1, 2, \ldots) \tag{5.2}$$

can be considered as extensions of Gregory's series and of (4.1). These series admit exact analogues to Propositions 1 and 2; one only has to replace $f(x) = 1/x$ by $f(x) = x^{-(2n+1)}$, respectively $x^{-(2n)}$, in the proofs.

We note that the Euler-MacLaurin summation formula leads to similar results for

$$\sum_{k=1}^{\infty} k^{-2n} = \frac{|B_{2n}|2^{2n-1}}{(2n)!}\pi^{2n}, \tag{5.3}$$

where multiples of the Bernoulli numbers B_{2n} take the place of the E_n and T_n in Propositions 1 and 2.

ii). A generalization of the Euler-MacLaurin and Boole summation formulas was derived by Berndt [3]. This can be applied to character analogues of the series (5.1)–(5.3). The roles of the E_n and T_n in Proposition 1 and 2 are then played by generalized Bernoulli numbers or by related numbers.

6. Additional Comments. The phenomenon observed in the introduction results from taking N to be a power of ten; taking $N = 2 \cdot 10^m$ also leads to "clean" expressions. References [1], [5], [6], and [9] include the basic material on Bernoulli and Euler numbers, while [8] deals extensively with their calculation, and [2] describes an entertaining analogue of Pascal's triangle. Much on the calculation of pi and related matters may be found in [4]. Euler summation is treated in [5], [6], and [9], while Boole summation is treated in [9]. Related material on the computation and acceleration of alternating series is given in [7].

Added in Proof.   A version of the phenomeon was observed by M. R. Powell and various explanations were offered (see *The Mathematical Gazette*, 66 (1982) 220–221, and 67(1983) 171–188).

REFERENCES

1.   M. Abramowitz and I. Stegun, Handbook of Mathematical Functions, Dover, N.Y., 1964.
2.   M. D. Atkinson, How to compute the series expansions of sec x and tan x, *Amer. Math. Monthly*, 93 (1986) 387-388.
3.   B. C. Berndt, Character analogues of the Poisson and Euler-MacLaurin summation formulas with applications, *J. Number Theory*, 7 (1975 413-445.
4.   J. M. Borwein and P. B. Borwein, Pi and the AGM - A Study in Analytic Number Theory and Computational Complexity, Wiley, N.Y., 1987.
5.   T. J. I'a Bromwich, An Introduction to the Theory of Infinite Series, 2nd ed., MacMillan, London, 1926.
6.   R. L. Graham, D. E. Knuth, and O. Patashnik, Concrete Mathematics, Addison Wesley, Reading, Mass., 1989.
7.   R. Johnsonbaugh, Summing an alternating series, this MONTHLY, 86 (1979) 637-648.
8.   D. E. Knuth and T. J. Buckholtz, Computation of Tangent, Euler, and Bernoulli numbers, *Math. Comput.*, 21 (1967) 663-688.
9.   N. Nörlund, *Vorlesungen über Differenzenrechnung*, Springer-Verlag, Berlin, 1924.
10.   R. D. North, personal communications, 1988.
11.   N. J. A. Sloane, A Handbook of Integer Sequences, Academic Press, New York, 1973.

13. A spigot algorithm for the digits of π (1995)

Paper 13: Stanley Rabinowitz and Stan Wagon, "A spigot algorithm for the digits of π," *American Mathematical Monthly*, vol. 102 (March 1995), p. 195–203. Copyright 1995 Mathematical Association of America. All Rights Reserved.

Synopsis:

In this paper, Rabinowitz and Wagon introduce a very interesting "spigot algorithm" for the digits of π. In particular, the algorithm, which acts on a two-dimensional table of data, generates the decimal digits of π one by one. The algorithm acts only on modest-sized integer data — no floating-point arithmetic is involved. The process was discovered by some experimentation using the *Mathematica* software.

This scheme is different in nature from the BBP algorithm, to be presented in the next chapter, in that it cannot directly generate digits of π at an arbitrary starting point. But it is very simple, and, unlike the BBP formula and algorithm, it works in base ten.

Keywords: Algorithms

© Springer International Publishing Switzerland 2016

D.H. Bailey, J.M. Borwein, *Pi: The Next Generation*,

DOI 10.1007/978-3-319-32377-0_13

A Spigot Algorithm for the Digits of π

Stanley Rabinowitz and Stan Wagon

It is remarkable that the algorithm illustrated in Table 1, which uses no floating-point arithmetic, produces the digits of π. The algorithm starts with some 2s, in columns headed by the fractions shown. Each entry is multiplied by 10. Then, starting from the right, the entries are reduced modulo *den*, where the head of the column is *num/den*, producing a quotient q and remainder r. The remainder is left in place and $q \times num$ is carried one column left. This reduce-and-carry is continued all the way left. The tens digit of the leftmost result is the next digit of π. The process continues with the multiplication of the remainders by 10, the reductions modulo the denominators, and the augmented carrying.

TABLE 1. The workings of an algorithm that produces digits of π. The dashed line indicates the key step: starting from the right, entries are reduced modulo the denominator of the column head $(25, 23, 21, \ldots,$ resp.), with the quotients, after multiplication by the numerator $(12, 11, 10, \ldots)$, carried left. For example, the 20 in the $\frac{9}{19}$'s column yields a remainder of 1 and a left carry of $1 \cdot 9 = 9$. After the leftmost carries, the tens digits are $3, 1, 4, 1$. To get more digits of π one must start with a longer string of 2s.

	Digits of π		$\frac{1}{3}$	$\frac{2}{5}$	$\frac{3}{7}$	$\frac{4}{9}$	$\frac{5}{11}$	$\frac{6}{13}$	$\frac{7}{15}$	$\frac{8}{17}$	$\frac{9}{19}$	$\frac{10}{21}$	$\frac{11}{23}$	$\frac{12}{25}$
Initialize		2	2	2	2	2	2	2	2	2	2	2	2	2
×10		20	20	20	20	20	20	20	20	20	20	20	20	20
Carry	3	+10	+12	+12	+12	+10	+12	+7	+8	+9	+0	+0	+0	=
		30	32	32	32	30	32	27	28	29	20	20	20	20
Remainders		0	2	2	4	3	10	1	13	12	1	20	20	20
×10		0	20	20	40	30	100	10	130	120	10	200	200	200
Carry	1	+13	+20	+33	+40	+65	+48	+98	+88	+72	+150	+132	+96	=
		13	40	53	80	95	148	108	218	192	160	332	296	200
Remainders		3	1	3	3	5	5	4	8	5	8	17	20	0
×10		30	10	30	30	50	50	40	80	50	80	170	200	0
Carry	4	+11	+24	+30	+40	+40	+42	+63	+64	+90	+120	+88	+0	=
		41	34	60	70	90	92	103	144	140	200	258	200	0
Remainders		1	1	0	0	0	4	12	9	4	10	6	16	0
×10		10	10	0	0	0	40	120	90	40	100	60	160	0
Carry	1	+4	+2	+9	+24	+55	+84	+63	+48	+72	+60	+66	+0	=
		14	12	9	24	55	124	183	138	112	160	126	160	0

This algorithm is a "spigot" algorithm: it pumps out digits one at a time and does not use the digits after they are computed. Moreover, the digits are generated without any use of high-precision (or low-precision) operations on floating-point real numbers; the entire algorithm uses only ordinary integer arithmetic on relatively small integers. For example, to obtain the first 5,000 digits of π requires only arithmetic operations on integers less than 600,000,000. Although high-precision floating-point routines are built up from integer operations, the algorithms in this paper are quite simple and do not simulate floating-point computations.

In order to motivate the π-algorithm, we first discuss the much simpler case of e, for which a spigot algorithm was discovered by Sale [**Sale**]. His algorithm is the basis of the discussion in §1.

1. A NUMBER SYSTEM IN WHICH e's DIGITS ARE PERIODIC. A real number's decimal representation may be interpreted as an infinitely nested expression; for example:

$$\sqrt{2} = 1.41421356\ldots = 1 + \tfrac{1}{10}\big(4 + \tfrac{1}{10}\big(1 + \tfrac{1}{10}\big(4 + \tfrac{1}{10}\big(2 + \tfrac{1}{10}\big(1 + \cdots\big)\big)\big)\big)\big).$$

Some interesting and useful representations may be obtained if we change the base-sequence, which in the case above is $(\tfrac{1}{10}, \tfrac{1}{10}, \tfrac{1}{10}, \tfrac{1}{10}\ldots)$. For example, using the base $\mathbf{b} = (\tfrac{1}{2}, \tfrac{1}{3}, \tfrac{1}{4}, \tfrac{1}{5}, \ldots)$ yields the following form, called a *mixed-radix* representation (see [**Knu**, §4.1]):

$$a_0 + \tfrac{1}{2}\big(a_1 + \tfrac{1}{3}\big(a_2 + \tfrac{1}{4}\big(a_3 + \tfrac{1}{5}\big(a_4 + \tfrac{1}{6}\big(a_5 + \ldots\big)\big)\big)\big)\big),$$

where the a_i (the *digits*) are nonnegative integers. If $0 \le a_i \le i$ for $i \ge 1$, the representation is called *regular*. Mixed-radix representations will be denoted by $(a_0; a_1, a_2, a_3, a_4, \ldots)_{\mathbf{b}}$. For base $\mathbf{b}$, every positive real number has a regular representation and representations are unique provided we exclude representations that terminate with maximal digits (otherwise, for example, $\tfrac{1}{2} = (0; 1, 0, 0, \ldots)_{\mathbf{b}} = (0; 0, 2, 3, 4, 5, 6, \ldots)_{\mathbf{b}}$); from now on and for all bases, we exclude such representations. The proof of the following Lemma is in Appendix 1.

Lemma 1(a). *If* $i \ge 1$, $(0; 0, 0, \ldots, 0, a_i, a_{i+1}, \ldots)_{\mathbf{b}} < \tfrac{1}{i!}$; *in particular,* $(0; a_1, a_2, a_3, a_4, \ldots)_{\mathbf{b}} < 1$.

(b). *Representations using the mixed-radix base* $\mathbf{b}$ *are unique.*

(c). *The integer part of* $(a_0; a_1, a_2, a_3, a_4, \ldots)_{\mathbf{b}}$ *is* a_0 *and the fractional part is* $(0; a_1, a_2, a_3, a_4, \ldots)_{\mathbf{b}}$.

In this number system some irrationals become periodic. For example, $e = (2; 1, 1, 1, 1, \ldots)_{\mathbf{b}}$; this is just a restatement of the infinite series $\sum \tfrac{1}{i!}$ as $1 + \tfrac{1}{1}(1 + \tfrac{1}{2}(1 + \tfrac{1}{3}(1 + \tfrac{1}{4}(1 + \tfrac{1}{5}(1 + \ldots))))))$. Rational numbers in this system correspond to digit-sequences that terminate (Appendix 1, Lemma 2).

The decimal digits of a real number x in $[0, 10)$ can be obtained by taking the integer part of x, multiplying its fractional part by 10, taking the integer part of the result, multiply the resulting fractional part by 10, and so on. In some mixed-radix bases, this is especially simple. If $x = (a_0; a_1, a_2, \ldots, a_n)_{\mathbf{b}}$, then $10x = (10a_0; 10a_1, 10a_2, 10a_3, \ldots, 10a_n)_{\mathbf{b}}$. The latter may not be a regular expression: some digits may be too big. But we can decrease digits by reducing them modulo i, where i is the denominator of the corresponding element of $\mathbf{b}$. Starting these reductions at the right end, we carry the quotients left, eventually getting the regular representation of $10x$. Thus multiplying by 10 is algorithmically straightforward. Taking the integer and fractional parts for $\mathbf{b}$-representations is also easy, thanks to Lemma 1(c).

We can now give the algorithm to get the first n base-10 digits of e. A proof of correctness—the error analysis showing that $n + 2$ mixed-radix digits suffice[1] to get n base-10 digits—is given as Lemma 3 in Appendix 1.

Algorithm *e-spigot*

1. *Initialize:* Let the first digit be 2 and initialize an array A of length $n + 1$ to $(1, 1, 1, \ldots, 1)$.
2. Repeat $n - 1$ times:
   Multiply by 10: Multiply each entry of A by 10.
   Take the fractional part: Starting from the right, reduce the ith entry of A modulo $i + 1$, carrying the quotient one place left.
   Output the next digit: The final quotient is the next digit of e.

The first few steps of this algorithm, starting with an array of 10 1s (this corresponds to 11 mixed-radix digits, good for 9 digits of e; only 5 are shown), are displayed in Table 2.

TABLE 2. The workings of a spigot algorithm for the digits of e (in bold). The reductions in the column headed $\frac{1}{i}$ are performed modulo i. The leftmost base-10 real numbers are the values of the rows viewed as mixed-radix representations. Since only 11 mixed-radix digits start the algorithm, the first base-10 number is only an approximation to e.

Base 10		$^1/_2$	$^1/_3$	$^1/_4$	$^1/_5$	$^1/_6$	$^1/_7$	$^1/_8$	$^1/_9$	$^1/_{10}$	$^1/_{11}$
2.718281826...	**2**	1	1	1	1	1	1	1	1	1	1
7.18281826...		10	10	10	10	10	10	10	10	10	10
carries	**7**	±3	±3	±2	±1	±1	±1	±1	±1	±0	==
		14	13	12	11	11	11	11	11	10	10
0.18281826...		0	1	0	1	5	4	3	2	0	10
1.8281826...		0	10	0	10	50	40	30	20	0	100
carries	**1**	±3	±0	±3	±9	±6	±4	±2	±0	±9	==
		3	10	3	19	56	44	32	20	9	100
0.8281826...		1	1	3	4	2	2	0	2	9	1
8.281826...		10	10	30	40	20	20	0	20	90	10
carries	**8**	±6	±9	±8	±3	±2	±0	±3	±9	±0	==
		16	19	38	43	22	20	3	29	90	10
0.281826...		0	1	2	3	4	6	3	2	0	10
2.81826...		0	10	20	30	40	60	30	20	0	100
carries	**2**	±5	±6	±7	±8	±9	±4	±2	±0	±9	==
		5	16	27	38	49	64	32	20	9	100
0.81826...		1	1	3	3	1	1	0	2	9	1

2. A SPIGOT FOR DIGITS OF π. The ideas of §1 lead to a spigot algorithm for π, but there are additional complexities and additional interesting questions that distinguish π from e. Our starting point is the following moderately well-known

---

[1] Any digit-producing algorithm for a presumed-normal number x suffers from a drawback that, although unlikely, can impinge on the result. If x is between 1 and 10 and the algorithm says that the first 100 digits of x are, say, $4, 6, 5, 0, 7, \ldots, 3, 9, 9, 9, 9, 9$ then one cannot be sure that the last 6 digits are correct. They will be the digits of a certain approximation to x that is within $5 \cdot 10^{-100}$ of the true value. One cannot simply go farther until a non-9 is reached, because memory allocations must be made in advance. The user must realize that a terminating string of 9s is a red flag concerning those digits and even with no 9s, the last digit might be incorrect. In practice, one might ask for, say, 6 extra digits, reducing the odds of this problem to one in a million.

series:

$$\pi = \sum_{i=0}^{\infty} \frac{(i!)^2 2^{i+1}}{(2i+1)!}.$$

This series can be derived from the Wallis product for π; another approach uses an acceleration technique called Euler's transform applied to the series $\pi = 4 - \frac{4}{3} + \frac{4}{5} - \frac{4}{7} + \ldots$. These proofs, together with three others and references to earlier sources, may be found in [Li]. We let $k!!$ denote the product $1 \cdot 3 \cdot 5 \cdots k$ for odd integers k; then the series is equivalent to

$$\frac{\pi}{2} = \sum_{i=0}^{\infty} \frac{i!}{(2i+1)!!} = 1 + \frac{1}{3} + \frac{1 \cdot 2}{3 \cdot 5} + \frac{1 \cdot 2 \cdot 3}{3 \cdot 5 \cdot 7} + \cdots,$$

which expands to become

$$\frac{\pi}{2} = 1 + \frac{1}{3}\left(1 + \frac{2}{5}\left(1 + \frac{3}{7}\left(1 + \frac{4}{9}(1 + \cdots)\right)\right)\right).$$

This last expression leads to the mixed-radix base $\mathbf{c} = (\frac{1}{3}, \frac{2}{5}, \frac{3}{7}, \frac{4}{9}, \ldots)$, with respect to which π is simply $(2; 2, 2, 2, 2, 2, \ldots)_{\mathbf{c}}$. For a regular representation in base $\mathbf{c}$, the digit in the ith place must lie in the interval $[0, 2i]$. Unfortunately, base $\mathbf{c}$ is less accommodating than $\mathbf{b}$.

Lemma 4 (Proof in Appendix 1). *The base-$\mathbf{c}$ number with maximal digits, $(0; 2, 4, 6, 8, \ldots)$, represents 2; hence regular representations of the form $(0; a, b, c, \ldots)_{\mathbf{c}}$ lie between 0 and 2.*

Lemma 4 implies that $\mathbf{c}$-representations are not unique. For example, $(0; 0, 4, 6, 8, \ldots)_{\mathbf{c}} = 2 - \frac{2}{3} = \frac{4}{3}$, whence $(0; 0, 2, 3, 4, \ldots)_{\mathbf{c}} = \frac{2}{3} = (0; 2, 0, 0, 0, \ldots)_{\mathbf{c}}$. More relevant algorithmically, integer and fractional parts using $\mathbf{c}$ are not straightforward, as they are for $\mathbf{b}$. The integer part of $(a_0; a_1, a_2, \ldots)_{\mathbf{c}}$ is either a_0 or $a_0 + 1$ according as $(0; a_1, a_2, \ldots)$ is in $[0, 1)$ or $[1, 2)$. This problem is surmounted by leaving the units digit of a_0 in place during the next iteration and calling the tens digit of a_0 a *predigit*. The predigits must be temporarily held because occasionally (once every 20 iterations, roughly) the next predigit is a 10; this will happen when the carry, which is between 0 and 19, is greater than 10 and, simultaneously, the leftover units digit of a_0 is 9, which becomes 90 in the multiply-by-10 step. This event requires that the held number be increased by 1 before being released. Specific details of the algorithm follow; the presentation at the beginning of this paper sidestepped the problem of the occasional 10. The proof that $\lfloor 10n/3 \rfloor$ mixed-radix digits suffice for n digits of π is in Appendix 1 (Lemma 5). Appendix 2 contains a Pascal implementation of this algorithm.

Algorithm π-spigot

1. *Initialize:* Let $A = (2, 2, 2, 2, \ldots, 2)$ be an array of length $\lfloor 10n/3 \rfloor$.
2. Repeat n times:
   Multiply by 10: Multiply each entry of A by 10.
   Put A into regular form: Starting from the right, reduce the ith element of A (corresponding to $\mathbf{c}$-entry $(i-1)/(2i-1)$) modulo $2i - 1$, to get a quotient q and a remainder r. Leave r in place and carry $q(i-1)$ one place left. The last integer carried (from the position where $i - 1 = 2$) may be as large as 19.

Get the next predigit: Reduce the leftmost entry of A (which is at most $109[= 9 \cdot 10 + 19]$) modulo 10. The quotient, q, is the new predigit of π, the remainder staying in place.

Adjust the predigits: If q is neither 9 nor 10, release the held predigits as true digits of π and hold q. If q is 9, add q to the queue of held predigits. If q is 10 then:

- set the current predigit to 0 and hold it;
- increase all other held predigits by 1 (9 becomes 0);
- release as true digits of π all but the current held predigit.

This algorithm uses only integer arithmetic and is easy to program. The table at the beginning of the paper shows it in action, starting with 13 mixed-radix digits of π (good for 4 base-10 digits). To clarify the working of the algorithm, note that the (finite) first row of Table 1 is a mixed-radix representation of $3.1414796\ldots$, the second row represents $31.414796\ldots$, the fifth row represents $1.414796\ldots$, the sixth row is $14.14796\ldots$, the ninth row is $4.14796\ldots$, and so on. Table 3 shows the result of a computation using a larger initial array; the holding aspect does not become relevant until the 32nd digit.

TABLE 3. The actual digits of π (bottom) compared to the sequence of leftmost base-c digits for 35 iterations with a starting array of 116 2s (good for 35 digits). At the 32nd iteration a 102 shows up, yielding a predigit of 10.

```
                                                              ↓
30 13 41 15 58 92 26 64 53 35 58 89 97 78 92 32 23 38 84 45 62 26 63 42 33 38 82 32 27 78 94 49 102 28 87
 3  1  4  1  5  9  2  6  5  3  5  8  9  7  9  3  2  3  8  4  6  2  6  4  3  3  8  3  2  7  9  5     0  2  8
                                                                                            ↑
```

We repeat that the algorithm uses only integer operations. To get 5,000 digits of π requires only integer arithmetic on numbers less than 600,000,000. The algorithm leads naturally to the question of improving it to one that is essentially as simple as *e-spigot*.

Question. Is there a base **d** of rationals such that π has a **d**-representation that is periodic, or an arithmetic progression, and such that a_0 is always the integer part of $(a_0; a_1, a_2, \ldots)_\mathbf{d}$?

Gosper [**Gos**, p. 32] has discovered a series for π that brings us tantalizingly close to spigot-perfection:

$$\pi = 3 + \frac{1}{60}8 + \frac{1}{60}\frac{2 \cdot 3}{7 \cdot 8 \cdot 3}13 + \frac{1}{60}\frac{2 \cdot 3}{7 \cdot 8 \cdot 3}\frac{3 \cdot 5}{10 \cdot 11 \cdot 3}18 +$$

$$\frac{1}{60}\frac{2 \cdot 3}{7 \cdot 8 \cdot 3}\frac{3 \cdot 5}{10 \cdot 11 \cdot 3}\frac{4 \cdot 7}{13 \cdot 14 \cdot 3}23 + \cdots.$$

He obtained this series by using a refinement of the Euler transform on $4 - \frac{4}{3} + \frac{4}{5} - \frac{4}{7} + \ldots$. Gosper's series leads to the base $\mathbf{d} = (\frac{1}{60}, \frac{6}{168}, \frac{15}{330}, \frac{28}{546}, \ldots)$, with respect to which π is $(3; 8, 13, 18, \ldots)$. A computation shows that $(0; 59, 167, 329, 545, \ldots)_\mathbf{d} = 1.092\ldots$, a substantial improvement over the 2 that arose for **c**. Under the usual randomness assumption for π's digits, the odds of a bad predigit in base **c** are 1 in 20, while in base **d** they decrease to less than 1 in 110; this is because a **d**-predigit of 10 occurs only when the remainder is a 9 (which becomes 90) and the carry is a 10. The former happens 10% of the time, while the latter happens no more than once in 11 iterations because the carry is the integer part of a real between 0 and 10.93. So base **d** is within 1% of spigot-perfection. Because Gosper's series converges more quickly than the one we used, it has less memory requirements: n digits of π require an initial array of length n; however, the arithmetic on the array will involve integers larger than those in an array of the same size using base **c**.

One way to improve the Gosper-series approach is to reduce the fractions in **d** to lowest terms. Then the regular number with maximal digits is $(0; 59, 27, 21, 38, \ldots)_\mathbf{d}$, which equals $1.0000476468\ldots$. It is not hard to see that the regular representation of π is unchanged in this new base. However, the work expended in reducing to lowest terms outweighs the gain made in reducing the number of times a 10 appears as a predigit. Thus it is likely that an affirmative answer to the question above is of more theoretical than practical interest.

The spigot algorithm for π is by no means competitive with the recently discovered fast algorithms (due to the Borwein brothers, the Chudnovsky brothers, and others) that have been used to compute hundreds of millions of digits of π (see [**BBB**]). But the spigot algorithm does have the advantage of avoiding all floating-point computations; thus it is easily implemented on a home computer where it can produce thousands of digits in a few minutes. Moreover, it gives the result directly in base 10 (most other π-algorithms produce the result in binary or some internal format and a second pass must be made to obtain decimal digits).

The algorithm given here can be made to run faster by outputting multiple digits at a time. For example, to get five decimal digits at a time, simply compute the digits of π using base 100,000. This can be done by multiplying by 100,000 instead of 10 in the main step. The integer part is then the next "digit" in base 100,000.

If one is working in base 100,000 and knows in advance that the portion of digits to be computed does not contain the string 00000, then one can omit the lengthy part of the algorithm that adjusts the predigits. This can lead to an exceedingly short computer program. For example, Rabinowitz [**Rab**] used this idea to exhibit a 14-line Fortran program that outputs 1,000 decimal digits of π.

Finally, we mention that the algorithm can be parallelized, in which case it becomes blindingly fast up to about 10,000 digits.

For examples of spigot algorithms for other functions, see [**Abd**].

APPENDIX 1. FIVE LEMMAS

Lemma 1(a). *If $i \geq 1$, $(0; 0, 0, \ldots, 0, a_i, a_{i+1}, \ldots)_\mathbf{b} < \frac{1}{i}$; in particular, $(0; a_1, a_2, a_3, a_4, \ldots)_\mathbf{b} < 1$. Representations using mixed-radix base **b** are unique.*

(b). *Representations using mixed-radix base **b** are unique.*

(c). *The integer part of $(a_0; a_1, a_2, a_3, a_4, \ldots)_\mathbf{b}$ is a_0 and the fractional part is $(0; a_1, a_2, a_3, a_4, \ldots)_\mathbf{b}$.*

Proof: (a). It suffices to prove that $\sum_{k=i+1}^{\infty}(k-1)/k! = 1/i!$, which follows from the fact that the series telescopes to:

$$\left(\frac{1}{i!} - \frac{1}{(i+1)!}\right) + \left(\frac{1}{(i+1)!} - \frac{1}{(i+2)!}\right) + \left(\frac{1}{(i+2)!} - \frac{1}{(i+3)!}\right) + \cdots.$$

(b). Suppose $(a_0; a_1, a_2, a_3, a_4, \ldots)_b$ and $(c_0; c_1, c_2, c_3, c_4, \ldots)_b$ represent the same real number. Then, for some i, $0 = \sum_{k=i}^{\infty} d_k / k!$, where $|d_k| < k$ and $d_i \neq 0$. But then $|d_i|/i! \leq \sum_{k=i+1}^{\infty} |d_k|/k!$, contradicting (a).

(c). This follows from (a).

Lemma 2. *A positive number is rational iff its digits using the mixed-radix base* **b** *are eventually* 0.

Proof: The reverse direction is obvious. For the forward direction we use a sublemma.

Sublemma. *For any integers t and n, with $0 \leq n < t!$, there are integers d_i in $[0, i]$ such that* $n = d_1 t(t - 1)(t - 2) \cdots 4 \cdot 3 + d_2 t(t - 1)(t - 2) \cdots 5 \cdot 4 + \cdots + d_{t-3} t(t - 1) + d_{t-2} t + d_{t-1}.$

Proof: By induction on t. If $n < t!$ write n as $qt + r$ with $0 \leq r < t$ and $0 \leq q < (t - 1)!$. By induction there is a sequence $(d_1, d_2, \ldots d_{t-3}, d_{t-2})$ that is a solution for q with respect to terms $(t - 1)(t - 2) \cdots 4 \cdot 3$, and the like, whence $(d_1, d_2, \ldots d_{t-3}, d_{t-2}, r)$ is a solution for n w.r.t. the terms $t(t - 1)(t - 2) \cdots 4 \cdot 3$, and the like.

Returning to Lemma 2's proof, suppose a positive rational s/t is given. Use the sublemma to express $s(t - 1)!$ in the form $d_1 t(t - 1)(t - 2) \cdots 4 \cdot 3 + d_2 t (t - 1)(t - 2) \cdots 5 \cdot 4 + \cdots + d_{t-3} t(t - 1) + d_{t-2} t + d_{t-1}$. Dividing by $t!$ then yields a representation of s/t as a sum of reciprocals of factorials with appropriately small coefficients, which is the same as a terminating representation in the mixed-radix base **b**.

Lemma 3. *The algorithm for digits of e is correct.*

Proof: It must be shown that $n + 2$ mixed-radix digits of e suffice to get n base-10 digits of e. We first prove that if $n \geq 28$ $(= \lceil 10e \rceil)$, then n mixed-radix digits suffice for n base-10 digits. Using n mixed-radix digits means we are actually getting the base-10 digits of $e_n = (2; 1, 1, 1, \ldots, 1) = \sum_{i=0}^{n} 1/i!$. Thus we must show that
$e - e_n \leq 5 \cdot 10^{-n}$ (see footnote at beginning of paper). A geometric series estimation of the tail of the series shows that $e - e_n < 2/(n + 1)!$, and then Stirling's formula yields

$$\frac{2}{(n + 1)!} < \frac{1}{n!} < \left(\frac{e}{n}\right)^n < \left(\frac{1}{10}\right)^n.$$

If $n < 28$ then a direct computation of the digits shows that $n + 2$ mixed-radix digits suffice.

Lemma 4. *The base-c number with maximal digits, $(0; 2, 4, 6, 8, \ldots)$, represents 2; hence regular representations of the form $(0; a, b, c, \ldots)_c$ lie between 0 and 2.*

Proof: Instead of giving a formal proof, we show how some *Mathematica* computations led to the result (and a proof). In terms of series, the lemma states that

$$\sum_{i=0}^{\infty} \frac{(2i)i!}{(2i + 1)!!} = 2.$$

A rough calculation showed that the sum is near 2. Then a rational computation of the remainders—the differences between the partial sums and 2—yielded the following sequence.

$$\tfrac{4}{3}, \tfrac{4}{5}, \tfrac{16}{35}, \tfrac{16}{63}, \tfrac{32}{231}, \tfrac{32}{429}, \tfrac{256}{6435}, \tfrac{256}{12155}, \tfrac{512}{46189}, \tfrac{512}{88179}.$$

The pattern in these remainders was found by dividing each by the preceding one, which yielded:

$$\tfrac{3}{5}, \tfrac{4}{7}, \tfrac{5}{9}, \tfrac{6}{11}, \tfrac{7}{13}, \tfrac{8}{15}, \tfrac{9}{17}, \tfrac{10}{19}, \tfrac{11}{21}.$$

Induction proves the pattern to be valid in general; it follows that the remainders have the closed form $2^{n+1}/\binom{2n+1}{n}$, which converges to 0, as claimed.

Lemma 5. *The algorithm for digits of π is correct.*

Proof: As for e, we look at $\pi - \pi_m$, where $\pi_m = (2; 2, 2, \ldots, 2)_c$. This error is the tail of our main series for π: $\sum_{i=m}^{\infty}(i!)^2 2^{i+1}/(2i+1)!$. This tail is less than twice its first term since each subsequent term is less than half its predecessor, leading us to study $m!^2 2^{m+2}/(2m+1)!$. Splitting the denominator into evens and odds turns this into: $m! \, 2^2/(3 \cdot 5 \cdots (2m+1))$, which is less than $\tfrac{2}{3}m! \, 2^2/(2 \cdot 4 \cdots (2m))$, or $1/(3 \cdot 2^{m-1})$. It is easy to see (using the fact that $\tfrac{3}{10} < \log_{10} 2$) that this last is less than $5 \cdot 10^{-n}$ when $m = \lfloor 10n/3 \rfloor$, as claimed.

APPENDIX 2. PASCAL CODE

The following program, for which we are grateful to Macalester student Simeon Simeonov, implements the algorithm π-*spigot*. This code makes use of the fact that the queue of predigits always has a pile of 9s to the right of its leftmost member, and so only this leftmost predigit and the number of 9s need be remembered. The program computes 1000 digits of π and requires a version of Pascal with a longint data type (32-bit integer).

```
Program Pi_Spigot;
const n      = 1000;
len          = 10*n div 3;
var   i, j, k, q, x, nines, predigit : integer;
      a : array[1..len] of longint;
begin
  for j := 1 to len do a[j] := 2;              {Start with 2s}
  nines := 0; predigit := 0          {First predigit is a 0}
  for j := 1 to n do
  begin q := 0;
    for i := len downto 1 do                   {Work backwards}
    begin
      x := 10*a[i] + q*i;
      a[i] := x mod (2*i - 1);
      q := x div (2*i - 1);
    end;
    a[1] := q mod 10; q := q div 10;
    if q = 9 then nines := nines + 1
    else if q = 10 then
```

```
begin write(predigit + 1);
  for k := 1 to nines do write(0);          {zeros}
  predigit := 0; nines := 0
end
else begin
  write(predigit); predigit := q;
  if nines <> 0 then
  begin
    for k := 1 to nines do write(9);
    nines := 0
  end
end
end;
writeln(predigit);
end.
```

ADDED IN PROOF. The latest version of *Mathematica* (2.3) can sum many of the series that occur in this paper. It takes only a second or so to get $\pi/2$ as the sum of the crucial series at the beginning of section 2, to get $1/i!$ for the series in Lemma 1's proof, and to get 2 as the sum of the series in Lemma 4's proof.

REFERENCES

[Abd]   S. Kamal Abdali, Algorithm 393—Special series summation with arbitrary precision, *Comm. ACM* **13** (1970) 570.
[BBB]   J. M. Borwein, P. B. Borwein, and D. H. Bailey, Ramanujan, modular equations, and approximations to pi, or How to compute one billion digits of pi, this *Monthly* **96** (1989) 201–219
[Gos]   R. W. Gosper, Acceleration of series, Memo no. 304, M.I.T. Artificial Intelligence Laboratory, Cambridge, Mass., 1974.
[Knu]   D. E. Knuth, *The Art of Computer Programming*, volume 2, Reading, Mass., Addison-Wesley, 1981.
[Li]   J. C. R. Li, Problem E854, this *Monthly* **56** (1949) 633–635.
[Rab]   S. Rabinowitz, Abstract 863-11-482: A spigot algorithm for pi, *Abstracts Amer. Math. Society* **12** (1991) 30.
[Sale] A. H. J. Sale, The calculation of *e* to many significant digits, *Comput. J.* **11** (1968) 229–230.

MathPro Press
P.O. Box 713
Westford, MA 01886
72717.3515@compuserve.com

Department of Mathematics
Macalester College
St. Paul, MN 55105
wagon@macalstr.edu

14. On the rapid computation of various polylogarithmic constants (1997)

Paper 14: David H. Bailey, Peter B. Borwein and Simon Plouffe, "On the rapid computation of various polylogarithmic constants," *Mathematics of Computation*, vol. 66 (1997), p. 903–913. Reprinted by permission of the American Mathematical Society.

Synopsis:

For many years, it had been widely presumed impossible to calculate, say, the millionth binary digit of π any faster than simply computing all one million digits. Thus it was with considerable interest that in 1996 the above authors announced that they could do just that. The authors presented a surprisingly simple algorithm, based on the formula

$$\pi = \sum_{k=0}^{\infty} \frac{1}{16^k} \left(\frac{4}{8k+1} - \frac{2}{8k+4} - \frac{1}{8k+5} - \frac{1}{8k+6} \right),$$

that permits one to directly calculate a string of binary or hexadecimal digits of π, beginning at position n, without needing to compute any of the first $n-1$ digits. The authors further presented formulas of this type for numerous other mathematical constants.

However, the real story of interest here is that the above formula (now known as the BBP formula) was discovered by a computer program, running Helaman Ferguson's PSLQ integer relation algorithm. This was the first notable discovery by the PSLQ algorithm; since 1997, a large number of other previously unknown mathematical identities have been found in this process. Indeed, PSLQ is now a premier tool of the rapidly expanding field of experimental mathematics.

Keywords: Algorithms, Computation

© Springer International Publishing Switzerland 2016
D.H. Bailey, J.M. Borwein, *Pi: The Next Generation*,
DOI 10.1007/978-3-319-32377-0_14

MATHEMATICS OF COMPUTATION
Volume 66, Number 218, April 1997, Pages 903–913
S 0025-5718(97)00856-9

ON THE RAPID COMPUTATION OF VARIOUS POLYLOGARITHMIC CONSTANTS

DAVID BAILEY, PETER BORWEIN, AND SIMON PLOUFFE

ABSTRACT. We give algorithms for the computation of the d-th digit of certain transcendental numbers in various bases. These algorithms can be easily implemented (multiple precision arithmetic is not needed), require virtually no memory, and feature run times that scale nearly linearly with the order of the digit desired. They make it feasible to compute, for example, the billionth binary digit of $\log (2)$ or π on a modest work station in a few hours run time.

We demonstrate this technique by computing the ten billionth hexadecimal digit of π, the billionth hexadecimal digits of π^2, $\log(2)$ and $\log^2(2)$, and the ten billionth decimal digit of $\log(9/10)$.

These calculations rest on the observation that very special types of identities exist for certain numbers like π, π^2, $\log(2)$ and $\log^2(2)$. These are essentially polylogarithmic ladders in an integer base. A number of these identities that we derive in this work appear to be new, for example the critical identity for π:

$$\pi = \sum_{i=0}^{\infty} \frac{1}{16^i} \left(\frac{4}{8i+1} - \frac{2}{8i+4} - \frac{1}{8i+5} - \frac{1}{8i+6} \right).$$

1. INTRODUCTION

It is widely believed that computing just the d-th digit of a number like π is really no easier than computing all of the first d digits. From a bit complexity point of view this may well be true, although it is probably very hard to prove. What we will show is that it is possible to compute just the d-th digit of many transcendentals in (essentially) linear time and logarithmic space. So while this is not of fundamentally lower complexity than the best known algorithms (for say π or $\log(2)$), this makes such calculations feasible on modest workstations without needing to implement arbitrary precision arithmetic.

We illustrate this by computing the ten billionth hexadecimal digit of π, the billionth hexadecimal digits of π^2, $\log(2)$ and $\log^2(2)$, and the ten billionth decimal digit of $\log(9/10)$. Details are given in Section 4. A previous result in this same spirit is the Rabinowitz-Wagon "spigot" algorithm for π. In that scheme, however, the computation of the digit at position n depends on all digits preceding position n.

We are interested in computing in polynomially logarithmic space and polynomial time. This class is usually denoted SC (space $= \log^{O(1)}(d)$ and time $= d^{O(1)}$

Received by the editor October 11, 1995 and, in revised form, February 16, 1996.
1991 *Mathematics Subject Classification*. Primary 11A05, 11Y16, 68Q25.
Key words and phrases. Computation, digits, log, polylogarithms, SC, π, algorithm.
Research of the second author was supported in part by NSERC of Canada.

where d is the place of the "digit" to be computed). Actually we are most interested in the space we will denote by SC* of polynomially logarithmic space and (almost) linear time (here we want the time $= O(d \log^{O(1)}(d))$). There is always a possible ambiguity when computing a digit string base b in distinguishing a sequence of digits $a(b-1)(b-1)(b-1)$ from $(a+1)000$. In this particular case we consider either representation as an acceptable computation. In practice this problem does not arise.

It is not known whether division is possible in SC, similarly it is not known whether base change is possible in SC. The situation is even worse in SC*, where it is not even known whether multiplication is possible. If two numbers are in SC* (in the same base) then their product computes in time $= O(d^2 \log^{O(1)}(d))$ and is in SC but not obviously in SC*. The d^2 factor here is present because the logarithmic space requirement precludes the usage of advanced multiplication techniques, such as those based on FFTs.

We will not dwell on complexity issues except to point out that different algorithms are needed for different bases (at least given our current ignorance about base change) and very little closure exists on the class of numbers with d-th digit computable in SC. Various of the complexity related issues are discussed in [6], [8], [9], [11], [14].

As we will show in Section 3, the class of numbers we can compute in SC* in base b includes all numbers of the form

$$(1.1) \qquad \sum_{k=1}^{\infty} \frac{p(k)}{b^{ck} q(k)},$$

where p and q are polynomials with integer coefficients and c is a positive integer. Since addition is possible in SC*, integer linear combinations of such numbers are also feasible (provided the base is fixed).

The algorithm for the binary digits of π, which also shows that π is in SC* in base 2, rests on the following remarkable identity:

Theorem 1. *The following identity holds:*

$$(1.2) \qquad \pi = \sum_{i=0}^{\infty} \frac{1}{16^i} \left(\frac{4}{8i+1} - \frac{2}{8i+4} - \frac{1}{8i+5} - \frac{1}{8i+6} \right).$$

This can also be written as:

$$(1.3) \qquad \pi = \sum_{i=1}^{\infty} \frac{p_i}{16^{\lfloor \frac{1}{8} \rfloor i}}, \qquad [p_i] = \overline{[4,0,0,-2,-1,-1,0,0]}$$

where the overbar notation indicates that the sequence is periodic.

Proof. This identity is equivalent to:

$$(1.4) \qquad \pi = \int_0^{1/\sqrt{2}} \frac{4\sqrt{2} - 8x^3 - 4\sqrt{2}x^4 - 8x^5}{1 - x^8} dx,$$

which on substituting $y := \sqrt{2}x$ becomes

$$\pi = \int_0^1 \frac{16y - 16}{y^4 - 2y^3 + 4y - 4} dy.$$

The equivalence of (1.2) and (1.4) is straightforward. It follows from the identity

$$\int_0^{1/\sqrt{2}} \frac{x^{k-1}}{1-x^8}\, dx = \int_0^{1/\sqrt{2}} \sum_{i=0}^{\infty} x^{k-1+8i}\, dx$$

$$= \frac{1}{\sqrt{2}^k} \sum_{i=0}^{\infty} \frac{1}{16^i (8i+k)}.$$

That the integral (1.4) evaluates to π is an exercise in partial fractions most easily done in Maple or Mathematica.   □

This proof entirely conceals the route to discovery. We found the identity (1.2) by a combination of inspired guessing and extensive searching using the PSLQ integer relation algorithm [3],[12].

Shortly after the authors originally announced the result (1.2), several colleagues, including Helaman Ferguson, Tom Hales, Victor Adamchik, Stan Wagon, Donald Knuth and Robert Harley, pointed out to us other formulas for π of this type. One intriguing example is

$$\pi = \sum_{i=0}^{\infty} \frac{1}{16^i} \Big(\frac{2}{8i+1} + \frac{2}{4i+2} + \frac{1}{4i+3} - \frac{1/2}{4i+5} - \frac{1/2}{4i+6} - \frac{1/4}{4i+7} \Big),$$

which can be written more compactly as

$$\pi = \sum_{i=0}^{\infty} \frac{(-1)^i}{4^i} \Big(\frac{2}{4i+1} + \frac{2}{4i+2} + \frac{1}{4i+3} \Big).$$

In [2], this and some related identities are derived using Mathematica.

As it turns out, these other formulas for π can all be written as formula (1.2) plus a rational multiple of the identity

$$0 = \sum_{i=0}^{\infty} \frac{1}{16^i} \Big(\frac{-8}{8i+1} + \frac{8}{8i+2} + \frac{4}{8i+3} + \frac{8}{8i+4} + \frac{2}{8i+5} + \frac{2}{8i+6} - \frac{1}{8i+7} \Big).$$

The proof of this identity is similar to that of Theorem 1.

The identities of the next section and Section 5 show that, in base 2, π^2, $\log^2(2)$ and various other constants, including $\{\log(2), \log(3), \dots, \log(22)\}$ are in SC*. (We don't know however if $\log(23)$ is even in SC.)

We will describe the algorithm in Section 3. Complexity issues are discussed in [3], [5], [6], [7], [8], [9], [14], [19], [21] and algorithmic issues in [5], [6], [7], [8], [14]. The requisite special function theory may be found in [1], [5], [15], [16], [17], [20].

2. IDENTITIES

As usual, we define the m-th polylogarithm L_m by

$$(2.1) \qquad\qquad L_m(z) := \sum_{i=1}^{\infty} \frac{z^i}{i^m}, \qquad |z| < 1.$$

The most basic identity is

$$(2.2) \qquad\qquad -\log(1 - 2^{-n}) = L_1(1/2^n)$$

which shows that $\log(1 - 2^{-n})$ is in SC* base 2 for integer n. (See also section 5.)

Much less obvious are the identities

$$(2.3) \qquad \pi^2 = 36L_2(1/2) - 36L_2(1/4) - 12L_2(1/8) + 6L_2(1/64)$$

and

$$(2.4) \qquad \log^2(2) = 4L_2(1/2) - 6L_2(1/4) - 2L_2(1/8) + L_2(1/64) \,.$$

These can be written as

$$(2.5) \qquad \pi^2 = 36 \sum_{i=1}^{\infty} \frac{a_i}{2^i i^2}, \qquad [a_i] = \overline{[1, -3, -2, -3, 1, 0]},$$

$$(2.6) \qquad \log^2(2) = 2 \sum_{i=1}^{\infty} \frac{b_i}{2^i i^2}, \qquad [b_i] = \overline{[2, -10, -7, -10, 2, -1]} \,.$$

Here the overline notation indicates that the sequences repeat. Thus we see that π^2 and $\log^2(2)$ are in SC* in base 2. These two formulas can alternately be written

$$\pi^2 = \frac{9}{8} \sum_{i=0}^{\infty} \frac{1}{64^i} \left(\frac{16}{(6i+1)^2} - \frac{24}{(6i+2)^2} - \frac{8}{(6i+3)^2} - \frac{6}{(6i+4)^2} + \frac{1}{(6i+5)^2} \right),$$

$$\log^2(2) = \frac{1}{8} \sum_{i=0}^{\infty} \frac{1}{64^i} \left(\frac{-16}{(6i)^2} + \frac{16}{(6i+1)^2} - \frac{40}{(6i+2)^2} - \frac{14}{(6i+3)^2} \right.$$
$$\left. - \frac{10}{(6i+4)^2} + \frac{1}{(6i+5)^2} \right).$$

Identities (2.3)-(2-6) are examples of polylogarithmic ladders in the base 1/2 in the sense of [16]. As with (1.2) we found them by searching for identities of this type using an integer relation algorithm. We have not found them directly in print. However (2.5) follows from equation (4.70) of [15] with $\alpha = \pi/3, \beta = \pi/2$ and $\gamma = \pi/3$. Identity (2.6) now follows from the well known identity

$$(2.7) \qquad 12L_2(1/2) = \pi^2 - 6\log^2(2) \,.$$

A distinct but similar formula that we have found for π^2 is

$$\pi^2 = \sum_{i=0}^{\infty} \frac{1}{16^i} \left(\frac{16}{(8i+1)^2} - \frac{16}{(8i+2)^2} - \frac{8}{(8i+3)^2} - \frac{16}{(8i+4)^2} \right.$$
$$\left. - \frac{4}{(8i+5)^2} - \frac{4}{(8i+6)^2} + \frac{2}{(8i+7)^2} \right),$$

which can be derived from the methods of section 1.

There are several ladder identities involving L_3:

$$(2.8) \quad 35/2\zeta(3) - \pi^2 \log(2) = 36L_3(1/2) - 18L_3(1/4) - 4L_3(1/8) + L_3(1/64) \,,$$

$$(2.9) \quad 2\log^3(2) - 7\zeta(3) = -24L_3(1/2) + 18L_3(1/4) + 4L_3(1/8) - L_3(1/64) \,,$$

(2.10)
$$10\log^3(2) - 2\pi^2\log(2) = -48L_3(1/2) + 54L_3(1/4) + 12L_3(1/8) - 3L_3(1/64) \,.$$

The favored algorithms for π of the last centuries involved some variant of Machin's 1706 formula:

$$(2.11) \qquad\qquad \frac{\pi}{4} = 4\arctan\frac{1}{5} - \arctan\frac{1}{239} \,.$$

There are many related formulas [15], [16], [17], [20] but to be useful to us all the arguments of the arctans have to be a power of a common base, and we have not discovered any such formula for π . One can however write

(2.12)
$$\frac{\pi}{2} = 2 \arctan \frac{1}{\sqrt{2}} + \arctan \frac{1}{\sqrt{8}}.$$

This can be written as

(2.13) $\sqrt{2}\pi = 4f(1/2) + f(1/8)$ where $f(x) := \displaystyle\sum_{i=1}^{\infty} \frac{(-1)^i x^i}{2i+1}$

and allows for the calculation of $\sqrt{2}\pi$ in SC*.

Another two identities involving Catalan's constant G, π and $\log(2)$ are:

(2.14) $G - \dfrac{\pi \log(2)}{8} = \displaystyle\sum_{i=1}^{\infty} \frac{c_i}{2^{\lfloor \frac{i+1}{2} \rfloor} i^2}$, $[c_i] = \overline{[1,1,1,0,-1,-1,-1,0]}$

and

(2.15) $\dfrac{5}{96}\pi^2 - \dfrac{\log^2(2)}{8} = \displaystyle\sum_{i=1}^{\infty} \frac{d_i}{2^{\lfloor \frac{i+1}{2} \rfloor} i^2}$, $[d_i] = \overline{[1,0,-1,-1,-1,0,1,1]}$.

These may be found in [17, p. 105, p. 151]. Thus $8G - \pi \log(2)$ is also in SC* in base 2, but it is open and interesting as to whether G is itself in SC* in base 2.

A family of base 2 ladder identities exist:

(2.16) $\dfrac{L_m(1/64)}{6^{m-1}} - \dfrac{L_m(1/8)}{3^{m-1}} - \dfrac{2\,L_m(1/4)}{2^{m-1}} + \dfrac{4\,L_m(1/2)}{9} - \dfrac{5\,(-\log(2))^m}{9\,m!}$

$+ \dfrac{\pi^2\,(-\log(2))^{m-2}}{54\,(m-2)!} - \dfrac{\pi^4\,(-\log(2))^{m-4}}{486\,(m-4)!} - \dfrac{403\,\zeta(5)\,(-\log(2))^{m-5}}{1296\,(m-5)!} = 0.$

The above identity holds for $1 \le m \le 5$; when the arguments to factorials are negative they are taken to be infinite so the corresponding terms disappear. See [16, p. 45].

As in the case of formula (1.2) for π, colleagues of the authors have subsequently pointed out several other formulas of this type for various constants. Three examples reported by Knuth, which are based on formulas in [13, p. 17, 18, 22, 47, 139], are

$$\sqrt{2}\ln(1+\sqrt{2}) = \sum_{i=0}^{\infty} \frac{1}{16^i}\Big(\frac{1}{8i+1} + \frac{1/2}{8i+3} + \frac{1/4}{8i+5} + \frac{1/8}{8i+7}\Big),$$

$$\sqrt{2}\arctan(1/\sqrt{2}) = \sum_{i=0}^{\infty} \frac{1}{16^i}\Big(\frac{1}{8i+1} - \frac{1/2}{8i+3} + \frac{1/4}{8i+5} - \frac{1/8}{8i+7}\Big),$$

$$\arctan(1/3) = \sum_{i=0}^{\infty} \frac{1}{16^i}\Big(\frac{1}{8i+1} - \frac{1}{8i+2} - \frac{1/2}{8i+4} - \frac{1/4}{8i+5}\Big).$$

Thus these constants are also in class SC*. Some other examples can be found in [18].

3. THE ALGORITHM

Our algorithm to compute individual base-b digits of certain constants is based on the binary scheme for exponentiation, wherein one evaluates x^n rapidly by successive squaring and multiplication. This reduces the number of multiplications to less than $2\log_2(n)$. According to Knuth [14], where details are given, this trick goes back at least to 200 B.C. In our application, we need to perform exponentiation modulo a positive integer c, but the overall scheme is the same — one merely performs all operations modulo c. An efficient formulation of this algorithm is as follows.

To compute $r = b^n \bmod c$, first set t to be the largest power of two $\leq n$, and set $r = 1$. Then

> A: if $n \geq t$ then $r \leftarrow br \bmod c$; $n \leftarrow n - t$;   endif
> $t \leftarrow t/2$
> if $t \geq 1$ then $r \leftarrow r^2 \bmod c$;   go to A;   endif

Here and in what follows, "mod" is used in the binary operator sense, namely as the binary function defined by $x \bmod y := x - \lfloor x/y \rfloor y$. Note that the above algorithm is entirely performed with positive integers that do not exceed c^2 in size. Thus it can be correctly performed, without round-off error, provided a numeric precision of at least $1 + 2\log_2 c$ bits is used.

Consider now a constant defined by a series of the form

$$S = \sum_{k=0}^{\infty} \frac{1}{b^{ck}p(k)},$$

where b and c are positive integers and $p(k)$ is a polynomial with integer coefficients. First observe that the digits in the base b expansion of S beginning at position $n+1$ can be obtained from the fractional part of $b^n S$. Thus we can write

$$(3.4) \qquad b^n S \bmod 1 = \sum_{k=0}^{\infty} \frac{b^{n-ck}}{p(k)} \bmod 1$$

$$= \sum_{k=0}^{\lfloor n/c \rfloor} \frac{b^{n-ck} \bmod p(k)}{p(k)} \bmod 1 + \sum_{k=\lfloor n/c \rfloor + 1}^{\infty} \frac{b^{n-ck}}{p(k)} \bmod 1.$$

For each term of the first summation, the binary exponentiation scheme is used to evaluate the numerator. Then floating-point arithmetic is used to perform the division and add the result to the sum mod 1. The second summation, where the exponent of b is negative, may be evaluated as written using floating-point arithmetic. It is only necessary to compute a few terms of this second summation, just enough to insure that the remaining terms sum to less than the "epsilon" of the floating-point arithmetic being used. The final result, a fraction between 0 and 1, is then converted to the desired base b.

Since floating-point arithmetic is used here in divisions and in addition modulo 1, the result is of course subject to round-off error. If the floating-point arithmetic system being used has the property that the result of each individual floating-point operation is in error by at most one bit (as in systems implementing the IEEE arithmetic standard), then no more than $\log_2(2n)$ bits of the final result will be corrupted. This is actually a generous estimate, since it does not assume any cancelation of errors, which would yield a lower estimate. In any event, it is clear that ordinary IEEE 64-bit arithmetic is sufficient to obtain a numerically significant result for even a large computation, and "quad precision" (i.e. 128-bit) arithmetic, if available, can insure that the final result is accurate to several digits beyond the one desired. One can check the significance of a computed result beginning at position n by also performing a computation at position $n + 1$ or $n - 1$ and comparing the trailing digits produced.

The most basic interesting constant whose digits can be computed using this scheme is

$$\log(2) = \sum_{k=1}^{\infty} \frac{1}{k 2^k}$$

in base 2. Using this scheme to compute hexadecimal digits of π from identity (1.2) is only marginally more complicated, since one can rewrite formula (1.2) using four sums of the required form. Details are given in the next section. In both cases, in order to compute the n-th binary digit (or a fixed number of binary digits at the n-th place) we must sum $O(n)$ terms of the series. Each term requires $O(\log(n))$ arithmetic operations and the required precision is $O(\log(n))$ digits. This gives a total bit complexity of $O(n \log(n) M(\log(n)))$ where $M(j)$ is the complexity of multiplying j bit integers. So even with ordinary multiplication the bit complexity is $O(n \log^3(n))$.

This algorithm is, by a factor of $\log(\log(\log(n)))$, asymptotically slower than the fastest known algorithms for generating the n-th digit by generating all of the first n digits of $\log(2)$ or π [7]. The asymptotically fastest algorithms for all the first n digits known requires a Strassen-Schönhage multiplication [19]; the algorithms actually employed use an FFT based multiplication and are marginally slower than our algorithm, from a complexity point of view, for computing just the n-th digit. Of course this complexity analysis is totally misleading: the strength of our algorithm rests mostly on its easy implementation in standard precision without requiring FFT methods to accelerate the computation.

It is clear that the above methods can easily be extended to evaluate digits of contstants defined by a formula of the form

$$S = \sum_{k=0}^{\infty} \frac{p(k)}{b^{ck} q(k)},$$

where p and q are polynomials with integer coefficients and c is a positive integer. Similarly if p and q are slowly growing analytic functions of various types the method extends.

4. COMPUTATIONS

We report here computations of π, $\log(2)$, $\log^2(2)$, π^2 and $\log(9/10)$, based on the formulas (1.1), (2.2), (2.5), (2.6) and the identity $\log(9/10) = -L_1(1/10)$, respectively.

Each of our computations employed quad precision floating-point arithmetic for division and sum mod 1 operations. Quad precision is supported from Fortran on the IBM RS6000/590 and the SGI Power Challenge (R8000), which were employed by the authors in these computations. We were able to avoid the usage of explicit quad precision in the exponentiation scheme by exploiting a hardware feature common to these two systems, namely the 106-bit internal registers in the multiply-add operation. This saved considerable time, because quad precision operations are significantly more expensive than 64-bit operations.

Computation of π^2 and $\log^2(2)$ presented a special challenge, because one must perform the exponentiation algorithm modulo k^2 instead of k. When n is larger than only 2^{13}, some terms of the series (2.5) and (2.6) must be computed with a modulus k^2 that is greater than 2^{26}. Squares that appear in the exponentiation algorithm will then exceed 2^{52}, which is nearly the maximum precision of IEEE 64-bit floating-point numbers. When n is larger than 2^{26}, then squares in the exponentiation algorithm will exceed 2^{104}, which is nearly the limit of quad precision.

This difficulty can be remedied using a method which has been employed for example in searches for Wieferich primes [10]. Represent the running value r in the exponentiation algorithm by the ordered pair (r_1, r_2), where $r = r_1 + kr_2$, and where r_1 and r_2 are positive integers less than k. Then one can write

$$r^2 = (r_1 + kr_2)^2 = r_1^2 + 2r_1r_2k + r_2^2k^2.$$

When this is reduced mod k^2, the last term disappears. The remaining expression is of the required ordered pair form, provided that r_1^2 is first reduced mod k, the carry from this reduction is added to $2r_1r_2$, and this sum is also reduced mod k. Note that this scheme can be implemented with integers of size not exceeding $2k^2$. Since the computation of r^2 mod k^2 is the key operation of the binary exponentiation algorithm, this means that ordinary IEEE 64-bit floating-point arithmetic can be used to compute the n-th hexadecimal digit of π^2 or $\log^2(2)$ for n up to about 2^{24}. For larger n, we still used this basic scheme, but we employed the multiply-add "trick" mentioned above to avoid the need for explicit quad precision in this section of code.

Our results are given below. The first entry, for example, gives the 10^6-th through $10^6 + 13$-th hexadecimal digits of π after the "decimal" point. In all cases we did the calculations twice — the second calculation was similar to the first, except shifted back one position. Since this changes all the arithmetic performed, it is a highly rigorous validity check. Thus we believe that all the digits shown below are correct.

These computations were done at NASA Ames Research Center, using workstation cycles that otherwise would have been idle.

5. LOGS IN BASE 2

It is easy to compute, in base 2, the d-th binary digit of

$$(5.1) \qquad\qquad \log(1 - 2^{-n}) = L_1(1/2^n).$$

So it is easy to compute $\log(m)$ for any integer m that can be written as

$$(5.2) \qquad\qquad m := \frac{(2^{a_1} - 1)(2^{a_2} - 1)\cdots(2^{a_h} - 1)}{(2^{b_1} - 1)(2^{b_2} - 1)\cdots(2^{b_j} - 1)}.$$

In particular the n-th cyclotomic polynomial evaluated at 2 is so computable. A check shows that all primes less than 19 are of this form. The beginning of this list is:

$$\{2, 3, 5, 7, 11, 13, 17, 31, 43, 57, 73, 127, 151, 205, 257\}.$$

Since

$$2^{18} - 1 = 7 \cdot 9 \cdot 19 \cdot 73.$$

THE RAPID COMPUTATION OF VARIOUS POLYLOGARITHMIC CONSTANTS 911

Constant:	Base:	Position:	Digits from Position:
π	16	10^6	26C65E52CB4593
		10^7	17AF5863EFED8D
		10^8	ECB840E21926EC
		10^9	85895585A0428B
		10^{10}	921C73C6838FB2
$\log(2)$	16	10^6	418489A9406EC9
		10^7	815F479E2B9102
		10^8	E648F40940E13E
		10^9	B1EEF1252297EC
π^2	16	10^6	685554E1228505
		10^7	9862837AD8AABF
		10^8	4861AAF8F861BE
		10^9	437A2BA4A13591
$\log^2(2)$	16	10^6	2EC7EDB82B2DF7
		10^7	33374B47882B32
		10^8	3F55150F1AB3DC
		10^9	8BA7C885CEFCE8
$\log(9/10)$	10	10^6	80174212190900
		10^7	21093001236414
		10^8	01309302330968
		10^9	44066397959215
		10^{10}	82528693381274

and since 7, $\sqrt{9}$ and 73 are all on the above list we can compute $\log(19)$ in SC^* from

$$\log(19) = \log(2^{18} - 1) - \log(7) - \log(9) - \log(73).$$

Note that $2^{11} - 1 = 23 \cdot 89$ so either both $\log(23)$ and $\log(89)$ are in SC^* or neither is.

We would like to thank Carl Pomerance for showing that an identity of type (5.2) does not exist for 23. This is a consequence of the fact that each cyclotomic polynomial evaluated at two has a new distinct prime factor. We would also like to thank Robert Harley for pointing out that 29 and 37 are in SC^* in base 2 via consideration of the Aurefeuillian factors $2^{2n-1} + 2^n + 1$ and $2^{2n-1} - 2^n + 1$.

6. RELATION BOUNDS

One of the first questions that arises in the wake of the above study is whether there exists a scheme of this type to compute decimal digits of π. At present we know of no identity like (1.2) in base 10. The chances that there is such an identity are dimmed by some numerical results that we have obtained using the PSLQ integer relation algorithm [3], [12]. These computations establish (with the usual provisos of computer "proofs") that there are no identities (except for the case $n = 16$) of the form

$$\pi = \frac{a_1}{a_0} + \frac{1}{a_0} \sum_{k=0}^{\infty} \frac{1}{n^k} \left[\frac{a_2}{mk+1} + \frac{a_3}{mk+2} + \cdots + \frac{a_{m+1}}{mk+m} \right],$$

where n ranges from 2 to 128, where m ranges from 1 to $\min(n, 32)$, and where the Euclidean norm of the integer vector $(a_0, a_1, \cdots, a_{m+1})$ is 10^{12} or less. These results of course do not have any bearing on the possibility that there is a formula not of this form which permits computation of π in some non-binary base.

In fact, J. P. Buhler has reported a proof that any identity for π of the above form must have $n = 2^K$ or $n = \sqrt{2}^K$. This also does not exclude more complicated formulae for the computation of π base 10.

7. Questions

As mentioned in the previous section, we cannot at present compute decimal digits of π by our methods because we know of no identity like (1.2) in base 10. But it seems unlikely that it is fundamentally impossible to do so. This raises the following obvious problem:

1] Find an algorithm for the n-th decimal digit of π in SC*. It is not even clear that π is in SC in base 10 but it ought to be possible to show this.

2] Show that π is in SC in all bases.

3] Are e and $\sqrt{2}$ in SC (SC*) in any base?

Similarly the treatment of log is incomplete:

4] Is $\log(2)$ in SC* in base 10?

5] Is $\log(23)$ in SC* in base 2?

8. Acknowledgments

The authors wish to acknowledge the following for their helpful comments: V. Adamchik, J. Borwein, J. Buhler, R. Crandall, H. Ferguson, T. Hales, R. Harley, D. Knuth, C. Pomerance and S. Wagon.

References

1. M. Abramowitz and I.A. Stegun, *Handbook of Mathematical Functions*, Dover, New York, NY, 1966. MR **34**:8606

2. V. Adamchik and S. Wagon, *Pi: A 2000-year search changes direction (preprint)*.

3. A. V. Aho, J.E. Hopcroft, and J. D. Ullman, *The Design and Analysis of Computer Algorithms*, Addison-Wesley, Reading, MA, 1975. MR **54**:1706

4. D. H. Bailey, J. Borwein and R. Girgensohn, *Experimental evaluation of Euler sums*, Experimental Mathematics **3** (1994), 17–30. MR **96e**:11168

5. J. Borwein, and P Borwein, *Pi and the AGM – A Study in Analytic Number Theory and Computational Complexity*, Wiley, New York, NY, 1987. MR **89a**:11134

6. J. Borwein and P. Borwein, *On the complexity of familiar functions and numbers*, SIAM Review **30** (1988), 589–601. MR **89k**:68061

7. J. Borwein, P. Borwein and D. H. Bailey, *Ramanujan, modular equations and approximations to pi*, Amer. Math. Monthly **96** (1989), 201–219. MR **90d**:11143

8. R. Brent, *The parallel evaluation of general arithmetic expressions*, J. Assoc. Comput. Mach. **21** (1974), 201–206. MR **58**:31996

9. S. Cook, *A taxonomy of problems with fast parallel algorithms*, Information and Control **64** (1985), 2–22. MR **87k**:68043

10. R. Crandall, K. Dilcher, and C. Pomerance, *A search for Wieferich and Wilson primes*, Math. Comp. **66** (1997), 433–449. CMP 96:07

THE RAPID COMPUTATION OF VARIOUS POLYLOGARITHMIC CONSTANTS   913

11. R. Crandall and J. Buhler, *On the evaluation of Euler sums*, Experimental Mathematics **3**, (1995), 275–285. MR **96e**:11113
12. H. R. P. Ferguson and D. H. Bailey, *Analysis of PSLQ, an integer relation algorithm (preprint)*.
13. E. R. Hansen, *A Table of Series and Products*, Prentice-Hall, Englewood Cliffs, NJ, 1975.
14. D. E. Knuth, *The Art of Computer Programming. Vol. 2: Seminumerical Algorithms*, Addison-Wesley, Reading, MA, 1981. MR **83i**:68003
15. L. Lewin, *Polylogarithms and Associated Functions*, North Holland, New York, 1981. MR **83b**:33019
16. L. Lewin, *Structural Properties of Polylogarithms*, Amer. Math. Soc., RI., 1991. MR **93b**:11158
17. N. Nielsen, *Der Eulersche Dilogarithmus*, Halle, Leipzig, 1909.
18. S. D. Rabinowitz and S. Wagon, *A spigot algorithm for the digits of pi*, Amer. Math. Monthly **102** (1995), 195–203. MR **96a**:11152
19. A. Schönhage, *Asymptotically fast algorithms for the numerical multiplication and division of polynomials with complex coefficients*, in: EUROCAM (1982) Marseille, Springer Lecture Notes in Computer Science, vol. 144, 1982, pp. 3–15. MR **83m**:68064
20. J. Todd, *A problem on arc tangent relations*, Amer. Math. Monthly **56** (1949), 517–528. MR **11**:159d
21. H. S. Wilf, *Algorithms and Complexity*, Prentice Hall, Englewood Cliffs, NJ, 1986. MR **88j**:68073

NASA AMES RESEARCH CENTER, MAIL STOP T27A-1, MOFFETT FIELD, CALIFORNIA 94035-1000
E-mail address: dbailey@nas.nasa.gov

DEPARTMENT OF MATHEMATICS AND STATISTICS, SIMON FRASER UNIVERSITY, BURNABY, B.C., CANADA V5A 1S6
E-mail address: pborwein@cecm.sfu.ca

DEPARTMENT OF MATHEMATICS AND STATISTICS, SIMON FRASER UNIVERSITY, BURNABY, B.C., CANADA V5A 1S6
E-mail address: plouffe@cecm.sfu.ca

15. Similarities in irrationality proofs for π, $\ln 2$, $\zeta(2)$, and $\zeta(3)$ (2001)

Paper 15: Dirk Huylebrouck, "Similarities in irrational proofs for π, $\ln 2$, $\zeta(2)$ and $\zeta(3)$," *American Mathematical Monthly*, vol. 108 (2001), p. 222–231. Copyright 2001 Mathematical Association of America. All Rights Reserved.

Synopsis:

Ever since antiquity, mathematicians have wondered whether numbers such as π are rational, the quotient of two integers. In fact, one prime motivations in computations of π through the centuries was the hope that the digits produced might provide some insight into this question. In 1761, Lambert settled the question for π, but the irrationality of some other constants remains in question.

In this paper, the author presents, in a single proof, irrationality proofs for π, $\ln 2$, $\zeta(2) = \sum_{k=1}^{\infty} 1/k^2$ and $\zeta(3) = \sum_{k=1}^{\infty} 1/k^3$. These proofs are surprisingly accessible — for the most part they involve only operations of calculus, together with some properties of Legendre polynomials.

Keywords: Irrationality

© Springer International Publishing Switzerland 2016
D.H. Bailey, J.M. Borwein, *Pi: The Next Generation*,
DOI 10.1007/978-3-319-32377-0_15

Similarities in Irrationality Proofs for π, ln2, $\zeta(2)$, and $\zeta(3)$

Dirk Huylebrouck

1. FOUR REMARKABLE NUMBERS. The first two numbers, π and ln2, are familiar to high school graduates. Their expressions as series are standard:

$$\pi = 1 - \frac{1}{3} + \frac{1}{5} - \frac{1}{7} + \cdots = \sum_{i \geq 0} (-1)^i \cdot \frac{1}{2i+1} \quad \text{(Leibniz' series)}$$

$$\ln2 = 1 - \frac{1}{2} + \frac{1}{3} - \frac{1}{4} + \cdots = \sum_{i \geq 1} (-1)^{i-1} \cdot \frac{1}{i} = 0.693147\ldots.$$

Similar expressions define the less familiar $\zeta(2)$ and $\zeta(3)$:

$$\zeta(2) = 1 + \frac{1}{2^2} + \frac{1}{3^2} + \cdots + \frac{1}{i^2} + \cdots = \sum_{i \geq 1} \frac{1}{i^2} = 1.64493\ldots$$

$$\text{and} \quad \zeta(3) = 1 + \frac{1}{2^3} + \frac{1}{3^3} + \cdots + \frac{1}{i^3} + \cdots = \sum_{i \geq 1} \frac{1}{i^3} = 1.20205\ldots.$$

The irrationality of π dominated a good 2000 years of mathematical history, starting with the closely related circle-squaring problem of the ancient Greeks. In 1761 Lambert proved the irrationality of π (Lindemann would complete the transcendence proof in 1882 [**2**, pp. 52 and 172]). The interest in π's younger brother $\zeta(3)$ started only a few centuries ago, but the number resisted until 1978, when R. Apéry presented his 'miraculous' proof [**14**]. Even after Apéry's lecture, scepticism remained general, until Beukers' simplified version confirmed it [**3**]. The character of the ζ-numbers still fascinates the mathematical community, and even very recently it was upset by results of Tanguy Rivoal (communication J. Van Geel, University of Ghent).

Essential in the simplified proofs are the representations of $\zeta(2)$ and $\zeta(3)$ as integrals. Since

$$\iint_0^1 \frac{1}{1-xy} dxdy = \iint_0^1 \sum_{i \geq 0} x^i y^i dxdy$$

$$= \sum_{i \geq 0} \int_0^1 x^i dx \int_0^1 y^i dy$$

$$= \sum_{i \geq 0} \left[\frac{x^{i+1}}{i+1}\right]_0^1 \cdot \left[\frac{y^{i+1}}{i+1}\right]_0^1 = \sum_{i \geq 0} \frac{1}{(i+1)^2},$$

we have

$$\zeta(2) = \iint_0^1 \frac{1}{1-xy} dxdy.$$

Similarly for $\zeta(3)$:

$$\int\int_0^1 \frac{1}{1-xy}\ln xy\,dxdy = \int\int_0^1 \sum_{i\geq 0} x^i y^i \ln xy\,dxdy = 2\int\int_0^1 \sum_{i\geq 0} x^i y^i \ln x\,dxdy$$

$$= 2\sum_{i\geq 0}\int_0^1 x^i \ln x\,dx \int_0^1 y^i\,dy$$

$$= 2\sum_{i\geq 0}\frac{1}{i+1}\left([x^{i+1}\ln x]_0^1 - \int_0^1 x^{i+1}\frac{1}{x}dx\right)\frac{1}{i+1}$$

$$= 2\sum_{i\geq 0}\frac{1}{i+1}\left(-\frac{1}{i+1}\right)\frac{1}{i+1} = -2\sum_{i\geq 0}\frac{1}{(i+1)^3} = -2\zeta(3).$$

Thus,

$$\zeta(3) = -\frac{1}{2}\int\int_0^1 \frac{1}{1-xy}\ln xy\,dxdy.$$

Incidentally, the integral for $\zeta(2)$ can be evaluated very easily. This computation is not really needed here, but it provides an illustration of a calculation involving a zeta expression. The easiest components of Apostol's, Beuker's, and Kalman's results are combined, and except for some elementary knowledge about double integrals, no other prerequisites seem needed: see [1], [3], [10]. The combination provides a proof at graduate level for Euler's $\zeta(2)$ result; see [4] for a more rigorous and general proof.

First, rewrite the difference of two integrals using the substitution $X = x^2$ and $Y = y^2$:

$$\int\int_0^1 \left(\frac{1}{1-xy} - \frac{1}{1+xy}\right)dxdy = \int\int_0^1 \left(\frac{2xy}{1-x^2y^2}\right)dxdy$$

$$= \frac{1}{2}\int\int_0^1 \left(\frac{1}{1-XY}\right)dXdY. \tag{1}$$

Next, obtain their sum:

$$\int\int_0^1 \left(\frac{1}{1-xy} + \frac{1}{1+xy}\right)dxdy = 2\int\int_0^1 \frac{1}{1-x^2y^2}dxdy. \tag{2}$$

Adding (1) and (2) gives

$$2\int\int_0^1 \frac{1}{1-xy}dxdy = \frac{1}{2}\int\int_0^1 \frac{1}{1-XY}dXdY + 2\int\int_0^1 \frac{1}{1-x^2y^2}dxdy,$$

or

$$2\zeta(2) = \frac{1}{2}\zeta(2) + 2\int\int_0^1 \frac{1}{1-x^2y^2}dxdy.$$

Thus,

$$\frac{3}{4}\zeta(2) = \int\int_0^1 \frac{1}{1-x^2y^2}dxdy.$$

Substituting $x = \sin\theta/\cos\phi$ and $y = \sin\phi/\cos\theta$ yields a Jacobian equal to $1 - \tan^2\theta\tan^2\phi$ and this is the denominator of the integrand.

$$\zeta(2) = \frac{4}{3}\int_0^{\frac{\pi}{2}} d\theta \int_0^{\frac{\pi}{2}-\theta} \frac{1}{1 - \left(\frac{\sin\theta}{\cos\varphi}\right)^2\left(\frac{\sin\varphi}{\cos\theta}\right)^2}\left(1 - \tan^2\theta\tan^2\varphi\right) d\varphi$$

$$= \frac{4}{3}\int_0^{\frac{\pi}{2}} d\theta \int_0^{\frac{\pi}{2}-\theta} d\varphi = \frac{4}{3}\frac{\pi^2}{8} = \frac{\pi^2}{6}.$$

This is Euler's well-known result: $\zeta(2) = \pi^2/6$.

2. FOUR PROOFS IN ONE. The Borweins collected irrationality proofs for these four numbers [5, pp. 353, 366, 369, and 370], in a very rigorous treatise. In order to make their similarity more evident, we first summarise the highlights of the demonstrations in general terms.

Suppose the irrationality of a number ξ must be shown. In the four cases we present here, a *family of integrals* ($j \in \mathbb{N}$) concerning that number is proposed:

$$\int_0^1 x^j f(x)dx = R_j + S_j\xi,$$

where R_j and $S_j \in \mathbb{Q}$; f is an unknown function.

Now if ξ were a fraction a/b, this family of integrals would yield rational expressions $\int_0^1 x^j f(x)dx = C_j/D_j$, where C_j and $D_j \in \mathbb{Z}$. Integer multiples of these integrals and their sums would again exhibit this property. Thus, if $p_{nj} \in \mathbb{Z}$ and $n \in \mathbb{N}$:

$$\int_0^1 \sum_{j=0}^n p_{nj}x^j f(x)dx = \sum_{j=0}^n p_{nj}\int_0^1 x^j f(x)dx = \sum_{j=0}^n p_{nj}\frac{C_j}{D_j} = \frac{E_n}{F_n},$$

with again E_n and $F_n \in \mathbb{Z}$.

Apply this property to the Legendre polynomials:

$$P_n(x) = \frac{1}{n!}\frac{d^n}{dx^n}\left(x^n(1-x)^n\right) = \sum_{j=0}^n p_{nj}x^j.$$

For example, $P_0(x) = 0$, $P_1(x) = 1 - 2x$, $P^2(x) = 2 - 12x - 12x^2$. Note that $p_{nj} \in \mathbb{Z}$.

Thus, for the given family of integrals,

$$\int_0^1 P_n(x)f(x)dx = \frac{A_n}{B_n},$$

with A_n, $B_n \in \mathbb{Z}$.

The choice of Legendre polynomials is inspired by the possibility of performing integrations by parts easily. Indeed, since $P_n(0) = 0 = P_n(1)$, and similarly for the derivatives of $x^n(1-x)^n$ up to the order n, many terms can be simplified:

$$\int_0^1 P_n(x)f(x)dx = \int_0^1 \frac{1}{n!}\frac{d}{dx}\left(\frac{d^{n-1}}{dx^{n-1}}\left(x^n(1-x)^n\right)\right)f(x)dx$$

 [Monthly 108

$$= \left[\frac{1}{n!} \frac{d^{n-1}}{dx^{n-1}} \left(x^n (1-x)^n \right) f(x) \right]_0^1$$

$$- \int_0^1 \frac{1}{n!} \frac{d}{dx} \frac{d^{n-1}}{dx^{n-1}} \left(x^n (1-x)^n \right) \frac{df(x)}{dx} dx$$

$$= 0 - \int_0^1 \frac{1}{n!} \frac{d^{n-1}}{dx^{n-1}} \left(x^n (1-x)^n \right) \frac{df(x)}{dx} dx.$$

Integrating by parts n times leads to

$$\int_0^1 \frac{1}{n!} x^n (1-x)^n \frac{d^n f(x)}{dx^n} dx.$$

In the four encountered cases the function $f(x)$ happens to be such that

$$\left| \int_0^1 \frac{1}{n!} x^n (1-x)^n \frac{d^n f(x)}{dx^n} dx \right| = \left| \int_0^1 \frac{1}{n!} (g(x))^n h(x) dx \right|,$$

where the maximum value M of $g(x)$ is small enough to ensure that

$$\left| B_n M^n \int_0^1 h(x) dx \right| \to 0.$$

In addition, all the integrals $\int_0^1 P_n(x) f(x) dx$ are non-zero, and this immediately implies the irrationality of ξ:

$$0 < |A_n| = \left| B_n \int_0^1 P_n(x) f(x) dx \right| \le \left| B_n M^n \int_0^1 h(x) dx \right| \to 0.$$

Indeed, for any $n \in \mathbb{N}$, $|A_n|$ is a positive integer so this is impossible; ξ cannot be rational.

Of course, the difficulty in the proofs lies in the appropriate choice of $f(x)$. It must yield a family of non-zero integrals whose members are easily expressed as a combination rational number and ξ. In addition, it should simultaneously be possible to maximize their product with an integer that becomes larger and larger, and still ensure the indicated convergence to 0.

3–I. Irrationality of π Take $f(x) = \sin(\pi x)$. It is a standard calculus exercise to show that the members of family of integrals of the form $\int_0^1 x^j \sin(\pi x) dx$ are polynomials in π of degree at most j, divided by π^j. The linear combinations $\int_0^1 P_n(x) \sin(\pi x) dx$ are non-zero, and thus, if π were the rational number a/b:

$$0 < |A_n| = \left| a^n \int_0^1 P_n(x) \sin(\pi x) dx \right|$$

$$= \left| a^n \int_0^1 \frac{1}{n!} x^n (1-x)^n \frac{d^n}{dx^n} (\sin \pi x) dx \right| \le \left| a^n \int_0^1 \frac{1}{n!} x^n (1-x)^n \pi^n dx \right|,$$

since the n-th order derivative of $\sin(\pi x)$ is $\pm \pi^n \sin(\pi x)$ or $\pm \pi^n \cos(\pi x)$. The maximum value of $x(1-x)$ is $1/4$, attained at $x = 1/2$. Thus, the final expression is less than

$$\left| a^n \frac{1}{n!} \left(\frac{1}{4} \right)^n \pi^n 1 \right|,$$

which is arbitrary small for large values of n; see [13].

3–II. Irrationality of ln2 The choice is $f(x) = 1/(1+x)$. An Euclidian division of x^j by $1+x$ allows us to compute the family of integrals

$$\int_0^1 \frac{x^j}{1+x} dx = \frac{1}{j} - \frac{1}{j-1} + \cdots \mp 1 \pm \ln 2.$$

If ln2 were a/b, then

$$0 < |A_n| = \left| (bd_n) \int_0^1 P_n(x) \frac{1}{1+x} dx \right| \qquad \text{(where } d_n = \text{LCM}\{1, 2, 3, \ldots, n\})$$

$$= \left| (bd_n) \int_0^1 \frac{1}{n!} x^n (1-x)^n \left[\frac{d^n}{dx^n} \left(\frac{1}{1+x} \right) \right] dx \right|$$

$$\leq \left| (bd_n) \int_0^1 \left(\frac{x(1-x)}{1+x} \right)^n \frac{1}{1+x} dx \right|.$$

since the n-th order derivative of $1/(1+x)$ is $(-1)\ldots(-n)(1/(1+x))^{n+1}$. Now on $[0, 1]$, the maximum value of $x(1-x)/(1+x)$ is $3 - 2\sqrt{2}$, achieved at $x = -1 + \sqrt{2}$. A rough inequality from number theory is $d_n = \text{LCM}\{1, \ldots, n\} \leq 3^n$. Finally, since $(3(3 - 2\sqrt{2}))^n < 1$, the irrationality of ln2 is established; see [5, p. 370].

3–III. Irrationality of $\zeta(2)$ The choice is

$$f(x) = \int_0^1 \frac{(1-y)^n}{1-xy} dy.$$

Each member of the family of integrals

$$\int_0^1 x^j \left[\int_0^1 \frac{(1-y)^n}{1-xy} dy \right] dx$$

is a sum of integrals of the form

$$\iint_0^1 \frac{x^r y^s}{1-xy} dy dx, \text{ with } r, s \in \mathbb{N}.$$

These can again be computed through an Euclidian division of $x^j y^k$ by $1-xy$, which gives a sum of integrals of the form

$$\iint_0^1 x^p y^q dy dx, \quad \iint_0^1 \frac{x^p}{1-xy} dy dx, \quad \iint_0^1 \frac{y^q}{1-xy} dy dx \quad \text{or} \quad \iint_0^1 \frac{1}{1-xy} dy dx.$$

The latter is $\zeta(2)$, while the others are sums of fractions, which can be computed using partial integration for the integral $\int x^m \ln x \, dx$. Thus when $r \neq s$,

$$\iint_0^1 \frac{x^r y^s}{1 - xy} dy dx$$

is a sum of fractions whose common denominator is the square of least common multiple (LCM) of the first $n + 1$ integers. When $r = s$, the integral equals

$$\sum_{i>r} \frac{1}{i^2} = \zeta(2) - \left(1 + \cdots + \frac{1}{r^2}\right).$$

Thus,

$$\left|\int_0^1 P_n(x) f(x) dx\right| = \frac{|A_n|}{d_{n+1}^2},$$

where d_{n+1} is the LCM of the first $n + 1$ natural numbers and $A_n \in \mathbb{Z}_0$.

Now

$$0 < |A_n| = \left|d_{n+1}^2 \int_0^1 P_n(x) f(x) dx\right|$$

$$= \left|d_{n+1}^2 \int_0^1 \frac{1}{n!} x^n (1-x)^n \frac{d^n}{dx^n} \left(\int_0^1 \frac{(1-y)^n}{1-xy} dy\right) dx\right|$$

$$= \left|d_{n+1}^2 \int_0^1 \frac{1}{n!} x^n (1-x)^n (-1) \cdots (-n) \int_0^1 \frac{y^n (1-y)^n}{(1-xy)^{n+1}} dy dx\right|$$

$$= \left|d_{n+1}^2 \int_0^1 \left(\frac{x(1-x)y(1-y)}{(1-xy)}\right)^n \frac{1}{1-xy} dy\right| \qquad (3)$$

On [0, 1], the maximum value of $x(1-x)y(1-y)/(1-xy)$ is $((-1+\sqrt{5})/2)^5$, and is attained for $x = y = (-1+\sqrt{5})/2$. Together with $d_{n+1} \leq 3^{n+1}$, this shows that the expression (3) is less than

$$\left|(3^{n+1})^2 \int_0^1 \left(\left(\frac{-1+\sqrt{5}}{2}\right)^5\right)^n \frac{1}{1-xy} dx dy\right| = \left|9 \left(\frac{9\left(\frac{-1+\sqrt{5}}{2}\right)^5}{2^5}\right)^n \zeta(2)\right| < 1,$$

which establishes the irrationality of $\zeta(2)$; see [3].

3–IV. Irrationality of $\zeta(3)$ Take

$$f(x) = \int_0^1 \frac{P_n(y)}{1-xy} \ln xy \, dx dy.$$

The members of the family of integrals

$$\int_0^1 x^j \left[\int_0^1 \frac{P_n(y)}{1-xy} \ln xy \, dx dy\right] dx$$

are computed through the derivative of

$$\iint_0^1 \frac{x^{r+t}y^{s+t}}{1-xy}dxdy$$

with respect to t, which is

$$\iint_0^1 \frac{x^{r+t}y^{s+t}}{1-xy}\ln xy\,dxdy.$$

If $r \neq s$, this is a sum of fractions since $d(r+t)^{-m}/dt = -m/(r+t)^{-m-1}$, and the LCM of the denominators is $(d_{r+t})^3$.

When $r = s$, the result is

$$\sum_{i>r+t} \frac{d}{dt}\frac{1}{(r+t)^2} = \sum_{i>r+t} \frac{-2}{(r+t)^3}$$

$$= -2\left(\zeta(3) - \left(1 + \cdots + \frac{1}{r^3} + \frac{1}{(r+1)^3} + \cdots + \frac{1}{(r+t)^3}\right)\right).$$

Thus,

$$\left|\int_0^1 P_n(x)f(x)dx\right| = \frac{|A_n|}{|d_{n+1}^3|},$$

where $A_n \in \mathbb{Z}$ and d_{n+1} is the LCM of the first $n+1$ natural numbers.

Now

$$|A_n| = \left|d_{n+1}^3 \int_0^1 P_n(x)\left[\int_0^1 \frac{P_n(y)}{1-xy}\ln xy\,dy\right]dx\right|$$

$$= \left|d_{n+1}^3 \iiint_0^1 \frac{P_n(x)P_n(y)}{1-(1-xy)z}dxdydz\right| \qquad \text{(integration by parts)}$$

$$= \left|d_{n+1}^3 \iiint_0^1 \frac{x^n(1-x)^n P_n(y)y^n z^n}{(1-(1-xy)z)^{n+1}}dxdydz\right| \qquad \left(\text{put } w = \frac{1-z}{1-(1-xy)z}\right)$$

$$= \left|d_{n+1}^3 \iiint_0^1 \frac{(1-x)^n P_n(y)(1-w)^n}{1-(1-xy)w}dxdydw\right| \qquad \text{(integration by parts)}$$

$$= \left|d_{n+1}^3 \iiint_0^1 \frac{x^n(1-x)^n y^n(1-y)^n w^n(1-w)^n}{(1-(1-xy)w)^{n+1}}dxdydw\right|$$

$$= \left|d_{n+1}^3 \cdot \iiint_0^1 \left(\frac{w(1-x)y(1-y)w(1-w)}{1-(1-xy)w}\right)^n \frac{1}{1-(1-xy)w}dxdydw\right| \qquad (4)$$

The maximum value of $x(1-x)y(1-y)w(1-w)/(1-(1-xy)w)$ on $[0,1]$ is $(\sqrt{2}-1)^4$, and is attained for $x = y = -1 + \sqrt{2}$ and $z = 1/\sqrt{2}$. Together with $d_{n+1} \leq 3^{n+1}$, this shows that the expression (4) is less than

$$\left|(3^{n+1})^3 \iiint_0^1 (\sqrt{2}-1)^{4n} \frac{1}{1-(1-xy)w}dxdydw\right| = \left|27\left(27\left(\sqrt{2}-1\right)^4\right)^n \zeta(3)\right|$$

$$< 1,$$

which establishes the irrationality of $\zeta(3)$, see [3].

4. THE GOLDEN SECTION. In our proofs, the maximum values of certain functions play an important role. They were attained at the points $1/2$, $(-1 + \sqrt{5})/2$, and $-1 + \sqrt{2}$. There is a coincidental link since they are all related to the golden section.

Classically, the *golden section* ϕ arises when a line segment of length x greater than 1 is divided into two parts. This could be done by cutting it into two halves (recall that $1/2$ popped up in the irrationality proof of π). Or, if unequal segments are desired, one could look for two pieces of lengths 1 and $x - 1$, such that the ratio $x/1$ equals the ratio $1/(x - 1)$. This equality produces the quadratic equation $x^2 - x - 1 = 0$, of which $1.6180\ldots = \phi$ is the positive solution.

More generally, the positive roots of $x^2 - nx - 1 = 0$ yield the family of *metallic means*, for various values of $n \in \mathbb{N}$. For $n = 2$, we get the *silver mean* $\sigma_{Ag} = 1 + \sqrt{2}$, for $n = 3$ the *bronze mean* $\sigma_{Br} = (3 + \sqrt{13})/2$, etc. The properties of these numbers have been described in numerous publications; for a comprehensive survey, see [6], while [8] and [12] pointed out that some authors often had too much enthusiasm. A common misconception is that a rectangle of width 1 and length ϕ would be the "most elegant" one and thus is used in various designs. However, no reliable statistical studies confirm this statement about the optimal choice provided by the golden number [7].

A statement that comes close is the fact that ϕ would be "the most irrational of all irrational numbers" because its representation as a continued fraction contains only 1s:

$$\phi = 1 + \cfrac{1}{1 + \frac{1}{1+\cdots}} = [1, 1, \ldots].$$

The silver mean would be "the second most irrational number" since $\sigma_{Ag} = [2, 2, \ldots]$, etc.; see [6]. Yet, this again does not provide an interpretation of the golden section an optimal solution, in the standard mathematical sense.

However, the various irrationality proofs lead to a property of these metallic means where the expression "optimal solution" has its common mathematical meaning. Table 1 illustrates some interesting facts:

TABLE 1.

Proof	Function used	Maximum	Attained at	Name
π	$x(1 - x)$	$x = \dfrac{1}{4}$	$x = \dfrac{1}{2}$	2^{-1}
$\ln 2$	$\dfrac{x(1 - x)}{(1 + x)}$	$3 - 2\sqrt{2}$	$x = -1 + \sqrt{2}$	$-\sigma_{Ag}^{-1}$
$\zeta(2)$	$\dfrac{x(1 - x)y(1 - y)}{(1 - xy)}$	$\left(\dfrac{(-1 + \sqrt{5})}{2}\right)^5$	$x = y = \dfrac{(-1 + \sqrt{5})}{2}$	$-\phi^{-1}$
$\zeta(3)$	$\dfrac{x(1 - x)y(1 - y)w(1 - w)}{(1 - (1 - xy)w)}$	$(-1 + \sqrt{2})^4$	$x = y = -1 + \sqrt{2}; \ z = \dfrac{1}{\sqrt{2}}$	$-\sigma_{Ag}^{-1}$

A substitution of $X = -1/x$ and $Y = -1/y$ in $x(1 - x)y(1 - y)/(1 - xy)$ changes the expression into $(1 + x)(1 + y)/(xy(xy - 1))$. Its extremum is obtained at $X = Y = \phi$. Similarly, σ_{Ag} provides the optimal solution to $(X - 1)/((X + 1)X)$. In these cases, the word "optimal" is used in the usual mathematical way, in contrast to the loose terms often used in golden section papers. The geometric interpretation of these facts is developed in a forthcoming text [9].

There are other links between $\zeta(2)$ and $\zeta(3)$, and the metallic means. For example, in the easy proof for Euler's $\zeta(2)$ result, hyperbolic sines and cosines ($x = \sinh\theta/\cosh\phi$ and $y = \sinh\phi/\cosh$) can be substituted instead of the similar expressions with circular sines and cosines. In that case,

$$\zeta(2) = \left(\ln\left(1 + \sqrt{2}\right)\right)^2 + 2\int_{\ln(1+\sqrt{2})}^{+\infty} (t - \arg\cosh(\sinh t))\, dt.$$

Now the silver section $1 + \sqrt{2}$ appears, while in Table 1, $\zeta(2)$ was already linked to the golden section. More involved computations relate $\zeta(3)$ to the golden section, too [11, p. 156]:

$$\zeta(3) = 10\int_0^{\ln(\frac{1+\sqrt{5}}{2})} t^2 \cosh t\, dt.$$

These relations did not inform us about the "optimal" properties of the metallic numbers, and we give them here for sake of completeness.

Incidentally, since

$$\zeta(4) = \iiiint_0^1 \frac{(1 - xy)}{(1 - (1 - xy)w)(1 - (1 - xy)v)}dx\,dy\,dw\,dv = 1 + \frac{1}{2^4} + \frac{1}{3^4} + \cdots,$$

a natural generalization of the functions used in the study of the irrationality of $\zeta(2)$ and $\zeta(3)$ would be

$$\frac{x(1 - x)y(1 - y)w(1 - w)v(1 - v)(1 - xy)}{(1 - (1 - xy)w)(1 - (1 - xy)v)}.$$

The maximum of this function is

$$\frac{\left(5 - \sqrt{13}\right)^4 \left(-7 + 2\sqrt{13}\right)^2}{54(-3 + \sqrt{13})^4},$$

obtained for $z = w = (1 - \sqrt{(xy)})/(1 - xy)$ and $x = y = (-3 + \sqrt{13})/2$. Here again a metallic mean is found, and it is the next one, $-\sigma_{Br}^{-1}$. Unfortunately, it does not provide a proof for the irrationality of $\zeta(4)$ (and by extension for $\zeta(5)$) since the members of the family of integrals are not combinations of $\zeta(4)$ with rational numbers. The quest for these proofs remains open.

REFERENCES

1. Tom M. Apostol, A Proof that Euler Missed: Evaluating $\zeta(2)$ the Easy Way, *Math. Intelligencer* 5 (1983) 59–60.
2. P. Beckman, *A History of Pi*, 4th ed. Golem Press, Boulder, Colorado, 1977.
3. F. Beukers, A Note on the irrationality of $\zeta(2)$ and $\zeta(3)$, *Bull. London Math. Soc.* 11 (1979) 268–272.
4. F. Beukers, J. A. C. Kolk, and E. Calabi, Sums of generalized harmonic series and volumes, *Nieuw Arch. Wisk.* (4) 11 (1993) 217–224.
5. J. M. Borwein and P. B. Borwein, *Pi and the AGM: A Study in Analytic Number Theory and Computational Complexity*, Wiley, New York, 1987.
6. Vera W. de Spinadel, *From the Golden Mean to Chaos*, Nueva Libreria S.R.L. Buenos Aires, Argentina 1998.
7. Christopher D. Green, All that Glitters: A Review of Psychological Research on the Aesthetics of the Golden Section, *Perception* 24 (1995) 937–968.

8. Roger Herz-Fischler, *A Mathematical History of Division in Extreme and Mean Ratio*, Wilfrid Laurier University Press, Waterloo, Canada, 1987.

9. Dirk Huylebrouck, the golden section as an optimal solution, to appear.

10. Dan Kalman, Six Ways to Sum a Series, *College Math. J.* 24 (1993) 402–421.

11. L. Lewin, *Polylogarithms and Associated Functions*, North Holland, New York, 1981.

12. George Markowsky, Misconceptions about the Golden Ratio, *College Math. J.* 23 (1992) 2–19.

13. I. Niven, A Simple Proof that π is Irrational, *Bull. Amer. Math. Soc.* 53 (1947) 509.

14. A. van der Poorten, A Proof that Euler Missed...Apéry's Proof of the Irrationality of $\zeta(3)$, An informal report, *Math. Intelligencer* 1 (4) (1978/79) 195–203.

D. HUYLEBROUCK obtained his Ph.D. in linear algebra at the University of Ghent (Belgium). Several years of teaching in Congo came to a sudden end after a conflict between former president Mobutu and Belgium. He taught in Portugal, and later at University of Maryland for American GIs in Europe. His fascination for the history of mathematics led him to produce a poster of 100,000 coloured decimals of pi.

Another stay in Africa again came to an abrupt end after a coup in Burundi (by someone else), and so he settled at the Sint-Lucas Institute for Architecture in Brussels, Belgium. He still has an African dream: to let its oldest mathematical artifact one day reach space.
Aartshertogstraat 42, 8400 Oostende, Belgium
dirk.huylebrouck@pi.be

Solution to Sherwood Forest Puzzle on p. 143 of the February issue:

The idea of 'singer of a person's song to another' can be taken as an operation $*$, that is, "$x * y = z$" denotes "x's song is sung to y by z". Thus:

1) we have closure of $*$ (everyone's song is sung to everyone by a singer),

2) we have an 'identity' (the priest),

3) we have 'inverse' for everyone (mates),

4) the (unavoidably) cryptic third paragraph is the associativity of the operation $*$.

Thus we have a group structure. But, $10,201 = 101 \times 101$ and 101 is a prime, and any group of order of square of a prime is abelian.

Thus, it was Marian who sang Little John's song to Robin. For the second question, it suffices that we have a group structure since we have Marian = Robin $*$ Little John. We should recall Marian and Robin are mates and so we can left multiply the equation by Robin's inverse to get Marian $*$ Marian = Little John. And so Little John sang to Marian her song.

---

16. Unbounded spigot algorithms for the digits of pi (2006)

Paper 16: Jeremy Gibbons, "Unbounded spigot algorithms for the digits of pi," *American Mathematical Monthly*, vol. 113 (2006), p. 318–328. Copyright 2006 Mathematical Association of America. All Rights Reserved.

Synopsis:

In 1995, Rabinowitz and Wagon presented a "spigot" algorithm for π, in a paper included earlier in this volume. One limitation of the Rabinowitz-Wagon algorithm was that the computation was inherently bounded — one has to decide in advance that one will compute up to a certain number of digits.

In this paper, Gibbons presents a different spigot algorithm, based on the same infinite series that lies behind the Rabinowitz-Wagon algorithm, but which avoids this limitation. One does not need to commit in advance to compute a certain maximum number of digits. In theory, the algorithm could continue to generate decimal digits of π indefinitely. Gibbon's algorithm is not as simple as the Rabinowitz-Wagon algorithm, but it still can be stated (and implemented) very concisely.

Keywords: Algorithms

© Springer International Publishing Switzerland 2016
D.H. Bailey, J.M. Borwein, *Pi: The Next Generation*,
DOI 10.1007/978-3-319-32377-0_16

Unbounded Spigot Algorithms for the Digits of Pi

Jeremy Gibbons

1. INTRODUCTION. Rabinowitz and Wagon [8] present a "remarkable" algorithm for computing the decimal digits of π, based on the expansion

$$\pi = \sum_{i=0}^{\infty} \frac{(i!)^2 2^{i+1}}{(2i+1)!}. \tag{1}$$

Their algorithm uses only bounded integer arithmetic and is surprisingly efficient. Moreover, it admits extremely concise implementations. Witness, for example, the following (deliberately obfuscated) C program due to Dik Winter and Achim Flammenkamp [1, p. 37], which produces the first 15,000 decimal digits of π:

```
a[52514],b,c=52514,d,e,f=1e4,g,h;
main(){for(;b=c-=14;h=printf("%04d", e+d/f))
  for(e=d%=f;g=--b*2;d/=g)d=d*b+f*(h?a[b]:f/5),a[b]=d%--g;}
```

Rabinowitz and Wagon call their algorithm a *spigot algorithm*, because it yields digits incrementally and does not reuse digits after they have been computed. The digits drip out one by one, as if from a leaky tap. In contrast, most algorithms for computing the digits of π execute inscrutably, delivering no output until the whole computation is completed.

However, the Rabinowitz-Wagon algorithm has its weaknesses. In particular, the computation is inherently bounded: one has to commit in advance to computing a certain number of digits. Based on this commitment, the computation proceeds on an appropriate finite prefix of the infinite series (1). In fact, it is essentially impossible to determine in advance how big that finite prefix should be for a given number of digits—specifically, a computation that terminates with nines for the last few digits of the output is inconclusive, because there may be a "carry" from the first few truncated terms. Rabinowitz and Wagon suggest that "in practice, one might ask for, say, six extra digits, reducing the odds of this problem to one in a million" [8, p. 197], a not entirely satisfactory recommendation. Indeed, the implementation printed at the end of their paper is not quite right [1, p. 82], sometimes printing an incorrect last digit because the finite approximation of the infinite series is one term too short.

We propose a different algorithm, based on the same series (1) for π but avoiding these problems. We also show the same technique applied to other characterizations of π. No commitment need be made in advance to the number of digits to be computed; given enough memory, the programs will generate digits ad infinitum. Once more (necessarily, in fact, given the previous property), the programs are spigot algorithms in Rabinowitz and Wagon's sense: they yield digits incrementally and do not reuse them after producing them. Of course, no algorithm using a bounded amount of memory can generate a nonrepeating sequence such as the digits of π indefinitely, so we have to allow arbitrary-precision arithmetic, or some other manifestation of dynamic memory allocation. Like Rabinowitz and Wagon's algorithm, our proposals are not competitive with state-of-the-art *arithmetic-geometric mean* algorithms for computing π [2], [9]. Nevertheless, our algorithms are simple to understand and admit almost as concise an implementation. As evidence to support the second claim, here is a (deliberately obscure) program that will generate as many digits of π as memory will allow:

```
> pi = g(1,0,1,1,3,3) where
>   g(q,r,t,k,n,l) = if 4*q+r-t<n*t
>       then n : g(10*q,10*(r-n*t),t,k,div(10*(3*q+r))t-10*n,l)
>       else g(q*k,(2*q+r)*l,t*l,k+1,div(q*(7*k+2)+r*l)(t*l),l+2)
```

The remainder of this paper provides a justification for the foregoing program, and some others like it.

These algorithms exhibit a pattern that we call *streaming* [**3**]. Informally, a streaming algorithm consumes a (potentially infinite) sequence of inputs and generates a (possibly infinite) sequence of outputs, maintaining some state as it goes. Based on the current state, at each step there is a choice between producing an element of the output and consuming an element of the input. Streaming seems to be a common pattern for various kinds of *representation changers*, including several data compression and number conversion algorithms.

The program under discussion is written in Haskell [**5**], a lazy functional programming language. As a secondary point of this paper, we hope to convince the reader that such languages are excellent vehicles for expressing mathematical computations, certainly when compared with other general-purpose programming languages such as Java, C, and Pascal, and arguably even when compared with computer algebra systems such as Mathematica. In particular, a lazy language allows direct computations with infinite data structures, which require some kind of indirect representation in most other languages. The Haskell program presented earlier has been compressed to compete with the C program for conciseness, so we do not argue that this particular one is easy to follow—but we do claim that the later Haskell programs are.

2. LAZY FUNCTIONAL PROGRAMMING IN HASKELL. To aid the reader's understanding, we start with a brief (and necessarily incomplete) description of the concepts of functional programming (henceforth FP) and laziness and their manifestation in Haskell [**5**], the de facto standard lazy FP language. Further resources, including pointers to tutorials and free implementations for many platforms, can be found at the Haskell website [**6**].

FP is programming with *expressions* rather than *statements*. Everything is a value, and there are no assignments or other state-changing commands. Therefore, a pure FP language is *referentially transparent*: an expression may always be substituted for one with an equal value, without changing the meaning of the surrounding context. This makes reasoning in FP languages just like reasoning in high-school algebra.

Here is a simple Haskell program:

```
> square :: Integer -> Integer
> square x = x * x
```

Program text is marked with a ">" in the left-hand column, and comments are unmarked. This is a simple form of *literate programming*, in which the emphasis is placed on making the program easy for people rather than computers to read. Code and documentation are freely interspersed; indeed, the manuscript for this article is simultaneously an executable Haskell program.

The first line in the `square` program is a type declaration; the symbol "`::`" should be read "has type." Thus, `square` has type `Integer -> Integer`, that is, it is a function from `Integer`s to `Integer`s. (Type declarations in Haskell are nearly always optional, because they can be inferred, but we often specify them anyway for clarity.) The second line gives a definition, as an equation: in any context, a subexpression of the form square x for any x may be replaced safely with x * x.

Lists are central to FP. Haskell uses square brackets for lists, so [1,2,3] has three elements, and [] is the empty list. The operator ":" prefixes an element; so [1,2,3] = 1 : (2 : (3 : [])). For any type a, there is a corresponding type [a] of lists with elements drawn from type a, so, for example, [1,2,3] :: [Integer]. Haskell has a very convenient *list comprehension* notation, analogous to set comprehensions; for instance, [square x | x <- [1,2,3]] denotes the list of squares [1,4,9].

For the "substitution of equals for equals" property to be universally valid, it is important not to evaluate expressions unless their values are needed. For example, consider the following Haskell program:

```
> three :: Integer -> Integer
> three x = 3

> nonsense :: Integer
> nonsense = 1 + nonsense
```

The first definition is of a function that ignores its argument x and always returns the integer 3; the second is of a value of type Integer, but one whose evaluation never terminates. For substitutivity to hold, and in particular for three nonsense to evaluate to 3 as the equation suggests, it is important not to evaluate the function argument nonsense in the function application three nonsense. With *lazy evaluation*, in which evaluation is demand-driven, no expression is evaluated unless and until its value is needed to make further progress.

A useful by-product of lazy evaluation is the ability to handle infinite data structures: they are evaluated only as far as is necessary. We illustrate this with a definition of the infinite sequence of Fibonacci numbers:

```
> fibs :: [Integer]
> fibs = f (0,1) where f (a,b) = a : f (b,a+b)
```

(Here, (0,1) is a pair of Integers, and the where clause introduces a local definition.) Evaluating fibs never terminates, of course, but computing a finite prefix of it does. For example, with

```
> take :: Integer -> [Integer] -> [Integer]
> take 0 xs = []
> take (n+1) [] = []
> take (n+1) (x:xs) = x : take n xs
```

(so take takes two arguments, an Integer and a [Integer], and returns another [Integer]), we have take 10 fibs = [0,1,1,2,3,5,8,13,21,34], which terminates normally.

Functions may be *polymorphic*, defined for arbitrary types. Thus, the function take in fact works for lists of any element type, since elements are merely copied and not further analyzed. The most general type assignable to take is Integer -> [a] -> [a] for an arbitrary type a.

Because this is FP, functions are "first-class citizens" of the language, with all the rights of any other type. Among other things, they may be passed as arguments to and returned as results from *higher-order functions*. For example, with the definition

```
> map :: (a->b) -> [a] -> [b]
> map f [] = []
> map f (x:xs) = f x : map f xs
```

the list comprehension [square x | x <- [1,2,3]] is equal to map square [1,2,3]. Note that map and take are *curried*. In fact, the function type former -> associates to the right, so the type of map is equivalent to (a->b) -> ([a]->[b]), which one might read as saying that map takes a function of type a->b and transforms it into one of type [a]->[b].

3. RABINOWITZ AND WAGON'S SPIGOT ALGORITHM.
Rabinowitz and Wagon's algorithm is based on the series (1) for π, which expands out to the expression

$$\pi = 2 + \frac{1}{3}\left(2 + \frac{2}{5}\left(2 + \frac{3}{7}\left(\cdots\left(2 + \frac{i}{2i+1}\left(\cdots\right)\right)\right)\right)\right). \tag{2}$$

This expression for π can be derived from the well-known Leibniz series

$$\frac{\pi}{4} = \sum_{i=0}^{\infty} \frac{(-1)^i}{2i+1} \tag{3}$$

using Euler's convergence-accelerating transform, among several other methods [7]. One can view expression (2) as representing a number $(2; 2, 2, 2, \ldots)$ in a mixed-radix base $\mathcal{B} = (\frac{1}{3}, \frac{2}{5}, \frac{3}{7}, \ldots)$, in the same way that the usual decimal expansion

$$\pi = 3 + \frac{1}{10}\left(1 + \frac{1}{10}\left(4 + \frac{1}{10}\left(1 + \frac{1}{10}5 + \cdots\right)\right)\right)$$

represents $(3; 1, 4, 1, 5, \ldots)$ in the fixed-radix base $\mathcal{F}_{10}$, where $\mathcal{F}_m = (\frac{1}{m}, \frac{1}{m}, \frac{1}{m}, \ldots)$. The task of computing the decimal digits of π is then simply a matter of converting from base $\mathcal{B}$ to base $\mathcal{F}_{10}$.

We consider *regular* representations. For a regular representation in decimal, every digit after the decimal point is in the range $[0, 9]$. The decimal number $(0; 9, 9, 9, \ldots)$ with a zero before the point and maximal digits afterwards represents 1. Accordingly, regular decimal representations with a zero before the point lie between 0 and 1. By analogy, for a regular representation in base $\mathcal{B}$, the digit in position i after the point (the first after the point being in position 1) is in the range $[0, 2i]$. The number $(0; 2, 4, 6, \ldots)$ with zero before the point and maximal digits afterwards represents 2. Accordingly, regular base $\mathcal{B}$ representations with zero before the point lie between 0 and 2. (We call the representations "regular" rather than "normal" because they are not unique.)

Conversion from base $\mathcal{B}$ to decimal proceeds as one might expect. The integer part of the input becomes the integer part of the output. The fractional part of the input is multiplied by ten; the integer part of this becomes the first output digit after the decimal point, and the fractional part is retained. This is again multiplied by ten; the integer part of this becomes the second output digit after the point; and so on.

Multiplying a number in base $\mathcal{B}$ by ten is achieved simply by multiplying each digit by ten. However, that yields an irregular result, because some of the resulting digits may be too big. Regularization proceeds from right to left, reducing each digit as necessary and propagating any carry leftwards.

The only remaining problem is in computing the integer part of a number $(a_0; a_1, a_2, a_3)$ in base $\mathcal{B}$. This is either a_0 or $a_0 + 1$, depending on whether the remainder $(0; a_1, a_2, a_3)$ is in $[0, 1)$ or $[1, 2)$. (In principle, the remainder could equal 2; but, in practice, this cannot happen in the computation of an irrational such as π.) Therefore, Rabinowitz and Wagon's algorithm temporarily buffers any nines that are produced, until it is clear whether or not there could be a carry that would invalidate them.

This whole conversion is performed on a *finite* number $(2; 2, 2, 2, \ldots, 2)$ in base $\mathcal{B}$—necessarily, as regularization proceeds from right to left. Rabinowitz and Wagon provide a bound on the number of base $\mathcal{B}$ digits needed to yield a given number of decimal digits. In fact, as mentioned earlier, they underestimate by one in some cases: $\lfloor 10n/3 \rfloor$ digits is usually sufficient, but sometimes $\lfloor 10n/3 \rfloor + 1$ input digits is necessary (and sufficient) for n decimal digits. (Here, "$\lfloor x \rfloor$" denotes the greatest integer not larger than x.) Again, as noted, this does not mean that those n decimal digits are all correct digits of π, only that the n-digit decimal number produced is within 5×10^{-n} of the desired result.

4. STREAMING ALGORITHMS.

We turn now to streaming algorithms, by way of a simpler example than computing the digits of π. Consider the problem of converting a fraction in the interval $[0, 1]$ from one base to another. We represent fractions as digit sequences, and for simplicity (but without loss of generality) we consider only infinite sequences. For this reason, we cannot consume all the input before producing any output; we must alternate between consumption and production. The computation will therefore maintain some state depending on the inputs consumed and the outputs produced thus far. Based on that state, it will either produce another output, if that is possible given the available information, or consume another input if it is not. This pattern is captured by the following higher-order function:

```
> stream :: (b->c) -> (b->c->Bool) -> (b->c->b) -> (b->a->b)
>                  -> b -> [a] -> [c]
> stream next safe prod cons z (x:xs)
>   = if safe z y
>   then y : stream next safe prod cons (prod z y) (x:xs)
>   else stream next safe prod cons (cons z x) xs
>   where y = next z
```

This defines a function `stream` taking six arguments. The result of applying `stream` is an infinite list of output terms, each of type c. The last argument `x:xs` is a list of input terms, each of type a; the first element or "head" is `x`, and the infinite remainder or "tail" is `xs`. The penultimate argument `z` is the state, of type b. The other four arguments (`next` of type b->c, `safe` of type b->c->Bool, `prod` of type b->c->b, and `cons` of type b->a->b) are all functions. From the state `z` the function produces a provisional output term `y = next z` of type c. If `y` is `safe` to commit to from the current state `z` (whatever input terms may come next), then it is produced, and the state adjusted accordingly using `prod`; otherwise, the next term `x` of the input is consumed into the state. This process continues indefinitely: the input is assumed never to run out, and, if the process is productive, the output never terminates.

In the case of conversion from an infinite digit sequence in base $\mathcal{F}_m$ to an infinite sequence in base $\mathcal{F}_n$, clearly both the input and output elements are of type `Integer`. The state maintained is a pair (u, v) of `Rational`s, satisfying the "invariant" (that is, a property that is established before a loop commences and is maintained by each iteration of that loop) that the original input

$$\frac{1}{m}\left(x_0 + \frac{1}{m}\left(x_1 + \cdots\right)\right)$$

is equal to

$$\frac{1}{n}\left(y_0 + \frac{1}{n}\left(y_1 + \cdots + \frac{1}{n}\left(y_{j-1} + v \times \left(u + \frac{1}{m}\left(x_i + \frac{1}{m}x_{i+1} + \cdots\right)\right)\right)\right)\right)$$

when i input terms $x_0, x_1, \ldots, x_{i-1}$ have been consumed and j output terms $y_0, y_1, \ldots, y_{j-1}$ have been produced. Initially i and j are zero, so the invariant is established with $u = 0$ and $v = 1$. In order to maintain the invariant, the state (u, v) should be transformed to $(u - y/(n\,v), n\,v)$ when producing an additional output term y and to $(x + u\,m, v/m)$ when consuming an additional input term x. The value of the remaining input

$$\frac{1}{m}\left(x_i + \frac{1}{m}\left(x_{i+1} + \cdots\right)\right)$$

ranges between 0 and 1, so the next output term is determined provided that nvu and $nv(u + 1)$ have the same integer part or floor. This justifies the following streaming algorithm:

```
> convert :: (Integer,Integer) -> [Integer] -> [Integer]
> convert (m,n) xs = stream next safe prod cons init xs
>   where
>     init = (0%1, 1%1)
>     next (u,v) = floor (u*v*n')
>     safe (u,v) y = (y == floor ((u+1)*v*n'))
      prod (u,v) y = (u - fromInteger y/(v*n'), v*n')
>     cons (u,v) x = (fromInteger x + u*m', v/m')
>     (m',n')      = (fromInteger m, fromInteger n)
```

(Here, "%" constructs a Rational from two Integers, "==" is the comparison operator, and the function fromInteger coerces Integers to Rationals.)

For example, $1/e$ is $0.1002210112\ldots$ in base 3 and $0.2401164352\ldots$ in base 7. Therefore, applying the function convert (3,7) to the infinite list [1,0,0,2,2,1, 0,1,1,2... should yield the infinite list [2,4,0,1,1,6,4,3,5,2.... The first few states through which execution of this conversion proceeds are illustrated in the following table:

input		1	0	0		2		2		1			
state	$\frac{0}{1},\frac{1}{1}$	$\frac{1}{1},\frac{1}{3}$	$\frac{3}{1},\frac{1}{9}$	$\frac{9}{1},\frac{1}{27}$	$\frac{9}{7},\frac{7}{27}$	$\frac{41}{7},\frac{7}{81}$	$\frac{137}{7},\frac{7}{243}$	$\frac{418}{7},\frac{7}{729}$	$\frac{10}{49},\frac{49}{729}$	$\frac{10}{49},\frac{343}{729}$	$\cdots$
output				2					4	0		

The middle row shows consecutive values of the state (u, v). The upper row shows input digits consumed, above the state resulting from their consumption. The lower row shows output digits produced, below the state resulting from their production. Notice that outputs are produced precisely when the corresponding state (u, v), which is in the previous column, is safe; that is, when $\lfloor 7uv \rfloor = \lfloor 7(u + 1)v \rfloor$. For example, the fourth state $(\frac{9}{1}, \frac{1}{27})$ is safe, because $\lfloor 7 \times \frac{9}{1} \times \frac{1}{27} \rfloor = 2 = \lfloor 7 \times (\frac{9}{1} + 1) \times \frac{1}{27} \rfloor$.

This paper is not the place to make a more formal justification for the correctness of this program, although it is not hard to establish from the invariant stated. Nevertheless, it is possible to *derive* the streaming program from a specification expressed in terms of independent operations for expanding and collapsing digit sequences, using a general theory of such algorithms [**3**]. (In the general case, either the input or the output or both may be finite. We have stuck to the simple case of necessarily-infinite lists here, because that is all that is needed for computing the digits of π.) We have found this pattern of computation in numerous problems concerning *changes of data representation*, of which conversions between number formats are a representative example. Consequently, we have been calling such algorithms *metamorphisms*.

5. A STREAMING ALGORITHM FOR THE DIGITS OF π.

The main problem with Rabinowitz and Wagon's spigot algorithm is that it is bounded: one must make a commitment in advance to the number of terms of the series (1) to use. This commitment arises because the process of regularizing a number in base B proceeds from right to left, hence works only for finite numbers in that base.

It turns out that there is a rather simple *streaming algorithm* for regularizing infinite numbers in base B. This means that we can make Rabinowitz and Wagon's algorithm unbounded: there is no longer any need to make a prior commitment to a particular finite prefix of the expansion of π. However, we will not say any more about this approach, for there is a more direct way of computing the digits of π from the expression (2), to which we now turn.

One can view the expansion (2) as the composition

$$\pi = \left(2 + \frac{1}{3} \times\right)\left(2 + \frac{2}{5} \times\right)\left(2 + \frac{3}{7} \times\right) \cdots \left(2 + \frac{i}{2i+1} \times\right) \cdots \quad (4)$$

of an infinite series of *linear fractional transformations* or *Möbius transformations*. These are functions taking x to $(qx + r)/(sx + t)$ for integers q, r, s, and t with $qt - rs \neq 0$—that is, yielding a ratio of integer-coefficient linear transformations of x. Such a transformation can be represented by the four coefficients q, r, s, and t, and if they are arranged as a matrix $\left(\begin{smallmatrix} q & r \\ s & t \end{smallmatrix}\right)$ then function composition corresponds to matrix multiplication.

```
> type LFT = (Integer, Integer, Integer, Integer)

> extr :: LFT -> Integer -> Rational
> extr (q,r,s,t) x = ((fromInteger q) * x + (fromInteger r)) /
>                     ((fromInteger s) * x + (fromInteger t))
> unit :: LFT
> unit = (1,0,0,1)
> comp :: LFT -> LFT -> LFT
> comp (q,r,s,t) (u,v,w,x) = (q*u+r*w,q*v+r*x,s*u+t*w,s*v+t*x)
```

(The first line introduces the abbreviation LFT for the type of four-tuples of Integers.)

The infinite composition of transformations in (4) converges, in the following sense. Although the products of finite prefixes of the composition have coefficients that grow without bound, the transformations represented by these products map the interval [3, 4] onto converging subintervals of itself. (This is easy to see, as each term is a monotonic transformation, reduces the width of an interval by at least a factor of two, and maps [3, 4] onto a subinterval of itself. Indeed, the same also holds for any tail of the infinite composition.)

Therefore, equation (4) can be thought of as the representation of some real number, and computing the decimal digits of this number is a change of representation, effectable by a streaming algorithm. The streaming process maintains as its state an additional linear fractional transformation, representing the required function from the inputs yet to be consumed to the outputs yet to be produced. This state is initially the identity matrix; consumption of another input term is matrix multiplication; production of a digit n is multiplication by $\begin{pmatrix} 10 & -10n \\ 0 & 1 \end{pmatrix}$, the inverse of the linear fractional transformation taking x to $n + \frac{x}{10}$. If the current state is the transformation z, then the next digit to be produced lies somewhere in the image under z of the interval $[3, 4]$; if the two endpoints of this image have the same integer part, then that next digit is completely determined and it is safe to commit to it.

```
> pi = stream next safe prod cons init lfts where
>    init      = unit
>    lfts      = [(k, 4*k+2, 0, 2*k+1) | k<-[1..]]
>    next z    = floor (extr z 3)
>    safe z n  = (n == floor (extr z 4))
>    prod z n  = comp (10, -10*n, 0, 1) z
>    cons z z' = comp z z'
```

The definition of lfts uses a list comprehension, with generator [1..], the infinite list of Integers from 1 upwards. The list consists of the expression (k, 4*k+2, 0, 2*k+1) to the left of the vertical bar, evaluated for each value of k from 1 upwards. The first few terms are

$$\left[\begin{pmatrix} 1 & 6 \\ 0 & 3 \end{pmatrix}, \begin{pmatrix} 2 & 10 \\ 0 & 5 \end{pmatrix}, \begin{pmatrix} 3 & 14 \\ 0 & 7 \end{pmatrix}, \dots \right].$$

For example, the first term $\begin{pmatrix} 1 & 6 \\ 0 & 3 \end{pmatrix}$ represents the transformation taking x to

$$\frac{1 \times x + 6}{0 \times x + 3},$$

or $2 + \frac{1}{3}x$.

The condensed program shown in section 1 can be obtained from this program by making various simple optimizations. These include: unfolding intermediate definitions; exploiting the invariant that the bottom left element s of every linear fractional transformation $\begin{pmatrix} q & r \\ s & t \end{pmatrix}$ be 0; constructing the input transformations in place; representing the sequence of remaining transformations simply by the index k; and simplifying away one of the divisions.

Another optimization that can be performed is the elimination of any factors common to all four entries resulting from a matrix multiplication. This optimization is valid, since linear fractional transformations are invariant under scaling of the matrix, and helpful, as it keeps the numbers small. (In fact, it is better still to perform this cancellation less frequently than every iteration.)

6. MORE STREAMING ALGORITHMS FOR THE DIGITS OF π. The expression (2) turns out not to be a very efficient one for computation. Each term shrinks the range by a factor of about a half, so more than three terms are required on average for every digit of the output. Better sequences are known; the book π *Unleashed* [1] presents many. We conclude this paper with two more applications of the streaming technique from section 4 to computing π, using the same approach but based on two of these different expressions.

Lambert's expression. Here is a more efficient expression for π, due to Lambert in 1770 [**1**, eq. (16.99)], that yields two decimal digits for every three terms:

$$\pi = \cfrac{4}{1 + \cfrac{1^2}{3 + \cfrac{2^2}{5 + \cfrac{3^2}{7 + \cdots}}}}. \tag{5}$$

Again, one can view this as an infinite composition of linear fractional transformations:

$$\pi = (4 \div)(1 + 1^2 \div)(3 + 2^2 \div)(5 + 3^2 \div) \cdots (2i - 1 + i^2 \div) \cdots.$$

After consuming i terms of the input, the remaining terms represent the composition

$$(2i - 1 + i^2 \div)(2i + 1 + (i + 1)^2 \div)(2i + 3 + (i + 2)^2 \div) \cdots,$$

which denotes a value in the range $[2i - 1, 2i - 1 + \frac{i}{2}]$. As before, we subject this infinite sequence to a streaming process. This time, however, we maintain a state consisting of not just a linear fractional transformation $\left(\begin{smallmatrix} q & r \\ s & t \end{smallmatrix}\right)$, but also the number i of terms consumed thus far (needed in order to determine the next digit to produce, which lies between $\left(\begin{smallmatrix} q & r \\ s & t \end{smallmatrix}\right)(2i - 1)$ and $\left(\begin{smallmatrix} q & r \\ s & t \end{smallmatrix}\right)(2i - 1 + \frac{i}{2})$).

This reasoning justifies the following program:

```
> piL = stream next safe prod cons init lfts where
>    init                = ((0,4,1,0), 1)
>    lfts                = [(2*i-1, i*i, 1, 0) | i<-[1..]]
>    next ((q,r,s,t),i)  = floor ((q*x+r) % (s*x+t))
                           where x=2*i-1
>    safe ((q,r,s,t),i) n = (n == floor ((q*x+2*r) % (s*x+2*t)))
>                           where x=5*i-2
>    prod (z,i) n        = (comp (10, -10*n, 0, 1) z, i)
>    cons (z,i) z'       = (comp z z', i+1)
```

Gosper's series. An even more efficient series for π, yielding more than one decimal digit for each term, is due to Gosper [**4**]:

$$\pi = 3 + \frac{1 \times 1}{3 \times 4 \times 5} \times \left(8 + \frac{2 \times 3}{3 \times 7 \times 8} \times \left(\cdots 5i - 2 + \frac{i(2i - 1)}{3(3i + 1)(3i + 2)} \times \cdots\right)\right). \tag{6}$$

Once more, we can view this as an infinite composition of linear fractional transformations, namely,

$$\pi = \left(3 + \frac{1 \times 1}{3 \times 4 \times 5} \times\right)\left(8 + \frac{2 \times 3}{3 \times 7 \times 8} \times\right)$$
$$\cdots \left(5i - 2 + \frac{i(2i - 1)}{3(3i + 1)(3i + 2)} \times\right) \cdots.$$

 [Monthly 113

It is not hard to show that, after consuming $i - 1$ terms of the input, the remaining terms denote a value in the range $[\frac{27}{5}i - \frac{12}{5}, \frac{27}{5}i - \frac{2^3 3^3}{5^3}]$, which gives rise to the following program:

```
> piG = stream next safe prod cons init lfts where
>    init                = ((1,0,0,1), 1)
>    lfts                = [let j = 3*(3*i+1)*(3*i+2)
>                          in (i*(2*i-1),j*(5*i-2),0,j) |
>                          i<-[1..]]
>    next ((q,r,s,t),i)   = div (q*x+5*r) (s*x+5*t) where
>                          x = 27*i+15
>    safe ((q,r,s,t),i) n = (n == div (q*x+125*r) (s*x+125*t))
>                          where x=675*i-216
>    prod (z,i) n         = (comp (10, -10*n, 0, 1) z, i)
>    cons (z,i) z'        = (comp z z', i+1)
```

(The let here is another form of local definition.)

A challenge. Gosper's series (6) yields more than one digit per term on average, since the scaling factors approach $\frac{2}{27}$ (which is less than $\frac{1}{10}$) from below. This suggests that we could dispense with the test altogether and strictly alternate between consumption and production, as expressed by the following conjecture. Eliminating the test would speed up the algorithm considerably.

Conjecture 1. *Define the following functions:*

$$n(z, i) = \left\lfloor z \left(\frac{27i + 15}{5} \right) \right\rfloor$$

$$p((z, i), n) = \left(z \begin{pmatrix} 10 & -10n \\ 0 & 1 \end{pmatrix}, i \right)$$

$$c((z, i), z') = (z\,z', i + 1)$$

$$s((z, i), n) = \left(n = \left\lfloor z \left(\frac{675i - 216}{125} \right) \right\rfloor \right).$$

For $i = 1, 2 \ldots$, let

$$x_i = \begin{pmatrix} i(2i - 1) & 3(3i + 1)(3i + 2)(5i - 2) \\ 0 & 3(3i + 1)(3i + 2) \end{pmatrix}$$

$$u_0 = \left(\begin{pmatrix} 1 & 0 \\ 0 & 1 \end{pmatrix}, 1 \right)$$

$$u_i = p(v_i, y_i)$$

$$v_i = c(u_{i-1}, x_i)$$

$$y_i = n(v_i).$$

Then $s(v_i, y_i)$ holds for all i.

If this conjecture holds, then, in the following program, every value taken by the variable v satisfies the condition safe v (next v).

```
> piG2 = process next prod cons init lfts
> process next prod cons u (x:xs)
>   = y : process next prod cons (prod v y) xs
>     where v = cons u x
>           y = next v
```

(Here, next, prod, cons, init, and lfts, as well as the predicate safe mentioned in the claim, are as in section 6.)

We have not been able to prove Conjecture 1, although we have verified it for the first thousand terms. Perhaps some diligent reader can provide enlightenment. If it is valid, then piG2 does indeed produce the digits of π, and the optimizations outlined at the end of section 5 can be applied to piG2, yielding the following program:

```
> piG3 = g(1,180,60,2) where
>   g(q,r,t,i) =
>     let (u,y)=(3*(3*i+1)*(3*i+2),div(q*(27*i-12)+5*r)(5*t))
>     in y : g(10*q*i*(2*i-1),10*u*(q*(5*i-2)+r-y*t),t*u,i+1)
```

This is of comparable length to the compressed program given in section 1, but approximately five times faster.

ACKNOWLEDGMENTS Thanks are due to the anonymous referees, the Algebra of Programming research group at Oxford, Stan Wagon, Sue Gibbons, and especially to Christoph Haenel, who suggested Conjecture 1. All of them have made suggestions that have improved the presentation of this paper.

REFERENCES

1. J. Arndt and C. Haenel, π *Unleashed*, 2nd ed., Springer-Verlag, Berlin, 2001.
2. R. P. Brent, Fast multiple-precision evaluation of elementary functions, *J. ACM* **23** (1976) 242–251.
3. J. Gibbons. Streaming representation-changers, in *Mathematics of Program Construction*, D. Kozen, ed., Springer-Verlag, Berlin, 2004, pp. 142–168.
4. R. W. Gosper, Acceleration of series, Technical Report AIM-304, AI Laboratory, MIT (March 1974); available at ftp://publications.ai.mit.edu/ai-publications/pdf/AIM-304.pdf.
5. S. Peyton Jones, ed., *Haskell 98 Language and Libraries: The Revised Report*, Cambridge University Press, Cambridge, 2003.
6. Haskell web site, http://www.haskell.org/.
7. J. C. R. Li et al., Solutions to Problem E854: A series for π, this MONTHLY **56** (1949) 633–635.
8. S. Rabinowitz and S. Wagon, A spigot algorithm for the digits of π, this MONTHLY **102** (1995) 195–203.
9. E. Salamin, Computation of π using arithmetic-geometric mean, *Math. Comp.* **30** (1976) 565–570.

JEREMY GIBBONS is a university lecturer in software engineering and continuing education at the University of Oxford. He obtained his doctorate at Oxford in 1991, then spent five years at the University of Auckland in New Zealand and three years at Oxford Brookes University, before returning to the University of Oxford. He now teaches on the professional postgraduate Software Engineering Programme (http://www.softeng.ox.ac.uk/). His research interests are in programming languages and methods, particularly in the functional and object-oriented paradigms, and in recurring patterns in the structure of computer programs.
Oxford University Computing Laboratory, Wolfson Building, Parks Road, Oxford OX1 3QD, UK
jeremy.gibbons@comlab.ox.ac.uk
http://www.comlab.ox.ac.uk/jeremy.gibbons/

17. Mathematics by experiment: Plausible reasoning in the 21st Century (2008)

Paper 17: David H. Bailey and Jonathan M. Borwein, "Pi and its friends," and "Normality: A stubborn question," from *Mathematics by Experiment: Plausible Reasoning in the 21st Century*, A. K. Peters, Natick, MA, 2nd edition, 2008. Reproduced with permission of AK Peters.

Synopsis:

We present here excerpts from two chapters of the second edition of our book *Mathematics by Experiment*. The first chapter presents an overview of computations of π through the ages, followed by a summary of computer-age developments, including a detailed analysis of the BBP formula and algorithm for computing binary digits of π beginning at an arbitrary starting position.

The second selection from our book takes a look at a question that has puzzled mathematicians from time immemorial (and which has spurred many computations of π): whether and why the digits of π and other well-known mathematical constants are "normal," meaning that every m-long string of base-b digits appears, in the limit, with frequency $1/b^m$. Included here are some details of some recent results in this area, such as a proof of normality of the Stoneham numbers, namely constants of the form

$$\alpha_{b,c} = \sum_{k=0}^{\infty} \frac{1}{c^k b^{c^k}},$$

where $b \geq 2$ and $c \geq 2$ are relatively prime.

Keywords: Algorithms, Computation, History, Normality

© Springer International Publishing Switzerland 2016
D.H. Bailey, J.M. Borwein, *Pi: The Next Generation*,
DOI 10.1007/978-3-319-32377-0_17

Nowadays, this is almost trivial: A "Minpoly" calculation immediately returns $29 - 80x - 24x^2 + 16x^4 = 0$ and this has the surd above as its smallest positive root. At this point, the authors could use known results only to prove the value of $\alpha(1), \alpha(2)$ and $\alpha(3)$. Those for $\alpha(5)$ and $\alpha(7)$ remained conjectural. There was, however, an empirical family of algorithms for π: let $\alpha_0 = \alpha(N)$ and $k_0 = k'_N$ (where $k' = \sqrt{1 - k^2}$) and iterate

$$k_{n+1} = \frac{1 - k'_n}{1 + k'_n} \tag{3.22}$$

$$\alpha_{n+1} = (1 + k_{n+1})^2 \alpha_n - \sqrt{N}\, 2^{n+1} k_{n+1}. \tag{3.23}$$

Then

$$\lim_{n \to \infty} \alpha_n^{-1} = \pi. \tag{3.24}$$

Again, (3.24) was provable for $N = 1, 2, 3$ and only conjectured for $N = 5, 7$. In each case the algorithm *appeared* to converge quadratically to π. On closer inspection while the provable cases were correct to $5,000$ digits, the empirical ones agreed with π to roughly 100 places only. Now, in many ways to have discovered a "natural" number that agreed with π to that level—and no more—would have been more interesting than the alternative. That seemed unlikely, but recoding and rerunning the iterations kept producing identical results.

Twenty years ago, very high-precision calculation was less accessible, and the code was being run in a Berkeley Unix integer package. After about six weeks of effort, it was found that the square root algorithm in the package was badly flawed, but only if run with an odd precision of more than 60 digits! And for idiosyncratic reasons that had only been the case in the two unproven cases. Needless to say, tracing the bug was a salutary and somewhat chastening experience.

3.4   Computing Individual Digits of Pi

An outsider might be forgiven for thinking that essentially everything of interest with regards to π has been discovered. For example, this sentiment is suggested in the closing chapters of Beckmann's 1971 book on the history of π [48, pg. 172]. Ironically, the Salamin–Brent quadratically convergent iteration was discovered only five years later, and the higher-order convergent algorithms followed in the 1980s. In 1990, Rabinowitz and Wagon discovered a "spigot" algorithm for π, which permits successive digits of π (in any desired base) to be computed with a relatively simple recursive

3.4. Computing Individual Digits of Pi

algorithm based on the previously generated digits (see [239] and Item 15 at the end of this chapter).

But even insiders are sometimes surprised by a new discovery. Prior to 1996, almost all mathematicians believed that if you want to determine the d-th digit of π, you have to generate the entire sequence of the first d digits. (For all of their sophistication and efficiency, the schemes described above all have this property.) But it turns out that this is not true, at least for hexadecimal (base 16) or binary (base 2) digits of π. In 1996, Peter Borwein, Simon Plouffe, and one of the present authors (Bailey) found an algorithm for computing individual hexadecimal or binary digits of π [33]. To be precise, this algorithm:

(1) directly produces a modest-length string of digits in the hexadecimal or binary expansion of π, beginning at an arbitrary position, without needing to compute any of the previous digits;

(2) can be implemented easily on any modern computer;

(3) does not require multiple precision arithmetic software;

(4) requires very little memory; and

(5) has a computational cost that grows only slightly faster than the digit position.

Using this algorithm, for example, the one millionth hexadecimal digit (or the four millionth binary digit) of π can be computed in less than a minute on a 2003-era computer. The new algorithm is not fundamentally faster than best-known schemes for computing all digits of π up to some position, but its elegance and simplicity are nonetheless of considerable interest. This scheme is based on the following remarkable new formula for π:

Theorem 3.1.

$$\pi = \sum_{i=0}^{\infty} \frac{1}{16^i} \left(\frac{4}{8i+1} - \frac{2}{8i+4} - \frac{1}{8i+5} - \frac{1}{8i+6} \right). \quad (3.25)$$

Proof. First note that for any $k < 8$,

$$\int_0^{1/\sqrt{2}} \frac{x^{k-1}}{1-x^8} \, dx = \int_0^{1/\sqrt{2}} \sum_{i=0}^{\infty} x^{k-1+8i} \, dx$$

$$= \frac{1}{2^{k/2}} \sum_{i=0}^{\infty} \frac{1}{16^i(8i+k)}. \quad (3.26)$$

120   3. Pi and Its Friends

Thus one can write

$$\sum_{i=0}^{\infty} \frac{1}{16^i}\left(\frac{4}{8i+1} - \frac{2}{8i+4} - \frac{1}{8i+5} - \frac{1}{8i+6}\right)$$
$$= \int_0^{1/\sqrt{2}} \frac{4\sqrt{2} - 8x^3 - 4\sqrt{2}x^4 - 8x^5}{1-x^8}\,dx, \qquad (3.27)$$

which on substituting $y = \sqrt{2}x$ becomes

$$\int_0^1 \frac{16\,y - 16}{y^4 - 2\,y^3 + 4\,y - 4}\,dy \;=\; \int_0^1 \frac{4y}{y^2 - 2}\,dy - \int_0^1 \frac{4y-8}{y^2 - 2y + 2}\,dy$$

$$= \; \pi. \qquad (3.28)$$

$\square$

However, in presenting this formal derivation, we are disguising the actual route taken to the discovery of this formula. This route is a superb example of experimental mathematics in action.

It all began in 1995, when Peter Borwein and Simon Plouffe of Simon Fraser University observed that the following well-known formula for $\log 2$ permits one to calculate isolated digits in the binary expansion of $\log 2$:

$$\log 2 \;=\; \sum_{k=0}^{\infty} \frac{1}{k2^k}. \qquad (3.29)$$

This scheme is as follows. Suppose we wish to compute a few binary digits beginning at position $d+1$ for some integer $d > 0$. This is equivalent to calculating $\{2^d \log 2\}$, where $\{\cdot\}$ denotes fractional part. Thus we can write

$$\{2^d \log 2\} \;=\; \left\{\left\{\sum_{k=0}^{d} \frac{2^{d-k}}{k}\right\} + \sum_{k=d+1}^{\infty} \frac{2^{d-k}}{k}\right\}$$
$$= \left\{\left\{\sum_{k=0}^{d} \frac{2^{d-k} \bmod k}{k}\right\} + \sum_{k=d+1}^{\infty} \frac{2^{d-k}}{k}\right\}. \qquad (3.30)$$

We are justified in inserting "mod k" in the numerator of the first summation, because we are only interested in the fractional part of the quotient when divided by k.

Now the key observation is this: The numerator of the first sum in Equation (3.30), namely $2^{d-k} \bmod k$, can be calculated very rapidly by

3.4. Computing Individual Digits of Pi   121

means of the binary algorithm for exponentiation, performed modulo k. The binary algorithm for exponentiation is merely the formal name for the observation that exponentiation can be economically performed by means of a factorization based on the binary expansion of the exponent. For example, we can write $3^{17} = ((((3^2)^2)^2)^2) \cdot 3$, thus producing the result in only 5 multiplications, instead of the usual 16. According to Knuth, this technique dates back at least to 200 BCE [188, pg. 461]. In our application, we need to obtain the exponentiation result modulo a positive integer k. This can be done very efficiently by reducing modulo k the intermediate multiplication result at each step of the binary algorithm for exponentiation. A formal statement of this scheme is as follows:

Algorithm 3.2. Binary algorithm for exponentiation modulo k.

To compute $r = b^n \bmod k$, where r, b, n and k are positive integers: First set t to be the largest power of two such that $t \leq n$, and set $r = 1$. Then
>   A: if $n \geq t$ then $r \leftarrow br \bmod k$; $n \leftarrow n - t$;   endif
>
> $t \leftarrow t/2$
>
>   if $t \geq 1$ then $r \leftarrow r^2 \bmod k$;   go to A;   endif   □

Note that the above algorithm is performed entirely with positive integers that do not exceed k^2 in size. Thus ordinary 64-bit floating-point or integer arithmetic, available on almost all modern computers, suffices for even rather large calculations. 128-bit floating-point arithmetic (double-double or quad precision), available at least in software on many systems (see Section 6.2.1), suffices for the largest computations currently feasible.

We can now present the algorithm for computing individual binary digits of $\log 2$.

Algorithm 3.3. Individual digit algorithm for $\log 2$.

To compute the $(d + 1)$-th binary digit of $\log 2$: Given an integer $d > 0$, (1) calculate each numerator of the first sum in Equation (3.30), using Algorithm 3.2, implemented using ordinary 64-bit or 128-bit floating-point arithmetic; (2) divide each numerator by the respective value of k, again using ordinary floating-point arithmetic; (3) sum the terms of the first summation, while discarding any integer parts; (4) evaluate the second summation as written using floating-point arithmetic—only a few terms are necessary since it rapidly converges; and (5) add the result of the first and second summations, discarding any integer part. The resulting fraction, when expressed in binary, gives the first few digits of the binary expansion of $\log 2$ beginning at position $d + 1$.   □

As soon as Borwein and Plouffe found this algorithm, they began seeking other mathematical constants that shared this property. It was clear that any constant α of the form

$$\alpha \;=\; \sum_{k=0}^{\infty} \frac{p(k)}{q(k)2^k}, \qquad (3.31)$$

where $p(k)$ and $q(k)$ are integer polynomials, with $\deg p < \deg q$ and q having no zeroes at nonnegative integer arguments, is in this class. Further, any rational linear combination of such constants also shares this property. Checks of various mathematical references eventually uncovered about 25 constants that possessed series expansions of the form given by Equation (3.31).

As you might suppose, the question of whether π also shares this property did not escape these researchers. Unfortunately, exhaustive searches of the mathematical literature did not uncover any formula for π of the requisite form. But given the fact that any rational linear combination of constants with this property also shares this property, Borwein and Plouffe performed integer relation searches to see if a formula of this type existed for π. This was done, using computer programs written by one of the present authors (Bailey), which implement the "PSLQ" integer relation algorithm in high-precision, floating-point arithmetic [16, 140]. We will discuss the PSLQ algorithm and related techniques more in Section 6.3.

In particular, these three researchers sought an integer relation for the real vector $(\alpha_1, \alpha_2, \cdots, \alpha_n)$, where $\alpha_1 = \pi$ and $(\alpha_i, \; 2 \le i \le n)$ is the collection of constants of the requisite form gleaned from the literature, each computed to several hundred decimal digit precision. To be precise, they sought an n-long vector of integers (a_i) such that $\sum_i a_i \alpha_i = 0$, to within a very small "epsilon." After a month or two of computation, with numerous restarts using new α vectors (when additional formulas were found in the literature) the identity (3.25) was finally uncovered. The actual formula found by the computation was:

$$\pi \;=\; 4F(1/4, 5/4; 1; -1/4) + 2\arctan(1/2) - \log 5, \qquad (3.32)$$

where $F(1/4, 5/4; 1; -1/4) = 0.955933837\ldots$ is a hypergeometric function evaluation. Reducing this expression to summation form yields the new π formula:

$$\pi \;=\; \sum_{i=0}^{\infty} \frac{1}{16^i}\left(\frac{4}{8i+1} - \frac{2}{8i+4} - \frac{1}{8i+5} - \frac{1}{8i+6}\right). \qquad (3.33)$$

To return briefly to the derivation of Formula (3.33), let us point out that it was discovered not by formal reasoning, or even by computer-based

3.4. Computing Individual Digits of Pi   123

symbolic processing, but instead by numerical computations using a high-precision implementation of the PSLQ integer relation algorithm. It is most likely the first instance in history of the discovery of a new formula for π by a computer. We might mention that, in retrospect, Formula (3.33) could be found much more quickly, by seeking integer relations in the vector $(\pi, S_1, S_2, \cdots, S_8)$, where

$$S_j \;=\; \sum_{k=0}^{\infty} \frac{1}{16^k(8k+j)}. \tag{3.34}$$

Such a calculation could be done in a few seconds on a computer, even if one did not know in advance to use 16 in the denominator and 9 terms in the search, but instead had to stumble on these parameters by trial and error. But this observation is, as they say, 20-20 hindsight. The process of real mathematical discovery is often far more tortuous and less elegant than the polished version typically presented in textbooks and research journals.

It should be clear at this point that the scheme for computing individual hexadecimal digits of π is very similar to Algorithm 3.3. For completeness, we state it as follows:

Algorithm 3.4. Individual digit algorithm for π.

To compute the $(d+1)$-th hexadecimal digit of π: Given an integer $d > 0$, we can write

$$\{16^d\pi\} \;=\; \{4\{16^dS_1\} - 2\{16^dS_4\} - \{16^dS_5\} - \{16^dS_6\}\}, \tag{3.35}$$

using the S_j notation of Equation (3.34). Now apply Algorithm 3.3, with

$$\{16^dS_j\} \;=\; \left\{\left\{\sum_{k=0}^{d}\frac{16^{d-k}}{8k+j}\right\} + \sum_{k=d+1}^{\infty}\frac{16^{d-k}}{8k+j}\right\}$$

$$=\; \left\{\left\{\sum_{k=0}^{d}\frac{16^{d-k} \bmod 8k+j}{8k+j}\right\} + \sum_{k=d+1}^{\infty}\frac{16^{d-k}}{8k+j}\right\} \tag{3.36}$$

instead of Equation (3.30), to compute $\{16^dS_j\}$ for $j = 1, 4, 5, 6$. Combine these four results, discarding integer parts, as shown in (3.35). The resulting fraction, when expressed in hexadecimal notation, gives the hex digit of π in position $d+1$, plus a few more correct digits.   □

As with Algorithm 3.3, multiple-precision arithmetic software is not required—ordinary 64-bit or 128-bit floating-point arithmetic suffices even

124   3. Pi and Its Friends

for some rather large computations. We have omitted here some numerical details for large computations—see [33]. Sample implementations in both C and Fortran-90 are available from http://www.experimentalmath .info.

One mystery that remains unanswered is why Formula (3.33) was not discovered long ago. As you can see from the above proof, there is nothing very sophisticated about its derivation. There is no fundamental reason why Euler, for example, or Gauss or Ramanujan, could not have discovered it. Perhaps the answer is that its discovery was a case of "reverse mathematical engineering." Lacking a motivation to find such a formula, mathematicians of previous eras had no reason to derive one. But this still doesn't answer the question of why the algorithm for computing individual digits of $\log 2$ had not been discovered before—it is based on a formula, namely Equation (3.29), that has been known for centuries.

Needless to say, Algorithm 3.4 has been implemented by numerous researchers. In 1997, Fabrice Bellard of INRIA computed 152 binary digits of π starting at the trillionth binary digit position. The computation took 12 days on 20 workstations working in parallel over the Internet. His scheme is actually based on the following variant of 3.33:

$$
\pi = 4 \sum_{k=0}^{\infty} \frac{(-1)^k}{4^k (2k+1)}
$$

$$
-\frac{1}{64} \sum_{k=0}^{\infty} \frac{(-1)^k}{1024^k} \left(\frac{32}{4k+1} + \frac{8}{4k+2} + \frac{1}{4k+3} \right). \qquad (3.37)
$$

This formula permits individual hex or binary digits of π to be calculated roughly 43% faster than (3.25).

A year later, Colin Percival, then a 17-year-old student at Simon Fraser University, utilized a network of 25 machines to calculate binary digits in the neighborhood of position 5 trillion, and then in the neighborhood of 40 trillion. In September 2000, he found that the quadrillionth binary digit is "0," based on a computation that required 250 CPU-years of run time, carried out using 1,734 machines in 56 countries. Table 3.4 gives some results known as of this writing.

One question that immediately arises in the wake of this discovery is whether or not there is a formula of this type and an associated computational scheme to compute individual *decimal* digits of π. Searches conducted by numerous researchers have been unfruitful. Now it appears that there is no nonbinary formula of this type—this is ruled out by a new result co-authored by one of the present authors (see Section 3.7) [73]. However, none of this removes the possibility that there exists some completely different approach that permits rapid computation of individual decimal digits

3.5. Unpacking the BBP Formula for Pi   125

Position	Hex Digits Beginning at This Position
10^6	26C65E52CB4593
10^7	17AF5863EFED8D
10^8	ECB840E21926EC
10^9	85895585A0428B
10^{10}	921C73C6838FB2
10^{11}	9C381872D27596
1.25×10^{12}	07E45733CC790B
2.5×10^{14}	E6216B069CB6C1

Table 3.4. Computed hexadecimal digits of π.

of π. Also, as we will see in the next section, there do exist formulas for certain other constants that admit individual digit calculation schemes in various nonbinary bases (including base ten).

3.5   Unpacking the BBP Formula for Pi

It is worth asking "why" the formula

$$\pi \;=\; \sum_{i=0}^{\infty} \frac{1}{16^i}\left(\frac{4}{8i+1} - \frac{2}{8i+4} - \frac{1}{8i+5} - \frac{1}{8i+6}\right) \qquad (3.38)$$

exists. As observed above, this identity is equivalent to, and can be proved by establishing:

$$\pi = \int_0^{1/\sqrt{2}} \frac{4\sqrt{2} - 8x^3 - 4\sqrt{2}x^4 - 8x^5}{1 - x^8}\,dx.$$

The present version of *Maple* evaluates this integral to

$$-2\log 2 + 2\log(2 - \sqrt{2}) + \pi + 2\log(2 + \sqrt{2}), \qquad (3.39)$$

which simplifies to π. In any event, one can ask what the individual series in (3.38) comprise. So consider

$$S_b \;=\; \sum_{k=0}^{\infty} \frac{1}{16^k(8k+b)}$$

126   3. Pi and Its Friends

for $1 \leq b \leq 8$ and the corresponding normalized integrals

$$I(b) \quad = \quad 2^{b/2} \int_0^{1/\sqrt{2}} \frac{x^{b-1}}{1 - x^8}\, dx. \tag{3.40}$$

Again, *Maple* provides closed forms for $I(b)$ in which the basic quantities seem to be the following: $\arctan(2), \arctan(1/2), \sqrt{2}\,\arctan(1/\sqrt{2})$, $\log(2), \log(3), \log(5)$, and $\log(\sqrt{2} \pm 1)$. At this point one may use integer relation methods and obtain:

$$S_1 = \frac{\pi}{8} + \frac{\log 5}{8} - \frac{\sqrt{2}\log(\sqrt{2}-1)}{4} - \frac{\arctan(1/2)}{4} + \frac{\sqrt{2}\arctan(\sqrt{2}/2)}{4}$$

$$S_2 = \frac{\log(3)}{4} + \frac{\arctan(1/2)}{2}$$

$$S_3 = \frac{\pi}{4} - \frac{\sqrt{2}\log(\sqrt{2}-1)}{2} - \frac{\arctan(1/2)}{2} - \frac{\sqrt{2}\arctan(\sqrt{2}/2)}{2} - \frac{\log 5}{4}$$

$$S_4 = \frac{\log 5}{2} - \frac{\log 3}{2}$$

$$S_5 = -\frac{\pi}{2} - \sqrt{2}\log(\sqrt{2}-1) + \arctan(1/2) + \sqrt{2}\arctan(\sqrt{2}/2) - \frac{\log 5}{2}$$

$$S_6 = \log 3 - 2\arctan(1/2)$$

$$S_7 = -\pi + \log 5 - 2\sqrt{2}\log(\sqrt{2}-1) + 2\arctan(1/2) - 2\sqrt{2}\arctan(\sqrt{2}/2)$$

$$S_8 = 8\log 2 - 2\log 5 - 2\log 3. \tag{3.41}$$

Thus the "simple" hexadecimal formula (3.38) is actually a molecule made up of more subtle hexadecimal atoms: with the final bond coming from the simple identity $\arctan 2 + \arctan(1/2) = \pi/2$. As an immediate consequence, one obtains the formula $\arctan(1/2) = S_2 - S_6/4$.

   Furthermore, the facts that

$$\mathrm{Im}\left(\log\left(1 - \frac{1-i}{x}\right)\right) \quad = \quad \arctan\left(\frac{1}{1-x}\right)$$

$$2\arctan(1/3) + \arctan(1/7) \quad = \quad \arctan(1/2) + \arctan(1/3)$$

$$= \quad \arctan 1 \;=\; \pi/4 \tag{3.42}$$

allow one to write directly a base-64 series for $\arctan(1/3)$ (using $x = 4$) and a base-1024 series for $\arctan(1/7)$ (using $x = 8$). This yields the identity

$$\frac{\pi}{4} \quad = \quad \frac{1}{16}\sum_{n=0}^{\infty} \frac{(-1)^n}{64^n}\left(\frac{8}{4n+1} + \frac{4}{4n+2} + \frac{1}{4n+3}\right)$$

$$+ \quad \frac{1}{256}\sum_{n=0}^{\infty}\frac{(-1)^n}{1024^n}\left(\frac{32}{4n+1} + \frac{8}{4n+2} + \frac{1}{4n+3}\right), \tag{3.43}$$

which is similar to, although distinct from, the identity used by Bellard and Percival in their computations.

3.6 Other BBP-Type Formulas

A formula of the type mentioned in the previous sections, namely

$$\alpha \;\; = \;\; \sum_{k=0}^{\infty} \frac{p(k)}{b^k q(k)}, \tag{3.44}$$

is now referred to as a BBP-type formula, named after the initials of the authors of the 1997 paper where the π hex digit algorithm appeared [33]. For a constant α given by a formula of this type, it is clear that individual base-b digits can be calculated, using the scheme similar to the ones outlined in the previous section. The paper [33] includes formulas of this type for several other constants. Since then, a large number of other BBP-type formulas have been discovered.

Most of these identities were discovered using an experimental approach, using PSLQ searches. Others were found as the result of educated guesses based on experimentally obtained results. In each case, these formulas have been formally established, although the proofs are not always as simple as the proof of Theorem 3.1. We present these results, in part, to underscore the fact that the approach used to find the new formula for π has very broad applicability.

A sampling of the known binary BBP-type formulas (i.e., formulas with a base $b = 2^p$ for some integer p) is shown in Table 3.5. Some nonbinary BBP-type formulas are shown in Table 3.6. These formulas are derived from several sources: [33,93,94]. An updated collection is available at [19]. The constant G that appears in Table 3.5 is Catalan's constant, namely $G = 1 - 1/3^2 + 1/5^2 - 1/7^2 + \cdots = 0.9159655941\ldots$

In addition to the formulas in Tables 3.5 and 3.6, there are two other classes of constants known to possess binary BBP-type formulas. The first is logarithms of certain integers. Clearly, $\log n$ can be written with a binary BBP formula (i.e. a formula with $b = 2^m$ for some integer m) provided n factors completely using primes whose logarithms have binary BBP formulas—one merely combines the individual series for the different primes into a single binary BBP formula. We have seen that the logarithm of the prime 2 possesses a binary BBP formula, and so does $\log 3$, by the following reasoning:

$$\log 3 \;=\; 2\log 2 + \log\left(1 - \frac{1}{4}\right) = 2\sum_{k=1}^{\infty}\frac{1}{k2^k} - \sum_{k=1}^{\infty}\frac{1}{k4^k}$$

$$=\; \frac{1}{2}\sum_{k=0}^{\infty}\frac{1}{4^k}\left(\frac{2}{2k+1} + \frac{1}{2k+2}\right) - \frac{1}{4}\sum_{k=0}^{\infty}\frac{1}{4^k}\left(\frac{2}{2k+2}\right)$$

$$=\; \sum_{k=0}^{\infty}\frac{1}{4^k}\left(\frac{1}{2k+1}\right). \tag{3.46}$$

$$\pi\sqrt{3} \;=\; \frac{9}{32}\sum_{k=0}^{\infty}\frac{1}{64^k}\left(\frac{16}{6k+1} + \frac{8}{6k+2} - \frac{2}{6k+4} - \frac{1}{6k+5}\right)$$

$$\pi^2 \;=\; \frac{9}{8}\sum_{k=0}^{\infty}\frac{1}{64^k}\left(\frac{16}{(6k+1)^2} - \frac{24}{(6k+2)^2} - \frac{8}{(6k+3)^2} - \frac{6}{(6k+4)^2}\right.$$
$$\left. + \frac{1}{(6k+5)^2}\right)$$

$$\log^2 2 \;=\; \frac{1}{32}\sum_{k=0}^{\infty}\frac{1}{64^k}\left(\frac{64}{(6k+1)^2} - \frac{160}{(6k+2)^2} - \frac{56}{(6k+3)^2} - \frac{40}{(6k+4)^2}\right.$$
$$\left. + \frac{4}{(6k+5)^2} - \frac{1}{(6k+6)^2}\right)$$

$$\pi\log 2 \;=\; \frac{1}{256}\sum_{k=0}^{\infty}\frac{1}{4096^k}\left(\frac{4096}{(24k+1)^2} - \frac{8192}{(24k+2)^2} - \frac{26112}{(24k+3)^2} + \frac{15360}{(24k+4)^2}\right.$$
$$- \frac{1024}{(24k+5)^2} + \frac{9984}{(24k+6)^2} + \frac{11520}{(24k+8)^2} + \frac{2368}{(24k+9)^2} - \frac{512}{(24k+10)^2}$$
$$+ \frac{768}{(24k+12)^2} - \frac{64}{(24k+13)^2} + \frac{408}{(24k+15)^2} + \frac{720}{(24k+16)^2}$$
$$\left. + \frac{16}{(24k+17)^2} + \frac{196}{(24k+18)^2} + \frac{60}{(24k+20)^2} - \frac{37}{(24k+21)^2}\right)$$

$$G \;=\; \frac{1}{1024}\sum_{k=0}^{\infty}\frac{1}{4096^k}\left(\frac{3072}{(24k+1)^2} - \frac{3072}{(24k+2)^2} - \frac{23040}{(24k+3)^2} + \frac{12288}{(24k+4)^2}\right.$$
$$- \frac{768}{(24k+5)^2} + \frac{9216}{(24k+6)^2} + \frac{10368}{(24k+8)^2} + \frac{2496}{(24k+9)^2} - \frac{192}{(24k+10)^2}$$
$$+ \frac{768}{(24k+12)^2} - \frac{48}{(24k+13)^2} + \frac{360}{(24k+15)^2} + \frac{648}{(24k+16)^2}$$
$$\left. + \frac{12}{(24k+17)^2} + \frac{168}{(24k+18)^2} + \frac{48}{(24k+20)^2} - \frac{39}{(24k+21)^2}\right)$$

Table 3.5. Binary BBP-type formulas.

3.6. Other BBP-Type Formulas　　　　129

$$\log 2 = \frac{2}{3} \sum_{k=0}^{\infty} \frac{1}{9^k (2k+1)}$$

$$\pi\sqrt{3} = \frac{1}{9} \sum_{k=0}^{\infty} \frac{1}{729^k} \left(\frac{81}{12k+1} - \frac{54}{12k+2} - \frac{9}{12k+4} - \frac{12}{12k+6} \right.$$
$$\left. - \frac{3}{12k+7} - \frac{2}{12k+8} - \frac{1}{12k+10} \right)$$

$$\log 3 = \frac{1}{729} \sum_{k=0}^{\infty} \frac{1}{729^k} \left(\frac{729}{6k+1} + \frac{81}{6k+2} + \frac{81}{6k+3} + \frac{9}{6k+4} \right.$$
$$\left. + \frac{9}{6k+5} + \frac{1}{6k+6} \right)$$

$$\pi^2 = \frac{2}{27} \sum_{k=0}^{\infty} \frac{1}{729^k} \left(\frac{243}{(12k+1)^2} - \frac{405}{(12k+2)^2} - \frac{81}{(12k+4)^2} \right.$$
$$- \frac{27}{(12k+5)^2} - \frac{72}{(12k+6)^2} - \frac{9}{(12k+7)^2} - \frac{9}{(12k+8)^2}$$
$$\left. - \frac{5}{(12k+10)^2} + \frac{1}{(12k+11)^2} \right)$$

$$\log\left(\frac{9}{10}\right) = \frac{-1}{10} \sum_{k=1}^{\infty} \frac{1}{k 10^k}$$

$$\log\left(\frac{1111111111}{387420489}\right) = \frac{1}{10^8} \sum_{k=0}^{\infty} \frac{1}{10^{10k}} \left(\frac{10^8}{10k+1} + \frac{10^7}{10k+2} + \frac{10^6}{10k+3} \right.$$
$$+ \frac{10^5}{10k+4} + \frac{10^4}{10k+5} + \frac{10^3}{10k+6} + \frac{10^2}{10k+7}$$
$$\left. + \frac{10}{10k+8} + \frac{1}{10k+9} \right)$$

$$\frac{25}{2} \log\left(\frac{781}{256} \left(\frac{57 - 5\sqrt{5}}{57 + 5\sqrt{5}} \right)^{\sqrt{5}} \right) = \sum_{k=0}^{\infty} \frac{1}{5^{5k}} \left(\frac{5}{5k+2} + \frac{1}{5k+3} \right) \tag{3.45}$$

Table 3.6. Nonbinary BBP-type formulas.

In a similar manner, it can be shown, by examining the factorization of $2^n + 1$ and $2^n - 1$, where n is an integer, that numerous other primes have this property. Some additional primes can be obtained by noting that the real part of the Taylor series expansion of

$$\alpha = \log\left(1 \pm \frac{(1+i)^k}{2^n} \right) \tag{3.47}$$

yields a BBP-type formula. See [19] for details.

130   3. Pi and Its Friends

The logarithms of the following primes are now known to possess binary BBP formulas [103]:

2, 3, 5, 7, 11, 13, 17, 19, 29, 31, 37, 41, 43, 61, 73, 109, 113, 127, 151,
241, 257, 331, 337, 397, 683, 1321, 1429, 1613, 2113, 2731, 5419, 8191,
14449, 26317, 38737, 43691, 61681, 65537, 87211, 131071, 174763,
246241, 262657, 268501, 279073, 312709, 524287, 525313, 599479,
2796203, 4327489, 7416361, 15790321, 18837001, 22366891   (3.48)

This list is certainly not complete, and it is unknown whether or not all primes have this property, or even whether the list of such primes is finite or infinite. One can also obtain BBP-type formulas in nonbinary bases for the logarithms of certain integers and rational numbers. One example is given by the base ten formula for $\log(9/10)$ in Table 3.6. This has been used to compute the ten billionth decimal digit of $\log(9/10)$ [33].

One additional class of binary BBP-type formulas that we will mention here is arctangents of certain rational numbers. We present here the results of experimental searches, using the PSLQ integer relation algorithm, which we have subsequently established formally. The formal derivation of these results proceeds as follows. Consider the set of rationals given by $q = |\mathrm{Im}(T)/\mathrm{Re}(T)|$ or $|\mathrm{Re}(T)/\mathrm{Im}(T)|$, where

$$T = \prod_{k=1}^{m} \left(1 \pm \frac{i}{2^{t_k}}\right)^{u_k} \left(1 \pm \frac{1+i}{2^{v_k}}\right)^{w_k} \qquad (3.49)$$

for various m-long nonnegative integer vectors t, u, v, w and choices of signs as shown [74, pg. 344]. For example, setting $t = (1,1)$, $u = (1,1)$, $v = (1,3)$, $w = (1,1)$, with signs $(1,-1,-1,1)$, gives the result $T = 25/32 - 5i/8$, which yields $q = 4/5$. Indeed, one can obtain the formula

$$\arctan\left(\frac{4}{5}\right) = \frac{1}{2^{17}} \sum_{k=0}^{\infty} \frac{1}{2^{20k}} \left(\frac{524288}{40k+2} - \frac{393216}{40k+4} - \frac{491520}{40k+5} + \frac{163840}{40k+8} \right.$$
$$+ \frac{32768}{40k+10} - \frac{24576}{40k+12} + \frac{5120}{40k+15} + \frac{10240}{40k+16}$$
$$+ \frac{2048}{40k+18} + \frac{1024}{40k+20} + \frac{640}{40k+24} + \frac{480}{40k+25}$$
$$+ \frac{128}{40k+26} - \frac{96}{40k+28} + \frac{40}{40k+32} + \frac{8}{40k+34}$$
$$\left. - \frac{5}{40k+35} - \frac{6}{40k+36} \right). \qquad (3.50)$$

The set of rationals for which BBP formulas can be obtained in this way can be further expanded by applying the formula

$$\tan(r + s) \; = \; \frac{\tan r + \tan s}{1 - \tan r \tan s}, \tag{3.51}$$

for rationals r and s for which binary BBP-type formulas are found. By applying these methods, it can be shown that binary BBP formulas exist for the arctangents of the following rational numbers. Only those rationals with numerators $<$ denominators ≤ 25 are listed here.

1/2, 1/3, 2/3, 1/4, 3/4, 1/5, 2/5, 3/5, 4/5, 1/6, 5/6, 1/7, 3/7, 4/7, 5/7, 6/7, 1/8, 7/8, 1/9, 2/9, 7/9, 8/9, 3/10, 1/11, 2/11, 3/11, 7/11, 8/11, 10/11, 1/12, 5/12, 1/13, 4/13, 6/13, 7/13, 9/13, 11/13, 12/13, 3/14, 5/14, 1/15, 4/15, 8/15, 1/16, 7/16, 11/16, 13/16, 15/16, 1/17, 4/17, 6/17, 7/17, 9/17, 11/17, 15/17, 16/17, 1/18, 13/18, 3/19, 4/19, 6/19, 7/19, 8/19, 9/19, 11/19, 17/19, 9/20, 1/21, 13/21, 16/21, 20/21, 3/22, 7/22, 9/22, 19/22, 21/22, 2/23, 4/23, 6/23, 7/23, 9/23, 10/23, 11/23, 14/23, 15/23, 7/24, 11/24, 23/24, 1/25, 2/25, 13/25, 19/25, 21/25   (3.52)

Note that not all "small" rationals appear in this list. As it turns out, by applying the methods given in the paper [73] (see the next section), one can rule out the possibility of Machin-type BBP formulas (as described in Section 3.6) for the arctangents of 2/7, 3/8, 5/8, 4/9, and 5/9. Thus we believe the above list to be complete for rationals with numerators and denominators up to ten. Beyond this level, we do not know for sure whether this list is complete, or whether applying formula (3.49), together with addition and subtraction formulas, generates all possible rationals possessing binary BBP-type formulas.

One can obtain BBP formulas in nonbinary bases for the arctangents of certain rational numbers by employing an appropriate variant of formula (3.49).

3.7 Does Pi Have a Nonbinary BBP Formula?

As we mentioned above, from the day that the BBP-formula for π was discovered, many researchers have wondered whether there exist BBP-type formulas that would permit computation of individual digits in bases other

132   3. Pi and Its Friends

than powers of two (such as base ten). This is not such a far-fetched possibility, because both base-2 and base-3 formulas are known for π^2, as well as for $\log 2$ (see Tables 3.5 and 3.6). But extensive computations failed to find any nonbinary formulas for π.

Recently one of the present authors, together with David Borwein (Jon's father) and William Galway, established that there are no nonbinary Machin-type arctangent formulas for π. We believe that if there is no nonbinary Machin-type arctangent formula for π, then there is no nonbinary BBP-type formula of any form for π. We will summarize this result here. Full details and other related results can be found in [73].

We say that the integer $b > 1$ is not a proper power if it cannot be written as c^m for any integers c and $m > 1$. We will use the notation $\mathrm{ord}_p(z)$ to denote the p-adic order of the rational $z \in Q$. In particular, $\mathrm{ord}_p(p) = 1$ for prime p, while $\mathrm{ord}_p(q) = 0$ for primes $q \neq p$, and $\mathrm{ord}_p(wz) = \mathrm{ord}_p(w) + \mathrm{ord}_p(z)$. The notation $\nu_b(p)$ will mean the order of the integer b in the multiplicative group of the integers modulo p. We will say that p is a primitive prime factor of $b^m - 1$ if m is the least integer such that $p|(b^m - 1)$. Thus p is a primitive prime factor of $b^m - 1$ provided $\nu_b(p) = m$. Given the Gaussian integer $z \in Q[i]$ and the rational prime $p \equiv 1 \pmod 4$, let $\theta_p(z)$ denote $\mathrm{ord}_{\mathfrak{p}}(z) - \mathrm{ord}_{\overline{\mathfrak{p}}}(z)$, where $\mathfrak{p}$ and $\overline{\mathfrak{p}}$ are the two conjugate Gaussian primes dividing p, and where we require $0 < \Im(\mathfrak{p}) < \mathrm{R}(\mathfrak{p})$ to make the definition of θ_p unambiguous. Note that

$$\theta_p(wz) \quad = \quad \theta_p(w) + \theta_p(z). \tag{3.53}$$

Given $\kappa \in R$, with $2 \leq b \in Z$ and b not a proper power, we say that κ has a Z-linear or Q-linear *Machin-type BBP arctangent formula* to the base b if and only if κ can be written as a Z-linear or Q-linear combination (respectively) of generators of the form

$$\arctan\left(\frac{1}{b^m}\right) \quad = \quad \Im \log\left(1 + \frac{i}{b^m}\right) = b^m \sum_{k=0}^{\infty} \frac{(-1)^k}{b^{2mk}(2k+1)}. \tag{3.54}$$

We will also use the following theorem, first proved by Bang in 1886:

Theorem 3.5. *The only cases where $b^m - 1$ has no primitive prime factor(s) are when $b = 2$, $m = 6$, $b^m - 1 = 3^2 \cdot 7$; and when $b = 2^N - 1$, $N \in Z$, $m = 2$, $b^m - 1 = 2^{N+1}(2^{N-1} - 1)$.*

We can now state the main result of this section:

Theorem 3.6. *Given $b > 2$ and not a proper power, then there is no Q-linear Machin-type BBP arctangent formula for π.*

3.8. Commentary and Additional Examples   133

Proof. It follows immediately from the definition of a Q-linear Machin-type BBP arctangent formula that any such formula has the form

$$\pi \;=\; \frac{1}{n}\sum_{m=1}^{M} n_m \Im \log(b^m - i) \tag{3.55}$$

where $n > 0 \in Z$, $n_m \in Z$, and $M \geq 1$, $n_M \neq 0$. This implies that

$$\prod_{m=1}^{M}(b^m - i)^{n_m} \in e^{ni\pi}Q^{\times} \;=\; Q^{\times} \tag{3.56}$$

For any $b > 2$ and not a proper power we have $M_b \leq 2$, so it follows from Bang's Theorem that $b^{4M} - 1$ has a primitive prime factor, say p. Furthermore, p must be odd, since $p = 2$ can only be a *primitive* prime factor of $b^m - 1$ when b is odd and $m = 1$. Since p is a primitive prime factor, it does not divide $b^{2M} - 1$, and so p must divide $b^{2M} + 1 = (b^M + i)(b^M - i)$. We cannot have both $p|b^M + i$ and $p|b^M - i$, since this would give the contradiction that $p|(b^M + i) - (b^M - i) = 2i$. It follows that $p \equiv 1$ (mod 4), and that p factors as $p = \mathfrak{p}\bar{\mathfrak{p}}$ over $Z[i]$, with exactly one of $\mathfrak{p}$, $\bar{\mathfrak{p}}$ dividing $b^M - i$. Referring to the definition of θ, we see that we must have $\theta_p(b^M - i) \neq 0$. Furthermore, for any $m < M$, neither $\mathfrak{p}$ nor $\bar{\mathfrak{p}}$ can divide $b^m - i$ since this would imply $p \mid b^{4m} - 1$, $4m < 4M$, contradicting the fact that p is a primitive prime factor of $b^{4M} - 1$. So for $m < M$, we have $\theta_p(b^m - i) = 0$. Referring to equation (3.55), using Equation (3.53) and the fact that $n_M \neq 0$, we get the contradiction

$$0 \;\neq\; n_M\theta_p(b^M - i) \;=\; \sum_{m=1}^{M} n_m\theta_p(b^m - i) \;=\; \theta_p(Q^{\times}) = 0. \tag{3.57}$$

Thus our assumption that there was a b-ary Machin-type BBP arctangent formula for π must be false. $\qquad\qquad\square$

3.8 Commentary and Additional Examples

1. **The ENIAC Integrator and Calculator.** ENIAC, built in 1946 at the University of Pennsylvania, had 18,000 vacuum tubes, 6,000 switches, 10,000 capacitors, 70,000 resistors, 1,500 relays, was 10 feet tall, occupied 1,800 square feet, and weighed 30 tons. ENIAC could perform 5,000 arithmetic operations per second—1,000 times faster

Figure 3.3. The ENIAC computer. Courtesy of Smithsonian Institution.

than any earlier machine, but a far cry from today's leading-edge microprocessors, which can perform more than four billion operations per second. The first stored-memory computer, ENIAC could store 200 digits, which again is a far cry from the hundreds of megabytes in a modern personal computer system. Data flowed from one accumulator to the next, and after each accumulator finished a calculation, it communicated its results to the next in line. The accumulators were connected to each other manually. A photo is shown in Figure 3.3.

2. **Four approximations to pi.** Here are two well known, but fascinating, approximations to π:

$$\pi \approx \frac{3}{\sqrt{163}} \log{(640320)},$$

correct to 15 decimal places, and

$$\pi \approx \frac{3}{\sqrt{67}} \log{(5280)},$$

3.8. Commentary and Additional Examples   135

correct to 9 decimal places. Both rely on somewhat deeper number theory (see Section 1.4 in the second volume). Here are two nice algebraic π approximations:

$$\pi \approx 66 \, \frac{\sqrt{2}}{33 \sqrt{29} - 148}$$

and

$$\pi \approx \frac{63}{25} \, \frac{17 + 15\sqrt{5}}{7 + 15\sqrt{5}}.$$

3. **An arctan series for pi.** Find rational coefficients a_i such that the identity

$$
\begin{aligned}
\pi \;=\; & a_1 \arctan \frac{1}{390112} + a_2 \arctan \frac{1}{485298} \\
& + a_3 \arctan \frac{1}{683982} + a_4 \arctan \frac{1}{1984933} \\
& + a_5 \arctan \frac{1}{2478328} + a_6 \arctan \frac{1}{3449051} \\
& + a_7 \arctan \frac{1}{18975991} + a_8 \arctan \frac{1}{22709274} \\
& + a_9 \arctan \frac{1}{24208144} + a_{10} \arctan \frac{1}{201229582} \\
& + a_{11} \arctan \frac{1}{2189376182}
\end{aligned}
$$

holds [10, pg. 75]. Also show that an identity with even simpler coefficients exists if $\arctan 1/239$ is included as one of the terms on the RHS. *Hint*: Use an integer relation program (see Section 6.3), or try the tools at one of these sites: http://oldweb.cecm.sfu.ca/projects/IntegerRelations or http://www.experimentalmath.info.

4. **Ballantine's series for pi.** A formula of Euler for arccot is

$$x \sum_{n=0}^{\infty} \frac{(n!)^2 \, 4^n}{(2\,n+1)! \, (x^2+1)^{n+1}} \;=\; \arctan\left(\frac{1}{x}\right). \qquad (3.58)$$

As observed by Ballantine in 1939, ([49]) this allows one to rewrite the variant of Machin's formula, used by Guilloud and Bouyer in 1973 to compute a million digits of π,

$$\frac{\pi}{4} \;=\; 12 \arctan \frac{1}{18} + 8 \arctan \frac{1}{57} - 5 \arctan \frac{1}{239} \qquad (3.59)$$

in the neat form

$$\pi = 864 \sum_{n=0}^{\infty} \frac{(n!)^2 \, 4^n}{(2\,n+1)! \, 325^{n+1}} + 1824 \sum_{n=0}^{\infty} \frac{(n!)^2 \, 4^n}{(2\,n+1)! \, 3250^{n+1}}$$
$$- \; 20 \arctan \frac{1}{239}, \tag{3.60}$$

where the terms of the second series are just decimal shifts of the terms of the first.

5. **Convergence rates for pi formulas.** Analyze the rates of convergence of Archimedes iteration (3.1), the Salamin-Brent iteration (3.16), the Borwein cubic iteration (3.17) and the Borwein quartic iteration (3.18), by means of explicit computations. Use the high-precision arithmetic facility built into *Maple* or *Mathematica*, or write your own C++ or Fortran-90 code using the ARPREC arbitrary precision software available at http://www.experimentalmath.info, or the GNU multiprecision software available at http://www.gnu.org /software/gmp/gmp.html. Such iterations are discussed more in Sections 5.6.2 and 5.6.3.

6. **Biblical pi.** As noted in Section 3.1, the Biblical passages 1 Kings 7:23 and 2 Chronicles 4:2 indicate that $\pi = 3$. In spite of the fact the context of these verses clearly suggests an informal approximation, not a precise statement of mathematical fact, this discrepancy has been a source of consternation among Biblical literalists for centuries. For example, an 18th-century German Bible commentary attempted to explain away this discrepancy using the imaginative (if pathetic) suggestion that the circular pool in Solomon's temple (clearly described in 2 Chron. 4:2 as "round in compass") was instead hexagonal in shape [48, pg. 75–76]. Even today, some are still unwilling to accept that the Bible could simply be mistaken here. One evangelical scholar, for example, writes:

> However, the recorded dimensions are still no problem if we consider the shape of the vessel. In 1 Kings 7:26, we read that its "brim was made like the brim of a cup, as a lily blossom." Hence, the sea was not a regular cylinder, but had an outward curving rim. Although we do not know the exact points on the vessel where the measurements were taken, the main part of the sea always will be somewhat smaller than the 10 cubits measured "from brim to brim." [213]

3.8. Commentary and Additional Examples 137

Another geometric difficulty in these Biblical passages is that 1 Kings 7:26 (three verses after it gives its dimensions) gives a volume of 2,000 "baths" for the basin, while 2 Chron. 4:5 gives the figure 3,000 baths. 2 Chron. 4:2 gives the height of the basin as five cubits. Using the accepted conversions that one "cubit" is roughly 46 cm, and that one "bath" is roughly 23 liters, then assuming Solomon's pool was cylindrical in shape, we obtain an actual volume of roughly 1660 baths. If the basin was rounded on the bottom, then its volume was even lower than this.

7. **Exponentiation of pi.** Arguably the most accessible transcendental number to compute is e^{π}, which can be computed using the following iteration.

Algorithm 3.7. Computation of $\exp(\pi)$.

Set $k_0 = 1/\sqrt{2}$ and for $n < N = \lceil \log_2(D/1.36) \rceil$ iterate

$$k'_n = \sqrt{1 - k_n^2}, \qquad k_{n+1} = \frac{1 - k'_n}{1 + k'_n}.$$

Then return

$$\left(\frac{k_N}{4} \right)^{-1/2^{N-1}}.$$

Some care needs to be taken with guard digits. □

8. **Algorithms for Gamma values.** An algorithm for π may be viewed as an algorithm for $\Gamma\left(\frac{1}{2}\right)$, and there is a quite analogous iteration for Γ at the values 1/3, 2/3, 1/4, 3/4, 1/6, and 5/6. This, in turn, allows rapid computation of $\Gamma\left(\frac{k}{24}\right)$, for all integer k. We illustrate with:

Algorithm 3.8. Computation of $\Gamma\left(\frac{1}{4}\right)$.

Let $x_0 = 2^{1/2}, y_1 = 2^{1/4}$. Let

$$x_{n+1} = \frac{\sqrt{x_n} + 1/\sqrt{x_n}}{2} \tag{3.61}$$

$$y_{n+1} = \frac{y_n \sqrt{x_n} + 1/\sqrt{x_n}}{y_n + 1}. \tag{3.62}$$

3. Pi and Its Friends

Then

$$\Gamma^4\left(\frac{1}{4}\right) = 16\,(1+\sqrt{2})^3 \prod_{n=1}^{\infty} x_n^{-1} \left(\frac{1+x_n}{1+y_n}\right)^3.$$

This yields a quadratically convergent iteration for $\Gamma\left(\frac{1}{4}\right)$. $\square$

No such iteration is known for $\Gamma\left(\frac{1}{5}\right)$; see [68, 74].

9. **An integral representation of Euler's constant.** While it is known that $\Gamma\left(\frac{1}{3}\right)$ and $\Gamma\left(\frac{1}{4}\right)$ are transcendental, the status of *Euler s constant*

$$\gamma = \lim_{n\to\infty} \sum_{k=1}^{n} \frac{1}{k} - \log(n) \tag{3.63}$$

is unsettled.

Problem: Show that

$$\gamma = \int_0^{\infty} \left(\frac{1}{e^t - 1} - \frac{1}{t\,e^t}\right)\,dt.$$

10. **Computation of Euler's constant.** Perhaps the most efficient method of computation, due to Brent and MacMillan ([74, pg. 336]), is based on Bessel function identities. It allows one to show that if γ is rational it must have a denominator with millions of digits. The underlying identity, known to Euler, is

$$\gamma + \log(z/2) = \frac{S_0(z) - K_0(z)}{I_0(z)} \tag{3.64}$$

where $I_\nu(z) = \sum_{k=0}^{\infty} (z/2)^{2k+\nu}/(k!\,\Gamma(k+\nu+1))$, while $K_0(z) = \partial I_0(z)/\partial\nu$ and $S_0(z) = \sum_{k=0}^{\infty}(\sum_{j=1}^{k} 1/j)\,(z/2)^{2k}/(k!)^2$.

An algorithm follows from knowing the first terms of the asymptotic expansion for K_0 and I_0. It is

Algorithm 3.9. Computation of K_0 and I_0:

$A_0 = -\log(n), B_0 = 1, U_0 = A_0, V_0 = 1$, and for $k = 1, 2, \cdots$,

$$\begin{aligned} B_k &= B_{k-1}n^2/k^2, & A_k &= (A_{k-1}n^2/k + B_k)/k, \\ U_k &= U_{k-1} + A_k, & V_k &= V_{k-1} + B_k. \end{aligned} \tag{3.65}$$

3.8. Commentary and Additional Examples   139

Terminate when U_k and V_k no longer change, and return $\gamma \approx U_k/V_k$. With $\log(n)$ computed efficiently, this scheme takes $O(D)$ storage and approximately $2.07\,D$ steps to compute γ to D decimal places.   □

11. **Buffon's needle.** Suppose we have a lined sheet of paper, and a needle that is precisely as long as the distance between the lines. Compute the probability that the needle "thrown at random" on the sheet of paper will lie on a line. Answer: $2/\pi$. Although this is certainly not a good way to calculate π (millions of trials would be required to obtain just a few digits), it is an instructive example of how π arises in unlikely settings (see also the next two exercises). Some additional discussion of this problem, plus a computer-based tool that allows one to perform these trials, is available at the URL http://www.mste.uiuc.edu/reese/buffon/buffon.html.

12. **Putnam problem 1993-B3.** If two real numbers x and y are generated uniformly at random in $(0, 1)$, what is the probability that the nearest integer to x/y is even? *Hint*: Ignoring negligible events, for this to occur either $0 < x/y < 1/2$ or $(4n-1)/2 < x/y < (4n+1)/2$. The first occurs in a triangle of area $1/4$ and the subsequent in triangles of area $1/(4n-1) - 1/(4n+1)$. Now apply the Gregory-Leibniz formula. Answer: The probability is $(5 - \pi)/4 \approx 0.4646018$.

13. **Number-theory probabilities.** Prove (a) The probability that an integer is square-free is $6/\pi^2$. (b) The probability that two integers are relatively prime is also $6/\pi^2$. This is a good example of π appearing in a number-theory setting. See [163].

14. **The irrationality of pi.** We reproduce in extenso Ivan Niven's 1947 very concise proof that π is irrational [224].

    Let $\pi = a/b$, the quotient of positive integers. We define the polynomials
    $$f(x) = \frac{x^n(a - bx)^n}{n!}$$
    $$F(x) = f(x) - f^{(2)}(x) + f^{(4)}(x) - \cdots + (-1)^n f^{(2n)}(x),$$
    the positive integer n being specified later. Since $n!f(x)$ has integral coefficients and terms in x of degree not less than n, $f(x)$ and its derivatives $f^{(j)}(x)$ have integral values for $x = 0$; also for $x = \pi = a/b$, since $f(x) = f(a/b - x)$. By elementary calculus we have
    $$\frac{d}{dx}\{F'(x)\sin x - F(x)\cos x\} = F''(x)\sin x + F(x)\sin x = f(x)\sin x$$

and

$$\int_0^\pi f(x)\sin x\, dx = [F'(x)\sin x - F(x)\cos x]_0^\pi = F(\pi) + F(0). \ (3.66)$$

Now $F(\pi) + F(0)$ is an *integer*, since $f^{(j)}(0)$ and $f^{(j)}(\pi)$ are integers. But for $0 < x < \pi$,

$$0 < f(x)\sin x < \frac{\pi^n a^n}{n!},$$

so that the integral in (3.66) is *positive but arbitrarily small* for n sufficiently large. Thus (3.66) is false, and so is our assumption that π is rational. $\qquad\qquad\square$

This proof gives a good taste of the ingredients of more subtle irrationality and transcendence proofs.

15. **A spigot algorithm for e and pi.** A spigot method for a numerical constant is one that can produce digits one by one ("drop by drop") [239]. This is especially easy for e as carries are not a big issue.

   (a) The following algorithm, due to Rabinowitz and Wagon, generates successive digits of e. Initialize an array A of length $n + 1$ to 1. Then repeat the following $n - 1$ times: (a) multiply each entry in A by ten; (b) Starting from the right, reduce the i-th entry of A modulo $i + 1$, carrying the quotient of the division one place left. The final quotient produced is the next digit of e. This algorithm is based on the following formula, which is simply a restatement of $e = \sum 1/i!$.

   $$e = 1 + \frac{1}{1}\left(1 + \frac{1}{2}\left(1 + \frac{1}{3}\left(1 + \frac{1}{4}\left(1 + \frac{1}{5}(1 + \cdots)\right)\right)\right)\right).$$

   (b) Implement a parallel spigot algorithm for π, based on showing that:

   $$\pi = 2 + \frac{1}{3}\left(2 + \frac{2}{5}\left(2 + \frac{3}{7}\left(2 + \cdots \left(2 + \frac{k}{2k+1}\cdots\right)\right)\right)\right).$$

   The last term can be approximated by $2 + 4k/(2k + 1)$ where $k = \log_2(10)n$ to produce n digits of π "drop by drop." If one wishes to run the algorithm without a specified end, one must take more care.

3.8. Commentary and Additional Examples   141

One may view the iteration (4.7), which we will study in Chapter 4, as a spectacular (albeit unproven) spigot algorithm for π base 16.

16. **Wagon's BBP identity.** Determine the range of validity of the following identity, which is due to Stan Wagon:

$$\pi + 4\arctan z + 2\log\left(\frac{1 - 2z - z^2}{z^2 + 1}\right) =$$
$$\sum_{k=0}^{\infty} \frac{1}{16^k}\left(\frac{4(z+1)^{1+8k}}{1+8k} - \frac{2(z+1)^{4+8k}}{4+8k} - \frac{(z+1)^{5+8k}}{5+8k} - \frac{(z+1)^{6+8k}}{6+8k}\right).$$

17. **Monte Carlo calculation of pi.** Monte Carlo simulation was pioneered during the Manhattan project by Stanislaw Ulam and others, who recognized that this scheme permitted simulations beyond the reach of conventional methods on the systems then available. We illustrate here a Monte Carlo calculation of π, which is a poor method to compute π, but illustrative of this general class of computation. Nowadays, Monte Carlo methods are quite popular because they are well suited to parallel computation on systems such as "Beowulf" clusters.

   (a) Design and implement a Monte Carlo simulation for π, based on generating pairs of uniformly distributed numbers in the unit square and testing whether they lie inside the unit circle. Use the pseudorandom number generator $x_0 = 314159$ and $x_n = cx_{n-1} \bmod 2^{32}$, where $c = 5^9 = 1953125$. This generator is of the well known class of linear congruential generators and has period 2^{30} [188, pg. 21]. It can be easily implemented on a computer using IEEE 64-bit "double" datatype, since the largest integer that can arise here is less than 2^{53}. Variations with longer periods can easily be designed, although the implementation is not as convenient. The results of this generator are normalized, by 2^{32} in this case, to produce results in the unit interval.

   (b) Extend your program to run on a parallel computer system, with the property that your parallel program generates the same overall scheme of pseudorandom numbers, and thus gets the same result for π, as a serial implementation (you may for convenience assume that n, the total number of pseudorandom numbers generated, is evenly divisible by p, the number of processors). This is a very desirable feature of a parallel program, because it allows you to certify your parallel results by comparing them with a conventional single-processor run, and it permits you to take advantage of a range of system sizes. *Hint*: Design the program so that processor k (where processors are numbered from 0 to $p - 1$) generates the $m = n/p$ members of the sequence $(x_{km}, x_{km+1}, x_{km+2}, \cdots, x_{km+m-1})$. Note that the starting value x_{km} for processor k can be directly computed as $x_{km} = 5^{9km} x_0 \bmod 2^{32}$. This exponentiation modulo 2^{32} may be performed by using Algorithm 3.2, implemented with 128-bit floating-point or "double-double" arithmetic (see Section 6.2.1).

(c) Generate the following sequence of pseudorandom numbers and experimentally determine what distribution they satisfy (i.e., by computing means, standard deviations, graphs, etc.): Let x_1 and x_2 be a pair of uniform $(0, 1)$ pseudorandom numbers generated as described above. Set $v = \sqrt{x_1^2 + x_2^2}$ and $w = \sqrt{-2 \log v / v}$. Then produce the results $y_1 = w x_1$ and $y_2 = w x_2$.

18. **Life of Pi.** At the end of his story, Piscine (Pi) Molitor [215, pp.316–7] writes

> I am a person who believes in form, in harmony of order. Where we can, we must give things a meaningful shape. For example—I wonder—could you tell my jumbled story in exactly one hundred chapters, not one more, not one less? I'll tell you, that's one thing I hate about my nickname, the way that number runs on forever. It's important in life to conclude things properly. Only then can you let go.

We may not share the sentiment, but we should celebrate that Pi knows π to be irrational.

4 | Normality of Numbers

Anyone who wants to make a name for himself can examine the major
issue of whether π is *normal*, or perhaps more accurately, whether π
is *not normal*.

> – Jörg Arndt and Christoph Haenel, *Pi Unleashed*, 2001

In this chapter, we address a fundamental problem of mathematics, a para-
dox of sorts: Whereas on one hand it can be proven that "almost all" real
numbers are normal, and whereas it appears from experimental analysis
that many of the fundamental constants of mathematics are normal to
commonly used number bases, as yet there are no proofs, nor even any
solid reason why we should observe this behavior. What we shall show
here is that the theory of BBP constants, which as we have seen is a classic
case study of experimental mathematics in action, opens a pathway into the
investigation of normality, and in fact has already yielded some intriguing
results.

4.1   Normality: A Stubborn Question

Given a real number α and an integer $b > 2$, we say that α is *b-normal*
or *normal base b* if *every* sequence of k consecutive digits in the base-b
expansion of α appears with limiting frequency b^{-k}. In other words, if
a constant is 10-normal, then the limiting frequency of "3" (or any other
single digit) in its decimal expansion is 1/10, the limiting frequency of
"58" (or any other two-digit pair) is 1/100, and so forth. We say that a
real number α is *absolutely normal* if it is b-normal for all integers $b > 1$
simultaneously.

In spite of these strong conditions, it is well known from measure theory that the set of absolutely normal real numbers in the unit interval has
measure one, or in other words, that almost all real numbers are absolutely normal (see Exercise 1 at the end of this chapter). Further, from

numerous analyses of computed digits, it appears that many of the fundamental constants of mathematics are normal to commonly used number bases. By "fundamental constants," we include π, e, $\sqrt{2}$, the golden mean $\tau = (1 + \sqrt{5})/2$, as well as $\log n$ and the Riemann zeta function $\zeta(n)$ for positive integers $n > 1$, and many others. For example, it is a reasonable conjecture that *every* irrational algebraic number is absolutely normal, since there is no known example of an irrational algebraic number whose decimal expansion (or expansion in any other base) appears to have skewed digit-string frequencies.

Decimal values are given for a variety of well known mathematical constants in Table 4.1 [99, 142]. In addition to the widely recognized constants such as π and e, we have listed Catalan's constant (G), Euler's constant (γ), an evaluation of the elliptic integral of the first kind $K(1/\sqrt{2})$, an evaluation of an elliptic integral of the second kind $E(1/\sqrt{2})$, Feigenbaum's α and δ constants, Khintchine's constant $\mathcal{K}$, and Madelung's constant $\mathcal{M}_3$. Binary values for some of these constants, as well as Chaitin's Ω constant (from the field of computational complexity) [99], are given in Table 4.2. As you can see, none of the expansions in either table exhibits any evident "pattern."

The digits of π have been studied more than any other single constant, in part because of the widespread fascination with π. Along this line, Yasumasa Kanada of the University of Tokyo has tabulated the number of occurrences of the ten decimal digits "0" through "9" in the first one trillion decimal digits of π. These counts are shown in Table 4.3. For reasons given in Section 3.4, binary (or hexadecimal) digits of π are also of considerable interest. To that end, Kanada has also tabulated the number of occurrences of the 16 hexadecimal digits "0" through "F," as they appear in the first one trillion hexadecimal digits. These counts are shown in Table 4.4. As you can see, both the decimal and hexadecimal single-digit counts are entirely reasonable.

Some readers may be amused by the LBNL PiSearch utility, which is available at http://pisearch.lbl.gov. This online tool permits one to enter one's name (or any other modest-length alphabetic string, or any modest-length hexadecimal string) and see if it appears encoded in the first several billion binary digits of π. Along this line, a graphic based on a random walk of the first million decimal digits of π, courtesy of David and Gregory Chudnovsky, is shown in Figure 4.1 (see Color Plate VIII). It maps the digit stream to a surface in ways similar to those used by Mandelbrot and others.

As we mentioned in Section 3.2, the question of whether π, in particular, or, say, $\sqrt{2}$, is normal or not has intrigued mathematicians for centuries. But in spite of centuries of effort, not a single one of the fundamental

4.1. Normality: A Stubborn Question   145

Constant	Value
$\sqrt{2}$	$1.4142135623730950488\ldots$
$\sqrt{3}$	$1.7320508075688772935\ldots$
$\sqrt{5}$	$2.2360679774997896964\ldots$
$\phi = \frac{\sqrt{5}-1}{2}$	$0.6180339887498948482\ldots$
π	$3.1415926535897932384\ldots$
$1/\pi$	$0.3183098861837906715\ldots$
e	$2.7182818284590452353\ldots$
$1/e$	$0.3678794411714423215\ldots$
e^{π}	$23.140692632779269005\ldots$
$\log 2$	$0.6931471805599453094\ldots$
$\log 10$	$2.3025850929940456840\ldots$
$\log_2 10$	$3.3219280948873623478\ldots$
$\log_{10} 2$	$0.3010299956639811952\ldots$
$\log_2 3$	$1.5849625007211561814\ldots$
$\zeta(2)$	$1.6449340668482264364\ldots$
$\zeta(3)$	$1.2020569031595942854\ldots$
$\zeta(5)$	$1.0369277551433699263\ldots$
G	$0.9159655941772190150\ldots$
γ	$0.5772156649015328606\ldots$
$\Gamma(1/2) = \sqrt{\pi}$	$1.7724538509055160272\ldots$
$\Gamma(1/3)$	$2.6789385347077476336\ldots$
$\Gamma(1/4)$	$3.6256099082219083119\ldots$
$K(1/\sqrt{2})$	$1.8540746773013719184\ldots$
$E(1/\sqrt{2})$	$1.3506438810476755025\ldots$
α_f	$4.6692016091029906718\ldots$
δ_f	$2.5029078750958928222\ldots$
$\mathcal{K}$	$2.6854520010653064453\ldots$
$\mathcal{M}_3$	$1.7475645946331821903\ldots$

Table 4.1. Decimal values of various mathematical constants.

Constant	Value
π	$11.0010010000111111011010101000100010000101101000110000010001\ldots$
e	$10.1011011111000010101000101100010100010101110110100101010100\ldots$
$\sqrt{2}$	$1.0110101000001001111001100110011111111001110111001100100100\ldots$
$\sqrt{3}$	$1.1011101101100111101011101000010110000100110010101010011100\ldots$
$\log 2$	$0.1011000101110010000101111111011111101000111001111011110011 0\ldots$
$\log 3$	$1.0001100100111110101001111010101011010000001100001010100101\ldots$
Ω	$0.0000001000000100001000001000011101110011001001111000100100\ldots$

Table 4.2. Binary values of various mathematical constants.

4. Normality of Numbers

Digit	Occurrences
0	99999485134
1	99999945664
2	100000480057
3	99999787805
4	100000357857
5	99999671008
6	99999807503
7	99999818723
8	100000791469
9	99999854780
Total	1000000000000

Table 4.3. Statistics for the first trillion decimal digits of π.

Digit	Occurrences
0	62499881108
1	62500212206
2	62499924780
3	62500188844
4	62499807368
5	62500007205
6	62499925426
7	62499878794
8	62500216752
9	62500120671
A	62500266095
B	62499955595
C	62500188610
D	62499613666
E	62499875079
F	62499937801
Total	1000000000000

Table 4.4. Statistics for the first trillion hexadecimal digits of π.

4.1. Normality: A Stubborn Question

Figure 4.1. A random walk based on one million digits of π (see Color Plate VIII). Courtesy of David and Gregory Chudnovsky.

constants of mathematics has ever been proven to be b-normal for any integer b, much less for all integer bases simultaneously. And this is not for lack of trying—some very good mathematicians have seriously investigated this problem, but to no avail. Even much weaker results, such as the digit "1" appears with nonzero limiting frequency in the binary expansion of π, and the digit "5" appears infinitely often in the decimal expansion of $\sqrt{2}$, have heretofore remained beyond the reach of modern mathematics.

One result in this area is the following. Let $f(n) = \sum_{1 \le j \le n} \lfloor \log_{10} j \rfloor$. Then the Champernowne number,

$$\sum_{n=1}^{\infty} \frac{n}{10^{n+f(n)}} = 0.12345678910111213141516171819202122232425\ldots,$$

where the positive integers are concatenated in a decimal value, is known to be 10-normal (See Exercise 5). There are similar constants and normality results for other number bases. However, no one, to the authors' knowledge, has ever argued that this constant and its relatives are "natural" or "fundamental" constants.

Consequences of a proof in this area would definitely be interesting. For starters, such a proof would immediately provide an inexhaustible source of provably reliable pseudorandom numbers for numerical or scientific experimentation. We also would obtain the mind-boggling but uncontestable consequence that if π, for example, is shown to be 2-normal, then the entire text of the Bible, the Koran and the works of William Shakespeare, as well as the full LATEX source text for this book, must all be contained somewhere in the binary expansion of π, where consecutive blocks of eight bits (two hexadecimal digits) each represent one ASCII character. Unfortunately, this would not be much help to librarians or archivists, since every conceivable misprint of each of these books would also be contained in the binary digits of π.

Before continuing, we should mention the "first digit" principle, also known as Benford's principle. In the 1880s, Simon Newcomb observed a pattern in the first digits of logarithm tables: A "1" is significantly more likely to occur than "2," a "2" more than a "3," and so on. In other words, the collection of first digits of data in logarithm tables certainly does not reflect the statistics expected of 10-normal numbers. In the 20th century, Frank Benford rediscovered this phenomenon, noting that it applies to many types of numerical data, ranging from values of physical constants to census data to the stock market. One can deduce this principle by observing that natural laws surely cannot be dependent on our choice of units, and thus must be scale-independent. This suggests that we view numerical data on a logarithmic scale. In the logarithmic sense, a leading "1" appears roughly 30% of the time (since $\log_{10} 2 - \log_{10} 1 = 0.30102999\ldots$), a "2" appears roughly 17.6% of the time (since $\log_{10} 3 - \log_{10} 2 = 0.1760912\ldots$), and so on. More recently scientists have applied Benford's principle in diverse ways, including fraud detection in business accounting [172].

4.2   BBP Constants and Normality

Until recently, the BBP formulas mentioned in Sections 3.4 and 3.6 were assigned by some to the realm of "recreational" mathematics—interesting but of no serious consequence. But the history of mathematics has seen many instances where results once thought to be idle curiosities were later found to have significant consequences. This now appears to be the case with the theory of BBP-type constants.

What we shall establish below, in a nutshell, is that the 16-normality of π (which, of course, is equivalent to the 2-normality of π), as well as the normality of numerous other irrational constants that possess BBP-type formulas, can be reduced to a certain plausible conjecture in the theory of chaotic sequences. At this time we do not know the full implications of this result. It may be the first salvo in the resolution of this age-old mathematical question, or it may be merely a case of reducing one very difficult mathematical problem to another. But at the least, this result

18. Approximations to π derived from integrals with nonnegative integrands (2009)

Paper 18: Stephen K. Lucas, "Approximations to π derived from integrals with nonnegative integrands," *American Mathematical Monthly*, vol. 116 (2009), p. 166–172. Copyright 2009 Mathematical Association of America. All Rights Reserved.

Synopsis:

In this article, Stephen Lucas addresses the classic grade school approximation $\pi \approx 22/7$. He shows how this venerable approximation can be seen in context with approximations based on integrals, such as

$$\int_0^1 \frac{x^4(1-x)^4}{1+x^2}\mathrm{d}x = \frac{22}{7} - \pi.$$

Lucas then exhibits other integrals that yield even better approximations, such as

$$\int_0^1 \frac{x^8(1-x)^8(25+816x^2)}{3164(1+x^2)}\mathrm{d}x = \frac{355}{113} - \pi.$$

Keywords: Approximations

NOTES

Edited by **Ed Scheinerman**

Approximations to π Derived from Integrals with Nonnegative Integrands

Stephen K. Lucas

One of the more intriguing results related to approximating π is the integral

$$I = \int_0^1 \frac{x^4(1-x)^4}{1+x^2}\,dx = \frac{22}{7} - \pi. \tag{1}$$

Since the integrand is nonnegative and "small" (as we shall shortly see) on the interval $[0, 1]$, it shows that π is smaller than $22/7$, and that this rational approximation is inded a good approximation to it.

The earliest statement of (1) that we are aware of is Dalzell [6] in 1944. Proving (1) was a question on a University of Sydney examination in November 1960 (Borwein et al. [4]), and it was apparently shown by Kurt Mahler to his students in the mid 1960s. Proving (1) was also the first question in the William Lowell Putnam mathematical competition of December 1968, as published by McKay [10] in 1969. In 1971, Dalzell [7] again derived (1) in a larger work published in the Cambridge student journal *Eureka*. This paper is the one most often cited in connection with the result (1) (e.g., Backhouse [2] and Borwein et al. [4]). It was also presented without reference in Cornwell [5] in 1980. A more recent reference is Medina [11].

In this article we shall look at some features of this integral, including error bounds and a related series expansion. Then, we present a number of generalizations, including a new series approximation to π where each term adds as many digits of accuracy as desired. We conclude by presenting a number of related integral results for other continued fraction convergents of π.

1. THE CLASSIC INTEGRAL. Proving (1) is not difficult, if perhaps somewhat tedious. A partial fraction decomposition leads to

$$\frac{x^4(1-x)^4}{1+x^2} = x^6 - 4x^5 + 5x^4 - 4x^2 + 4 - \frac{4}{1+x^2}, \tag{2}$$

and (1) immediately follows by integration. An alternative is to use the substitution $x = \tan\theta$, leading to

$$\int_0^1 \frac{x^4(1-x)^4}{1+x^2}\,dx = \int_0^{\pi/4} \frac{\tan^4\theta(1-\tan\theta)^4}{\sec^2\theta}\sec^2\theta\,d\theta$$

$$= \int_0^{\pi/4} \tan^4\theta - 4\tan^5\theta + 6\tan^6\theta - 4\tan^7\theta + \tan^8\theta\,d\theta.$$

This can be solved using the recurrence relation ($\tan^n\theta = \tan^{n-2}\theta(\sec^2\theta - 1)$)

$$\int_0^{\pi/4} \tan^n\theta\,d\theta = \frac{1}{n-1} - \int_0^{\pi/4} \tan^{n-2}\theta\,d\theta,$$

 [Monthly 116

with

$$\int_0^{\pi/4} d\theta = \frac{\pi}{4}, \qquad \int_0^{\pi/4} \tan\theta\, d\theta = \ln\sqrt{2},$$

which returns the required result after some algebra. Of course, the simplest approach today is to verify (1) using a symbolic manipulation package.

1.1. Error estimation. As well as showing $22/7 > \pi$ we can use (1) to get bounds on the error. One approach, following Nield [**12**], is to note that since $x(1-x) \leq 1/4$ and $1 + x^2 \geq 1$ on [0, 1] with equality only at the endpoints, the integrand takes maximum value $(1/4)^4 = 1/256$. Combined with the fact that $22/7 - \pi$ is positive, we get the error bound

$$\frac{5625}{1792} = \frac{22}{7} - \frac{1}{256} < \pi < \frac{22}{7}.$$

However, a better bound can be found by noting that $1 < 1 + x^2 < 2$ for $0 < x < 1$, and $\int_0^1 x^4(1-x)^4\, dx = 1/630$ (as in Dalzell [**6, 7**] and Nield [**13**]). Then

$$\frac{1}{1260} < \frac{22}{7} - \pi < \frac{1}{630} \quad \text{or} \quad \frac{1979}{630} = \frac{22}{7} - \frac{1}{630} < \pi < \frac{22}{7} - \frac{1}{1260} = \frac{3959}{1260}.$$

The interval $[1979/630, 3959/1260]$ is of width 7.94×10^{-4}, and is not centered at π.

1.2. Series expansion. Dalzell [**6**] also provides a series expansion for π based upon (2), which is included in Borwein et al. [**4**] as an example. After gathering the two pieces with $1 + x^2$ as the denominator, we can write

$$\frac{1}{1+x^2} = \frac{x^6 - 4x^5 + 5x^4 - 4x^2 + 4}{4 + x^4(1-x)^4} \quad \text{or} \quad \frac{4}{1+x^2} = \frac{x^6 - 4x^5 + 5x^4 - 4x^2 + 4}{1 + x^4(1-x)^4/4}.$$

Integrating both sides between 0 and 1 and using the Taylor series expansion for $1/(1+t)$ leads to

$$\pi = \sum_{k=0}^{\infty} \left(-\frac{1}{4}\right)^k \int_0^1 (x^6 - 4x^5 + 5x^4 - 4x^2 + 4)x^{4k}(1-x)^{4k}\, dx. \qquad (3)$$

Applying integration by parts n times reducing the exponent of $(1-x)$, we have

$$\int_0^1 x^m(1-x)^n\, dx = \frac{m!\, n!}{(m+n+1)!}, \qquad (4)$$

where m and n are nonnegative integers, which when applied to (3) gives

$$\pi = \sum_{k=0}^{\infty} \left(-\frac{1}{4}\right)^k \left[\frac{(4k)!\,(4k+6)!}{(8k+7)!} - \frac{4(4k)!\,(4k+5)!}{(8k+6)!} + \frac{5(4k)!\,(4k+4)!}{(8k+5)!}\right.$$

$$\left. - \frac{4(4k)!\,(4k+2)!}{(8k+3)!} + \frac{4(4k)!^2}{(8k+1)!}\right]. \qquad (5)$$

The series (5) is equivalent to that derived in Dalzell [6, 7] and Borwein et al. [4], but written in a different form. There the sixth order polynomial in (3) is recognized to be unchanged when x is replaced by $1 - x$, and so additional algebra is performed to reformulate the integral in (3) as

$$\int_0^1 \left(3 + x(1 - x) - \frac{1}{2}x^2(1 - x)^2 - x^3(1 - x)^3 \right) x^{4k}(1 - x)^{4k}\, dx$$
$$= \frac{3(4k)!^2}{(8k + 1)!} + \frac{(4k + 1)!^2}{(8k + 3)!} - \frac{(4k + 2)!^2}{2(8k + 5)!} - \frac{(4k + 3)!^2}{(8k + 7)!}, \tag{6}$$

where we have applied (4) with $m = n$ several times. To show that the bracketed expression in (5) is equivalent to (6), it is easiest to factor both of them, and incidentally get a cleaner solution. In both cases, start by giving them the common denominator $(8k + 7)!$, and take out the common factor $(4k)!^2$. It quickly becomes apparent that $(4k + 1)(4k + 2)(4k + 3)$ is also a common factor, and both expressions lead to

$$\pi = \sum_{k=0}^{\infty} (-1)^k 4^{2-k} \frac{(4k)!\,(4k + 3)!}{(8k + 7)!} (820k^3 + 1533k^2 + 902k + 165)$$
$$= \frac{22}{7} - \frac{19}{15015} + \frac{543}{594914320} - \frac{77}{104187267600} + \cdots . \tag{7}$$

The convergence rate can be found by applying Stirling's approximation for the factorials (see for example [1, 6.1.37]) and taking the ratios of successive terms, giving that each term has magnitude roughly $1/1024$ of the previous term, or roughly 3 decimal digits of accuracy are added per term. Using just the first two terms, and knowing that the error when truncating an alternating series with terms decreasing in magnitude is less than or equal to the absolute value of the first term in the truncated part, we can form the bound

$$\frac{22}{7} - \frac{19}{15015} \le \pi \le \frac{22}{7} + \frac{19}{15015}, \tag{8}$$

which is of width 2.53×10^{-3}. This result is poorer than the bound from the previous section, but if we use three terms, then

$$\frac{22}{7} - \frac{19}{15015} - \frac{543}{594914320} \le \pi \le \frac{22}{7} - \frac{19}{15015} + \frac{543}{594914320}, \tag{9}$$

which is a bound of width 1.83×10^{-6}, an improvement.

2. RELATED FAMILIES OF INTEGRALS.
There turn out to be a number of families of integrals that are similar in style to (1). The most obvious is originally due to Nield [12] in 1982, who introduces

$$I_{4n} = \int_0^1 \frac{x^{4n}(1 - x)^{4n}}{1 + x^2}\, dx \tag{10}$$

for positive integers n. Then I from (1) is equivalent to I_4. Medina [11] has investigated this set of integrals in detail, where the upper bound in the integral has been replaced by x, and he has used these integrals to develop polynomial approximations to $\arctan(x)$

     [Monthly 116

with rational coefficients. From our perspective, one of the most useful results he gives is the closed form expression

$$\frac{x^{4n}(1-x)^{4n}}{1+x^2} = (x^6 - 4x^5 + 5x^4 - 4x^2 + 4)\sum_{k=0}^{n-1}(-4)^{n-1-k}x^{4k}(1-x)^{4k} + \frac{(-4)^n}{1+x^2}.$$
(11)

While this equation is not proven explicitly in [**11**], a proof by mathematical induction is straightforward. Integrating (11) using (4) and simplifying leads to

$$\frac{(-1)^n}{4^{n-1}}\int_0^1 \frac{x^{4n}(1-x)^{4n}}{1+x^2}\,dx$$

$$= \pi - \sum_{k=0}^{n-1}(-1)^k\frac{2^{4-2k}(4k)!\,(4k+3)!}{(8k+7)!}(820k^3 + 1533k^2 + 902k + 165),$$

where the integration and simplification were already done earlier for Dalzell's series expansion. So in fact the closed form expression for $(-1)^n I_{4n}/4^{n-1}$ is equivalent to the error when approximating π by a truncated version of Dalzell's series expansion!

In 1995, Backhouse [**2**] generalized (1) to

$$I_{m,n} = \int_0^1 \frac{x^m(1-x)^n}{1+x^2}\,dx = a + b\pi + c\ln(2),$$
(12)

where a, b, and c are rationals that depend on the positive integers m and n, and a and b have opposite sign. In this case, $I \equiv I_{4,4}$. Backhouse [**2**] showed that if $2m - n \equiv 0$ (mod 4), then $c = 0$ and approximations to π result. In what follows we shall assume that this is the case. An integral equal to $a + b\pi$ gives the approximation $-a/b$ for π. As m and n increase, the integrand becomes increasingly flat (Backhouse calls them "pancake functions") and the approximations to π improve as well. Unfortunately, there is no straightforward formula relating a and b directly to m and n as in the I_{4n} case. However, Weisstein [**14**] at least states the result

$$I_{m,n} = 2^{-(m+n+1)}\sqrt{\pi}\,\Gamma(m+1)\Gamma(n+1)$$

$$\times {}_3F_2\left(1, \frac{m+1}{2}, \frac{m+2}{2}; \frac{m+n+2}{2}; \frac{m+n+3}{2}; -1\right).$$

We previously saw that error bounds for I could be found using the bounds $1 \le 1 + x^2 \le 2$. The same approach for $I_{m,n}$ directly leads to

$$\frac{m!\,n!}{2(m+n+1)!} < a + b\pi = \int_0^1 \frac{x^m(1-x)^n}{1+x^2}\,dx < \frac{m!\,n!}{(m+n+1)!},$$

where a and b are the rationals depending on m and n. As m and n increase, the bounds on the error decrease with reasonable rapidity.

2.1. Series expansion. Given the closed form expression (11), we can follow the same process as Dalzell to produce series expansions for π, with a specific value of n

leading to

$$\pi = \sum_{m=0}^{\infty} \sum_{k=0}^{n-1} (-4)^{-nm-k} \int_0^1 (x^6 - 4x^5 + 5x^4 - 4x^2 + 4)x^{4(k+nm)}(1-x)^{4(k+nm)}\, dx,$$

(13)

which generalizes (3). Evaluating the integrals as before leads to

$$\pi = \sum_{m=0}^{\infty} \sum_{k=0}^{n-1} (-1)^{\alpha} 4^{2-\alpha} \frac{(4\alpha)!\,(4\alpha+3)!}{(8\alpha+7)!} (820\alpha^3 + 1533\alpha^2 + 902\alpha + 165),$$

(14)

where $\alpha = k + nm$. With $n = 1$, (14) is exactly Dalzell's expansion (7). With $n = 2$ we have

$$\pi = \sum_{m=0}^{\infty} 4^{2-2m} \left[\frac{(8m)!\,(8m+3)!}{(16m+7)!} (6560m^3 + 6132m^2 + 1804m + 165) \right.$$

$$\left. - \frac{(8m+4)!\,(8m+7)!}{(16m+15)!} (1640m^3 + 3993m^2 + 3214m + 855) \right]$$

$$= \frac{47171}{15015} + \frac{16553}{18150270600} + \frac{64615651}{102659859353904652800} + \cdots.$$

Note that this is not an alternating series, and each term is roughly $1/2^{20}$ of the previous, or roughly 6 digits of accuracy are added per term.

There is no reason why we can't take n as large as we like. While this increases the amount of work to find each term in the series, each term is roughly $1/2^{10n}$ the size of the previous term, or roughly $3n$ digits of accuracy are added with each term. In principle, there is no reason why a series where each term adds one hundred digits or more of accuracy cannot be explicitly written down from (14) with $n \geq 33$.

3. INTEGRALS LEADING TO CONVERGENTS.

The main reason for appreciating the elegance of (1) surely is that it approximates π by 22/7, the classic and most well-known rational approximation both within and outside the mathematics community. The number 22/7 is particularly good because it is better than other rational approximation p/q for $q < 57$. In fact it is one of the *convergents* of the continued fraction approximation to π, the first few of which are 3, 22/7, 333/106, 355/113, 103993/33102, and 104348/33215. There are many excellent texts on continued fractions, including Chapter 10 of Hardy and Wright [8]. A natural question, then, is whether there are integrals similar to the ones shown here that lead to other convergents of π.

Unfortunately, none of the integrals considered so far lead to approximations to π related to the other convergents of π. The 22/7 in (1) must be considered a happy coincidence. However, Lucas [9] developed a set of integrals with nonnegative integrands that equalled $355/113 - \pi$, where 355/113 is the next particularly good approximation to π. Here, we generalize those results, and show how integrals with nonnegative integrands can be formed for $z - \pi$ (if $z > \pi$) or $\pi - z$ (if $z < \pi$), with any real z.

We begin by noting that $I_{m,n}$ from (12) is a combination of multiples of 1, π, and $\ln 2$. Now consider the related integral

$$I'_{m,n} = \int_0^1 \frac{x^m (1-x)^n (a + bx + cx^2)}{1 + x^2}\, dx,$$

(15)

which can be evaluated as a combination of 1, π, and $\ln 2$, where the coefficients depend on a, b, and c. If we want I' to equal $z - \pi$, this leads to a set of three linear equations in a, b, and c, which is easily solved. However, there is no guarantee that our resulting integrand is nonnegative. We need to ensure m and n are large enough that $a + bx + cx^2 \geq 0$ for $x \in [0, 1]$. The closer z is to π, the larger m and n will need to be. As m and n increase, the coefficients a, b, and c become increasingly large, and so a "best" solution can be found, in the sense that the number of characters required to form the integrand is minimal. Using Maple code (available from the author) to list the various solutions verifies that (1) is the simplest integrand leading to 22/7, and that the simplest results for other continued fraction approximations to π are

$$\int_0^1 \frac{x^5(1-x)^6(197+462x^2)}{530(1+x^2)}\, dx = \pi - \frac{333}{106},$$

$$\int_0^1 \frac{x^8(1-x)^8(25+816x^2)}{3164(1+x^2)}\, dx = \frac{355}{113} - \pi,$$

$$\int_0^1 \frac{x^{14}(1-x)^{12}(124360+77159x^2)}{755216(1+x^2)}\, dx = \pi - \frac{103993}{33102},$$

and

$$\int_0^1 \frac{x^{12}(1-x)^{12}(1349-1060x^2)}{38544(1+x^2)}\, dx = \frac{104348}{33215} - \pi.$$

To conclude, there are a variety of further directions for experimentation. Are there other integrals with nonnegative integrands leading to approximations of π apart from (15) that are worth considering? Also, are there other constants that can be approximated, and hence new series approximations developed using these techniques? For example, (15) could be used to find approximations to $\ln 2$.

ACKNOWLEDGMENTS. This paper would not have been possible without input and discussion from Jonathan Borwein (Dalhousie), Grant Keady (Western Australia), Gerry Myerson (Macquarie), Garry Tee (Auckland), and the referees, whose contributions are gratefully acknowledged and appreciated.

REFERENCES

1. M. Abramowitz and I. A. Stegun, *Handbook of Mathematical Functions*, Dover, New York, 1972.
2. N. Backhouse, Note 79.36, Pancake functions and approximations to π, *Math. Gazette* **79** (1995) 371–374.
3. F. Beukers, A rational approach to π, *Nieuw Arch. Wiskd. (5)* **1** (2000) 372–379.
4. J. M. Borwein, D. H. Bailey, and R. Girgensohn, *Experimentation in Mathematics: Computational Paths to Discovery*, A.K. Peters, Wellesley, MA, 2004.
5. R. Cornwell, A proof that Pi is not 22/7, *New Zealand Math. Mag.* **17** (1980) 127.
6. D. P. Dalzell, On 22/7, *J. London Math. Soc.* **19** (1944) 133–134.
7. ———, On 22/7 and 355/113, *Eureka: The Archimedian's Journal* **34** (1971) 10–13.
8. G. H. Hardy and E. M. Wright, *An Introduction to the Theory of Numbers*, 5th ed., Oxford University Press, Oxford, 1979.
9. S. K. Lucas, Integral proofs that $355/113 > \pi$, *Gazette Aust. Math. Soc.* **32** (2005) 263–266.
10. J. H. McKay, The William Lowell Putnam mathematical competition, this MONTHLY **76** (1969) 909–915.
11. H. A. Medina, A sequence of polynomials for approximating inverse tangent, this MONTHLY **113** (2006) 156–161.
12. D. A. Nield (misspelled Neild in the article), Rational approximations to Pi, *New Zealand Math. Mag.* **18** (1982) 99–100.

13. ———, Rational approximations to Pi: A further comment, *New Zealand Math. Mag.* **19** (1982) 60–61.
14. E. W. Weisstein, Pi formulas—From MathWorld, A Wolfram Web Resource, `http://mathworld.wolfram.com/PiFormulas.html`.

Department of Mathematics and Statistics, James Madison University, Harrisonburg VA 22807
lucassk@jmu.edu

New Proofs of Euclid's and Euler's Theorems

Juan Pablo Pinasco

In this note we give a new proof of the existence of infinitely many prime numbers. There are several different proofs with many variants, and some of them can be found in [1, 3, 4, 5, 6]. This proof is based on a simple counting argument using the inclusion-exclusion principle combined with an explicit formula. A different proof based on counting arguments is due to Thue (1897) and can be found in [6] together with several generalizations, and a remarkable variant of it was given by Chaitin [2] using algorithmic information theory. Moreover, we prove that the series of reciprocals of the primes diverges. Our proofs arise from a connection between the inclusion-exclusion principle and the infinite product of Euler.

Let $\{p_i\}_i$ be the sequence of prime numbers, and let us define the following recurrence:

$$a_0 = 0, \qquad a_{k+1} = a_k + \frac{1 - a_k}{p_{k+1}}.$$

Let us note that the Nth term a_N generated by this recurrence coincides with

$$a_N = \sum_i \frac{1}{p_i} - \sum_{i<j} \frac{1}{p_i p_j} + \sum_{i<j<k} \frac{1}{p_i p_j p_k} - \cdots + (-1)^{N+1} \frac{1}{p_1 \cdots p_N},$$

and can be given in a closed form as

$$a_N = 1 - \prod_{i=1}^{N} \left(1 - \frac{1}{p_i}\right),$$

which implies that $0 < a_N < 1$, since each factor is strictly positive and less than one.

Now, we are ready to prove the classical Euclid's theorem:

Theorem 1. *There are infinitely many prime numbers.*

Proof. Let us suppose that $p_1 < p_2 < \cdots < p_N$ are all the primes. For any $x \geq 1$, and for $i = 1, \ldots, N$, let A_i be the set of integers in $[1, x]$ that are divisible by p_i. Then, the number of positive integers in $[1, x]$ is obtained by applying the inclusion-exclusion formula to find the cardinality of $\cup_{i=1}^{N} A_i$:

$$[x] = 1 + \sum_i \left[\frac{x}{p_i}\right] - \sum_{i<j} \left[\frac{x}{p_i p_j}\right] + \sum_{i<j<k} \left[\frac{x}{p_i p_j p_k}\right] - \cdots + (-1)^{N+1} \left[\frac{x}{p_1 \cdots p_N}\right],$$

19. Ramanujan's series for $1/\pi$: A survey (2009)

Paper 19: Nayandeep Deka Baruah, Bruce C. Berndt and Heng Huat Chan, "Ramanujan's series for $1/\pi$: A survey," *American Mathematical Monthly*, vol. 116 (2009), p. 567–587.

Synopsis:

In this piece, the authors discuss some formulas for $1/\pi$ originally discovered by Ramanujan. One of these is the formula

$$\frac{1}{\pi} = \frac{\sqrt{8}}{9801} \sum_{n=0}^{\infty} \frac{(4n)!(1103 + 26390n)}{(n!)^4 396^{4n}},$$

which was used by Gosper in 1985 to compute π to compute 17,526,100 decimal digits of π. Using the same mathematical approach, David and Gregory Chudnovsky subsequently deduced the formula

$$\frac{1}{\pi} = 12 \sum_{n=0}^{\infty} \frac{(-1)^n (6n)!}{(n!)^3 (3n)!} \frac{13591409 + 545140134n}{640320^{3n+3/2}},$$

which was then used by them to compute 2,260,331,336 digits. Results are presented here for numerous other series of this general form.

Keywords: Computation, History, Series

© Springer International Publishing Switzerland 2016
D.H. Bailey, J.M. Borwein, *Pi: The Next Generation*,
DOI 10.1007/978-3-319-32377-0_19

Ramanujan's Series for $1/\pi$: A Survey*

Nayandeep Deka Baruah, Bruce C. Berndt, and Heng Huat Chan

In Memory of V. Ramaswamy Aiyer,
Founder of the Indian Mathematical Society in 1907

When we pause to reflect on Ramanujan's life, we see that there were certain events that seemingly were necessary in order that Ramanujan and his mathematics be brought to posterity. One of these was V. Ramaswamy Aiyer's founding of the Indian Mathematical Society on 4 April 1907, for had he not launched the Indian Mathematical Society, then the next necessary episode, namely, Ramanujan's meeting with Ramaswamy Aiyer at his office in Tirtukkoilur in 1910, would also have not taken place. Ramanujan had carried with him one of his notebooks, and Ramaswamy Aiyer not only recognized the creative spirit that produced its contents, but he also had the wisdom to contact others, such as R. Ramachandra Rao, in order to bring Ramanujan's mathematics to others for appreciation and support. The large mathematical community that has thrived on Ramanujan's discoveries for nearly a century owes a huge debt to V. Ramaswamy Aiyer.

1. THE BEGINNING. Toward the end of the first paper [57], [58, p. 36] that Ramanujan published in England, at the beginning of Section 13, he writes, "I shall conclude this paper by giving a few series for $1/\pi$." (In fact, Ramanujan concluded his paper a couple of pages later with another topic: formulas and approximations for the perimeter of an ellipse.) After sketching his ideas, which we examine in detail in Sections 3 and 9, Ramanujan records three series representations for $1/\pi$. As is customary, set

$$(a)_0 := 1, \qquad (a)_n := a(a+1)\cdots(a+n-1), \qquad n \geq 1.$$

Let

$$A_n := \frac{(\frac{1}{2})_n^3}{n!^3}, \qquad n \geq 0. \tag{1.1}$$

Theorem 1.1. *If A_n is defined by (1.1), then*

$$\frac{4}{\pi} = \sum_{n=0}^{\infty}(6n+1)A_n\frac{1}{4^n}, \tag{1.2}$$

$$\frac{16}{\pi} = \sum_{n=0}^{\infty}(42n+5)A_n\frac{1}{2^{6n}}, \tag{1.3}$$

$$\frac{32}{\pi} = \sum_{n=0}^{\infty}\left((42\sqrt{5}+30)n+5\sqrt{5}-1\right)A_n\frac{1}{2^{6n}}\left(\frac{\sqrt{5}-1}{2}\right)^{8n}. \tag{1.4}$$

*This paper was originally solicited by the Editor of *Mathematics Student* to commemorate the founding of the Indian Mathematical Society in its centennial year. *Mathematics Student* is one of the two official journals published by the Indian Mathematical Society, with the other being the *Journal of the Indian Mathematical Society*. The authors thank the Editor of *Mathematics Student* for permission to reprint the article in this MONTHLY with minor changes from the original.

The first two formulas, (1.2) and (1.3), appeared in the Walt Disney film *High School Musical*, starring Vanessa Anne Hudgens, who plays an exceptionally bright high school student named Gabriella Montez. Gabriella points out to her teacher that she had incorrectly written the left-hand side of (1.3) as $8/\pi$ instead of $16/\pi$ on the blackboard. After first claiming that Gabriella is wrong, her teacher checks (possibly Ramanujan's *Collected Papers*?) and admits that Gabriella is correct. Formula (1.2) was correctly recorded on the blackboard.

After offering the three formulas for $1/\pi$ given above, at the beginning of Section 14 [**57**], [**58**, p. 37], Ramanujan claims, "There are corresponding theories in which q is replaced by one or other of the functions"

$$q_r := q_r(x) := \exp\left(-\pi \, \csc(\pi/r) \frac{{}_2F_1\left(\frac{1}{r}, \frac{r-1}{r}; 1; 1-x\right)}{{}_2F_1\left(\frac{1}{r}, \frac{r-1}{r}; 1; x\right)}\right), \tag{1.5}$$

where $r = 3$, 4, or 6, and where ${}_2F_1$ denotes one of the hypergeometric functions ${}_pF_{p-1}$, $p \geq 1$, which are defined by

$$ {}_pF_{p-1}\left(a_1, \ldots, a_p; b_1, \ldots, b_{p-1}; x\right) := \sum_{n=0}^{\infty} \frac{(a_1)_n \cdots (a_p)_n}{(b_1)_n \cdots (b_{p-1})_n} \frac{x^n}{n!}, \qquad |x| < 1. $$

(The meaning of q is explained in Section 3.) Ramanujan then offers 14 further series representations for $1/\pi$. Of these, 10 belong to the quartic theory, i.e., for $r = 4$; 2 belong to the cubic theory, i.e., for $r = 3$; and 2 belong to the sextic theory, i.e., for $r = 6$. Ramanujan never returned to the "corresponding theories" in his published papers, but six pages in his second notebook [**59**] are devoted to developing these theories, with all of the results on these six pages being proved in a paper [**16**] by Berndt, S. Bhargava, and F. G. Garvan. That the classical hypergeometric function ${}_2F_1(\frac{1}{2}, \frac{1}{2}; 1; x)$ in the classical theory of elliptic functions could be replaced by one of the three hypergeometric functions above and concomitant theories developed is one of the many incredibly ingenious and useful ideas bequeathed to us by Ramanujan. The development of these theories is far from easy and is an active area of contemporary research.

All 17 series for $1/\pi$ were discovered by Ramanujan in India before he arrived in England, for they can be found in his notebooks [**59**], which were written prior to Ramanujan's departure for England. In particular, (1.2), (1.3), and (1.4) can be found on page 355 in his second notebook and the remaining 14 series are found in his third notebook [**59**, p. 378]; see also [**14**, pp. 352–354]. It is interesting that (1.2), (1.3), and (1.4) are also located on a page published with Ramanujan's lost notebook [**60**, p. 370]; see also [**3**, Chapter 15].

2. THE MAIN ACTORS FOLLOWING IN THE FOOTSTEPS OF RAMANU-JAN.
Fourteen years after the publication of [**57**], the first mathematician to address Ramanujan's formulas was Sarvadaman Chowla [**37**], [**38**], [**39**, pp. 87–91, 116–119], who gave the first published proof of a general series representation for $1/\pi$ and used it to derive (1.2) of Ramanujan's series for $1/\pi$ [**57**, Eq. (28)]. We briefly discuss Chowla's ideas in Section 4.

Ramanujan's series were then forgotten by the mathematical community until November, 1985, when R. William Gosper, Jr. used one of Ramanujan's series, namely,

$$\frac{9801}{\pi\sqrt{8}} = \sum_{n=0}^{\infty} \frac{(4n)! \, (1103 + 26390n)}{(n!)^4 \, 396^{4n}}, \tag{2.1}$$

to calculate 17,526,100 digits of π, which at that time was a world record. There was only one problem with his calculation—(2.1) had not yet been proved. However, a comparison of Gosper's calculation of the digits of π with the previous world record held by Y. Kanada made it extremely unlikely that (2.1) was incorrect.

In 1987, Jonathan and Peter Borwein [23] succeeded in proving all 17 of Ramanujan's series for $1/\pi$. In a subsequent series of papers [24], [25], [29], they established several further series for $1/\pi$, with one of their series [29] yielding roughly fifty digits of π per term. The Borweins were also keen on calculating the digits of π, and accounts of their work can be found in [30], [28], and [26].

At about the same time as the Borweins were devising their proofs, David and Gregory Chudnovsky [40] also derived series representations for $1/\pi$ and, in particular, used their series

$$\frac{1}{\pi} = 12 \sum_{n=0}^{\infty} (-1)^n \frac{(6n)!}{(n!)^3 (3n)!} \frac{13591409 + 545140134n}{(640320)^{3n+3/2}} \tag{2.2}$$

to calculate a world record 2,260,331,336 digits of π. The series (2.2) yields 14 digits of π per term. A popular account of the Chudnovskys' calculations can be found in a paper written for *The New Yorker* [56].

The third author of the present paper and his coauthors (Berndt, S. H. Chan, A. Gee, W.-C. Liaw, Z.-G. Liu, V. Tan, and H. Verrill) in a series of papers [19], [31], [33], [34], [36] extended the ideas of the Borweins, in particular, without using Clausen's formula in [31] and [36], and derived general hypergeometric-like formulas for $1/\pi$. We devote Section 8 of our survey to discussing some of their results.

Stimulated by the work and suggestions of the third author, the first two authors [9], [7] systematically returned to Ramanujan's development in [57] and employed his ideas in order not only to prove most of Ramanujan's original representations for $1/\pi$ but also to establish a plethora of new such identities as well. In another paper [8], motivated by the work of Jesús Guillera [48]–[53], who both experimentally and rigorously discovered many new series for both $1/\pi$ and $1/\pi^2$, the first two authors continued to follow Ramanujan's ideas and devised series representations for $1/\pi^2$.

In the survey which follows, we delineate the main ideas in Sections 3, 6, 7, 8, and 9, where the ideas of Ramanujan, the Borwein brothers, the Chudnovsky brothers, Chan and his coauthors, and the present authors, respectively, are discussed.

3. RAMANUJAN'S IDEAS. To describe Ramanujan's ideas, we need several definitions from the classical theory of elliptic functions, which, in fact, we use throughout the paper. The complete elliptic integral of the first kind is defined by

$$K := K(k) := \int_0^{\pi/2} \frac{d\varphi}{\sqrt{1 - k^2 \sin^2 \varphi}}, \tag{3.1}$$

where k, $0 < k < 1$, denotes the *modulus*. Furthermore, $K' := K(k')$, where $k' := \sqrt{1 - k^2}$ is the *complementary modulus*. The complete elliptic integral of the second kind is defined by

$$E := E(k) := \int_0^{\pi/2} \sqrt{1 - k^2 \sin^2 \varphi} \, d\varphi. \tag{3.2}$$

If $q = \dot{e}xp(-\pi K'/K)$, then one of the central theorems in the theory of elliptic functions asserts that [13, p. 101, Entry 6]

$$\varphi^2(q) = \frac{2}{\pi}K(k) = {}_2F_1(\tfrac{1}{2}, \tfrac{1}{2}; 1; k^2), \tag{3.3}$$

where $\varphi(q)$ in Ramanujan's notation (or $\vartheta_3(q)$ in the classical notation) denotes the classical theta function defined by

$$\varphi(q) = \sum_{j=-\infty}^{\infty} q^{j^2}. \tag{3.4}$$

Note that, in the notation (1.5), $q = q_2$ and $x = k^2$. The second equality in (3.3) follows from expanding the integrand in a binomial series and integrating termwise. Conversely, it is also valuable to regard k as a function of q, and so we write $k = k(q)$.

Let K, K', L, and L' denote complete elliptic integrals of the first kind associated with the moduli k, k', ℓ, and ℓ', respectively. Suppose that, for some positive integer n,

$$n\frac{K'}{K} = \frac{L'}{L}. \tag{3.5}$$

A modular equation of degree n is an equation involving k and ℓ that is induced by (3.5). Modular equations are always algebraic equations. An example of a modular equation of degree 7 may be found later in (9.18). Alternatively, by (3.3), (3.5) can be expressed in terms of hypergeometric functions. We often say that ℓ has degree n over k. Derivations of modular equations ultimately rest on (3.3). If we set $K'/K = \sqrt{n}$, so that $q = e^{-\pi\sqrt{n}}$, then the corresponding value of k, which is denoted by $k_n := k(e^{-\pi\sqrt{n}})$, is called the *singular modulus*. The multiplier $m = m(q)$ is defined by

$$m := m(q) := \frac{{}_2F_1\left(\tfrac{1}{2}, \tfrac{1}{2}; 1; k^2(q)\right)}{{}_2F_1\left(\tfrac{1}{2}, \tfrac{1}{2}; 1; \ell^2(q)\right)}. \tag{3.6}$$

We note here that, by (3.3), (3.6), and [13, Entry 3, p. 98; Entry 25(vii), p. 40], $m(q)$ and $k^2(q)$ can be represented by

$$m(q) = \frac{\varphi^2(q)}{\varphi^2(q^n)} \quad \text{and} \quad k^2(q) = 16q\frac{\psi^4(q^2)}{\varphi^4(q)},$$

respectively, where $\varphi(q)$ is defined by (3.4) and

$$\psi(q) = \sum_{j=0}^{\infty} q^{j(j+1)/2}.$$

Thus, modular equations can also be written as theta function identities.

Ramanujan begins Section 13 of [57] with a special case of Clausen's formula [23, p. 178, Proposition 5.6(b)],

$$\frac{4K^2}{\pi^2} = \sum_{j=0}^{\infty} \frac{(\tfrac{1}{2})_j^3}{(j!)^3}(2kk')^{2j} = {}_3F_2\left(\tfrac{1}{2}, \tfrac{1}{2}, \tfrac{1}{2}; 1, 1; (2kk')^2\right), \tag{3.7}$$

which can be found as Entry 13 of Chapter 11 in his second notebook [59] [12, p. 58]. Except for economizing notation, we now quote Ramanujan. "Hence we have

$$q^{1/3}(q^2; q^2)^4_\infty = \left(\frac{1}{4}kk'\right)^{2/3} \sum_{j=0}^\infty \frac{(\frac{1}{2})^3_j}{(j!)^3}(2kk')^{2j},$$ (3.8)

where

$$(a; q)_\infty := (1 - a)(1 - aq)(1 - aq^2)\cdots.$$ (3.9)

Logarithmically differentiating both sides in (3.8) with respect to k, we can easily shew that

$$1 - 24\sum_{j=1}^\infty \frac{jq^{2j}}{1 - q^{2j}} = (1 - 2k^2)\sum_{j=0}^\infty (3j + 1)\frac{(\frac{1}{2})^3_j}{(j!)^3}(2kk')^{2j}.$$ (3.10)

But it follows from

$$1 - \frac{3}{\pi\sqrt{n}} - 24\sum_{j=1}^\infty \frac{j}{e^{2\pi j\sqrt{n}} - 1} = \left(\frac{K}{\pi}\right)^2 A(k)$$ (3.11)

where $A(k)$ is a certain type of algebraic number, that, when $q = e^{-\pi\sqrt{n}}$, n being a rational number, the left-hand side of (3.10) can be expressed in the form

$$A\left(\frac{2K}{\pi}\right)^2 + \frac{B}{\pi},$$

where A and B are algebraic numbers expressible by surds. Combining (3.7) and (3.10) in such a way as to eliminate the term $(2K/\pi)^2$, we are left with a series for $1/\pi$." He then gives the three examples (1.2)–(1.4).

Ramanujan's ideas will be described in more detail in Section 9. However, in closing this section, we note that the series on the left-hand sides of (3.10) and (3.11) is Ramanujan's Eisenstein series $P(q^2)$, with $q = e^{-\pi\sqrt{n}}$ in the latter instance, where

$$P(q) := 1 - 24\sum_{j=1}^\infty \frac{jq^j}{1 - q^j}, \qquad |q| < 1.$$ (3.12)

Ramanujan's derivation of (3.11) arises firstly from the transformation formula for $P(q)$, which in turn is an easy consequence of the transformation formula for the Dedekind eta-function, given in (9.11) below. The second ingredient in deriving (3.11) is an identity for $nP(q^{2n}) - P(q^2)$ in terms of the moduli k and ℓ, where ℓ has degree n over k. Formula (3.8) follows from a standard theorem in elliptic functions that Ramanujan also recorded in his notebooks [59], [13, p. 124, Entry 12(iii)].

4. SARVADAMAN CHOWLA. Chowla's ideas reside in the classical theory of elliptic functions and are not unlike those that the Borweins employed several years later. We now briefly describe Chowla's approach [37], [38], [39, pp. 87–91, 116–119]. Using classical formulas of Cayley and Legendre relating the complete elliptic integrals K and E, defined by (3.1) and (3.2), respectively, he specializes them by setting $K/K' = \sqrt{n}$. He then defines

$$S_r := \sum_{j=0}^{\infty} j^r \frac{(\frac{1}{2})_j^3}{(j!)^3} (2kk')^{2j} \tag{4.1}$$

and

$$T_r := \sum_{j=0}^{\infty} j^r \frac{(\frac{1}{4})_j^2}{(j!)^2} (2kk')^{2j}. \tag{4.2}$$

Chowla then writes "Then it is known that, when $k \leq 1/\sqrt{2}$,"

$$\frac{2K}{\pi} = 1 + T_0 \quad \text{and} \quad \frac{4K^2}{\pi^2} = 1 + S_0. \tag{4.3}$$

Chowla does not give his source for either formula, but the second formula in (4.3), as noted above, is a special case of Clausen's formula (3.7). The first formula is a special case of Kummer's quadratic transformation [**23**, pp. 179–180], which was also known to Ramanujan. Each of the formulas of (4.3) is differentiated twice with respect to k, and, without giving details, Chowla concludes that if $K/K' = \sqrt{n}$, then

$$\frac{1}{K} = a_1 + b_1 T_0 + c_1 T_1,$$

$$\frac{K}{\pi} = d_1 T_1 + e_1 T_2,$$

$$\frac{1}{\pi} = a_2 S_0 + b_2 S_1,$$

$$\frac{1}{K^2} = a_3 + b_3 S_0 + c_3 S_1 + d_3 S_2,$$

"where $a_1, b_1, \ldots$ are algebraic numbers." He then sets $n = 3$ and $k = \sin(\pi/12)$ in each of the four formulas above to deduce, in particular, identity (1.2) from the second formula above.

5. R. WILLIAM GOSPER, JR. As we indicated in the Introduction, in November, 1985, Gosper employed a lisp machine at Symbolics and Ramanujan's series (2.1) to calculate 17,526,100 digits of π, which at that time was a world record. (During the 1980s and 1990s, Symbolics made a lisp-based workstation running an object-oriented programming environment. Unfortunately, the machines were too expensive for the needs of most customers, and the company went bankrupt before it could squeeze the architecture onto a chip.) Of the 17 series found by Ramanujan, this one converges the fastest, giving about 8 digits of π per term. At the time of Gosper's calculation, the world record for digits of π was about 16 million digits calculated by Y. Kanada. Before the Borwein brothers had later found a "conventional" proof of Ramanujan's series (2.1), they had shown that either (2.1) yields an exact formula for π or that it differs from π by more than $10^{-3000000}$. Thus, by demonstrating that his calculation of π agreed with that of Kanada, Gosper effectively had completed the Borweins' first proof of (2.1). However, Gosper's primary goal was not to eclipse Kanada's record but to study the (simple) continued fraction expansion of π for which he calculated

17,001,303 terms. In email letters from February, 1992 and May, 1993, Gosper offered the following remarks on his calculations:

> Of course, what the scribblers always censor is that the digits were a by-product. I wanted to change the object of the game away from meaningless decimal digits.
>
> I used what I call a resumable matrix tower to exactly compute an enormous rational equal to the sum of a couple of million terms of Ramanujan's 99^{-4n} series. I then divided to form the binary integer = floor($\pi 2^{58,000,000}$). I converted this to decimal and sent a summary to Kanada for comparison, and converted the binary fraction to a cf using an fft based scheme.

Gosper's world record was short lived, as in January, 1986, D. H. Bailey used an algorithm of the Borweins arising from a fourth-order modular equation to compute 29,360,000 digits of π. Gosper's calculation of the continued fraction expansion of π was motivated by the fact that many important mathematical constants do not have interesting decimal expansions but do have interesting continued fraction expansions. That continued fraction expansions are considerably more interesting than decimal expansions is a view shared by the Chudnovsky brothers [43]. Continued fraction expansions can often be used to distinguish a constant from others, while decimal expansions likely will be unable to do so. For example, the simple continued fraction of e, namely,

$$e = 2 + \cfrac{1}{1} + \cfrac{1}{2} + \cfrac{1}{1} + \cfrac{1}{1} + \cfrac{1}{4} + \cfrac{1}{1} + \cfrac{1}{1} + \cfrac{1}{6} + \cfrac{1}{1} + \cdots$$

has a pattern. On the contrary, taking a large random string of digits of e would not help one identify e. It is an open problem if the simple continued fraction of π, namely,

$$\pi = 3 + \cfrac{1}{7} + \cfrac{1}{15} + \cfrac{1}{1} + \cfrac{1}{292} + \cfrac{1}{1} + \cfrac{1}{1} + \cfrac{1}{1} + \cfrac{1}{2} + \cfrac{1}{1} + \cfrac{1}{3} + \cfrac{1}{1} + \cfrac{1}{14}$$
$$+ \cfrac{1}{2} + \cfrac{1}{1} + \cfrac{1}{1} + \cfrac{1}{2} + \cfrac{1}{2} + \cfrac{1}{2} + \cfrac{1}{2} + \cfrac{1}{1} + \cfrac{1}{84} + \cfrac{1}{2} + \cfrac{1}{1} + \cfrac{1}{1} + \cfrac{1}{15}$$
$$+ \cfrac{1}{3} + \cfrac{1}{13} + \cfrac{1}{1} + \cfrac{1}{4} + \cfrac{1}{2} + \cfrac{1}{6} + \cdots$$

has a pattern.

Gosper also derived a hypergeometric-like series representation for π, namely,

$$\pi = \sum_{j=0}^{\infty} \frac{50j - 6}{\binom{3j}{j}2^j}, \tag{5.1}$$

which can be used to calculate any particular binary digit of π. See a paper by G. Almkvist, C. Krattenthaler, and J. Petersson [1] for a proof of (5.1) as well as generalizations, which include the following theorem.

Theorem 5.1. *For each integer $k \geq 1$, there exists a polynomial $S_k(j)$ in j of degree 4k with rational coefficients such that*

$$\pi = \sum_{j=0}^{\infty} \frac{S_k(j)}{\binom{8kj}{4kj}(-4)^{kj}}.$$

6. JONATHAN AND PETER BORWEIN. One key to the work of both Ramanujan and Chowla in their derivations of formulas for $1/\pi$ is Clausen's formula for the square of a complete elliptic integral of the first kind or, by (3.3), for the square of the hypergeometric function $_2F_1(\frac{1}{2}, \frac{1}{2}; 1; k^2)$. The aforementioned rendition (3.7) of Clausen's formula is not the most general version of Clausen's formula, namely,

$$_2F_1^2\left(a, b; a + b + \tfrac{1}{2}; z\right) = {}_3F_2\left(2a, 2b, a + b; a + b + \tfrac{1}{2}, 2a + 2b; z\right). \tag{6.1}$$

Indeed, the work of many authors who have proved Ramanujan-like series for $1/\pi$ ultimately rests on special cases of (6.1). In particular, squares of certain other hypergeometric functions lead one to Ramanujan's alternative theories of elliptic functions.

A second key step is to find another formula for $(K/\pi)^2$, which also contains another term involving $1/\pi$. Combining the two formulas to eliminate the term $(K/\pi)^2$ then produces a hypergeometric-type series representation for $1/\pi$. Evidently, unaware of Chowla's earlier work, the Borweins proceeded in a similar fashion and used Legendre's relation [**23**, p. 24]

$$E(k)K'(k) + E'(k)K(k) - K(k)K'(k) = \frac{\pi}{2}$$

and other relations between elliptic integrals to produce such formulas.

Having derived a series representation for $1/\pi$, one now faces the problem of evaluating the moduli and elliptic integrals that appear in the formulas. If $q = e^{-\pi\sqrt{n}}$, then for certain positive integers n one can evaluate the requisite quantities. This leads us to the definition of the Ramanujan–Weber class invariants. After Ramanujan, set

$$\chi(q) := (-q; q^2)_\infty, \qquad |q| < 1, \tag{6.2}$$

where $(a; q)_\infty$ is defined by (3.9). If n is a positive rational number and $q = e^{-\pi\sqrt{n}}$, then the class invariants G_n and g_n are defined by

$$G_n := 2^{-1/4} q^{-1/24} \chi(q) \qquad \text{and} \qquad g_n := 2^{-1/4} q^{-1/24} \chi(-q). \tag{6.3}$$

In the notation of H. Weber [**63**], $G_n = 2^{-1/4}\mathfrak{f}(\sqrt{-n})$ and $g_n = 2^{-1/4}\mathfrak{f}_1(\sqrt{-n})$. As mentioned in Section 3, $k_n := k(e^{-\pi\sqrt{n}})$ is called the *singular modulus*. In his voluminous work on modular equations, Ramanujan sets $\alpha := k^2$ and $\beta := \ell^2$. Accordingly, we set $\alpha_n := k_n^2$. Because [**13**, p. 124, Entries 12(v), (vi)]

$$\chi(q) = 2^{1/6}\{\alpha(1 - \alpha)/q\}^{-1/24} \qquad \text{and} \qquad \chi(-q) = 2^{1/6}\{\alpha(1 - \alpha)^{-2}/q\}^{-1/24},$$

it follows from (6.3) that

$$G_n = \{4\alpha_n(1 - \alpha_n)\}^{-1/24} \qquad \text{and} \qquad g_n = \{4\alpha_n(1 - \alpha_n)^{-2}\}^{-1/24}. \tag{6.4}$$

In the form (6.4), the class invariant G_n appears on the right-hand sides of (3.7) and (3.10), and consequently the values of G_n for several values of n are important in deriving certain series for $1/\pi$. It is known that if n is square-free and $n \equiv 1 \pmod{4}$ then G_n^4 is a real unit that generates the *Hilbert class field* of the quadratic field $\mathbb{Q}(\sqrt{-n})$ [**32**, Cor. 5.2], and this fact is very useful in evaluating G_n. When we say that a series for $1/\pi$ is associated with the imaginary quadratic field $\mathbb{Q}(\sqrt{-n})$, we mean that the constants involved in the series are related to the generators of the Hilbert class field of $\mathbb{Q}(\sqrt{-n})$.

Singular moduli and class invariants are actually algebraic numbers. In general, as n increases, the corresponding series for $1/\pi$ converges more rapidly. The series (2.2) is associated with the imaginary quadratic field $\mathbb{Q}(\sqrt{-163})$.

The Borweins' proofs of all 17 of Ramanujan's series for $1/\pi$ can be found in their book [23]. Their derivations arise from several general hypergeometric-like series representations for $1/\pi$ given in terms of singular moduli, class invariants, and complete elliptic integrals [23, pp. 181–184]. Another account of their work, but with fewer details, can be found in their paper [24] commemorating the centenary of Ramanujan's birth. Further celebrating the 100th anniversary of Ramanujan's birth, the Borweins derived further series for $1/\pi$ in [25]. The series in this paper correspond to imaginary quadratic fields with class number 2, with one of their series corresponding to $n = 427$ and yielding about 25 digits of π per term. In [29], the authors derived series for $1/\pi$ arising from fields with class number 3, with a series corresponding to $n = 907$ yielding about 37 or 38 digits of π per term. Their record is a series associated with a field of class number 4 giving about 50 digits of π per term; here, $n = 1555$ [28]. The latter paper gives the details of what we have written in this paragraph.

The Borweins have done an excellent job of communicating their work to a wide audience. Besides their paper [28], see their paper in this MONTHLY [30], with D. H. Bailey, on computing π, especially via work of Ramanujan, and their delightful paper in the *Scientific American* [26], which has been reprinted in [22, pp. 187–199] and [11, pp. 588–595].

7. DAVID AND GREGORY CHUDNOVSKY. In our Introduction we mentioned that the Chudnovsky brothers, Gregory and David, used (2.2) to calculate over 2 billion digits of π. They had first used (2.2) to calculate 1,130,160,664 digits of π in the fall of 1989 on a "borrowed" computer. They then built their own computer, "m zero," described colorfully in [56], and set a world record of 2,260,321,336 digits of π. The world record for digits of π has been broken several times since then, and since it is not the purpose of this paper to delineate this computational history, we refrain from mentioning further records.

The Chudnovskys, among others, have extensively examined their calculations for patterns. It is a long outstanding conjecture that π is *normal*. In particular, for each k, $0 \le k \le 9$,

$$\lim_{N \to \infty} \frac{\text{\# of appearances of } k \text{ in the first } N \text{ digits of } \pi}{N} = \frac{1}{10}.$$

The Chudnovskys' calculations, and all subsequent calculations of Kanada, lend credence to this conjecture. As a consequence, the average of the digits over a long interval should be approximately 4.5. The Chudnovskys found that for the first billion digits the average stays a bit on the high side, while for the next billion digits, the average hovers a bit on the low side. Their paper [43] gives an interesting statistical analysis of the digits up to one billion. For example, the maximal length of a string of identical digits, for each of the ten digits, is either 8, 9, or 10.

The Chudnovskys deduced (2.2) from a general series representation for $1/\pi$, which we will describe after making several definitions. For $\tau \in \mathcal{H} = \{\tau : \text{Im } \tau > 0\}$ and each positive integer k, the Eisenstein series $E_{2k}(\tau)$ is defined by

$$E_{2k}(\tau) := 1 - \frac{4k}{B_{2k}} \sum_{j=1}^{\infty} \sigma_{2k-1}(j) q^j, \qquad q = e^{\pi i \tau}, \tag{7.1}$$

where B_k, $k \geq 0$, denotes the kth Bernoulli number and $\sigma_k(n) = \sum_{d|n} d^k$. Klein's absolute modular J-invariant is defined by

$$J(\tau) := \frac{E_4^3(\tau)}{E_4^3(\tau) - E_6^2(\tau)}, \qquad \tau \in \mathcal{H}. \tag{7.2}$$

It is well known that if $\alpha(q) := k^2(q)$, where k is the modulus, then [13, pp. 126–127, Entry 13]

$$J(2\tau) = \frac{4\left(1 - \alpha(q) + \alpha^2(q)\right)^3}{27\alpha^2(q)(1 - \alpha(q))^2}. \tag{7.3}$$

Thus, (6.4) and (7.3) show that, when $q = e^{-\pi\sqrt{n}}$, singular moduli, class invariants, and the modular J-invariant are intimately related. Now define

$$s_2(\tau) := \frac{E_4(\tau)}{E_6(\tau)}\left(E_2(\tau) - \frac{3}{\pi \operatorname{Im}\tau}\right).$$

We are now ready to state the Chudnovskys' main formula [44, p. 122]. If $\tau = (1 + \sqrt{-n})/2$, then

$$\sum_{\mu=0}^{\infty}\left\{\frac{1}{6}(1 - s_2(\tau)) + \mu\right\}\frac{(6\mu)!}{(3\mu)!\,\mu!^3}\frac{1}{1728^\mu J^\mu(\tau)} = \frac{\sqrt{-J(\tau)}}{\pi}\frac{1}{\sqrt{n(1 - J(\tau))}}. \tag{7.4}$$

The Chudnovskys' series (2.2) is the special case $n = 163$ of (7.4).

The Chudnovsky brothers developed and extended Ramanujan's ideas in directions different from those of other authors. They obtained hypergeometric-like representations for other transcendental constants and proved, for example, that

$$\frac{\Gamma\left(\frac{1}{3}\right)}{\pi} \qquad \text{and} \qquad \frac{\Gamma^2\left(\frac{1}{24}\right)}{\Gamma\left(\frac{1}{3}\right)\Gamma\left(\frac{1}{4}\right)}$$

are transcendental [42]. Their advances involve the "second" solution of the hypergeometric differential equation. Recall from the theory of linear differential equations that $_2F_1(a, b; c; x)$ is a solution of a certain second-order linear differential equation with a regular singular point at the origin [5, p. 1]. A second linearly independent solution is generally not analytic at the origin, and in [43] and [44, pp. 124–126], the Chudnovsky brothers establish new hypergeometric series identities involving the latter function. Their identities lead to hypergeometric-like series representations for π, including Gosper's formula (5.1). In [45], the authors provide a lengthy list of such examples, including

$$45\pi + 644 = \sum_{j=0}^{\infty}\frac{8^j\left(430j^2 - 6240j - 520\right)}{\binom{4j}{j}}.$$

The Chudnovsky brothers have also employed series for $1/\pi$ to derive theorems on irrationality measures $\mu(\alpha)$, which are defined by

$$\mu(\alpha) := \inf\left\{\mu > 0 : 0 < \left|\alpha - \frac{p}{q}\right| < \frac{1}{q^\mu} \text{ has only finitely many solutions } \frac{p}{q} \in \mathbb{Q}\right\}.$$

By the famous Thue–Siegel–Roth Theorem [**6**, p. 66],

$$
\mu(\alpha) \quad
\begin{cases}
= 1, & \text{if } \alpha \text{ is rational,} \\
= 2, & \text{if } \alpha \text{ is algebraic but not rational,} \\
\geq 2, & \text{if } \alpha \text{ is transcendental.}
\end{cases}
$$

Although the Chudnovskys can obtain irrationality measures for various constants, the one they obtain for π is not as good as one would like. Currently, the world record for the irrationality measure of π is held by M. Hata [**54**], who proved that $\mu(\pi) \leq 8.016045\ldots$. Their methods are much better for obtaining irrationality measures for expressions, such as $\pi/\sqrt{640320}$, arising in their series (2.2). See also a paper by W. Zudilin [**64**].

8. RAMANUJAN'S CUBIC CLASS INVARIANT AND HIS ALTERNATIVE THEORIES.
In Sections 3, 4, 6, and 7, we emphasized how Ramanujan–Weber class invariants and singular moduli were of central importance for Ramanujan and others who followed in deriving series for $1/\pi$. We also stressed in Section 1, in particular, in the discourse after (1.5), that Ramanujan's remarkable idea of replacing the classical hypergeometric function $_2F_1(\frac{1}{2}, \frac{1}{2}; 1; x)$ by $_2F_1(\frac{1}{r}, \frac{r-1}{r}; 1; x)$, $r = 3, 4, 6$, leads to new and beautiful alternative theories. On the top of page 212 in his lost notebook [**60**], Ramanujan defines a cubic class invariant λ_n (i.e., $r = 3$ in (1.5)), which is an analogue of the Ramanujan–Weber classical invariants G_n and g_n defined in (6.3). Define Ramanujan's function

$$
f(-q) := (q; q)_\infty, \qquad |q| < 1, \tag{8.1}
$$

where $(a; q)_\infty$ is defined in (3.9), and the Dedekind eta-function $\eta(\tau)$

$$
\eta(\tau) := e^{2\pi i \tau/24} \prod_{j=1}^{\infty} (1 - e^{2\pi i j \tau}) =: q^{1/24} f(-q), \tag{8.2}
$$

where $q = e^{2\pi i \tau}$ and $\text{Im } \tau > 0$. Then Ramanujan's cubic class invariant λ_n is defined by

$$
\lambda_n = \frac{1}{3\sqrt{3}} \frac{f^6(q)}{\sqrt{q}\, f^6(q^3)} = \frac{1}{3\sqrt{3}} \left(\frac{\eta\left(\dfrac{1 + i\sqrt{n/3}}{2}\right)}{\eta\left(\dfrac{1 + i\sqrt{3n}}{2}\right)} \right)^6, \tag{8.3}
$$

where $q = e^{-\pi\sqrt{n/3}}$, i.e., $\tau = \frac{1}{2}i\sqrt{n/3}$.

Chan, Liaw, and Tan [**34**] established a general series representation for $1/\pi$ in terms of λ_n that is analogous to the general formulas of the Borweins and Chudnovskys in terms of the classical class invariants. To state this general formula, we first need some definitions. Define

$$
\frac{1}{\alpha^*(q)} := -\frac{1}{27q} \frac{f^{12}(q)}{f^{12}(q^3)} + 1. \tag{8.4}
$$

Thus, when $q = e^{-\pi\sqrt{n/3}}$ and $\alpha_n^* := \alpha^*(e^{-\pi\sqrt{n/3}})$, (8.3) and (8.4) imply that

$$
\frac{1}{\alpha_n^*} = 1 - \lambda_n^2.
$$

In analogy with (3.6), define the multiplier $\mathbf{m}(q)$ by

$$\mathbf{m}(q) := \mathbf{m}(\alpha^*, \beta^*) := \frac{{}_2F_1\left(\frac{1}{3}, \frac{2}{3}; 1; \alpha^*\right)}{{}_2F_1\left(\frac{1}{3}, \frac{2}{3}; 1; \beta^*\right)},$$

where $\beta^* = \alpha^*(q^n)$. We are now ready to state the general representation of $1/\pi$ derived by Chan, Liaw, and Tan [**34**, p. 102, Theorem 4.2].

Theorem 8.1. *For $n \geq 1$, let*

$$a_n = -\frac{\alpha_n^*(1 - \alpha_n^*)}{\sqrt{n}} \frac{d\mathbf{m}(\alpha^*, \beta^*)}{d\alpha^*}\Bigg|_{\alpha^*=1-\alpha_n^*, \beta^*=\alpha_n^*},$$

$$b_n = 1 - 2\alpha_n^*,$$

and

$$H_n = 4\alpha_n^*(1 - \alpha_n^*).$$

Then

$$\frac{1}{\pi}\sqrt{\frac{3}{n}} = \sum_{j=0}^{\infty}(a_n + b_n j)\frac{\left(\frac{1}{2}\right)_j \left(\frac{1}{3}\right)_j \left(\frac{2}{3}\right)_j}{(j!)^3}H_n^j. \tag{8.5}$$

We give one example. Let $n = 9$; then $\alpha_9^* = \frac{9}{8}$. Then, without providing further details,

$$\frac{4}{\pi\sqrt{3}} = \sum_{j=0}^{\infty}(5j + 1)\frac{\left(\frac{1}{2}\right)_j \left(\frac{1}{3}\right)_j \left(\frac{2}{3}\right)_j}{(j!)^3}\left(-\frac{9}{16}\right)^j,$$

which was discovered by Chan, Liaw, and Tan [**34**, p. 95].

Another general series representation for $1/\pi$ in the alternative theories of Ramanujan was devised by Berndt and Chan [**19**, p. 88, Eq. (5.80)]. We will not state this formula and all the requisite definitions, but let it suffice to say that the formula involves Ramanujan's Eisenstein series $P(q)$, $Q(q) = E_4(\tau)$, and $R(q) = E_6(\tau)$ at the argument $q = -e^{-\pi\sqrt{n}}$ and the modular j-invariant, defined by $j(\tau) = 1728J(\tau)$, where $J(\tau)$ is defined by (7.2). In particular,

$$j\left(\frac{3 + \sqrt{-3n}}{2}\right) = -27\frac{(\lambda_n^2 - 1)(9\lambda_n^2 - 1)^3}{\lambda_n^2},$$

a proof of which can be found in [**18**]. The hypergeometric terms are of the form

$$\frac{\left(\frac{1}{2}\right)_j \left(\frac{1}{6}\right)_j \left(\frac{5}{6}\right)_j}{(j!)^3}.$$

Berndt and Chan used their general formula to calculate a series for $1/\pi$ that yields about 73 or 74 digits of π per term.

       [Monthly 116

Lastly, we conclude this section by remarking that Berndt, Chan, and Liaw [20] have derived series representations for $1/\pi$ that fall under the umbrella of Ramanujan's quartic theory of elliptic functions. Because the quartic theory is intimately connected with the classical theory, their general formulas [20, p. 144, Theorem 4.1] involve the classical invariants G_n and g_n in their summands. Not surprisingly, the hypergeometric terms are of the form

$$B_j := \frac{\left(\frac{1}{2}\right)_j \left(\frac{1}{4}\right)_j \left(\frac{3}{4}\right)_j}{(j!)^3}.$$

The simplest example arising from their theory is given by [20, p. 145]

$$\frac{9}{2\pi} = \sum_{j=0}^{\infty} B_j(7j+1)\left(\frac{32}{81}\right)^j.$$

9. THE PRESENT AUTHORS AS DISCIPLES OF RAMANUJAN. As mentioned in Section 2, the first two authors were inspired by the third author to continue the development of Ramanujan's thoughts. In their first paper [9], Baruah and Berndt employed Ramanujan's ideas in the classical theory of elliptic functions to prove 13 of Ramanujan's original formulas and many new ones as well. In [7], they utilized Ramanujan's cubic and quartic theories to establish five of Ramanujan's 17 formulas in addition to some new representations. Lastly, in [8], motivated by the work of J. Guillera, described briefly in Section 10 below, the first two authors extended Ramanujan's ideas to derive hypergeometric-like series representations for $1/\pi^2$. For example,

$$\frac{24}{\pi^2} = \sum_{\mu=0}^{\infty} (44571654400\mu^2 + 5588768408\mu + 233588841) B_\mu \left(\frac{1}{99^4}\right)^{\mu+1},$$

where

$$B_\mu = \sum_{\nu=0}^{\mu} \frac{\left(\frac{1}{4}\right)_\nu \left(\frac{1}{4}\right)_{\mu-\nu} \left(\frac{1}{2}\right)_\nu \left(\frac{1}{2}\right)_{\mu-\nu} \left(\frac{3}{4}\right)_\nu \left(\frac{3}{4}\right)_{\mu-\nu}}{\nu!^3 (\mu-\nu)!^3}.$$

In Section 3, we defined Ramanujan's Eisenstein series $P(q)$ in (3.12) and offered several definitions from Ramanujan's theories of elliptic functions in giving a brief introduction to Ramanujan's ideas. Here we highlight the role of $P(q)$ in more detail before giving a complete proof of (1.3). Because these three series representations (1.2)–(1.4) can also be found in Ramanujan's lost notebook, our proof here is similar to that given in [3, Chapter 15].

Following Ramanujan, set

$$z := {}_2F_1\left(\tfrac{1}{2}, \tfrac{1}{2}; 1; x\right). \tag{9.1}$$

The two most important ingredients in our derivations are Ramanujan's representation for $P(q^2)$ given by [13, p. 120, Entry 9(iv)]

$$P(q^2) = (1 - 2x)z^2 + 6x(1-x)z\frac{dz}{dx} \tag{9.2}$$

and Clausen's formula (3.7), which, using (9.1), we restate in the form

$$z^2 = {}_3F_2(\tfrac{1}{2}, \tfrac{1}{2}, \tfrac{1}{2}; 1, 1; X) = \sum_{j=0}^{\infty} A_j X^j, \tag{9.3}$$

where, as in (1.1),

$$A_j := \frac{(\tfrac{1}{2})_j^3}{j!^3} \quad \text{and} \quad X := 4x(1-x). \tag{9.4}$$

From (9.3) and (9.4),

$$2z\frac{dz}{dx} = \sum_{j=0}^{\infty} A_j j X^{j-1} \cdot 4(1-2x). \tag{9.5}$$

Hence, from (9.2), (9.3), (9.5), and (9.4),

$$P(q^2) = (1-2x)\sum_{j=0}^{\infty} A_j X^j + 3(1-2x)\sum_{j=0}^{\infty} A_j j X^j$$

$$= \sum_{j=0}^{\infty} \{(1-2x) + 3(1-2x)j\} A_j X^j. \tag{9.6}$$

For $q := e^{-\pi\sqrt{n}}$, recall (9.1) and set

$$x_n = k^2(e^{-\pi\sqrt{n}}), \qquad z_n := {}_2F_1\left(\tfrac{1}{2}, \tfrac{1}{2}; 1; x_n\right) \tag{9.7}$$

and

$$X_n = 4x_n(1-x_n). \tag{9.8}$$

For later use, we note that [3, p. 375]

$$1 - x_n = x_{1/n} \quad \text{and} \quad z_{1/n} = \sqrt{n} z_n. \tag{9.9}$$

With the use of (9.7) and (9.8), (9.6) takes the form

$$P(e^{-2\pi\sqrt{n}}) = \sum_{j=0}^{\infty} \{(1-2x_n) + 3(1-2x_n)j\} A_j X_n^j$$

$$= (1-2x_n)z_n^2 + 3\sum_{j=0}^{\infty}(1-2x_n)j A_j X_n^j. \tag{9.10}$$

In order to utilize (9.10), we require two different formulas, each involving both $P(q^2)$ and $P(q^{2n})$, where n is a positive integer. The first comes from a transformation formula for $P(q)$, which in turn arises from the transformation formula for $f(-q)$ defined in (8.1) or the Dedekind eta function defined in (8.2), and is for general n. This transformation formula is given by [13, p. 43, Entry 27(iii)]

$$e^{-\alpha/12}\alpha^{1/4} f(-e^{-2\alpha}) = e^{-\beta/12}\beta^{1/4} f(-e^{-2\beta}), \tag{9.11}$$

where $\alpha\beta = \pi^2$, with α and β both positive. Taking the logarithm of both sides of (9.11), we find that

$$-\frac{\alpha}{12} + \frac{1}{4}\log\alpha + \sum_{j=1}^{\infty}\log(1 - e^{-2j\alpha}) = -\frac{\beta}{12} + \frac{1}{4}\log\beta + \sum_{j=1}^{\infty}\log(1 - e^{-2j\beta}).$$

(9.12)

Differentiating both sides of (9.12) with respect to α, we deduce that

$$-\frac{1}{12} + \frac{1}{4\alpha} + \sum_{j=1}^{\infty}\frac{2je^{-2j\alpha}}{1 - e^{-2j\alpha}} = \frac{\beta}{12\alpha} - \frac{1}{4\alpha} - \sum_{j=1}^{\infty}\frac{(2j\beta/\alpha)e^{-2j\beta}}{1 - e^{-2j\beta}}.$$

(9.13)

Multiplying both sides of (9.13) by 12α and rearranging, we arrive at

$$6 - \alpha\left(1 - 24\sum_{j=1}^{\infty}\frac{je^{-2j\alpha}}{1 - e^{-2j\alpha}}\right) = \beta\left(1 - 24\sum_{j=1}^{\infty}\frac{je^{-2j\beta}}{1 - e^{-2j\beta}}\right).$$

(9.14)

Setting $\alpha = \pi/\sqrt{n}$, so that $\beta = \pi\sqrt{n}$, recalling the definition (3.12) of $P(q)$, and rearranging slightly, we see that (9.14) takes the shape

$$\frac{6\sqrt{n}}{\pi} = P(e^{-2\pi/\sqrt{n}}) + nP(e^{-2\pi\sqrt{n}}).$$

(9.15)

This is the first desired formula.

The second gives representations for Ramanujan's function [57], [58, pp. 33–34]

$$f_n(q) := nP(q^{2n}) - P(q^2)$$

(9.16)

for *certain* positive integers n. (Ramanujan [57], [58, pp. 33–34] used the notation $f(n)$ instead of $f_n(q)$.) In [57], Ramanujan recorded representations for $f_n(q)$ for 12 values of n, but he gave no indication of how these might be proved. These formulas are also recorded in Chapter 21 of Ramanujan's second notebook [59], and proofs may be found in [13].

We now give the details for our proof of (1.3), which was clearly a favorite of Gabriella Montez, the precocious student in *High School Musical*. Unfortunately, we do not know whether she possessed a proof of her own. We restate (1.3) here for convenience.

Theorem 9.1. *If A_j, $j \geq 0$, is defined by (9.4), then*

$$\frac{16}{\pi} = \sum_{j=0}^{\infty}(42j + 5)A_j\frac{1}{2^{6j}}.$$

(9.17)

Proof. The identity (9.17) is connected with modular equations of degree 7. Thus, our first task is to calculate the singular modulus x_7. To that end, we begin with a modular equation of degree 7

$$\{x(q)x(q^7)\}^{1/8} + \{(1 - x(q))(1 - x(q^7))\}^{1/8} = 1,$$

(9.18)

due to C. Guetzlaff in 1834 but rediscovered by Ramanujan in Entry 19(i) of Chapter 19 of his second notebook [59], [13, p. 314]. In the notation of our definition of a modular equation after (3.5), we have set $x(q) = k^2(q)$, and so $x(q^7) = \ell^2(q)$. Set $q = e^{-\pi/\sqrt{7}}$ in (9.18) and use (9.9) and (9.8) to deduce that

$$2\left\{x_7(1 - x_7)\right\}^{1/8} = 1 \quad \text{and} \quad X_7 = \frac{1}{2^6}. \tag{9.19}$$

Ramanujan calculated the singular modulus x_7 in his first notebook [59], [15, p. 290], from which, or from (9.19), we easily can deduce that

$$1 - 2x_7 = \frac{3\sqrt{7}}{8}. \tag{9.20}$$

In the notation (9.16), from either [57], [58, p. 33], or [13, p. 468, Entry 5(iii)],

$$f_7(q) = 3z(q)z(q^7)\left(1 + \sqrt{x(q)x(q^7)} + \sqrt{(1 - x(q))(1 - x(q^7))}\right). \tag{9.21}$$

Putting $q = e^{-\pi/\sqrt{7}}$ in (9.21) and employing (9.9) and (9.19), we find that

$$f_7(e^{-\pi/\sqrt{7}}) = 3\sqrt{7}\left(1 + 2\sqrt{x_7(1 - x_7)}\right)z_7^2 = 3\sqrt{7} \cdot \frac{9}{8}z_7^2. \tag{9.22}$$

Letting $n = 7$ in (9.15) and (9.10), and using (9.20), we see that

$$\frac{6\sqrt{7}}{\pi} = P(e^{-2\pi/\sqrt{7}}) + 7P(e^{-2\pi\sqrt{7}}) \tag{9.23}$$

and

$$P(e^{-2\pi\sqrt{7}}) = (1 - 2x_7)z_7^2 + 3\sum_{j=0}^{\infty}(1 - 2x_7)jA_jX_7^j$$

$$= \frac{3\sqrt{7}}{8}z_7^2 + \frac{9\sqrt{7}}{8}\sum_{j=0}^{\infty}\frac{jA_j}{2^{6j}}, \tag{9.24}$$

respectively. Eliminating $P(e^{-2\pi/\sqrt{7}})$ from (9.22) and (9.23) and putting the resulting formula for $P(e^{-2\pi\sqrt{7}})$ in (9.24), we find that

$$\frac{3\sqrt{7}}{7\pi} + \frac{27\sqrt{7}}{16 \cdot 7}z_7^2 = \frac{3\sqrt{7}}{8}z_7^2 + \frac{9\sqrt{7}}{8}\sum_{j=0}^{\infty}\frac{jA_j}{2^{6j}},$$

which upon simplification with the use of (9.3) yields (9.17).   ∎

10. JESÚS GUILLERA.

A discrete function $A(n, k)$ is hypergeometric if

$$\frac{A(n + 1, k)}{A(n, k)} \quad \text{and} \quad \frac{A(n, k + 1)}{A(n, k)}$$

are both rational functions. A pair of functions $F(n, k)$ and $G(n, k)$ is said to be a WZ pair (after H. S. Wilf and D. Zeilberger) if F and G are hypergeometric and

$$F(n + 1, k) - F(n, k) = G(n, k + 1) - G(n, k).$$

In this case, H. Wilf and D. Zeilberger [55] showed that there exists a rational function $C(n, k)$ such that

$$G(n, k) = C(n, k)F(n, k).$$

The function $C(n, k)$ is called a certificate of (F, G). Defining

$$H(n, k) = F(n + k, n + 1) + G(n, n + k),$$

Wilf and Zeilberger showed that

$$\sum_{n=0}^{\infty} H(n, 0) = \sum_{n=0}^{\infty} G(n, 0).$$

Ekhad (Zeilberger's computer) and Zeilberger [46] were the first to use this method to derive a one-page proof of the representation

$$\frac{2}{\pi} = \sum_{j=0}^{\infty} (-1)^j (4j + 1) \frac{\left(\frac{1}{2}\right)_j^3}{(j!)^3}. \tag{10.1}$$

The identity (10.1) was first proved by G. Bauer in 1859 [10]. Ramanujan recorded (10.1) as Example 14 in Section 7 of Chapter 10 in his second notebook [59], [12, pp. 23–24]. Further references can be found in [9]. In 1905, generalizing Bauer's approach, J. W. L. Glaisher [47] found further series for $1/\pi$.

Motivated by this work, Guillera [48] found many new WZ-pairs (F, G) and derived new series not only for $1/\pi$ but for $1/\pi^2$ as well. One of his most elegant formulas is

$$\frac{128}{\pi^2} = \sum_{j=0}^{\infty} (-1)^j \binom{2j}{j}^5 \frac{(820j^2 + 180j + 13)}{2^{10j}}.$$

Subsequently, Guillera empirically discovered many series of the type

$$\frac{A}{\pi} = \sum_{j=0}^{\infty} c_j \frac{Bj^2 + Dj + E}{H^j}.$$

Most of the series he discovered cannot be proved by the WZ-method; it appears that the WZ-method is only applicable to those series for $1/\pi$ when H is a power of 2. An example of Guillera's series which remains to be proved is [49]

$$\frac{128\sqrt{5}}{\pi^2} = \sum_{j=0}^{\infty} (-1)^j \frac{\left(\frac{1}{2}\right)_j \left(\frac{1}{3}\right)_j \left(\frac{2}{3}\right)_j \left(\frac{1}{6}\right)_j \left(\frac{5}{6}\right)_j}{(j!)^5 803^{3j}} (5418j^2 + 693j + 29).$$

For further Ramanujan-like series for $1/\pi^2$, see Zudilin's papers [65], [66].

11. RECENT DEVELOPMENTS. We have emphasized in this paper that Clausen's formula (6.1) is an essential ingredient in most proofs of Ramanujan-type series representations for $1/\pi$. However, there are other kinds of series for $1/\pi$ that do not depend upon Clausen's formula. One such series discovered by Takeshi Sato [62] is given by

$$\frac{1}{\pi}\frac{\sqrt{15}}{120(4\sqrt{5}-9)} = \sum_{\mu=0}^{\infty}\sum_{\nu=0}^{\mu}\binom{\mu}{\nu}^2\binom{\mu+\nu}{\nu}^2\left(\frac{1}{2}-\frac{3}{20}\sqrt{5}+\mu\right)\left(\frac{\sqrt{5}-1}{2}\right)^{12\mu}.$$

(11.1)

In unpublished work (personal communication to the third author), Sato derived a more complicated series for $1/\pi$ that yields approximately 97 digits of π per term. A companion to (11.1), which was derived by a new method devised by the third author, S. H. Chan, and Z.-G. Liu [31], is given by

$$\frac{8}{\sqrt{3}\pi} = \sum_{\mu=0}^{\infty}a_\mu\frac{5\mu+1}{64^\mu},$$

(11.2)

where

$$a_\mu := \sum_{\nu=0}^{\mu}\binom{2\mu-2\nu}{\mu-\nu}\binom{2\nu}{\nu}\binom{\mu}{\nu}^2.$$

(11.3)

We cite three further new series arising from this new method. The first is another companion of (11.2), which arises from recent work of Chan and H. Verrill [36] (after the work of Almkvist and Zudilin [2]), and is given by

$$\frac{9}{2\sqrt{3}\pi} = \sum_{\mu=0}^{\infty}\sum_{\nu=0}^{\lceil\frac{\mu}{3}\rceil}(-1)^{\mu-\nu}3^{\mu-3\nu}\binom{\mu}{3\nu}\binom{\mu+\nu}{\nu}\frac{(3\nu)!}{(\nu!)^3}(4\mu+1)\left(\frac{1}{81}\right)^\mu.$$

The second is from a paper by Chan and K. P. Loo [35] and takes the form

$$\frac{2\sqrt{3}(3+2\sqrt{2})}{9\pi} = \sum_{\mu=0}^{\infty}C_\mu\left(\mu+1-\frac{2}{3}\sqrt{2}\right)\left(-1+\frac{3}{4}\sqrt{2}\right)^\mu,$$

where

$$C_\mu = \sum_{\nu=0}^{\mu}\left\{\sum_{j=0}^{\nu}\binom{\nu}{j}^3\sum_{i=0}^{\mu-\nu}\binom{\mu-\nu}{i}^3\right\}.$$

The third was derived by Y. Yang (personal communication) and takes the shape

$$\frac{18}{\pi\sqrt{15}} = \sum_{\mu=0}^{\infty}\sum_{\nu=0}^{\mu}\binom{\mu}{\nu}^4\frac{4\mu+1}{36^\mu}.$$

Motivated by his work with Mahler measures and new transformation formulas for $_5F_4$ series, M. D. Rogers [61, Corollary 3.2] has also discovered series for $1/\pi$ in the spirit of the formulas above. For example, if a_μ is defined by (11.3), then

$$\frac{2}{\pi} = \sum_{\mu=0}^{\infty}(-1)^\mu a_\mu\frac{3\mu+1}{32^\mu}.$$

This series was also independently discovered by Chan and Verrill [36].

According to W. Zudilin [67], G. Gourevich empirically discovered a hypergeometric-like series for $1/\pi^3$, namely,

$$\frac{32}{\pi^3} = \sum_{\mu=0}^{\infty} \frac{\left(\frac{1}{2}\right)_\mu^7}{(\mu!)^7 \, 2^{6\mu}} (168\mu^3 + 76\mu^2 + 14\mu + 1).$$

This series and the search for further series representations for $1/\pi^m$, $m \geq 2$, are described in a paper by D. H. Bailey and J. M. Borwein [4].

12. CONCLUSION. One test of "good" mathematics is that it should generate more "good" mathematics. Readers have undoubtedly concluded that Ramanujan's original series for $1/\pi$ have sown the seeds for an abundant crop of "good" mathematics.

ACKNOWLEDGMENTS. The authors are deeply grateful to R. William Gosper for carefully reading earlier versions of this paper, uncovering several errors, and offering excellent suggestions. We are pleased to thank Si Min Chan and Si Ya Chan for watching *High School Musical*, thereby making their father aware of Walt Disney Productions' interest in Ramanujan's formulas for $1/\pi$.

The research of Baruah was partially supported by BOYSCAST Fellowship grant SR/BY/M-03/05 from DST, Govt. of India. Berndt's research was partially supported by grant H98230-07-1-0088 from the National Security Agency. Lastly, Chan's research was partially supported by the National University of Singapore Academic Research Fund R-146-000-103-112.

REFERENCES

1. G. Almkvist, C. Krattenthaler, and J. Petersson, Some new formulas for π, *Experiment. Math.* **12** (2003) 441–456.
2. G. Almkvist and W. Zudilin, Differential equations, mirror maps and zeta values, in *Mirror Symmetry V*, N. Yui, S.-T. Yau, and J. D. Lewis, eds., AMS/IP Studies in Advanced Mathematics **38**, American Mathematical Society, Providence, RI, 2006, 481–515.
3. G. E. Andrews and B. C. Berndt, *Ramanujan's Lost Notebook*, Part II, Springer, New York, 2009.
4. D. H. Bailey and J. M. Borwein, Computer-assisted discovery and proof, in *Tapas in Experimental Mathematics*, T. Amdeberhan and V. H. Moll, eds., Contemporary Mathematics, vol. 457, American Mathematical Society, Providence, RI, 2008, 21–52.
5. W. N. Bailey, *Generalized Hypergeometric Series*, Cambridge University Press, London, 1935.
6. A. Baker, *Transcendental Number Theory*, Cambridge University Press, Cambridge, 1975.
7. N. D. Baruah and B. C. Berndt,, Ramanujan's series for $1/\pi$ arising from his cubic and quartic theories of elliptic functions, *J. Math. Anal. Appl.* **341** (2008) 357–371.
8. ———, Ramanujan's Eisenstein series and new hypergeometric-like series for $1/\pi^2$, *J. Approx. Theory* (to appear).
9. ———, Eisenstein series and Ramanujan-type series for $1/\pi$, *Ramanujan J.* (to appear).
10. G. Bauer, Von den Coefficienten der Reihen von Kugelfunctionen einer Variabeln, *J. Reine Angew. Math.* **56** (1859) 101–121.
11. L. Berggren, J. Borwein, and P. Borwein, *Pi: A Source Book*, Springer, New York, 1997.
12. B. C. Berndt, *Ramanujan's Notebooks*, Part II, Springer-Verlag, New York, 1989.
13. ———, *Ramanujan's Notebooks*, Part III, Springer-Verlag, New York, 1991.
14. ———, *Ramanujan's Notebooks*, Part IV, Springer-Verlag, New York, 1994.
15. ———, *Ramanujan's Notebooks*, Part V, Springer-Verlag, New York, 1998.
16. B. C. Berndt, S. Bhargava, and F. G. Garvan, Ramanujan's theories of elliptic functions to alternative bases, *Trans. Amer. Math. Soc.* **347** (1995) 4136–4244.
17. B. C. Berndt and H. H. Chan, Notes on Ramanujan's singular moduli, in *Number Theory*, Fifth Conference of the Canadian Number Theory Association, R. Gupta and K. S. Williams, eds., American Mathematical Society, Providence, RI, 1999, 7–16.
18. ———, Ramanujan and the modular j-invariant, *Canad. Math. Bull.* **42** (1999) 427–440.
19. ———, Eisenstein series and approximations to π, *Illinois J. Math.* **45** (2001) 75–90.
20. B. C. Berndt, H. H. Chan, and W.-C. Liaw, On Ramanujan's quartic theory of elliptic functions, *J. Number Theory* **88** (2001) 129–156.
21. B. C. Berndt, H. H. Chan, and L.-C. Zhang, Ramanujan's singular moduli, *Ramanujan J.* **1** (1997) 53–74.

22. B. C. Berndt and R. A. Rankin, *Ramanujan: Essays and Surveys*, American Mathematical Society, Providence, RI, 2001; London Mathematical Society, London, 2001.

23. J. M. Borwein and P. B. Borwein, *Pi and the AGM; A Study in Analytic Number Theory and Computational Complexity*, Wiley, New York, 1987.

24. ———, Ramanujan's rational and algebraic series for $1/\pi$, *J. Indian Math. Soc.* **51** (1987) 147–160.

25. ———, More Ramanujan-type series for $1/\pi$, in *Ramanujan Revisited*, G. E. Andrews, R. A. Askey, B. C. Berndt, K. G. Ramanathan, and R. A. Rankin, eds., Academic Press, Boston, 1988, 359–374.

26. ———, Ramanujan and pi, *Scientific American* **256**, February, 1988, 112–117; reprinted in [**22**, pp. 187–199]; reprinted in [**11**, pp. 588–595].

27. ———, A cubic counterpart of Jacobi's identity and the AGM, *Trans. Amer. Math. Soc.* **323** (1991) 691–701.

28. ———, Some observations on computer aided analysis, *Notices Amer. Math. Soc.* **39** (1992) 825–829.

29. ———, Class number three Ramanujan type series for $1/\pi$, *J. Comput. Appl. Math.* **46** (1993) 281–290.

30. J. M. Borwein, P. B. Borwein, and D. H. Bailey, Ramanujan, modular equations, and approximations to pi or how to compute one billion digits of pi, this MONTHLY **96** (1989) 201–219.

31. H. H. Chan, S. H. Chan, and Z. Liu, Domb's numbers and Ramanujan-Sato type series for $1/\pi$, *Adv. Math.* **186** (2004) 396–410.

32. H. H. Chan and S.-S. Huang, On the Ramanujan-Göllnitz-Gordon continued fraction, *Ramanujan J.* **1** (1997) 75–90.

33. H. H. Chan and W.-C. Liaw, Cubic modular equations and new Ramanujan-type series for $1/\pi$, *Pacific J. Math.* **192** (2000) 219–238.

34. H. H. Chan, W.-C. Liaw, and V. Tan, Ramanujan's class invariant λ_n and a new class of series for $1/\pi$, *J. London Math. Soc.* **64** (2001) 93–106.

35. H. H. Chan and K. P. Loo, Ramanujan's cubic continued fraction revisited, *Acta Arith.* **126** (2007) 305–313.

36. H. H. Chan and H. Verrill, The Apéry numbers, the Almkvist-Zudilin numbers and new series for $1/\pi$, *Math. Res. Lett.* (to appear).

37. S. Chowla, Series for $1/K$ and $1/K^2$, *J. London Math. Soc.* **3** (1928) 9–12.

38. ———, On the sum of a certain infinite series, *Tôhoku Math. J.* **29** (1928) 291–295.

39. ———, *The Collected Papers of Sarvadaman Chowla*, vol. 1, Les Publications Centre de Recherches Mathématiques, Montreal, 1999.

40. D. V. Chudnovsky and G. V. Chudnovsky, Approximation and complex multiplication according to Ramanujan, in *Ramanujan Revisited*, G. E. Andrews, R. A. Askey, B. C. Berndt, K. G. Ramanathan, and R. A. Rankin, eds., Academic Press, Boston, 1988, 375–472.

41. ———, The computation of classical constants, *Proc. Nat. Acad. Sci. USA* **86** (1989) 8178–8182.

42. ———, Computational problems in arithmetic of linear differential equations. Some diophantine applications, in *Number Theory: A Seminar held at the Graduate School and University Center of the City University of New York 1985–88*, Lecture Notes in Mathematics, vol. 1383, D. V. Chudnovsky, G. V. Chudnovsky, H. Cohn, and M. B. Nathanson, eds., Springer-Verlag, Berlin, 1989, 12–49.

43. ———, Classical constants and functions: Computations and continued fraction expansions, in *Number Theory: New York Seminar 1989/1990*, D. V. Chudnovsky, G. V. Chudnovsky, H. Cohn, and M. B. Nathanson, eds., Springer-Verlag, Berlin, 1991, 12–74.

44. ———, Hypergeometric and modular function identities, and new rational approximations to and continued fraction expansions of classical constants and functions, in *A Tribute to Emil Grosswald: Number Theory and Related Analysis*, Contemporary Mathematics, vol. 143, M. Knopp and M. Sheingorn, eds., American Mathematical Society, Providence, RI, 1993, 117–162.

45. ———, Generalized hypergeometric functions—classification of identities and explicit rational approximations, in *Algebraic Methods and q-Special Functions*, CRM Proceedings and Lecture Notes, vol. 22, American Mathematical Society, Providence, RI, 1993, 59–91.

46. S. Ekhad and D. Zeilberger, A WZ proof of Ramanujan's formula for π, in *Geometry, Analysis and Mechanics*, World Scientific, River Edge, NJ, 1994, 107–108.

47. J. W. L. Glaisher, On series for $1/\pi$ and $1/\pi^2$, *Quart. J. Pure Appl. Math.* **37** (1905) 173–198.

48. J. Guillera, Some binomial series obtained by the WZ-method, *Adv. in Appl. Math.* **29** (2002) 599–603.

49. ———, About a new kind of Ramanujan-type series, *Experiment. Math.* **12** (2003) 507–510.

50. ———, Generators of some Ramanujan formulas, *Ramanujan J.* **11** (2006) 41–48.

51. ———, A new method to obtain series for $1/\pi$ and $1/\pi^2$, *Experiment. Math.* **15** (2006) 83–89.

52. ———, A class of conjectured series representations for $1/\pi$, *Experiment. Math.* **15** (2006) 409–414.

53. ———, Hypergeometric identities for 10 extended Ramanujan type series, *Ramanujan J.* **15** (2008) 219–234.

54. M. Hata, Rational approximations to π and some other numbers, *Acta Arith.* **63** (1993) 335–349.

55. M. Petkovšek, H. S. Wilf, and D. Zeilberger, $A = B$, A K Peters, Wellesey, MA, 1996.

56. R. Preston, Profiles: The mountains of pi, *The New Yorker*, March 2, 1992, 36–67.
57. S. Ramanujan, Modular equations and approximations to π, *Quart. J. Math.* (Oxford) **45** (1914) 350–372.
58. ———, *Collected Papers*, Cambridge University Press, Cambridge, 1927; reprinted by Chelsea, New York, 1962; reprinted by the American Mathematical Society, Providence, RI, 2000.
59. ———, *Notebooks* (2 volumes), Tata Institute of Fundamental Research, Bombay, 1957.
60. ———, *The Lost Notebook and Other Unpublished Papers*, Narosa, New Delhi, 1988.
61. M. D. Rogers, New $_5F_4$ hypergeometric transformations, three-variable Mahler measures, and formulas for $1/\pi$, *Ramanujan J.* **18** (2009) 327–340.
62. T. Sato, Apéry numbers and Ramanujan's series for $1/\pi$, Abstract for a lecture presented at the annual meeting of the Mathematical Society of Japan, March 28–31, 2002.
63. H. Weber, *Lehrbuch der Algebra*, vol. 3, Chelsea, New York, 1961.
64. W. Zudilin, Ramanujan-type formulae and irrationality measures of certain multiples of π, *Mat. Sb.* **196** (2005) 51–66.
65. ———, Quadratic transformations and Guillera's formulae for $1/\pi^2$, *Math. Zametki* **81** (2007) 335–340 (Russian); *Math. Notes* **81** (2007) 297–301.
66. ———, More Ramanujan-type formulae for $1/\pi^2$, *Russian Math. Surveys* **62** (2007) 634–636.
67. ———, Ramanujan-type formulae for $1/\pi$: A second wind?, in *Modular Forms and String Duality*, N. Yui, H. Verrill, and C. F. Doran, eds., Fields Institute Communications, vol. 54, American Mathematical Society & The Fields Institute for Research in Mathematical Sciences, Providence, RI, 2008, pp. 179–188.

NAYANDEEP DEKA BARUAH is Reader of Mathematical Sciences at Tezpur University, Assam, India. He received his Ph.D. degree from this university in 2001. He was a recipient of an ISCA Young Scientist Award in 2004 and a BOYSCAST Fellowship in 2005–06 from the Department of Science and Technology, Government of India, under which he spent the year March 2006–March 2007 at the University of Illinois at Urbana-Champaign working with Professor Berndt. He has been a fan of π since the first year of his B.Sc. To learn more about π, he began reading about Ramanujan and his mathematics and became a huge fan of Ramanujan! He still loves to recite the first 2000 digits of π from memory in his leisure time.
Department of Mathematical Sciences, Tezpur University, Napaam-784028, Sonitpur, Assam, India
nayan@tezu.ernet.in

BRUCE C. BERNDT is Professor of Mathematics at the University of Illinois at Urbana-Champaign. Since 1977 he has devoted almost all of his research energy toward finding proofs for the claims made by Ramanujan in his (earlier) notebooks and lost notebook. Aiding him in this endeavor have been many of his 25 former and 6 current doctoral students, as well as several outstanding postdoctoral visitors. So far, he has published ten books on Ramanujan's work. Most recently, he and George Andrews have published their second of approximately four volumes on Ramanujan's lost notebook.
Department of Mathematics, University of Illinois, 1409 West Green St., Urbana, IL 61801, USA
berndt@illinois.edu

HENG HUAT CHAN is currently Professor of Mathematics at the National University of Singapore. He received his Ph.D. in 1995 under the supervision of Professor B. C. Berndt (the second author of this paper) and spent nine months at the Institute for Advanced Study after his graduation. He then took up a one-year visiting position at the National Chung Cheng University in Taiwan. While in Taiwan, he attempted without success to design a graduate course based on the book "Pi and the AGM" by J. M. Borwein and P. B. Borwein. However, it was during this period that he began his research on "Ramanujan-type series for $1/\pi$." Since then, he has published several papers on this topic with various mathematicians.
Department of Mathematics, National University of Singapore, 2 Science Drive 2, Singapore 117543, Republic of Singapore
matchh@nus.edu.sg

20. The computation of previously inaccessible digits of π^2 and Catalan's constant (2013)

Paper 20: David H. Bailey, Jonathan M. Borwein, Andrew Mattingly and Glenn Wightwick, "The computation of previously inaccessible digits of π^2 and Catalan's constant," *Notices of the American Mathematical Society*, vol. 60 (2013), p. 844–854. Reprinted by permission of the American Mathematical Society.

Synopsis:

An earlier selection (paper #14 in this collection) presented what is now known as the BBP formula for π, which permits one to calculate binary or base-16 digits of π beginning at an arbitrary starting point. The original BBP paper presented a similar formula for π^2, permitting arbitrary binary digits of π^2 to be calculated by this same general process. Since the publication of that paper, additional BBP-type formulas have also been found, among them one that permits arbitrary base-3 digits of π^2 to be calculated, and another that permits arbitrary binary digits of Catalan's constant $= \sum_{n=0}^{\infty}(-1)^n/(2n+1)^2 = 0.9159965594\ldots$ to be calculated.

This paper outlines the history of computing π and other constants through the ages, and then gives details on three new computations: base-64 digits of π^2, base-729 digits of π^2 and base-4096 digits of Catalan's constant, in each case beginning with position ten trillion. These computations, which required a total of approximately 1.5×10^{19} floating-point arithmetic operations, and which ran for tens of "rack-days" on an IBM Blue-Gene computer, are comparable in total cost, say, to that of generating a state-of-the-art animated movie.

Keywords: Computation, History, Normality, General Audience

© Springer International Publishing Switzerland 2016
D.H. Bailey, J.M. Borwein, *Pi: The Next Generation*,
DOI 10.1007/978-3-319-32377-0_20

The Computation of Previously Inaccessible Digits of π^2 and Catalan's Constant

David H. Bailey, Jonathan M. Borwein, Andrew Mattingly, and Glenn Wightwick

Introduction

We recently concluded a very large mathematical calculation, uncovering objects that until recently were widely considered to be forever inaccessible to computation. Our computations stem from the "BBP" formula for π, which was discovered in 1997 using a computer program implementing the "PSLQ" integer relation algorithm. This formula has the remarkable property that it permits one to directly calculate binary digits of π, beginning at an arbitrary position d, without needing to calculate any of the first $d-1$ digits. Since 1997

David H. Bailey is senior scientist at the Computation Research Department of the Lawrence Berkeley National Laboratory. His email address is dhbailey@lbl.gov.

Jonathan M. Borwein is professor of mathematics at the Centre for Computer Assisted Research Mathematics and its Applications (CARMA), University of Newcastle. His email address is jonathan.borwein@newcastle.edu.au.

Andrew Mattingly is senior information technology architect at IBM Australia. His email address is andrew_mattingly@au1.ibm.com.

Glenn Wightwick is director, IBM Research–Australia. His email address is glenn_wightwick@au.ibm.com.

The first author was supported in part by the Director, Office of Computational and Technology Research, Division of Mathematical, Information, and Computational Sciences of the U.S. Department of Energy, under contract number DE-AC02-05CH11231.

DOI: http://dx.doi.org/10.1090/noti1015

numerous other BBP-type formulas have been discovered for various mathematical constants, including formulas for π^2 (both in binary and ternary bases) and for Catalan's constant.

In this article we describe the computation of base-64 digits of π^2, base-729 digits of π^2, and base-4096 digits of Catalan's constant, in each case beginning at the ten trillionth place, computations that involved a total of approximately 1.549×10^{19} floating-point operations. We also discuss connections between BBP-type formulas and the age-old unsolved questions of whether and why constants such as $\pi, \pi^2, \log 2$, and Catalan's constant have "random" digits.

Historical Background

Since the dawn of civilization, mathematicians have been intrigued by the digits of π [6], more so than any other mathematical constant. In the third century BCE, Archimedes employed a brilliant scheme of inscribed and circumscribed $3 \cdot 2^n$-gons to compute π to two decimal digit accuracy. However, this and other numerical calculations of antiquity were severely hobbled by their reliance on primitive arithmetic systems.

One of the most significant scientific developments of history was the discovery of full positional decimal arithmetic with zero by an unknown mathematician or mathematicians in India at least by 500 CE and probably earlier. Some of the earliest documentation includes the *Aryab-hatiya*, the writings of the Indian mathematician Aryabhata dated to 499 CE; the *Lokavibhaga*, a cosmological work with astronomical observations that permit modern scholars to conclude that it was written on 25 August 458 CE [9]; and the Bakhshali manuscript, an ancient mathematical treatise that some scholars believe may be older still, but in any event is no later than the seventh century [7], [8], [2]. The Bakhshali manuscript includes, among other things, the following intriguing algorithm for computing the square root of q, starting with an approximation x_0:

$$a_n = \frac{q - x_n^2}{2x_n},$$

$$(1) \qquad x_{n+1} = x_n + a_n - \frac{a_n^2}{2(x_n + a_n)}.$$

This scheme is quartically convergent in that it approximately *quadruples* the number of correct digits with each iteration (although it was never iterated more than once in the examples given in the manuscript) [2].

In the tenth century, Gerbert of Aurillac, who later reigned as Pope Sylvester II, attempted to introduce decimal arithmetic in Europe, but little headway was made until the publication of Fibonacci's *Liber Abaci* in 1202. Several hundred more years would pass before the system finally gained universal, if belated, adoption in the West. The time of Sylvester's reign was a very turbulent one, and he died in 1003, shortly after the death of his protector, Emperor Otto III. It is interesting to speculate how history would have changed had he lived longer. A page from his mathematical treatise *De Geometria* is shown in Figure 1.

The Age of Newton

Armed with decimal arithmetic and spurred by the newly discovered methods of calculus, mathematicians computed with aplomb. Again, the numerical value of π was a favorite target. Isaac Newton devised an arcsine-like scheme to compute digits of π and recorded 15 digits, although he sheepishly acknowledged, "I am ashamed to tell you to how many figures I carried these computations, having no other business at the time." Newton wrote these words during the plague year 1666, when, ensconced in a country estate, he devised the fundamentals of calculus and the laws of motion and gravitation.

All large computations of π until 1980 relied on variations of Machin's formula:

$$(2) \qquad \frac{\pi}{4} = 4\arctan\left(\frac{1}{5}\right) - \arctan\left(\frac{1}{239}\right).$$

The culmination of these feats was a computation of π using (2) to 527 digits in 1853 by William Shanks,

Figure 1. Excerpt from *De Geometria* by Pope Sylvester II (reigned 999–1003 CE).

later (erroneously) extended to 707 digits. In the preface to the publication of this computation, Shanks wrote that his work "would add little or nothing to his fame as a Mathematician, though it might as a Computer" (until 1950 the word "computer" was used for a person, and the word "calculator" was used for a machine).

One motivation for such computations was to see whether the digits of π repeat, thus disclosing the fact that π is a ratio of two integers. This was settled in 1761, when Lambert proved that π is irrational, thus establishing that the digits of π do not repeat in any number base. In 1882 Lindemann established that π is transcendental, thus establishing that the digits of π^2 or any integer polynomial of π cannot repeat, and also settling once and for all the ancient Greek question of whether the circle could be squared—it cannot, because all numbers that can be formed by finite straightedge-and-compass constructions are necessarily algebraic.

The Computer Age

At the dawn of the computer age, John von Neumann suggested computing digits of prominent mathematical constants, including π and e, for statistical analysis. At his instigation, π was computed to 2,037 digits in 1949 on the Electronic

Division of Medicine & Science, National Museum of American History, Smithsonian Institution.

Figure 2. The ENIAC in the Smithsonian's National Museum of American History.

Table 1. Modern Computer-Era π Calculations.

Name	Year	Correct Digits
Miyoshi and Kanada	1981	2,000,036
Kanada-Yoshino-Tamura	1982	16,777,206
Gosper	1985	17,526,200
Bailey	Jan. 1986	29,360,111
Kanada and Tamura	Sep. 1986	33,554,414
Kanada and Tamura	Oct. 1986	67,108,839
Kanada et. al	Jan. 1987	134,217,700
Kanada and Tamura	Jan. 1988	201,326,551
Chudnovskys	May 1989	480,000,000
Kanada and Tamura	Jul. 1989	536,870,898
Kanada and Tamura	Nov. 1989	1,073,741,799
Chudnovskys	Aug. 1991	2,260,000,000
Chudnovskys	May 1994	4,044,000,000
Kanada and Takahashi	Oct. 1995	6,442,450,938
Kanada and Takahashi	Jul. 1997	51,539,600,000
Kanada and Takahashi	Sep. 1999	206,158,430,000
Kanada-Ushiro-Kuroda	Dec. 2002	1,241,100,000,000
Takahashi	Jan. 2009	1,649,000,000,000
Takahashi	Apr. 2009	2,576,980,377,524
Bellard	Dec. 2009	2,699,999,990,000
Kondo and Yee	Aug. 2010	5,000,000,000,000

Table 2. Computations of Other Mathematical Constants.

Constant	Decimal digits	Researcher	Date
$\sqrt{2}$	1,000,000,000,000	S. Kondo	2010
ϕ	1,000,000,000,000	A. Yee	2010
e	500,000,000,000	S. Kondo	2010
$\log 2$	100,000,000,000	S. Kondo	2011
$\log 10$	100,000,000,000	S. Kondo	2011
$\zeta(3)$	100,000,001,000	A. Yee	2011
G	31,026,000,000	A. Yee and R. Chan	2009
γ	29,844,489,545	A. Yee	2010

Numerical Integrator and Calculator (ENIAC); see Figure 2. In 1965 mathematicians realized that the newly discovered fast Fourier transform could be used to dramatically accelerate high-precision multiplication, thus facilitating not only large calculations of π and other mathematical constants but research in computational number theory as well.

In 1976 Eugene Salamin and Richard Brent independently discovered new algorithms for computing the elementary exponential and trigonometric functions (and thus constants such as π and e) much more rapidly than by using classical series expansions. Their schemes, based on elliptic integrals and the Gauss arithmetic-geometric mean iteration, approximately *double* the number of correct digits in the result with each iteration. Armed with such techniques, π was computed to over one million digits in 1973, to over one billion digits in 1989, to over one trillion digits in 2002, and to over five trillion digits at the present time; see Table 1.

Similarly, the constants $e, \phi, \sqrt{2}, \log 2, \log 10$, $\zeta(3)$, Catalan's constant $G = \sum_{n=0}^{\infty} (-1)^n / (2n+1)^2$, and Euler's γ constant have now been computed to impressive numbers of digits; see Table 2 [10].

One of the most intriguing aspects of this historical chronicle is the repeated assurances, often voiced by highly knowledgeable people, that future progress would be limited. As recently as 1963, Daniel Shanks, who himself calculated π to over 100,000 digits, told Philip Davis that computing one billion digits would be "forever impossible." Yet this feat was achieved less than thirty years later in 1989 by Yasumasa Kanada of Japan. Also in 1989, famous British physicist Roger Penrose, in the first edition of his bestselling book *The Emperor's New Mind*, declared that humankind likely will never know if a string of

ten consecutive 7s occurs in the decimal expansion of π. This string was found just eight years later, in 1997, also by Kanada, beginning at position 22,869,046,249. After being advised of this fact by one of the present authors, Penrose revised his second edition to specify twenty consecutive 7s.

Along this line, Brouwer and Heyting, exponents of the "intuitionist" school of mathematical logic, proposed, as a premier example of a hypothesis that could never be formally settled, the question of whether the string "0123456789" appears in the decimal expansion of π. Kanada found this at the 17,387,594,880-th position after the decimal point. Even astronomer Carl Sagan, whose lead character in his 1985 novel *Contact* (played by Jodi Foster in the movie version) sought confirmation in base-11 digits of π, expressed surprise to learn, shortly after the book's publication, that π had already been computed to many millions of digits.

The BBP Formula for pi

A 1997 paper [3], [5, Ch. 3] by one of the present authors (Bailey), Peter Borwein and Simon Plouffe presented the following unknown formula for π, now known as the "BBP" formula for π:

(3)
$$\pi = \sum_{k=0}^{\infty} \frac{1}{16^k} \left(\frac{4}{8k+1} - \frac{2}{8k+4} - \frac{1}{8k+5} - \frac{1}{8k+6} \right).$$

This formula has the remarkable property that it permits one to directly calculate binary or hexadecimal digits of π beginning at an arbitrary starting position without needing to calculate any of the preceding digits. The resulting simple algorithm requires only minimal memory, does not require multiple-precision arithmetic, and is very well suited to highly parallel computation. The cost of this scheme increases only slightly faster than the index of the starting position.

The proof of this formula is surprisingly elementary. First note that for any $k < 8$,

(4)
$$\int_0^{1/\sqrt{2}} \frac{x^{k-1}}{1-x^8}\,dx = \int_0^{1/\sqrt{2}} \sum_{i=0}^{\infty} x^{k-1+8i}\,dx$$
$$= \frac{1}{2^{k/2}} \sum_{i=0}^{\infty} \frac{1}{16^i(8i+k)}.$$

Thus one can write

(5)
$$\sum_{i=0}^{\infty} \frac{1}{16^i} \left(\frac{4}{8i+1} - \frac{2}{8i+4} - \frac{1}{8i+5} - \frac{1}{8i+6} \right)$$
$$= \int_0^{1/\sqrt{2}} \frac{4\sqrt{2} - 8x^3 - 4\sqrt{2}x^4 - 8x^5}{1-x^8}\,dx,$$

which on substituting $y := \sqrt{2}x$ becomes

(6)
$$\int_0^1 \frac{16\,y - 16}{y^4 - 2\,y^3 + 4\,y - 4}\,dy$$
$$= \int_0^1 \frac{4y}{y^2 - 2}\,dy - \int_0^1 \frac{4y-8}{y^2 - 2y + 2}\,dy = \pi,$$

reflecting a partial fraction decomposition of the integral on the left-hand side. In 1997 neither Maple nor Mathematica could evaluate (3) symbolically to produce the result π. Today both systems can do this easily.

Binary Digits of log 2

It is worth noting that the BBP formula (3) was not discovered by a conventional analytic derivation. Instead, it was discovered via a computer-based search using the PSLQ *integer relation detection algorithm* (see the section "Hunt for a pi Formula") of mathematician-sculptor Helaman Ferguson [4] in a process that some have described as an exercise in "reverse mathematical engineering". The motivation for this search was the earlier observation by the authors of [3] that log 2 also has this arbitrary position digit calculating property. This can be seen by analyzing the classic formula

(7)
$$\log 2 = \sum_{k=1}^{\infty} \frac{1}{k2^k},$$

which has been known at least since the time of Euler and which is closely related to the functional equation for the dilogarithm.

Let $r \bmod 1$ denote the fractional part of a nonnegative real number r, and let d be a nonnegative integer. Then the binary fraction of log 2 after the "decimal" point has been shifted to the right d places can be written as

(8)
$$(2^d \log 2) \bmod 1$$
$$= \left(\sum_{k=1}^{d} \frac{2^{d-k}}{k} \bmod 1 + \sum_{k=d+1}^{\infty} \frac{2^{d-k}}{k} \bmod 1 \right) \bmod 1$$
$$= \left(\sum_{k=1}^{d} \frac{2^{d-k} \bmod k}{k} \bmod 1 \right.$$
$$\left. + \sum_{k=d+1}^{\infty} \frac{2^{d-k}}{k} \bmod 1 \right) \bmod 1,$$

where "mod k" has been inserted in the numerator of the first term since we are only interested in the fractional part of the result after division.

The operation $2^{d-k} \bmod k$ can be performed very rapidly by means of the *binary algorithm for exponentiation*. This scheme is the simple observation that an exponentiation operation such as 3^{17} can be performed in only five multiplications instead of 16 by writing it as $3^{17} = (((3^2)^2)^2)^2) \cdot 3$. Additional savings can be realized by reducing all of the intermediate multiplication results modulo k at each step. This algorithm, together with the division and summation operations indicated in the first term, can be performed in ordinary double-precision floating-point arithmetic or for very large calculations by using quad- or oct-precision arithmetic.

Expressing the final fractional value in binary notation yields a string of digits corresponding to the binary digits of log 2 beginning immediately after the first d digits of log 2. Computed results can be easily checked by performing this operation for two slightly different positions, say $d - 1$ and d, then checking to see that resulting digit strings properly overlap.

Hunt for a pi Formula

In the wake of finding the above scheme for the binary digits of log 2, the authors of [3] immediately wondered if there was a similar formula for π (none was known at the time). Their approach was to collect a list of mathematical constants (α_i) for which formulas similar in structure to the formula for log 2 were known in the literature and then to determine by means of the PSLQ integer relation algorithm if there exists a nontrivial linear relation of the form

(9)
$$a_0\pi + a_1\alpha_1 + a_2\alpha_2 + \cdots + a_n\alpha_n = 0,$$

where a_i are integers (because such a relation could then be solved for π to yield the desired formula). After several months of false starts, the following relation was discovered:
(10)

$$\pi = 4 \cdot {}_2F_1 \left(\begin{matrix} 1, \frac{1}{4} \\ \frac{5}{4} \end{matrix} \,\middle|\, -\frac{1}{4} \right) + 2 \arctan \left(\frac{1}{2} \right) - \log 5,$$

where the first term is a Gauss hypergeometric function evaluation. After writing this formula explicitly in terms of summations, the BBP formula for π was uncovered:
(11)

$$\pi = \sum_{k=0}^{\infty} \frac{1}{16^k} \left(\frac{4}{8k+1} - \frac{2}{8k+4} - \frac{1}{8k+5} - \frac{1}{8k+6} \right).$$

One question that immediately arose in the wake of the discovery of the BBP formula for π was whether there are formulas of this type for π in other number bases—in other words, formulas where the 16 in the BBP formula is replaced by some other integer, such as 3 or 10. These computer searches were largely laid to rest in 2004, when one of the present authors (Jonathan Borwein), together with Will Galway and David Borwein, showed that there are no degree-1 BBP-type formulas of *Machin-type* for π, except those whose base is a power of two [5, pp. 131–133].

The BBP Formula in Action

Variants of the BBP formula have been used in numerous computations of high-index digits of π. In 1998 Colin Percival, then a 17-year-old undergraduate at Simon Fraser University in Canada, computed binary digits beginning at position one quadrillion (10^{15}). At the time, this was one of the largest, if not the largest, distributed computations ever done. More recently, in July 2010, Tsz-Wo Sze of *Yahoo! Cloud Computing*, in roughly 500 CPU-years of computing on *Apache Hadoop* clusters, found that the base-16 digits of π beginning at position 5×10^{14} (corresponding to binary position two quadrillion) are

0 E6C1294A ED40403F 56D2D764 026265BC A98511D0 FCFFAA10 F4D28B1B B5392B8

In an even more recent 2013 computation along this line, Ed Karrels of Santa Clara University used a system with NVIDIA graphics cards to compute 26 base-16 digits beginning at position one quadrillion. His result: 8353CB3F7F0C9ACCFA9AA215F2.

The BBP formulas have also been used to confirm other computations of π. For example, in August 2010, Shigeru Kondo (a hardware engineer) and Alexander Yee (an undergraduate software engineer) computed five trillion decimal digits of π on a home-built $18,000 machine. They found that the last thirty digits leading up to position five trillion are

Figure 3. (T) Shigeru Kondo and his π-computer. (B) Alex Yee and his elephant.

7497120374 4023826421 9484283852

Kondo and Yee (see Figure 3) used the following Chudnovsky-Ramanujan series:
(12)

$$\frac{1}{\pi} = 12 \sum_{k=0}^{\infty} \frac{(-1)^k (6k)!(13591409 + 545140134k)}{(3k)! (k!)^3 \, 640320^{3k+3/2}}.$$

They did not merely evaluate this formula as written but instead employed a clever quasi-symbolic scheme that mostly avoids the need for full-precision arithmetic.

Kondo and Yee first computed their result in hexadecimal (base-16) digits. Then, in a crucial verification step, they checked hex digits near the end against the same string of digits computed using the BBP formula for π. When this test passed, they converted their entire result to decimal. The entire computation took ninety days, including sixty-four hours for the BBP confirmation and eight days for base conversion to decimal. Note that the much lower time for the BBP confirmation, relative to the other two parts, greatly reduced the overall computational cost. A description of their work is available at [11].

BBP-Type Formulas for Other Constants

In the years since 1997, computer searches using the PSLQ algorithm, as well as conventional analytic investigations, have uncovered BBP-type formulas for numerous other mathematical constants, including $\pi^2, \log^2 2, \pi \log 2,$ $\zeta(3), \pi^3, \log^3 2, \pi^2 \log 2, \pi^4, \zeta(5)$ and Catalan's constant. BBP formulas are also known for many arctangents, as well as for $\log k,$ $2 \le k \le 22,$ although none is known for $\log 23$. These formulas and many others, together with references, are given in an online compendium [1].

One particularly intriguing fact is that, whereas only binary formulas exist for π, there are both binary and ternary (base-3) formulas for π^2:

(13)

$$\pi^2 = \frac{9}{8} \sum_{k=0}^{\infty} \frac{1}{64^k} \left(\frac{16}{(6k+1)^2} - \frac{24}{(6k+2)^2} - \frac{8}{(6k+3)^2} \right.$$
$$\left. - \frac{6}{(6k+4)^2} + \frac{1}{(6k+5)^2} \right),$$

(14)

$$\pi^2 = \frac{2}{27} \sum_{k=0}^{\infty} \frac{1}{729^k} \left(\frac{243}{(12k+1)^2} - \frac{405}{(12k+2)^2} \right.$$
$$- \frac{81}{(12k+4)^2} - \frac{27}{(12k+5)^2} - \frac{72}{(12k+6)^2}$$
$$- \frac{9}{(12k+7)^2} - \frac{9}{(12k+8)^2} - \frac{5}{(12k+10)^2}$$
$$\left. + \frac{1}{(12k+11)^2} \right).$$

Formula (13) appeared in [3], while formula (14) is due to Broadhurst. There are known binary BBP formulas for both $\zeta(3)$ and π^3, but no one has found a ternary formula for either.

Catalan's Constant

One other mathematical constant of central interest is Eugéne Charles Catalan's (1814–1894) constant,

(15)

$$G = \sum_{n=0}^{\infty} \frac{(-1)^n}{(2n+1)^2} = 0.91596559417722\ldots,$$

which is arguably the most basic constant whose irrationality and transcendence (though strongly suspected) remain unproven. Note the close connection to this formula for π^2:

(16)

$$\frac{\pi^2}{8} = \sum_{n=0}^{\infty} \frac{1}{(2n+1)^2} = 1.2337005501362\ldots.$$

Formulas (15) and (16) can be viewed as the simplest Dirichlet L-series values at 2. Such considerations were behind our decision to focus the computation described in this paper on these two constants.

Catalan's constant has already been the subject of some large computations. As mentioned above, in 2009 Alexander Yee and Raymond Chan

calculated G to 31.026 billion digits [10]. This computation employed two formulas, including this formula due to Ramanujan:

(17) $$G = \frac{3}{8} \sum_{n=0}^{\infty} \frac{1}{\binom{2n}{n}(2n+1)^2} + \frac{\pi}{8} \log(2 + \sqrt{3}),$$

which can be derived from the fact that

$$G = -T(\pi/4) = -3/2 \cdot T(\pi/12),$$

where $T(\theta) := \int_0^\theta \log \tan \sigma \, d\sigma$.

The BBP compendium lists two BBP-type formulas for G. The first was discovered numerically by Bailey, but both it and the second formula were subsequently proven by Kunle Adegoke, based in part on some results of Broadhurst.

For the present study, we sought a formula for G with as few terms as possible, because the run time for computing with a BBP-type formula increases roughly linearly with the number of nonzero coefficients. The two formulas in the compendium have twenty-two and eighteen nonzero coefficients, respectively. So we explored, by means of a computation involving the PSLQ algorithm, the linear space of formulas for G spanned by these two sets of coefficients, together with two known "zero relations" (BBP-type formulas whose sum is zero). These analyses and computations led to the following formula, which has only sixteen nonzero coefficients and which we believe to be the most economical BBP-type formula for computing Catalan's constant:

(18)

$$G = \frac{1}{4096} \sum_{k=0}^{\infty} \frac{1}{4096^k} \left(\frac{36864}{(24k+2)^2} - \frac{30720}{(24k+3)^2} \right.$$
$$- \frac{30720}{(24k+4)^2} - \frac{6144}{(24k+6)^2} - \frac{1536}{(24k+7)^2}$$
$$+ \frac{2304}{(24k+9)^2} + \frac{2304}{(24k+10)^2} + \frac{768}{(24k+14)^2}$$
$$+ \frac{480}{(24k+15)^2} + \frac{384}{(24k+11)^2} + \frac{1536}{(24k+12)^2}$$
$$+ \frac{24}{(24k+19)^2} - \frac{120}{(24k+20)^2} - \frac{36}{(24k+21)^2}$$
$$\left. + \frac{48}{(24k+22)^2} - \frac{6}{(24k+23)^2} \right).$$

BBP Formulas and Normality

One prime motivation in computing and analyzing digits of π and other well-known mathematical constants through the ages is to explore the age-old question of whether and why these digits appear "random". Numerous computer-based statistical checks of the digits of π—unlike those of e—so far have failed to disclose any deviation from reasonable statistical norms. See, for instance, Table 3, which presents the counts of individual hexadecimal digits among the first trillion hex digits, as obtained by Yasumasa Kanada.

Given some positive integer b, a real number α is said to be b-normal if every m-long string of base-b digits appears in the base-b expansion of α with precisely the expected limiting frequency $1/b^m$. It follows from basic probability theory that almost all real numbers are b-normal for any specific base b and even for all bases simultaneously. But proving normality for specific constants of interest in mathematics has proven remarkably difficult.

Interest in BBP-type formulas was heightened by the 2001 observation, by one of the present authors (Bailey) and Richard Crandall, that the normality of BBP-type constants such as $\pi, \pi^2, \log 2$ and G can be reduced to a certain hypothesis regarding the behavior of a class of chaotic iterations [5, pp. 141–173]. No proof is known for this general hypothesis, but even specific instances of this result would be quite interesting. For example, if it could be established that the iteration given by $w_0 = 0$ and

$$(19) \qquad w_n = \left(2w_{n-1} + \frac{1}{n}\right) \bmod 1$$

is equidistributed in $[0, 1)$ (i.e., is a "good" pseudorandom number generator), then, according to the Bailey-Crandall result, it would follow that $\log 2$ is 2-normal. In a similar vein, if it could be established that the iteration given by $x_0 = 0$ and

$$(20) \quad x_n = \Bigg(16x_{n-1}$$

$$+ \frac{120n^2 - 89n + 16}{512n^4 - 1024n^3 + 712n^2 - 206n + 21}\Bigg) \bmod 1$$

is equidistributed in $[0, 1)$, then it would follow that π is 2-normal.

Giving further hope to these studies is the recent extension of these methods to a rigorous proof of normality for an uncountably infinite class of real numbers. Given a real number r in $[0, 1)$, let r_k denote the k-th binary digit of r. Then the real number

$$(21) \qquad \alpha_{2,3}(r) = \sum_{k=0}^{\infty} \frac{1}{3^k 2^{3^k + r_k}}$$

is 2-normal. For example, the constant $\alpha_{2,3}(0) = \sum_{k\geq0} 1/(3^k 2^{3^k}) = 0.541883680831502985\ldots$ is provably 2-normal. A similar result applies if 2 and 3 in this formula are replaced by any pair of coprime integers (b, c) greater than one [5, pp. 141–173].

A Curious Hexadecimal Conjecture

It is tantalizing that if, using (20), one calculates the hexadecimal digit sequence

$$(22) \qquad y_n = \lfloor 16x_n \rfloor$$

Table 3. Digit counts in the first trillion hexadecimal (base-16) digits of π. Note that deviations from the average value 62,500,000,000 occur only after the first six digits, as expected.

Hex Digit	Occurrences
0	62499881108
1	62500212206
2	62499924780
3	62500188844
4	62499807368
5	62500007205
6	62499925426
7	62499878794
8	62500216752
9	62500120671
A	62500266095
B	62499955595
C	62500188610
D	62499613666
E	62499875079
F	62499937801
Total	1000000000000

(where $\lfloor \cdot \rfloor$ denotes greatest integer), then the sequence (y_n) appears to perfectly (not just approximately) produce the hexadecimal expansion of π. In explicit computations, we checked that the first 10,000,000 hexadecimal digits generated by this sequence are *identical* with the first 10,000,000 hexadecimal digits of $\pi - 3$. This is a fairly difficult computation, as it requires roughly n^2 bit-operations and is not easily performed on a parallel computer system. In our implementation, computing $2,000,000$ hex digits with (22) using Maple, required 17.3 hours on a laptop. Computing 4,100,000 using Mathematica, with a more refined implementation, required 46.5 hours. The full confirmation using a C++ program took 433,192 seconds (120.3 hours) on an IBM Power 780 (model: 9179-MHB, clock speed: 3.864 GHz). All these outputs were confirmed against stored hex digits of π in the software section of http://www.experimentalmath.info.

Conjecture 1. *The sequence $\lfloor 16x_n \rfloor$, where (x_n) is the sequence of iterates defined in equation (20), generates precisely the hexadecimal expansion of $\pi - 3$.*

We can learn more. Let $\|x - y\| = \min(|x - y|, |1 - (x - y)|)$ denote the "wrapped" distance between reals x and y in $[0, 1)$. The base-16

expansion of π, which we denote π_n, satisfies
(23)
$$||\pi_n - x_n||$$
$$\leq \sum_{k=n+1}^{\infty} \cdot \frac{120k^2 - 89k + 16}{16^{k-n}(512k^4 - 1024k^3 + 712k^2 - 206k + 21)}$$
$$\approx \frac{1}{64(n+1)^2},$$
so that ,upon summing from some N to infinity, we obtain the finite value

(24)
$$\sum_{n=N}^{\infty} ||\pi_n - x_n|| \leq \frac{1}{64(N+1)}.$$

Heuristically, let us assume that the π_n are independent, uniformly distributed random variables in $(0, 1)$, and let $\delta_n = ||\alpha_n - x_n||$. Note that an error (i.e., an instance where x_n lies in a subinterval of the unit interval different from π_n so that the corresponding hex digits don't match) can only occur when π_n is within δ_n of one of the points $(0, 1/16, 2/16, \ldots, 15/16)$. Since $x_n < \pi_n$ for all n (where $<$ is interpreted in the wrapped sense when x_n is slightly less than one), this event has probability $16\delta_n$. Then the fact that the sum (24) has a finite value implies that, by the first *Borel-Cantelli* lemma, there can only be finitely many errors. Further, the small value of the sum (24), even when $N = 1$, suggests that it is unlikely that any errors will be observed. If we set $N = 10,000,001$ in (24), since we know there are no errors in the first 10,000,000 elements, we obtain an upper bound of 1.563×10^{-9}, which suggests it is truly unlikely that errors will ever occur.

A similar correspondence can be seen between iterates of (19) and the binary digits of $\log 2$. In particular, let $z_n = \lfloor 2w_n \rfloor$, where w_n is given in (19). Then since the sum of the error terms for $\log 2$, corresponding to (24), is infinite, it follows by the second Borel-Cantelli lemma that discrepancies between (z_n) and the binary digits of $\log 2$ can be expected to appear indefinitely but with decreasing frequency. Indeed, in computations that we have done, we have found that the sequence (z_n) disagrees with ten of the first twenty binary digits of $\log 2$, but in only one position over the range 5,000 to 8,000.

Computing Digits of π^2 and Catalan's Constant

In illustration of this theory, we now present the results of computations of high-index binary digits of π^2, ternary digits of π^2, and binary digits of Catalan's constant, based on formulas (13), (14), and (18), respectively. These calculations were performed on a 4-rack *BlueGene/P* system at IBM's Benchmarking Center in Rochester, Minnesota (see Figure 4). This is a shared facility, so calculations were conducted over a several-month period during

Figure 4. Andrew Mattingly, Blue Gene/P, and Glenn Wightwick.

which time, none, some, or all of the system was available. It was programmed remotely from Australia, which permitted the system to be used off-hours. Sometimes it helps to be in a different time zone!

(1) *Base-64 digits of π^2 beginning at position 10 trillion.* The first run, which produced base-64 digits starting from position $10^{12} - 1$, required an average of 253,529 seconds per thread and was subdivided into seven partitions of 2048 threads each, so the total cost was $7 \cdot 2048 \cdot 253529 = 3.6 \times 10^9$ CPU-seconds. Each rack of the IBM system features 4096 cores, so the total cost is 10.3 "rack-days".

The second run, which produced base-64 digits starting from position 10^{12}, completed in nearly the same run time (within a few minutes). The two resulting base-8 digit strings appear in row A of Table X. (Each pair of base-8 digits corresponds to a base-64 digit.) Here the digits in agreement are delimited by |. Note that 53 consecutive base-8 digits (or, equivalently, 159 consecutive binary digits) are in perfect agreement.

(2) *Base-729 digits of π^2 beginning at position 10 trillion.* In this case the two runs each required an average of 795,773 seconds per thread, similarly subdivided as above, so that the total cost was 6.5×10^9 CPU-seconds, or 18.4 "rack-days". The two resulting base-9 digit strings are found in row B of Table X. (Each triplet of base-9 digits corresponds to one base-729 digit.) Note here that 47 consecutive base-9 digits (94 consecutive base-3 digits) are in perfect agreement.

(3) *Base-4096 digits of Catalan's constant beginning at position 10 trillion.* These two runs each required 707,857 seconds per thread, but in this case they were subdivided into eight partitions of 2048 threads each, so that the total cost was 1.2×10^{10} CPU-seconds, or 32.8 "rack-days". The two resulting base-8 digit strings are found in row C of Table X. (Each quadruplet of base-8 digits corresponds to one base-4096 digit.) Note that 47 consecutive base-8 digits (141 consecutive binary digits) are in perfect agreement.

These long strings of consecutively agreeing digits, beginning with the target digit, provide a compelling level of statistical confidence in the results. In the first case, for instance, note that the probability that thirty-two pairs of randomly chosen base-8 digits are in perfect agreement is roughly 1.2×10^{-29}. Even if one discards, say, the final six base-8 digits as a 1-in-262,144 statistical safeguard against numerical round-off error, one would still have twenty-four consecutive base-8 digits in perfect agreement, with a corresponding probability of 2.1×10^{-22}. Now strictly speaking, one cannot define a valid probability measure on digits of π^2, but nonetheless, from a practical point of view, such analysis provides a very high level of statistical confidence that the results have been correctly computed.

For this reason, computations of π and the like are a favorite tool for the integrity testing for computer system hardware and software. If either run of a paired computation of π succumbs to even a single fault in the course of the computation,

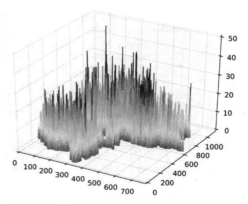

Figure 5. A "random" walk on a million digits of Catalan's constant.

then typically the final results will disagree almost completely. For example, in 1986 a similar pair of computations of π disclosed some subtle but substantial hardware errors in an early model of the Cray-2 supercomputer. Indeed, the calculations we have done arguably constitute the most strenuous integrity test ever performed on the BlueGene/P system. Table 4 gives some sense of the scale of the three record computations, which used more than 135 "rack-days", 1378 serial CPU-years, and more than 1.549×10^{19} floating point operations. This is comparable to the cost of the most sophisticated animated movies as of the present time (2011).

For the sake of completeness, in Table 5 we also record the one-, two-, and three-bit frequency counts from our Catalan computation.

Future Directions

It is ironic that, in an age when even pillars such as Fermat's Last Theorem and the Poincaré conjecture have succumbed to the brilliance of modern mathematics, one of the most elementary mathematical hypotheses, namely whether (and why) the digits of π or other constants such as $\log 2, \pi^2$, or G (see Figure 5) are "random", remains unanswered. In particular, proving that π (or $\log 2, \pi^2$, or G) is b-normal in some integer base b remains frustratingly elusive. Even much weaker results, for instance the simple assertion that a one appears in the binary expansion of π (or $\log 2, \pi^2$, or G) with limiting frequency $1/2$ (which assertion has been amply affirmed in numerous computations over the years), remain unproven and largely inaccessible at the present time.

Almost as much ignorance extends to simple algebraic irrationals such as $\sqrt{2}$. In this case it is now known that the number of ones in the

Table 4. (A) base-4 digits of π^2, (B) base-729 digits of π^2, and (C) base-4096 digits of Catalan's constant, in each case beginning at position 10 trillion.

A	75|60114505303236475724500005743262754530363052416350634|573227604 |60114505303236475724500005743262754530363052416350634|220210566
B	001|122644850645485831771111352101628560483234553468|10565567635862 |122644850645485831771111352101628560483234553468|04744867134524
C	0176|3470505377477705112261337162012525732721732452|6000177545727 |3470505377477705112261337162012525732721732452|5703510516602

Table 5. The scale of our computations. We estimate 4.5 quad-double operations per iteration and that each costs 266 single-precision operations. The total cost in single-precision operations is given in the last column. This total includes overhead which is largely due to a rounding operation that we implemented using bit-masking.

CONSTANT	n'	d	#ITERS ($\times 10^{15}$)	TIME/ITER (microsec)	TIME (yr)	WITH VERIFY	TOTAL (yr)	O'HEAD (%)	FLOPS ($\times 10^{18}$)
π^2 base-2^6	5	10^{13}	2.16	1.424	97.43	194.87	230.35	18.2	2.58
π^2 base-3^6	9	10^{13}	3.89	1.424	175.38	350.76	413.16	17.8	4.65
G base-4^6	16	10^{13}	6.91	1.424	311.79	623.58	735.02	17.9	8.26

Table 6. Base-4096 digits of G beginning at position 10 trillion: digit proportions.

Digit	0	1	2	3	4	5	6	7
base-2 (141)	0.454	0.546	-	-	-	-	-	-
base-4 (70)	0.171	0.329	0.229	0.271	-	-	-	-
base-8 (47)	0.085	0.128	0.213	0.128	0.064	0.128	0.043	0.213

first n binary digits of $\sqrt{2}$ must be at least of the order of $\sqrt{n}$, with similar results for other algebraic irrationals [5, pp. 141–173]. But this is a very weak result, given that this limiting ratio is almost certainly $1/2$, not only for $\sqrt{2}$ but more generally for all algebraic irrationals.

Nor can we prove much about continued fractions for various constants, except for a few well-known results for special cases such as quadratic irrationals, ratios of Bessel functions, and certain expressions involving exponential functions.

For these reasons there is continuing interest in the theory of BBP-type constants, since, as mentioned, there is an intriguing connection between BBP-type formulas and certain chaotic iterations that are akin to pseudorandom number generators. If these connections can be strengthened, then perhaps normality proofs could be obtained for a wide range of polylogarithmic constants, possibly including π, $\log 2$, π^2, and G.

As settings change, so do questions. Until the question of efficient single-digit extraction was asked, our ignorance about such issues was not exposed. The case of the exponential series

$$(25) \qquad e^x = \sum_{n=0}^{\infty} \frac{x^n}{n!}$$

is illustrative. For present purposes, the convergence rate in (25) is too good.

Conjecture 2. *There is no BBP formula for e. Moreover, there is no way to extract individual digits of e significantly more rapidly than by computing the first n digits.*

The same could be conjectured about other numbers, including Euler's constant $\gamma = 0.57721566490153\dots$. In short, until vastly stronger mathematical results are obtained in this area, there will doubtless be continuing interest in computing digits of these constants. In the present vacuum, that is perhaps all that we can do.

Acknowledgments

Thanks are due to many colleagues, but most explicitly to Prof. Mary-Anne Williams of University Technology Sydney, who conceived the idea of a π-related computation to conclude in conjunction with a public lecture at UTS on 3.14.2011 (see http://datasearch2.uts.edu.au/feit/news-events/event-detail.cfm?ItemId=25541). We also wish to thank Matthew Tam, who constructed the database version of [1].

Clay Mathematics Institute:
Clay Research Conference

Mathematical Institute Opening Conference

September 29 – October 4, 2013
University of Oxford
Mathematical Institute, Radcliffe Observatory Quarter

Clay Research Conference

Wednesday, October 2

- **Peter Constantin** (Princeton University)
- **Lance Fortnow** (Georgia Institute of Technology)
- **Fernando Rodriguez Villegas** (University of Texas at Austin)
- **Edward Witten** (Institute for Advanced Study)

Associated workshops will be held throughout the week of the conference:

29 Sept – 1 Oct:	The Navier-Stokes Equations and Related Topics
30 Sept – 4 Oct:	New Insights into Computational Intractability
30 Sept – 4 Oct:	Number Theory and Physics
30 Sept – 4 Oct:	Quantum Mathematics and Computation

For more information and to register, visit
www.claymath.org/CRC13/

Mathematical Institute Opening Conference

Thursday, October 3

To celebrate the opening of its new building, the Mathematical Institute of the University of Oxford is sponsoring a one-day conference.

- **Ingrid Daubechies** (Duke University)
- **Raymond Goldstein** (University of Cambridge)
- **Sir Andrew Wiles** (University of Oxford)

For more information and to register, visit
www.maths.ox.ac.uk/opening

www.claymath.org   www.maths.ox.ac.uk

References

1. DAVID H. BAILEY, BBP-type formulas, manuscript, 2011, available at http://davidhbailey.com/dhbpapers/bbp-formulas.pdf. A web database version is available at http://www.bbp.carma.newcastle.edu.au.
2. DAVID H. BAILEY and JONATHAN M. BORWEIN, Ancient Indian square roots: An exercise in forensic paleomathematics, *American Mathematical Monthly* **119**, no. 8 (Oct. 2012), 646–657.
3. DAVID H. BAILEY, PETER B. BORWEIN, and SIMON PLOUFFE, On the rapid computation of various polylogarithmic constants, *Mathematics of Computation* **66**, no. 218 (1997), 903–913.
4. DAVID H. BAILEY and DAVID J. BROADHURST, Parallel integer relation detection: Techniques and applications, *Mathematics of Computation* **70**, no. 236 (2000), 1719–1736.
5. J. M. BORWEIN and D. H. BAILEY, *Mathematics by Experiment: Plausible Reasoning in the 21st Century*, A K Peters Ltd., 2004. Expanded Second Edition, 2008.
6. L. BERGGREN, J. M. BORWEIN, and P. B. BORWEIN, *Pi: A Source Book*, Springer-Verlag, 1997, 2000, 2004. Fourth edition in preparation, 2011.
7. BIBHUTIBHUSAN DATTA, The Bakhshali mathematics, *Bulletin of the Calcutta Mathematical Society* **21**, no. 1 (1929), 1–60.
8. RUDOLF HOERNLE, *On the Bakhshali Manuscript*, Alfred Holder, Vienna, 1887.
9. GEORGES IFRAH, *The Universal History of Numbers: From Prehistory to the Invention of the Computer*, translated by David Vellos, E. F. Harding, Sophie Wood, and Ian Monk, John Wiley and Sons, New York, 2000.
10. ALEXANDER YEE, Large computations, 7 Mar 2011, available at http://www.numberworld.org/nagisa_runs/computations.html.
11. ALEXANDER YEE and SHIGERU KONDO, 5 trillion digits of pi—new world record, 7 Mar 2011, available at http://www.numberworld.org/misc_runs/pi-5t/details.html.

21. Walking on real numbers (2013)

Paper 21: Francisco J. Aragon Artacho, David H. Bailey, Jonathan M. Borwein and Peter B. Borwein, "Walking on real numbers," *Mathematical Intelligencer*, vol. 35 (2013), p. 42–60. With permission of Springer.

Synopsis:

As mentioned earlier, an age-old question (one that has been the motivation for both ancient and computer-age computations of π and other constants) is that of whether (and why) these digits are "random," which is usually taken to be the property of normality: a number α is said to be b-normal if every m-long string in the base-b expansion of α appears, in the limit, with frequency precisely $1/b^m$. It is straightforward to show, using probability and/or measure theory, that almost all real numbers must be normal, but it has been very difficult to prove normality for any of the classical constants of mathematics.

In this paper, the authors apply some new computer-based tools to address these questions. One of these is to cast, say, the base-4 digits of a constant, as a random walk, where at each step one moves a unit right, up, left or down depending on whether the digit at the given position is 0, 1, 2 or 3 (with similar extensions to other number bases). These analyses produce some surprising results, showing, for instance, that although the Stoneham numbers are provably normal, they exhibit self-similarity properties atypical of a truly "random" sequence. Numerous results along this line are presented.

Keywords: Curiosities, Graphical Representation, Normality, Random Walks

Walking on Real Numbers

Francisco J. Aragón Artacho, David H. Bailey, Jonathan M. Borwein, and Peter B. Borwein

The digit expansions of $\pi, e, \sqrt{2}$, and other mathematical constants have fascinated mathematicians from the dawn of history. Indeed, one prime motivation for computing and analyzing digits of π is to explore the age-old question of whether and why these digits appear "random." The first computation on ENIAC in 1949 of π to 2037 decimal places was proposed by John von Neumann so as to shed some light on the distribution of π (and of e) [15, pg. 277–281].

One key question of some significance is whether (and why) numbers such as π and e are "normal." A real constant α is *b*-normal if, given the positive integer $b \geq 2$, every *m*-long string of base-*b* digits appears in the base-*b* expansion of α with precisely the expected limiting frequency $1/b^m$. It is a well-established, albeit counterintuitive, fact that given an integer $b \geq 2$, almost all real numbers, in the measure theory sense, are *b*-normal. What's more, almost all real numbers are *b*-normal simultaneously for all positive integer bases (a property known as "absolutely normal").

Nonetheless, it has been surprisingly difficult to prove normality for well-known mathematical constants for any given base *b*, much less all bases simultaneously. The first constant to be proven 10-normal is the Champernowne number, namely the constant 0.123456789101112131415116..., produced by concatenating the decimal representation of all positive integers in order. Some additional results of this sort were established in the 1940s by Copeland and Erdős [26].

At present, normality proofs are not available for any well-known constant such as $\pi, e, \log 2, \sqrt{2}$. We do not even know, say, that a 1 appears one-half of the time, in the limit, in the binary expansion of $\sqrt{2}$ (although it certainly appears to), nor do we know for certain that a 1 appears infinitely often in the decimal expansion of $\sqrt{2}$. For that matter, it is widely believed

that *every* irrational algebraic number (i.e., every irrational root of an algebraic polynomial with integer coefficients) is *b*-normal to all positive integer bases *b*, but there is no proof, not for any specific algebraic number to any specific base.

In 2002, one of the present authors (Bailey) and Richard Crandall showed that given a real number r in $[0,1)$, with r_k denoting the *k*-th binary digit of r, the real number

$$\alpha_{2,3}(r) := \sum_{k=1}^{\infty} \frac{1}{3^k 2^{3^k + r_k}} \tag{1}$$

is 2-normal. It can be seen that if $r \neq s$, then $\alpha_{2,3}(r) \neq \alpha_{2,3}(s)$, so that these constants are all distinct. Since r can range over the unit interval, this class of constants is uncountable. So, for example, the constant $\alpha_{2,3} = \alpha_{2,3}(0) = \sum_{k \geq 1} 1/(3^k 2^{3^k}) = 0.0418836808315030\ldots$ is provably 2-normal (this special case was proven by Stoneham in 1973 [43]). A similar result applies if 2 and 3 in formula (1) are replaced by any pair of coprime integers (b, c) with $b \geq 2$ and $c \geq 2$ [10]. More recently, Bailey and Michal Misiurewicz were able to establish 2-normality of $\alpha_{2,3}$ by a simpler argument, by utilizing a "hot spot" lemma proven using ergodic theory methods [11].

In 2004, two of the present authors (Bailey and Jonathan Borwein), together with Richard Crandall and Carl Pomerance, proved the following: If a positive real y has algebraic degree $D > 1$, then the number #(y, N) of 1-bits in the binary expansion of y through bit position N satisfies #(y, N) $> C N^{1/D}$, for a positive number C (depending on y) and all sufficiently large N [5]. A related result has been obtained by Hajime Kaneko of Kyoto University in Japan [37]. However, these results fall far short of establishing *b*-normality for any irrational algebraic in any base *b*, even in the single-digit sense.

Supported in part by the Director, Office of Computational and Technology Research, Division of Mathematical, Information, and Computational Sciences of the U.S. Department of Energy, under contract number DE-AC02-05CH11231.

1 Twenty-First Century Approaches to the Normality Problem

In spite of such developments, there is a sense in the field that more powerful techniques must be brought to bear on this problem before additional substantial progress can be achieved. One idea along this line is to study directly the decimal expansions (or expansions in other number bases) of various mathematical constants by applying some techniques of scientific visualization and large-scale data analysis.

In a recent paper [4], by accessing the results of several extremely large recent computations [46, 47], the authors tested the first roughly four trillion hexadecimal digits of π by means of a Poisson process model: in this model, it is extraordinarily unlikely that π is not normal base 16, given its initial segment. During that work, the authors of [4], like many others, investigated visual methods of representing their large mathematical data sets. Their chosen tool was to represent these data as walks in the plane.

In this work, based in part on sources such as [22, 23, 21, 19, 14], we make a more rigorous and quantitative study of these walks on numbers. We pay particular attention to π, for which we have copious data and which—despite the fact that its digits can be generated by simple algorithms—behaves remarkably "randomly."

The organization of the article is as follows. In Section 2 we describe and exhibit uniform walks on various numbers, both rational and irrational, artificial and natural. In the next two sections, we look at quantifying two of the best-known features of random walks: the expected distance traveled after N steps (Section 3) and the number of sites visited (Section 4) In Section 5 we describe two classes for which normality and nonnormality results are known, and one for which we have only surmise. In Section 6 we show various examples and leave some open questions. Finally, in Appendix 7 we collect the numbers we have examined, with concise definitions and a few digits in various bases.

2 Walking on Numbers

2.1 Random and Deterministic Walks

One of our tasks is to compare deterministic walks (such as those generated by the digit expansion of a constant) with pseudorandom walks of the same length. For example, in Figure 1 we draw a uniform pseudorandom walk with one million base-4 steps, where at each step the path moves one unit east, north, west, or south, depending on the whether the pseudorandom iterate at that position is 0, 1, 2, or 3. The color indicates the path followed by the walk—it is shifted up the spectrum (red-orange-yellow-green-cyan-blue-purple-red) following an HSV scheme with S and V equal to one. The HSV (hue, saturation, and value) model is a cylindrical-coordinate representation that yields a rainbow-like range of colors.

Figure 1. A uniform pseudorandom walk.

FRANCISCO J. ARAGÓN ARTACHO earned his Ph.D. in optimization in 2007 at the University of Murcia, Spain. After working for a business in Madrid for a year, he took a postdoctoral position at the University of Alicante, supported by the program "Juan de la Cierva." In 2011, he became a Research Associate at the Priority Research Centre for Computer-Assisted Research Mathematics and its Applications (CARMA), University of Newcastle, Australia, under the direction of Jonathan Borwein, with whom he's currently collaborating on several projects.

Centre for Computer Assisted Research
Mathematics and its Applications
(CARMA)
University of Newcastle
Callaghan, NSW 2308
Australia
e-mail: francisco.aragon@ua.es

DAVID H. BAILEY is a Senior Scientist at Lawrence Berkeley National Laboratory. Before coming to the Berkeley Lab in 1998, he was at NASA's Ames Research Center for 14 years. Bailey has received the Chauvenet Prize from the Mathematical Association of America, the Sidney Fernbach Award from the IEEE Computer Society, and the Gordon Bell Prize from the Association for Computing Machinery. He is the author of five books, including *Mathematics by Experiment: Plausible Reasoning in the 21st Century*, coauthored with Jonathan Borwein.

Lawrence Berkeley National Laboratory
Berkeley, CA 94720
USA
e-mail: dhbailey@lbl.gov

Let us now compare this graph with that of some rational numbers. For instance, consider these two rational numbers $Q1$ and $Q2$:

But even more information is exhibited when we view a plot of the base-4 digits of $Q1$ and $Q2$ as deterministic walks, as shown in Figure 2. Here, as above, at each step

```
Q1=
104901227167749943748661928056544860161756735849156087616684838084314435844725287555162924702775955557045371567931305878324772977202177081818796590637365767487981422801328592027861019258140957135748704712290267465151312805954195399750420206138037382233895971339195419543/16122269626942909129404900662735492142298807557254685123533957184651913530173488143140175045399694454793530120643833272670970079330526292030350920973600450955456136596649325078391464772840162385651374295294530896122681527488756156580761624107880751845994219387748835

Q2=
727898485706687413042833612434773655776009792025799724606605332096715104161536221938098333730626479355955784966226331511063109122609667568778977976821682512653537303069288477901523227013159658247897670304354024902954394213109106393401484960281395/11187071843154281720476087474091733785438179364129161144313066289965259377090978187244251666337745459152093558288671765654061273733231787773611338297486163914262841526554379727447969242765226084470718753215525487295285372502631868599749526213466521
```

At first glance, these numbers look completely dissimilar. However, if we examine their digit expansions, we find that they are very close as real numbers: the first 240 decimal digits are the same, as are the first 400 base-4 digits.

the path moves one unit east, north, west, or south, depending on the whether the digit in the corresponding position is 0, 1, 2, or 3, and with color coded to indicate the overall position in the walk.

JONATHAN M. BORWEIN is currently Laureate Professor in the School of Mathematical and Physical Sciences and is Director of the Priority Research Centre in Computer Assisted Research Mathematics and its Applications at the University of Newcastle. An ISI highly cited scientist and former Chauvenet prize winner, he has published widely in various fields of mathematics. Two of his recent books are *Convex Functions*, with Jon Vanderwerff, and *Modern Mathematical Computation with Mathematica*, with Matt Skerritt.

Centre for Computer Assisted Research Mathematics and its Applications (CARMA)
University of Newcastle
Callaghan, NSW 2308
Australia
e-mail: jonathan.borwein@newcastle.edu.au

PETER B. BORWEIN is the founder and Executive Director of the IRMACS Research Center at Simon Fraser University. He holds a Burnaby Mountain Chair and is an award-winning mathematician (Chauvenet, Ford, and Hasse Prizes, CUFA BC Academic of the Year). His primary research interests are in analysis and number theory but always with an overarching interest in the computational and experimental aspects. He has authored several hundred research papers and more than a dozen books. For more than 30 years he has collaborated extensively with his brother Jon.

IRMACS
Simon Fraser University
Burnaby, BC V5A 1S6
Canada

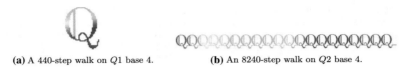

(a) A 440-step walk on $Q1$ base 4.   **(b)** An 8240-step walk on $Q2$ base 4.

Figure 2. Walks on the rational numbers $Q1$ and $Q2$.

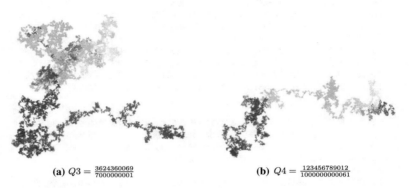

(a) $Q3 = \frac{3624360069}{7000000001}$ **(b)** $Q4 = \frac{123456789012}{1000000000061}$

Figure 3. Walks on the first million base-10 digits of the rationals $Q3$ and $Q4$ from [39].

The rational numbers $Q1$ and $Q2$ represent the two possibilities when one computes a walk on a rational number: either the walk is bounded as in Figure 2(a) (for any walk with more than 440 steps one obtains the same plot), or it is unbounded but repeating some pattern after a finite number of digits as in Figure 2(b).

Of course, not all rational numbers are that easily identified by plotting their walks. It is possible to create a rational number whose period is of any desired length. For example, the following rational numbers from [39],

$$Q3 = \frac{3624360069}{7000000001} \quad \text{and} \quad Q4 = \frac{123456789012}{1000000000061},$$

have base-10 periodic parts with length 1,750,000,000 and 1,000,000,000,060, respectively. A walk on the first million digits of both numbers is plotted in Figure 3. These huge periods derive from the fact that the numerators and denominators of $Q3$ and $Q4$ are relatively prime, and the denominators are not congruent to 2 or 5. In such cases, the period P is simply the discrete logarithm of the denominator D modulo 10; or, in other words, P is the smallest n such that $10^n \bmod D = 1$.

Graphical walks can be generated in a similar way for other constants in various bases—see Figures 2 through 7. Where the base $b \geq 3$, the base-b digits can be used to a select, as a direction, the corresponding base-b complex root of unity—a multiple of 120° for base three, a multiple of 90° for base four, a multiple of 72° for base 5, etc. We generally treat the case $b = 2$ as a base-4 walk, by grouping together pairs of base-2 digits (we could render a base-2 walk on a line, but the resulting images would be much less interesting). In Figure 4 the origin has been marked, but since this information is not that important for our purposes, and can be approximately deduced by the color in most cases, it is not indicated in the others. The color scheme for Figures 2 through 7 is the same as

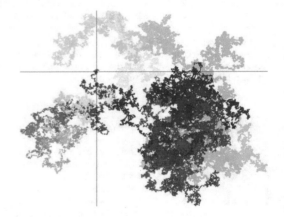

Figure 4. A million-step base-4 walk on e.

the above, except that Figure 6 is colored to indicate the number of returns to each point.

2.2 Normal Numbers as Walks

As noted previously, proving normality for specific constants of interest in mathematics has proven remarkably difficult. The tenor of current knowledge in this arena is illustrated by [45, 14, 34, 38, 40, 39, 44]. It is useful to know that, while small in measure, the "absolutely abnormal" or "absolutely non-normal" real numbers (namely those that are not b-normal for any integer b) are residual in the sense of topological category [1]. Moreover, the Hausdorff–Besicovitch dimension of the set of real numbers having no asymptotic frequencies is equal to 1. Likewise the set of Liouville numbers has measure zero but is of the second category [18, p. 352].

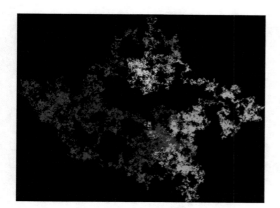

Figure 5. A walk on the first 100 billion base-4 digits of π (normal?).

Figure 6. A walk on the first 100 million base-4 digits of π, colored by number of returns (normal?). Color follows an HSV model (green-cyan-blue-purple-red) depending on the number of returns to each point (where the maxima show a tinge of pink/red).

One question that has possessed mathematicians for centuries is whether π is normal. Indeed, part of the original motivation of the present study was to develop new tools for investigating this age-old problem.

In Figure 5 we show a walk on the first 100 billion base-4 digits of π. This may be viewed dynamically in more detail online at http://gigapan.org/gigapans/106803, where the full-sized image has a resolution of 372,224 × 290,218 pixels (108.03 gigapixels in total). This must be one of the largest mathematical images ever produced. The computations for creating this image took roughly a month, where several parts of the algorithm were run in parallel with 20 threads on CARMA's MacPro cluster.

By contrast, Figure 6 exhibits a 100 million base-4 walk on π, where the color is coded by the number of returns to the point. In [4], the authors empirically tested the normality of its first roughly four trillion hexadecimal (base-16) digits using a Poisson process model, and they concluded that, according to this test, it is "extraordinarily unlikely" that π is not 16-normal (of course, this result does not pretend to be a proof).

In what follows, we propose various methods of analyzing real numbers and visualizing them as walks. Other methods widely used to visualize numbers include the matrix representations shown in Figure 8, where each pixel is colored

depending on the value of the digit to the right of the decimal point, following a left-to-right up-to-down direction (in base 4 the colors used for 0, 1, 2, and 3 are red, green, cyan, and purple, respectively). This method has been mainly used to visually test "randomness." In some cases, it clearly shows the features of some numbers; as for small periodic rationals, see Figure 8(c). This scheme also shows the nonnormality of the number $\alpha_{2,3}$—see Figure 8(d) (where the horizontal red bands correspond to the strings of zeroes)—and it captures some of the special peculiarities of the Champernowne's number C_4 (normal) in Figure 8(e). Nevertheless, it does not reveal the apparently nonrandom behavior of numbers such as the Erdős–Borwein constant; compare Figure 8(f) with Figure 7(e). See also Figure 21.

As we will see in what follows, the study of normal numbers and suspected normal numbers as walks will permit us to compare them with true random (or pseudorandom) walks, obtaining in this manner a new way to empirically test "randomness" in their digits.

3 Expected Distance to the Origin

Let $b \in \{3, 4, \ldots\}$ be a fixed base, and let $X_1, X_2, X_3, \ldots$ be a sequence of independent bivariate discrete random variables whose common probability distribution is given by

$$P\left(X = \begin{pmatrix} \cos\left(\frac{2\pi}{b}k\right) \\ \sin\left(\frac{2\pi}{b}k\right) \end{pmatrix}\right) = \frac{1}{b} \quad \text{for } k = 1, \ldots, b. \quad (2)$$

Then the random variable $S^N := \sum_{m=1}^{N} X_m$ represents a base-b random walk in the plane of N steps.

The following result on the asymptotic expectation of the distance to the origin of a base-b random walk is probably known, but being unable to find any reference in the literature, we provide a proof.

THEOREM 3.1 *The expected distance to the origin of a base-b random walk of N steps is asymptotically equal to $\sqrt{\pi N}/2$.*

PROOF. By the multivariate central limit theorem, the random variable $1/\sqrt{N} \sum_{m=1}^{N}(X_m - \mu)$ is asymptotically bivariate normal with mean $\begin{pmatrix} 0 \\ 0 \end{pmatrix}$ and covariance matrix M, where μ is the two-dimensional mean vector of X and M is its 2×2 covariance matrix. By applying Lagrange's trigonometric identities, one obtains

$$\mu = \begin{pmatrix} \frac{1}{b}\sum_{k=1}^{b}\cos\left(\frac{2\pi}{b}k\right) \\ \frac{1}{b}\sum_{k=1}^{b}\sin\left(\frac{2\pi}{b}k\right) \end{pmatrix} = \frac{1}{b}\begin{pmatrix} -\frac{1}{2} + \frac{\sin\left((b+1/2)\frac{2\pi}{b}\right)}{2\sin(\pi/b)} \\ \frac{1}{2}\cot(\pi/b) - \frac{\cos\left((b+1/2)\frac{2\pi}{b}\right)}{2\sin(\pi/b)} \end{pmatrix}$$

$$= \begin{pmatrix} 0 \\ 0 \end{pmatrix}. \quad (3)$$

Thus,

$$M = \frac{1}{b}\begin{bmatrix} \sum_{k=1}^{b}\cos^2\left(\frac{2\pi}{b}k\right) & \sum_{k=1}^{b}\cos\left(\frac{2\pi}{b}k\right)\sin\left(\frac{2\pi}{b}k\right) \\ \sum_{k=1}^{b}\cos\left(\frac{2\pi}{b}k\right)\sin\left(\frac{2\pi}{b}k\right) & \sum_{k=1}^{b}\sin^2\left(\frac{2\pi}{b}k\right) \end{bmatrix}. \quad (4)$$

(a) A million-step walk on $\alpha_{2,3}$ base 3 (normal?).

(b) A 100,000-step walk on $\alpha_{2,3}$ base 6 (nonnormal).

(c) A million-step walk on $\alpha_{2,3}$ base 2 (normal).

(d) A 100,000-step walk on Champernowne's number C_4 base 4 (normal).

(e) A million-step walk on $EB(2)$ base 4 (normal?).

(f) A million-step walk on $CE(10)$ base 4 (normal?).

Figure 7. Walks on various numbers in different bases.

Since

$$\sum_{k=1}^{b}\cos^2\left(\frac{2\pi}{b}k\right) = \sum_{k=1}^{b}\frac{1+\cos\left(\frac{4\pi}{b}k\right)}{2} = \frac{b}{2},$$

$$\sum_{k=1}^{b}\sin^2\left(\frac{2\pi}{b}k\right) = \sum_{k=1}^{b}\frac{1-\cos\left(\frac{4\pi}{b}k\right)}{2} = \frac{b}{2},$$

$$\sum_{k=1}^{b}\cos\left(\frac{2\pi}{b}k\right)\sin\left(\frac{2\pi}{b}k\right) = \sum_{k=1}^{b}\frac{\sin\left(\frac{4\pi}{b}k\right)}{2} = 0, \tag{5}$$

formula (4) reduces to

$$M = \begin{bmatrix} \frac{1}{2} & 0 \\ 0 & \frac{1}{2} \end{bmatrix}. \tag{6}$$

Hence, $S^N/\sqrt{N}$ is asymptotically bivariate normal with mean $\begin{pmatrix} 0 \\ 0 \end{pmatrix}$ and covariance matrix M. Because its components $(S_1^N/\sqrt{N},\ S_2^N/\sqrt{N})^T$ are uncorrelated, then they are independent random variables, whose distribution is (univariate)

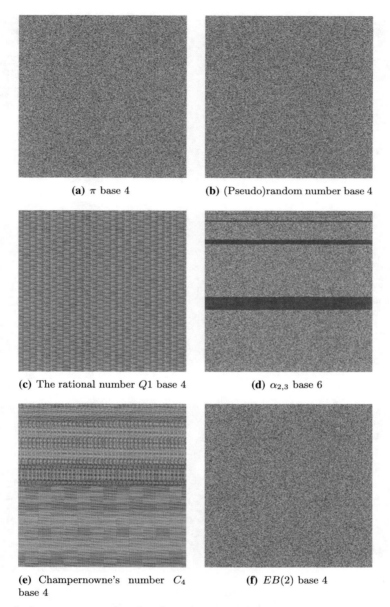

(a) π base 4

(b) (Pseudo)random number base 4

(c) The rational number $Q1$ base 4

(d) $\alpha_{2,3}$ base 6

(e) Champernowne's number C_4 base 4

(f) $EB(2)$ base 4

Figure 8. Horizontal color representation of a million digits of various numbers.

normal with mean 0 and variance 1/2. Therefore, the random variable

$$\sqrt{\left(\frac{\sqrt{2}}{\sqrt{N}}s_1^N\right)^2 + \left(\frac{\sqrt{2}}{\sqrt{N}}s_2^N\right)^2} \qquad (7)$$

converges in distribution to a χ random variable with two degrees of freedom. Then, for N sufficiently large,

$$E\left(\sqrt{(S_1^N)^2 + (S_2^N)^2}\right) = \frac{\sqrt{N}}{\sqrt{2}}E\left(\sqrt{\left(\frac{\sqrt{2}}{\sqrt{N}}S_1^N\right)^2 + \left(\frac{\sqrt{2}}{\sqrt{N}}S_2^N\right)^2}\right)$$

$$\approx \frac{\sqrt{N}}{\sqrt{2}}\frac{\Gamma(3/2)}{\Gamma(1)} = \frac{\sqrt{\pi N}}{2}, \qquad (8)$$

where $E(\cdot)$ stands for the expectation of a random variable. Therefore, the expected distance to the

Table 1. Average of the normalized distance to the origin of the walk of various constants in different bases

Number	Base	Steps	Average normalized distance to the origin	Normal
Mean of 10,000 random walks	4	1,000,000	1.00315	Yes
Mean of 10,000 walks on the digits of π	4	1,000,000	1.00083	?
$\alpha_{2,3}$	3	1,000,000	0.89275	?
$\alpha_{2,3}$	4	1,000,000	0.25901	Yes
$\alpha_{2,3}$	5	1,000,000	0.88104	?
$\alpha_{2,3}$	6	1,000,000	108.02218	No
$\alpha_{4,3}$	3	1,000,000	1.07223	?
$\alpha_{4,3}$	4	1,000,000	0.24268	Yes
$\alpha_{4,3}$	6	1,000,000	94.54563	No
$\alpha_{4,3}$	12	1,000,000	371.24694	No
$\alpha_{3,5}$	3	1,000,000	0.32511	Yes
$\alpha_{3,5}$	5	1,000,000	0.85258	?
$\alpha_{3,5}$	15	1,000,000	370.93128	No
π	4	1,000,000	0.84366	?
π	6	1,000,000	0.96458	?
π	10	1,000,000	0.82167	?
π	10	10,000,000	0.56856	?
π	10	100,000,000	0.94725	?
π	10	1,000,000,000	0.59824	?
e	4	1,000,000	0.59583	?
$\sqrt{2}$	4	1,000,000	0.72260	?
log 2	4	1,000,000	1.21113	?
Champernowne C_{10}	10	1,000,000	59.91143	Yes
$EB(2)$	4	1,000,000	6.95831	?
$CE(10)$	4	1,000,000	0.94964	?
Rational number Q_1	4	1,000,000	0.04105	No
Rational number Q_2	4	1,000,000	58.40222	No
Euler constant γ	10	1,000,000	1.17216	?
Fibonacci $\mathcal{F}$	10	1,000,000	1.24820	?
$\zeta(2) = \frac{\pi^2}{6}$	4	1,000,000	1.57571	?
$\zeta(3)$	4	1,000,000	1.04085	?
Catalan's constant G	4	1,000,000	0.53489	?
Thue–Morse $\mathcal{TM}_2$	4	1,000,000	531.92344	No
Paper-folding $\mathcal{P}$	4	1,000,000	0.01336	No

Table 2. Number of points visited in various N-step base-4 walks. The values of the two last columns are upper and lower bounds on the expectation of the number of distinct sites visited during an N-step random walk; see [31, Theorem 2] and [32]

Number	Steps	Sites visited	Bounds on the expectation of sites visited by a random walk	
			Lower bound	Upper bound
Mean of 10,000 random walks	1,000,000	202,684	199,256	203,060
Mean of 10,000 walks on the digits of π	1,000,000	202,385	199,256	203,060
$\alpha_{2,3}$	1,000,000	95,817	199,256	203,060
$\alpha_{4,3}$	1,000,000	68,613	199,256	203,060
$\alpha_{3,2}$	1,000,000	195,585	199,256	203,060
π	1,000,000	204,148	199,256	203,060
π	10,000,000	1,933,903	1,738,645	1,767,533
π	100,000,000	16,109,429	15,421,296	15,648,132
π	1,000,000,000	138,107,050	138,552,612	140,380,926
e	1,000,000	176,350	199,256	203,060
$\sqrt{2}$	1,000,000	200,733	199,256	203,060
log 2	1,000,000	214,508	199,256	203,060
Champernowne C_4	1,000,000	548,746	199,256	203,060
$EB(2)$	1,000,000	279,585	199,256	203,060
$CE(10)$	1,000,000	190,239	199,256	203,060
Rational number Q_1	1,000,000	378	199,256	203,060
Rational number Q_2	1,000,000	939,322	199,256	203,060
Euler constant γ	1,000,000	208,957	199,256	203,060
$\zeta(2)$	1,000,000	188,808	199,256	203,060
$\zeta(3)$	1,000,000	221,598	199,256	203,060
Catalan's constant G	1,000,000	195,853	199,256	203,060
$\mathcal{TM}_2$	1,000,000	1,000,000	199,256	203,060
Paper-folding $\mathcal{P}$	1,000,000	21	199,256	203,060

origin of the random walk is asymptotically equal to $\sqrt{\pi N}/2$.

As a consequence of this result, for any random walk of N steps in any given base, the expectation of the distance to the origin multiplied by $2/\sqrt{\pi N}$ (which we will call *normalized distance* to the origin) must approach 1 as N goes to infinity. Therefore, for a "sufficiently" big random walk, one would expect that the arithmetic mean of the normalized distances

(which will be called *average normalized distance* to the origin) should be close to 1.

We have created a sample of 10,000 (pseudo)random walks base-4 of one million points each in Python[1], and we have computed their average normalized distance to the origin. The arithmetic mean of these numbers for the mentioned sample of pseudorandom walks is 1.0031, whereas its standard deviation is 0.3676, so the asymptotic result fits quite well. We have also computed the normalized distance to the origin of 10,000 walks of one million steps each generated by the first ten billion digits of π. The resulting arithmetic mean is 1.0008, whereas the standard deviation is 0.3682. In Table 1 we show the average normalized distance to the origin of various numbers. There are several surprises in there data, such as the

---

[1]Python uses the Mersenne Twister as the core generator and produces 53-bit precision floats, with a period of $2^{19937} - 1 \approx 10^{6002}$. Compare the length of this period to the comoving distance from Earth to the edge of the observable universe in any direction, which is approximately $4.6 \cdot 10^{37}$ nanometers, or to the number of protons in the universe, which is approximately 10^{80}.

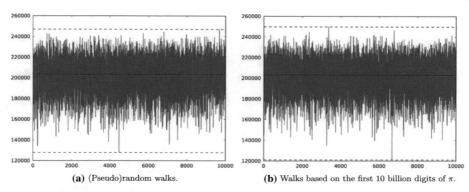

(a) (Pseudo)random walks.   (b) Walks based on the first 10 billion digits of π.

Figure 9. Number of points visited by 10^4 base-4 million-step walks.

fact that by this measure, Champernowne's number C_{10} is far from what is expected of a truly "random" number.

4 Number of Points Visited during an *N*-Step base-4 Walk

The number of distinct points visited during a walk of a given constant (on a lattice) can be also used as an indicator of how "random" the digits of that constant are. It is well known that the expectation of the number of distinct points visited by an *N*-step random walk on a two-dimensional lattice is asymptotically equal to $\pi N/\log(N)$; see, for example, [36, pg. 338] or [13, pg. 27]. This result was first proven by Dvoretzky and Erdős [33, Thm. 1]. The main practical problem with this asymptotic result is that its convergence is rather slow; specifically, it has order of $O(N \log \log N/(\log N)^2)$. In [31, 32], Downham and Fotopoulos show the following bounds on the expectation of the number of distinct points,

$$\left(\frac{\pi(N + 0.84)}{1.16\pi - 1 - \log 2 + \log(N + 2)}, \right.$$

$$\left. \frac{\pi(N + 1)}{1.066\pi - 1 - \log 2 + \log(N + 1)} \right), \tag{9}$$

which provide a tighter estimate on the expectation than the asymptotic limit $\pi N/\log(N)$. For example, for $N = 10^6$, these bounds are $(199256.1, 203059.5)$, whereas $\pi N/\log(N) = 227396$, which overestimates the expectation.

In Table 2 we have calculated the number of distinct points visited by the base-4 walks on several constants. One can see that the values for different step walks on π fit quite well the expectation. On the other hand, numbers that are known to be normal such as $\alpha_{2,3}$ or the base-4 Champernowne number substantially differ from the expectation of a random walk. These constants, despite being normal, do not have a "random" appearance when one draws the associated walk, see Figure 7(d).

At first look, the walk on $\alpha_{2,3}$ might seem random, see Figure 7(c). A closer look, shown in Figure 12, shows a more complex structure: the walk appears to be somehow self-repeating. This helps explain why the number of sites visited by the base-4 walk on $\alpha_{2,3}$ or $\alpha_{4,3}$ is smaller than the

expectation for a random walk. A detailed discussion of the Stoneham constants and their walks is provided in Section 5.2, where the precise structure of Figure 12 is conjectured.

5 Copeland–Erdős, Stoneham, and Erdős–Borwein Constants

As well as the classical numbers—such as e, π, γ—listed in the Appendix, we also considered various other constructions, which we describe in the next three subsections.

5.1 Champernowne Number and Its Concatenated Relatives

The first mathematical constant proven to be 10-normal is the *Champernowne number*, which is defined as the concatenation of the decimal values of the positive integers, that is, $C_{10} = 0.12345678910111213141516\ldots$ Champernowne proved that C_{10} is 10-normal in 1933 [24]. This was later extended to base-b normality (for base-b versions of the Champernowne constant) as in Theorem 5.1. In 1946, Copeland and Erdős established that the corresponding concatenation of primes $0.23571113171923\ldots$ and the concatenation of composites $0.46891012141516\ldots$, among others, are also 10-normal [26]. In general they proved that concatenation leads to normality if the sequence grows slowly enough. We call such numbers *concatenation numbers*:

THEOREM 5.1 ([26]). *If $a_1, a_2, \ldots$ is an increasing sequence of integers such that for every $\theta < 1$ the number of a_i's up to N exceeds N^θ provided N is sufficiently large, then the infinite decimal*

$$0.a_1 a_2 a_3 \cdots$$

is normal with respect to the base b in which these integers are expressed.

This result clearly applies to the Champernowne numbers (Fig. 7(d)), to the primes of the form $ak + c$ with a and c relatively prime, in any given base, and to the integers that are the sum of two squares (since every prime of the form $4k + 1$ is

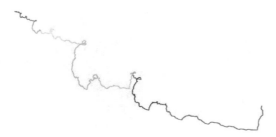

Figure 10. A walk on the first 100,000 bits of the primes ($CE(2)$) base two (normal).

included). In further illustration, using the primes in binary leads to normality in base 2 of the number

$$CE(2) = 0.101110111110111101100011001110111111011$$
$$111110010110100110011\ldots_2,$$

shown as a planar walk in Figure 10.

5.1.1 Strong Normality

In [14] it is shown that C_{10} fails the following stronger test of normality, which we now discuss. The test is is a simple one, in the spirit of Borel's test of normality, as opposed to other more statistical tests discussed in [14]. If the digits of a real number α are chosen at random in the base b, the *asymptotic frequency* $m_k(n)/n$ of each 1-string approaches $1/b$ with probability 1. However, the *discrepancy* $m_k(n) - n/b$ does not approach any limit, but fluctuates with an expected value equal to the standard deviation $\sqrt{(b-1)n/b}$. (Precisely $m_k(n) := \#\{i : a_i = k, i \leq n\}$ when α has fractional part $0.a_0 a_1 a_2 \cdots$ in base b.)

Kolmogorov's law of the iterated logarithm allows one to make a precise statement about the discrepancy of a random number. Belshaw and P. Borwein [14] use this to define their criterion and then to show that almost every number is absolutely strongly normal.

DEFINITION 5.2 (Strong normality [14]). For real α, and $m_k(n)$ as above, α is *simply strongly normal* in the base b if for each $0 \leq k \leq b - 1$ one has

$$\limsup_{n \to \infty} \frac{m_k(n) - n/b}{\sqrt{2n \log \log n}} = \frac{\sqrt{b-1}}{b} \quad \text{and}$$
$$\liminf_{n \to \infty} \frac{m_k(n) - n/b}{\sqrt{2n \log \log n}} = -\frac{\sqrt{b-1}}{b}. \quad (10)$$

A number is *strongly normal* in base b if it is simply strongly normal in each base $b^j, j = 1, 2, 3, \ldots$, and is *absolutely strongly normal* if it is strongly normal in every base.

In paraphrase (absolutely) strongly normal numbers are those that distributionally oscillate as much as is possible.

Belshaw and Borwein show that strongly normal numbers are indeed normal. They also make the important observation that Champernowne's base-b number is not strongly normal in base b. Indeed, there are b^{v-1} digits of length v and they all start with a digit between 1 and $b - 1$ whereas the following

$v - 1$ digits take values between 0 and $b - 1$ equally. In consequence, there is a dearth of zeroes. This is easiest to analyze in base 2. As illustrated below, the concatenated numbers start

$$1, 10, 11, 100, 101, 110, 111, 1000, 1001, 1010, 1011,$$
$$1100, 1101, 1110, 1111$$

For $v = 3$ there are 4 zeroes and 8 ones, for $v = 4$ there are 12 zeroes and 20 ones, and for $v = 5$ there are 32 zeroes and 48 ones.

Because the details were not provided in [14], we present them here.

THEOREM 5.3 (Belshaw and P. Borwein) *Champernowne's base-2 number is is not 2-strongly normal.*

PROOF. In general, let $n_k := 1 + (k - 1)2^k$ for $k \geq 1$. One has $m_0(n_k) = 1 + (k - 1)2^k$ and so

$$m_1(n_k) - m_0(n_k) = n_k - 2m_0(n_k) = 2^k - 1.$$

In fact $m_1(n) > m_0(n)$ for all n. To see this, suppose it true for $n \leq n_k$, and proceed by induction on k. Let us arrange the digits of the integers $2^k, 2^k + 1, \ldots, 2^k + 2^{k-1} - 1$ in a 2^{k-1} by $k + 1$ matrix, where the i-th row contains the digits of the integer $2^k + i - 1$. Each row begins 10, and if we delete the first two columns we obtain a matrix in which the i-th row is given by the digits of $i - 1$, possibly preceded by some zeroes. Neglecting the first row and the initial zeroes in each subsequent row, we see the first n_{k-1} digits of Champernowne's base-2 number, where by our induction hypothesis $m_1(n) > m_0(n)$ for $n \leq n_{k-1}$.

If we now count all the zeroes as we read the matrix in the natural order, any excess of zeroes must come from the initial zeroes, and there are exactly $2^{k-1} - 1$ of these. As shown above, $m_1(n_k) - m_0(n_k) = 2^k - 1$, so $m_1(n) > m_0(n) + 2^{k-1}$ for every $n \leq n_k + (k + 1) 2^{k-1}$. A similar argument for the integers from $2^k + 2^{k-1}$ to $2^{k+1} - 1$ shows that $m_1(n) > m_0(n)$ for every $n \leq n_{k+1}$. Therefore, $2m_1(n) > m_0(n) + m_1(n) = n$ for all n, and so

$$\liminf_{n \to \infty} \frac{m_1(n) - n/2}{\sqrt{2n \log \log n}} \geq 0 \neq -\frac{1}{2},$$

and, as asserted, Champernowne's base-2 number is not 2-strongly normal.

It seems likely that by appropriately shuffling the integers, one should be able to display a strongly normal variant. Along this line, Martin [40] has shown how to construct an explicit absolutely nonnormal number.

Finally, although the log log limiting behavior required by (10) appears difficult to test numerically to any significant level, it appears reasonably easy computationally to check whether other sequences, such as many of the concatenation sequences of Theorem 5.1, fail to be strongly normal for similar reasons.

Heuristically, we would expect the number $CE(2)$ above to fail to be strongly normal, because each prime of length k both starts and ends with a one, whereas intermediate bits should

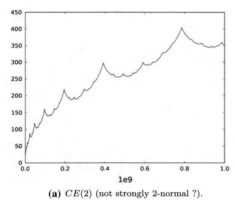

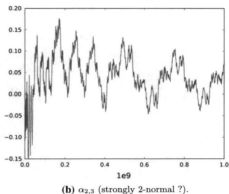

(a) $CE(2)$ (not strongly 2-normal ?).

(b) $\alpha_{2,3}$ (strongly 2-normal ?).

Figure 11. Plot of the first 10^9 values of $\frac{m_1(n)-n/2}{\sqrt{2n\log\log n}}$.

show no skewing. Indeed, for $CE(2)$ we have checked that $2m_1(n) > n$ for all $n \le 10^9$, see also Figure 11(a). Thus motivated, we are currently developing tests for strong normality of numbers such as $CE(2)$ and $\alpha_{2,3}$ below in binary.

For $\alpha_{2,3}$, the corresponding computation of the first 10^9 values of $\frac{m_1(n)-n/2}{\sqrt{2n\log\log n}}$ leads to the plot in Figure 11(b) and leads us to conjecture that it is 2-strongly normal.

5.2 Stoneham Numbers: A Class Containing Provably Normal and Nonnormal Constants

Giving further motivation for these studies is the recent provision of rigorous proofs of normality for the *Stoneham numbers*, which are defined by

$$\alpha_{b,c} := \sum_{m \ge 1} \frac{1}{c^m b^{c^m}}, \tag{11}$$

for relatively prime integers b, c [10].

Theorem 5.4 (Normality of Stoneham constants [3]). *For every coprime pair of integers (b, c) with $b \ge 2$ and $c \ge 2$, the constant $\alpha_{b,c} = \sum_{m \ge 1} 1/(c^m b^{c^m})$ is b-normal.*

So, for example, the constant $\alpha_{2,3} = \sum_{k \ge 1} 1/(3^k 2^{3^k}) = 0.0418836808315030\ldots$ is provably 2-normal. This special case was proven by Stoneham in 1973 [43]. More recently, Bailey and Misiurewicz were able to establish this normality result by a much simpler argument, based on techniques of ergodic theory [11] [16, pg. 141–173].

Equally interesting is the following result:

Theorem 5.5 (Nonnormality of Stoneham constants [3]). *Given coprime integers $b \ge 2$ and $c \ge 2$, and integers p, q, $r \ge 1$, with neither b nor c dividing r, let $B = b^p c^q r$. Assume that the condition $D = c^{q/p} r^{1/p}/b^{c-1} < 1$ is satisfied. Then the constant $\alpha_{b,c} = \sum_{k \ge 0} 1/(c^k b^{c^k})$ is B-nonnormal.*

In various of the Figures and Tables, we explore the striking differences of behavior—proven and unproven—for $\alpha_{b,c}$ as we vary the base. For instance, the nonnormality of $\alpha_{2,3}$ in

Figure 12. Zooming in on the base-4 walk on $\alpha_{2,3}$ of Figure 7(c) and Conjecture 5.6.

base-6 digits was proved just before we started to draw walks. Contrast Figure 7(b) to Figure 7(c) and Figure 7(a). Now compare the values presented in Table 1 and Table 2. Clearly, from this sort of visual and numeric data, the discovery of other cases of Theorem 5.5 is very easy.

As illustrated also in the "zoom" of Figure 12, we can use these images to discover more subtle structure. We conjecture the following relations on the digits of $\alpha_{2,3}$ in base 4 (which explain the values in Tables 1 and 2):

Conjecture 5.6 (Base-4 structure of $\alpha_{2,3}$). *Denote by a_k the kth digit of $\alpha_{2,3}$ in its base-4 expansion; that is, $\alpha_{2,3} = \sum_{k=1}^{\infty} a_k/4^k$, with $a_k \in \{0, 1, 2, 3\}$ for all k. Then, for all $n = 0, 1, 2, \ldots$ one has:*

(i) $\displaystyle\sum_{k=\frac{3}{2}(3^n+1)}^{\frac{3}{2}(3^n+1)+3^n} e^{a_k \pi i/2} = \frac{(-1)^{n+1}-1}{2} + \frac{(-1)^n-1}{2}i = -\begin{cases} i, & n\ odd \\ 1, & n\ even \end{cases};$

(ii) $a_k = a_{k+3^n} = a_{k+2\cdot 3^n}$ *for all* $k = \frac{3}{2}(3^n+1), \frac{3}{2}(3^n+1)+1, \ldots, \frac{3}{2}(3^n+1)+3^n-1.$

In Figure 13, we show the position of the walk after $\frac{3}{2}(3^n+1), \frac{3}{2}(3^n+1)+3^n$ and $\frac{3}{2}(3^n+1)+2\cdot 3^n$ steps for

Figure 13. A pattern in the digits of $\alpha_{2,3}$ base 4. We show only positions of the walk after $\frac{3}{2}(3^n + 1), \frac{3}{2}(3^n + 1) + 3^n$ and $\frac{3}{2}(3^n + 1) + 2 \cdot 3^n$ steps for $n = 0, 1, \ldots, 11$.

$n = 0, 1, \ldots, 11$, which, together with Figures 7(c) and 12, graphically explain Conjecture 5.6. Similar results seem to hold for other Stoneham constants in other bases. For instance, for $\alpha_{3,5}$ base 3 we conjecture the following.

CONJECTURE 5.7 (Base-3 structure of $\alpha_{3,5}$). *Denote by a_k the kth digit of $\alpha_{3,5}$ in its base-3 expansion; that is, $\alpha_{3,5} = \sum_{k=1}^{\infty} a_k/3^k$, with $a_k \in \{0, 1, 2\}$ for all k. Then, for all $n = 0, 1, 2, \ldots$ one has:*

(i) $\sum_{k=2+5^{n+1}}^{2+5^{n+1}+4 \cdot 5^n} e^{a_k \pi i/2} = (-1)^n \left(\frac{-1+\sqrt{3}i}{2}\right) = e^{(3n+2)\pi i/3}$;

(ii) $a_k = a_{k+4 \cdot 5^n} = a_{k+8 \cdot 5^n} = a_{k+12 \cdot 5^n} = a_{k+16 \cdot 5^n}$ *for* $k = 5^{n+1} + j, j = 2, \ldots, 2 + 4 \cdot 5^n$.

Along this line, Bailey and Crandall showed that, given a real number r in $[0, 1)$, and r_k denoting the k-th binary digit of r, the real number

$$\alpha_{2,3}(r) := \sum_{k=0}^{\infty} \frac{1}{3^k 2^{3^k + r_k}} \qquad (12)$$

is 2-normal. It can be seen that if $r \neq s$, then $\alpha_{2,3}(r) \neq \alpha_{2,3}(s)$, so that these constants are all distinct. Thus, this generalized class of Stoneham constants is uncountably infinite. A similar result applies if 2 and 3 in this formula are replaced by any pair of co-prime integers (b, c) greater than 1, [10] [16, pg. 141–173]. We have not yet studied this generalized class by graphical methods.

5.3 The Erdős–Borwein Constants
The constructions of the previous two subsections exhaust most of what is known for concrete irrational numbers. By contrast, we finish this section with a truly tantalizing case:

In a base $b \geq 2$, we define the *Erdős–(Peter) Borwein constant EB(b)* by the *Lambert series* [18]:

$$EB(b) := \sum_{n \geq 1} \frac{1}{b^n - 1} = \sum_{n \geq 1} \frac{\tau(n)}{b^n}, \qquad (13)$$

where $\tau(n)$ is the number of divisors of n. It is known that the numbers $\sum_{n \geq 1} 1/(b^n - r)$ are irrational for r a nonzero rational and $b = 2, 3, \ldots$ such that $r \neq b^n$ for all n [20].

Whence, as provably irrational numbers other than the standard examples are few and far between, it is interesting to consider their normality.

Crandall [27] has observed that the structure of (13) is analogous to the "BBP" formula for π (see [7, 16]) and used this, as well as some nontrivial knowledge of the arithmetic properties of τ, to establish results such as that the googol-th bit (namely, the bit in position 10^{100} to the right of the "decimal" point) of $EB(2)$ is a 1.

In [27] Crandall also computed the first 2^{43} bits (one Tbyte) of $EB(2)$, which required roughly 24 hours of computation, and found that there are **435**9105565638 zeroes and **443**6987456570 ones. There is a corresponding variation in the second and third place in the single-digit hex (base-16) distributions. This certainly leaves some doubt as to its normality. Likewise, Crandall finds that in the first 1,000 decimal positions after the quintillionth digit (10^{18}), the respective digit counts for digits 0, 1, 2, 3, 4, 5, 6, 7, 8, 9 are 104, 82, 87, 100, 73, 126, 87, 123, 114, 104. Our own more modest computations of $EB(10)$ base-10 again leave it far from clear that $EB(10)$ is 10-normal. See also Figure 7(e) but contrast it to Figure 8(f).

We should note that for computational purposes, we employed the identity

$$\sum_{n \geq 1} \frac{1}{b^n - 1} = \sum_{n \geq 1} \frac{b^n + 1}{b^n - 1} \frac{1}{b^{n^2}},$$

for $|b| > 1$, due to Clausen, as did Crandall [27].

6 Other Avenues and Concluding Remarks
Let us recall two further examples used in [14], that of $\Omega(n)$, the *Liouville function*, which counts the parity of the number of prime factors of n (see Figure 14), and the human genome taken from the *UCSC Genome Browser* at http://hgdownload. cse.ucsc.edu/goldenPath/hg19/chromosomes/ (see Fig. 15). Note the similarity of the genome walk to the those of concatenation sequences. We have explored a wide variety of walks on genomes, but we will reserve the results for a future study.

We should emphasize that, to the best of our knowledge, the normality and transcendence status of the numbers explored is unresolved other than in the cases indicated in sections 5.1 and 5.2 and indicated in Appendix 7. Although one of the clearly nonrandom numbers (say Stoneham or Copeland–Erdős) may pass muster on one or other measure of the walk, it is generally the case that it fails another. Thus, the Liouville number λ_2 (see Fig. 14) exhibits a much more structured drift than π or e, but looks more like them than like Figure 15(a).

This situation provides hope for more precise future analyses. We conclude by remarking on some unresolved issues and plans for future research.

6.1 Fractal and Box-Dimension
Another approach is to estimate the fractal dimensions of walks, which is an appropriate tool with which to measure the geometrical complexity of a set, characterizing its space-filling capacity (see, e.g., [6] for a nice introduction about fractals). The *box-counting dimension*, also known as the *Minkowski–*

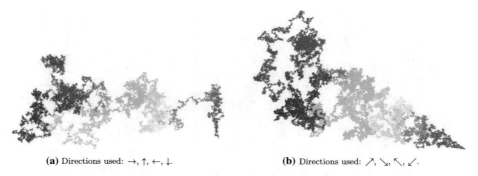

(a) Directions used: →, ↑, ←, ↓.   (b) Directions used: ↗, ↘, ↖, ↙.

Figure 14. Two different rules for plotting a base-2 walk on the first two million values of $\lambda(n)$ (the Liouville number λ_2).

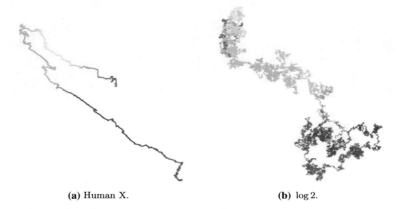

(a) Human X.   (b) log 2.

Figure 15. Base-4 walks on 10^6 bases of the X-chromosome and 10^6 digits of log 2.

Bouligand dimension, permits us to estimate the fractal dimension of a given set and often coincides with the fractal dimension. The box-dimension of the walk of numbers such as π turns out to be close to 2, whereas for nonrandom numbers as $\alpha_{2,3}$ in base 6 or Champernowne's number, the box-dimension is nearly 1.

6.2 Three Dimensions
We have also explored three-dimensional graphics—using base-6 for directions—both in perspective and in a large passive (glasses-free) three-dimensional viewer outside the CARMA laboratory; but we have not yet quantified these excursions.

6.3 Genome Comparison
Genomes are made up of so-called purine and pyrimidine nucleotides. In DNA, purine nucleotide bases are adenine and guanine (A and G), whereas the pyrimidine bases are thymine and cytosine (T and C). Thymine is replaced by uracyl in RNA. The haploid human genome (i.e., 23 chromosomes) is estimated to hold about 3.2 billion base pairs and so to contain 20,000-25,000 distinct genes. Hence there are many ways of representing a stretch of a chromosome as a walk, say as a base-4 uniform walk on the symbols (A, G, T, C) illustrated in Figure 15 (where A, G, T, and C draw the new point to the south, north, west, and east, respectively, and we have not plotted undecoded or unused portions), or as a three-dimensional logarithmic walk inside a tetrahedron.

We have also compared random *chaos games* in a square with genomes and numbers plotted by the same rules.[2] As an illustration, we show twelve games in Figure 16: four on a triangle, four on a square, and four on a hexagon. At each step we go from the current point halfway toward one of the vertices, chosen depending on the value of the digit. The color indicates the number of hits, in a similar manner as in Figure 6. The nonrandom behavior of the Champernowne numbers is apparent in the coloring patterns, as are the special features of the Stoneham numbers described in Section 5.2 (the non-normality of $\alpha_{2,3}$ and $\alpha_{3,2}$ in base 6 yields a paler color, whereas the repeating structure of $\alpha_{2,3}$ and $\alpha_{3,5}$ is the origin of the purple tone, see Conjectures 5.6 and 5.7).

---

[2]The idea of a chaos game was described by Barnsley in his 1988 book *Fractals Everywhere* [6]. Games on amino acids seem to originate with [35]. For a recent summary see [17, pp. 194–205].

Figure 16. Chaos games on various numbers, colored by frequency. Row 1: C_3, $\alpha_{3,5}$, a (pseudo)random number, and $\alpha_{2,3}$. Row 2: C_4, π, a (pseudo)random number, and $\alpha_{2,3}$. Row 3: C_6, $\alpha_{3,2}$, a (pseudo)random number, and $\alpha_{2,3}$.

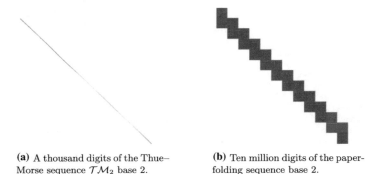

(a) A thousand digits of the Thue–Morse sequence $\mathcal{TM}_2$ base 2.

(b) Ten million digits of the paper-folding sequence base 2.

Figure 17. Walks on two automatic and nonnormal numbers.

6.4 Automatic Numbers

We have also explored numbers originating with finite state automata, such as those of the *paper-folding* and the *Thue–Morse* sequences, $\mathcal{P}$ and $\mathcal{TM}_2$, see [2] and Section 7. Automatic numbers are never normal and are typically transcendental; by comparison, the Liouville number λ_2 is not p-automatic for any prime p [25].

The walks on $\mathcal{P}$ and $\mathcal{TM}_2$ have a similar shape, see Figure 17, but while the Thue–Morse sequence generates very large pictures, the paper-folding sequence generates very small ones, because it is highly self-replicating; see also the values in Tables 1 and 2.

A *turtle plot* on these constants, where each binary digit corresponds to either "forward motion" of length 1 or "rotate the Logo turtle" in a fixed angle, exhibits some of their striking features (see Fig. 18). For instance, drawn with a rotating angle of $\pi/3$, $\mathcal{TM}_2$ converges to a Koch snowflake [41]; see Figure 18(c). We show a corresponding turtle graphic of π in Figure 18(d). Analogous features occur for the paper-folding sequence as described in [28, 29, 30], and two variants are shown in Figures 18(a) and 18(b).

6.5 Continued Fractions

Simple continued fractions often encode more information than base expansions about a real number. Basic facts are that a continued fraction terminates or repeats if and only if the

(a) Ten million digits of the paper-folding sequence with rotating angle $\pi/3$.

(b) Dragon curve from one million digits of the paper-folding sequence with rotating angle $2\pi/3$.

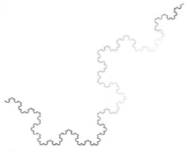

(c) Koch snowflake from 100,000 digits of the Thue–Morse sequence $\mathcal{TM}_2$ with rotating angle $\pi/3$.

(d) One million digits of π with rotating angle $\pi/3$.

Figure 18. Turtle plots on various constants with different rotating angles in base 2—where "0" gives forward motion and "1" rotation by a fixed angle.

number is rational or a quadratic irrational, respectively; see [16, 7]. By contrast, the simple continued fractions for π and e start as follows in the standard compact form:

$$\pi = [3, 7, 15, 1, 292, 1, 1, 1, 2, 1, 3, 1, 14, 2, 1,$$
$$1, 2, 2, 2, 2, 1, 84, 2, 1, 1, 15, 3, 13, 1, 4, \ldots]$$
$$e = [2, 1, 2, 1, 1, 4, 1, 1, 6, 1, 1, 8, 1, 1, 10, 1, 1, 12, 1, 1, 14, 1,$$
$$1, 16, 1, 1, 18, 1, 1, 20, 1, \ldots],$$

from which the surprising regularity of e and apparent irregularity of π as continued fractions is apparent. The counterpart to Borel's theorem—that almost all numbers are normal—is that almost all numbers have "normal" continued fractions $\alpha = [a_1, a_2, \ldots, a_n, \ldots]$, for which the *Gauss–Kuzmin distribution* holds [16]: for each $k = 1, 2, 3, \ldots$

$$\text{Prob}\{a_n = k\} = -\log_2\left(1 - \frac{1}{(k+1)^2}\right), \qquad (14)$$

so that roughly 41.5% of the terms are 1, 16.99% are 2, 9.31% are 3, etc.

In Figure 19, we show a histogram of the first 100 million terms, computed by Neil Bickford and accessible at http://neilbickford.com/picf.htm, of the continued fraction of π. We have not yet found a satisfactory way to embed this in a walk on a continued fraction, but in Figure 20 we show base-4 walks on π and e where we use the remainder modulo 4 to build the walk (with 0 being right, 1 being up, 2 being left, and 3 being down). We also show turtle plots on π, e.

Andrew Mattingly has observed that:

PROPOSITION 6.1 *With probability* 1, *a mod-4 random walk (with 0 being right, 1 being up, 2 being left, and 3 being down) on the simple continued fraction coefficients of a real number is asymptotic to a line making a positive angle with the x-axis of:*

$$\arctan\left(\frac{1}{2}\frac{\log_2(\pi/2) - 1}{\log_2(\pi/2) - 2\log_2(\Gamma(3/4))}\right) \approx 110.44°.$$

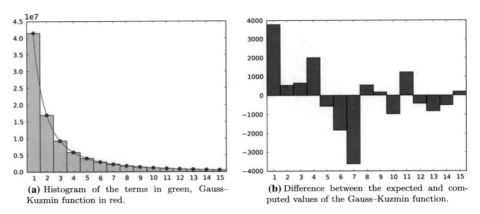

(a) Histogram of the terms in green, Gauss–Kuzmin function in red.

(b) Difference between the expected and computed values of the Gauss–Kuzmin function.

Figure 19. Expected values of the Gauss–Kuzmin distribution of (14) and the values of 100 million terms of the continued fraction of π.

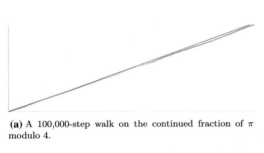

(a) A 100,000-step walk on the continued fraction of π modulo 4.

(b) A 100-step walk on the continued fraction of e modulo 4.

(c) A one million-step turtle walk on the continued fraction of π modulo 2 with rotating angle $\pi/3$.

(d) A 100-step turtle walk on the continued fraction of e modulo 2 with rotating angle $\pi/3$.

Figure 20. Uniform walks on π and e based on continued fractions.

PROOF. The result comes by summing the expected Gauss–Kuzmin probabilities of each step being taken as given by (14).

This is illustrated in Figure 20(a) with a 90° anticlockwise rotation; thus making the case that one must have some *a priori* knowledge before designing tools.

It is also instructive to compare digits and continued fractions of numbers as horizontal matrix plots of the form already used in Figure 8. In Figure 21, we show six pairs of million-term digit-strings and the corresponding continued fraction. In some cases both look random, in others one or the other does.

In conclusion, we have only scratched the surface of what is becoming possible in a period in which data—for example, five-hundred million terms of the continued fraction or five-trillion binary digits of π, full genomes, and much more—can be downloaded from the Internet, then rendered and visually mined, with fair rapidity.

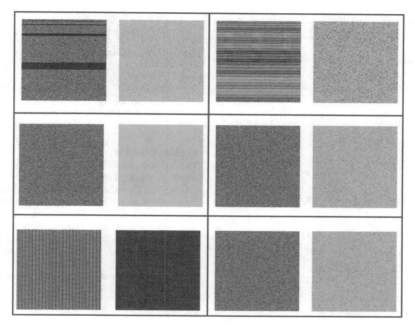

Figure 21. Million-step comparisons of base-4 digits and continued fractions. Row 1: $\alpha_{2,3}$ (*base* 6) and C_4. Row 2: e and π. Row 3: Q_1 and pseudorandom iterates; as listed from top left to bottom right.

7 Appendix Selected Numerical Constants

In what follows, $x := 0.a_1 a_2 a_3 a_4 \ldots_b$ denotes the base-b expansion of the number x, so that $x = \sum_{k=1}^{\infty} a_k b^{-k}$. Base-10 expansions are denoted without a subscript.

Catalan's constant (irrational?; normal?):

$$G := \sum_{k=0}^{\infty} \frac{(-1)^k}{(2k+1)^2} = 0.9159655941\ldots \quad (15)$$

Champernowne numbers (irrational; normal to corresponding base):

$$C_b := \sum_{k=1}^{\infty} \frac{\sum_{m=b^{k-1}}^{b^k - 1} mb^{-k\left[m-(b^{k-1}-1)\right]}}{b\sum_{m=0}^{k-1} m(b-1)b^{m-1}} \quad (16)$$

$$C_{10} = 0.123456789101112\ldots$$
$$C_4 = 0.1231011121320212223\ldots_4$$

Copeland–Erdős constants (irrational; normal to corresponding base):

$$CE(b) := \sum_{k=1}^{\infty} p_k b^{-\left(k+\sum_{m=1}^{k} \lfloor \log_b p_m \rfloor\right)}, \quad (17)$$

where p_k is the kth prime number

$$CE(10) = 0.2357111317\ldots$$
$$CE(2) = 0.1011101111\ldots_2$$

Exponential constant (transcendental; normal?):

$$e := \sum_{k=0}^{\infty} \frac{1}{k!} = 2.7182818284\ldots \quad (18)$$

Erdős–Borwein constants (irrational; normal?):

$$EB(b) := \sum_{k=1}^{\infty} \frac{1}{b^k - 1} \quad (19)$$

$$EB(2) = 1.6066951524\ldots = 1.212311001\ldots_4$$

Euler–Mascheroni constant (irrational?; normal?):

$$\gamma := \lim_{m\to\infty} \left(\sum_{k=1}^{m} \frac{1}{k} - \log m\right) = 0.5772156649\ldots \quad (20)$$

Fibonacci constant (irrational [12, Theorem 2]; normal?):

$$\mathcal{F} := \sum_{k=1}^{\infty} F_k 10^{-\left(1+k+\sum_{m=1}^{k} \lfloor \log_{10} F_m \rfloor\right)}, \text{ where}$$

$$F_k = \frac{\left(\frac{1+\sqrt{5}}{2}\right)^k - \left(\frac{1-\sqrt{5}}{2}\right)^k}{\sqrt{5}}$$

$$= 0.011235813213455\ldots$$

Liouville number (irrational; not p-automatic):

$$\lambda_2 := \sum_{k=1}^{\infty} \left(\frac{\lambda(k)+1}{2}\right) 2^{-k} \quad (22)$$

where $\lambda(k) := (-1)^{\Omega(k)}$ and $\Omega(k)$ counts prime factors of k

$$= 0.5811623188\ldots = 0.10010100110\ldots_2$$

Logarithmic constant (transcendental; normal?):

$$\log 2 := \sum_{k=1}^{\infty} \frac{1}{k2^k} \tag{23}$$

$$= 0.6931471806\ldots = 0.10110001011100100001\ldots_2$$

Pi (transcendental; normal?):

$$\pi := 2\int_{-1}^{1} \sqrt{1-x^2}\,dx = 4\sum_{k=0}^{\infty} \frac{(-1)^k}{2k+1} \tag{24}$$

$$= 3.1415926535\ldots = 11.00100100001111110110\ldots_2$$

Riemann zeta function at integer arguments (transcendental for n even; irrational for $n = 3$; unknown for $n \geq 5$ odd; normal?):

$$\zeta(s) := \sum_{k=1}^{\infty} \frac{1}{k^s} \tag{25}$$

In particular:

$$\zeta(2) = \frac{\pi^2}{6} = 1.6449340668\ldots$$

$$\zeta(2n) = (-1)^{n+1}\frac{(2\pi)^{2n}}{2(2n)!}B_{2n}$$

$$(\text{where } B_{2n} \text{ are Bernoulli numbers})$$

$$\zeta(3) = \text{Apery's constant} = \frac{5}{2}\sum_{k=1}^{\infty} \frac{(-1)^{k+1}}{k^3\binom{2k}{k}}$$

$$= 1.2020569031\ldots$$

Stoneham constants (irrational; normal in some bases; nonnormal in different bases; normality still is unknown other bases):

$$\alpha_{b,c} := \sum_{k=1}^{\infty} \frac{1}{b^{c^k}c^k} \tag{26}$$

$$\alpha_{2,3} = 0.0418836808\ldots = 0.0022232032\ldots_4$$
$$= 0.0130140430003334\ldots_6$$
$$\alpha_{4,3} = 0.0052087571\ldots = 0.0001111111301\ldots_4$$
$$= 0.0010430041343502130000\ldots_6$$
$$\alpha_{3,2} = 0.0586610287\ldots = 0.0011202021212121\ldots_3$$
$$= 0.0204005200030544000002\ldots_6$$
$$\alpha_{3,5} = 0.0008230452\ldots = 0.00000012101210121\ldots_3$$
$$= 0.002ba00000061d2\ldots_{15}$$

Thue–Morse constant (transcendental; 2-automatic, hence nonnormal):

$$\mathcal{TM}_2 := \sum_{k=1}^{\infty} \frac{1}{2^{t(n)}} \text{ where } t(0) = 0, \text{ while } t(2n) = t(n)$$
$$\text{and } t(2n+1) = 1 - t(n) \tag{27}$$

$$= 0.4124540336\ldots$$
$$= 0.011010011001011010010110011010001\ldots_2$$

Paper-folding constant (transcendental; 2-automatic, hence nonnormal):

$$\mathcal{P} := \sum_{k=0}^{\infty} \frac{8^{2^k}}{2^{2^{k+2}}-1} = 0.8507361882\ldots$$
$$= 0.1101100111001001\ldots_2 \tag{28}$$

ACKNOWLEDGMENTS

The authors thank the referee for a thoughtful and thorough report, David Allingham for his kind help in running the algorithms for plotting the "big walks" on π, Adrian Belshaw for his assistance with strong normality, Matt Skerritt for his 3D image of π, and Jake Fountain who produced a fine Python interface for us as a 2011-2012 Vacation Scholar at CARMA. We are also most grateful for several discussions with Andrew Mattingly (IBM) and Michael Coons (CARMA), who helped especially with continued fractions and automatic numbers, respectively.

REFERENCES

[1] S. Albeverioa, M. Pratsiovytyie, and G. Torbine G, "Topological and fractal properties of real numbers which are not normal". *Bulletin des Sciences Mathématiques*, **129** (2005), 615–630.

[2] J.-P. Allouche and J. Shallit, *Automatic Sequences: Theory, Applications, Generalizations*. Cambridge University Press, Cambridge, 2003.

[3] D. H. Bailey and J. M. Borwein, "Normal numbers and pseudorandom generators," *Proceedings of the Workshop on Computational and Analytical Mathematics in Honour of Jonathan Borwein's 60th Birthday*, Springer, 2012, in press.

[4] D. H. Bailey, J. M. Borwein, C. S. Calude, M. J. Dinneen, M. Dumitrescu, and A. Yee, "An empirical approach to the normality of pi". *Experimental Mathematics*, 2012; in press.

[5] D. H. Bailey, J. M. Borwein, R. E. Crandall, and C. Pomerance. "On the binary expansions of algebraic numbers". *Journal of Number Theory Bordeaux*, **16** (2004), 487–518.

[6] M. Barnsley, *Fractals Everywhere*, Academic Press, Inc., Boston, MA, 1988.

[7] D. H. Bailey, P. B. Borwein, and S. Plouffe, "On the rapid computation of various polylogarithmic constants". *Mathematics of Computation*, **66**, no. 218 (1997), 903–913.

[8] D. H. Bailey and D. J. Broadhurst, "Parallel integer relation detection: Techniques and applications". *Mathematics of Computation*, **70**, no. 236 (2000), 1719–1736.

[9] D. H. Bailey and R. E. Crandall, "On the random character of fundamental constant expansions". *Experimental Mathematics*, **10**, no. 2 (2001), 175–190.

[10] D. H. Bailey and R. E. Crandall, "Random generators and normal numbers," *Experimental Mathematics*, **11** (2002), no. 4, 527–546.

[11] D. H. Bailey and M. Misiurewicz, "A strong hot spot theorem," *Proceedings of the American Mathematical Society*, **134** (2006), no. 9, 2495–2501.

[12] G. Barat, R. F. Tichy, and R. Tijdeman, Digital blocks in linear numeration systems. *Number theory in progress*, **2** (Zakopane-Kościelisko, 1997), de Gruyter, Berlin (1999), 607–631.

[13] M. N. Barber and B. W. Ninham, *Random and Restricted Walks: Theory and Applications*, Gordon and Breach, New York, 1970.

[14] A. Belshaw and P. B. Borwein, "Champernowne's number, strong normality, and the X chromosome," *Proceedings of the Workshop on Computational and Analytical Mathematics in Honour of Jonathan Borwein's 60th Birthday*, Springer, 2012, in press.

[15] L. Berggren, J. M. Borwein, and P. B. Borwein, *Pi: a Source Book*, Springer-Verlag, Third Edition, 2004.

[16] J. M. Borwein and D. H. Bailey, *Mathematics by Experiment: Plausible Reasoning in the 21st Century*, 2nd ed., A. K. Peters, Natick, MA, 2008.

[17] J. Borwein, D. Bailey, N. Calkin, R. Girgensohn, R. Luke, V. Moll, *Experimental Mathematics in Action*. A. K. Peters, Natick, MA, 2007.

[18] J. M. Borwein and P. B. Borwein, *Pi and the AGM: A Study in Analytic Number Theory and Computational Complexity*, John Wiley, New York, 1987, paperback 1998.

[19] J. M. Borwein, P. B. Borwein, R. M. Corless, L. Jörgenson, and N. Sinclair, "What is organic mathematics?" *Organic mathematics* (Burnaby, BC, 1995), CMS Conf. Proc., **20**, Amer. Math. Soc., Providence, RI, 1997, 1–18.

[20] P. B. Borwein, "On the irrationality of certain series." *Math. Proc. Cambridge Philos. Soc.* **112** (1992) 141–146.

[21] P. B. Borwein and L. Jörgenson, " Visible structures in number theory," *Amer. Math. Monthly* **108** (2001), no. 10, 897–910.

[22] C. S. Calude, "Borel normality and algorithmic randomness," in G. Rozenberg, A. Salomaa (eds.), *Developments in Language Theory*, World Scientific, Singapore, 1994, 113–129.

[23] C.S. Calude, *Information and Randomness: An Algorithmic Perspective*, 2nd ed., Revised and Extended, Springer-Verlag, Berlin, 2002.

[24] D. G. Champernowne, "The construction of decimals normal in the scale of ten." *Journal of the London Mathematical Society*, **8** (1933) 254–260.

[25] M. Coons, "(Non)automaticity of number theoretic functions," *J. Théor. Nombres Bordeaux*, **22** (2010), no. (2), 339–352.

[26] A. H. Copeland and P. Erdős, "Note on normal numbers," *Bulletin of the American Mathematical Society*, **52** (1946), 857–860.

[27] R. E. Crandall, "The googol-th bit of the Erdős–Borwein constant," *Integers*, A23, 2012.

[28] M. Dekking, M. Mendès France, and A. van der Poorten, "Folds," *Math. Intelligencer* **4** (1982), no. 3, 130–138.

[29] M. Dekking, M. Mendès France, and A. van der Poorten, "Folds II," *Math. Intelligencer* **4** (1982), no. 4, 173–181.

[30] M. Dekking, M. Mendès France, and A. van der Poorten, "Folds III," *Math. Intelligencer* **4**(1982), no. (4), 190–195.

[31] D. Y. Downham and S. B. Fotopoulos, "The transient behaviour of the simple random walk in the plane," *J. Appl. Probab.* **25** (1988), no. 1, 58–69.

[32] D. Y. Downham and S. B. Fotopoulos, "A note on the simple random walk in the plane," *Statist. Probab. Lett.*, 17 (1993), no. 3, 221–224.

[33] A. Dvoretzky and P. Erdős, "Some problems on random walk in space," *Proceedings of the 2nd Berkeley Symposium on Mathematical Statistics and Probability*, (1951), 353–367.

[34] P. Hertling, "Simply normal numbers to different bases," *Journal of Universal Computer Science*, **8**, no. 2 (2002), 235–242.

[35] H. J. Jeffrey, Chaos game representation of gene structure, *Nucl. Acids Res.* **18** no 2, (1990) 2163–2170.

[36] B. D. Hughes, *Random Walks and Random Environments, Vol. 1. Random Walks*, Oxford Science Publications, New York, (1995).

[37] H. Kaneko, "On normal numbers and powers of algebraic numbers," *Integers*, **10** (2010), 31–64.

[38] D. Khoshnevisan, "Normal numbers are normal," *Clay Mathematics Institute Annual Report* (2006), 15 & 27–31.

[39] G. Marsaglia, "On the randomness of pi and other decimal expansions," preprint, 2010.

[40] G. Martin, "Absolutely abnormal numbers," *Amer. Math. Monthly*, **108** (2001), no. 8, 746-754.

[41] J. Mah and J. Holdener, "When Thue–Morse meets Koch," *Fractals*, **13** (2005), no. 3, 191–206.

[42] S. M. Ross, *Stochastic Processes*. John Wiley & Sons, New York, 1983.

[43] R. Stoneham, "On absolute (j, ε)-normality in the rational fractions with applications to normal numbers," *Acta Arithmetica*, **22** (1973), 277–286.

[44] M. Queffelec, "Old and new results on normality," *Lecture Notes – Monograph Series*, **48**, *Dynamics and Stochastics*, 2006, Institute of Mathematical Statistics, 225–236.

[45] W. Schmidt, "On normal numbers," *Pacific Journal of Mathematics*, **10** (1960), 661–672.

[46] A. J. Yee, "y-cruncher-multi-threaded pi program," http://www.numberworld.org/y-cruncher, 2010.

[47] A. J. Yee and S. Kondo, "10 trillion digits of pi: A case study of summing hypergeometric series to high precision on multicore systems," preprint, 2011, available at http://hdl.handle.net/2142/28348.

22. Birth, growth and computation of pi to ten trillion digits (2013)

Paper 22: Ravi Agarwal, Hans Agarwal and Syamal K. Sen, "Birth, growth and computation of pi to ten trillion digits," *Advances in Di erence Equations*, 2013:100, p. 1–59.

Synopsis:

This paper presents one of the most complete and up-to-date chronologies of the analysis and computation of π through the ages, from approximations used by Indian and Babylonian mathematicians, well before the time of Christ, to Archimedes of Syracuse, "who ranks with Newton and Gauss as one of the three greatest mathematicians who ever lived," to mathematicians in the Islamic world during the "dark ages," and on to mathematicians in Renaissance Europe, including Francois Viete, Ludolph van Ceulen, John Wallis, Isaac Newton, John Machin, Leonard Euler, William Shanks and many others.

In the twentieth century, the chronology continues with summaries of the work of Ramanujan, then computer-age calculations beginning with Wrench and Smith, Guilloud and Fillatre, and, in more recent years, the work of Gosper, Brent, Salamin and others who either development new mathematics related to π or performed prodigious computer-based calculations of π_m right up to and including the computation of π to ten trillion digits by Kondo and Yee, which, as of this writing, is almost the state-of-the-art (Kondo and Yee have subsequently computed twelve trillion digits).

The paper includes many formula and techniques used, all of which is spelled out in considerable detail.

Keywords: Algorithms, Computation, Curiosities, History, Series

Agarwal et al. *Advances in Difference Equations* 2013, **2013**:100
http://www.advancesindifferenceequations.com/content/2013/1/100

Advances in Difference Equations
a SpringerOpen Journal

RESEARCH   Open Access

Birth, growth and computation of pi to ten trillion digits

Ravi P Agarwal[1]*, Hans Agarwal[2] and Syamal K Sen[3]

*Correspondence:
Agarwal@tamuk.edu
[1]Department of Mathematics, Texas
A&M University-Kingsville, Kingsville,
TX, 78363, USA
Full list of author information is
available at the end of the article

Abstract

The universal real constant pi, the ratio of the circumference of any circle and its diameter, has no exact numerical representation in a finite number of digits in any number/radix system. It has conjured up tremendous interest in mathematicians and non-mathematicians alike, who spent countless hours over millennia to explore its beauty and varied applications in science and engineering. The article attempts to record the pi exploration over centuries including its successive computation to ever increasing number of digits and its remarkable usages, the list of which is not yet closed.

Keywords: circle; error-free; history of pi; Matlab; random sequence; stability of a computer; trillion digits

All circles have the same shape, and traditionally represent the infinite, immeasurable and even spiritual world. Some circles may be large and some small, but their 'circleness', their perfect roundness, is immediately evident. Mathematicians say that all circles are *similar*. Before dismissing this as an utterly trivial observation, we note by way of contrast that not all triangles have the same shape, nor all rectangles, nor all people. We can easily imagine tall narrow rectangles or tall narrow people, but a tall narrow circle is not a circle at all. Behind this unexciting observation, however, lies a profound fact of mathematics: that the *ratio* of circumference to diameter is the same for one circle as for another. Whether the circle is gigantic, with large circumference and large diameter, or minute, with tiny circumference and tiny diameter, the *relative* size of circumference to diameter will be exactly the same. In fact, the ratio of the circumference to the diameter of a circle produces, the most famous/studied/unlimited praised/intriguing/ubiquitous/external/mysterious mathematical number known to the human race. It is written as pi or as π [1–210], symbolically, and defined as

$$\text{pi} = \frac{\text{distance around a circle}}{\text{distance across and through the center of the circle}} = \frac{C}{D} = \pi.$$

Since the exact date of birth of π is unknown, one could imagine that π existed before the universe came into being and will exist after the universe is gone. Its appearance in the disks of the Moon and the Sun, makes it as one of the most ancient numbers known to humanity. It keeps on popping up inside as well as outside the scientific community, for example, in many formulas in geometry and trigonometry, physics, complex analysis, cosmology, number theory, general relativity, navigation, genetic engineering, statistics,

Agarwal et al. *Advances in Difference Equations* 2013, **2013**:100
http://www.advancesindifferenceequations.com/content/2013/1/100

fractals, thermodynamics, mechanics, and electromagnetism. Pi hides in the rainbow, and sits in the pupil of the eye, and when a raindrop falls into water π emerges in the spreading rings. Pi can be found in waves and ripples and spectra of all kinds and, therefore, π occurs in colors and music. The double helix of DNA revolves around π. Pi has lately turned up in super-strings, the hypothetical loops of energy vibrating inside subatomic particles. Pi has been used as a symbol for mathematical societies and mathematics in general, and built into calculators and programming languages. Pi is represented in the mosaic outside the mathematics building at the Technische Universität Berlin. Pi is also engraved on a mosaic at Delft University. Even a movie has been named after it. Pi is the secret code in Alfred Hitchcock's 'Torn Curtain' and in 'The Net' starring Sandra Bullock. Pi day is celebrated on March 14 (which was chosen because it resembles 3.14). The official celebration begins at 1:59 p.m., to make an appropriate 3.14159 when combined with the date. In 2009, the United States House of Representatives supported the designation of Pi Day. Albert Einstein was born on Pi Day (14 March 1879).

Throughout the history of π, which according to Beckmann (1971) 'is a quaint little mirror of the history of man', and James Glaisher (1848-1928) 'has engaged the attention of many mathematicians and calculators from the time of Archimedes to the present day, and has been computed from so many different formula, that a complete account of its calculation would almost amount to a history of mathematics', one of the enduring challenges for mathematicians has been to understand the nature of the number π (rational/irrational/transcendental), and to find its exact/approximate value. The quest, in fact, started during the pre-historic era and continues to the present day of supercomputers. The constant search by many including the greatest mathematical thinkers that the world produced, continues for new formulas/bounds based on geometry/algebra/analysis, relationship among them, relationship with other numbers such as $\pi = 5\cos^{-1}(\phi/2)$, $\pi \simeq 4/\sqrt{\phi}$, where ϕ is the Golden section (ratio), and $e^{i\pi} + 1 = 0$, which is due to Euler and contains 5 of the most important mathematical constants, and their merit in terms of computation of digits of π. Right from the beginning until modern times, attempts were made to exactly fix the value of π, but always failed, although hundreds constructed circle squares and claimed the success. These amateur mathematicians have been called the sufferers of *morbus cyclometricus*, the circle-squaring disease. Stories of these contributors are amusing and at times almost unbelievable. Many came close, some went to tens, hundreds, thousands, millions, billions, and now up to ten trillion (10^{13}) decimal places, but there is no exact solution. The American philosopher and psychologist William James (1842-1910) wrote in 1909 'the thousandth decimal of Pi sleeps there though no one may ever try to compute it'. Thanks to the twentieth and twenty-first century, mathematicians and computer scientists, it sleeps no more. In 1889, Hermann Schubert (1848-1911), a Hamburg mathematics professor, said 'there is no practical or scientific value in knowing more than the 17 decimal places used in the foregoing, already somewhat artificial, application', and according to Arndt and Haenel (2000), just 39 decimal places would be enough to compute the circumference of a circle surrounding the known universe to within the radius of a hydrogen atom. Further, an expansion of π to only 47 decimal places would be sufficiently precise to inscribe a circle around the visible universe that does not deviate from perfect circularity by more than the distance across a single proton. The question has been repeatedly asked why so many digits? Perhaps the primary motivation for these computations is the human desire to break records; the extensive calculations involved

Agarwal et al. *Advances in Difference Equations* 2013, **2013**:100
http://www.advancesindifferenceequations.com/content/2013/1/100

have been used to test supercomputers and high-precision multiplication algorithms (a stress test for a computer, a kind of 'digital cardiogram'), the statistical distribution of the digits, which is expected to be uniform, that is, the frequency with which the digits (0 to 9) appear in the result will tend to the same limit (1/10) as the number of decimal places increases beyond all bounds, and in recent years these digits are being used in applied problems as a random sequence. It appears experts in the field of π are looking for surprises in the digits of π. In fact, the Chudnovsky brothers once said: 'We are looking for the appearance of some rules that will distinguish the digits of π from other numbers. If you see a Russian sentence that extends for a whole page, with hardly a comma, it is definitely Tolstoy. If someone gave you a million digits from somewhere in π, could you tell it was from π'? Some interesting observations are: The first 144 digits of π add up to 666 (which many scholars say is 'the mark of the Beast'); Since there are 360 degrees in a circle, some mathematicians were delighted to discover that the number 360 is at the 359th digit position of π. A mysterious 2008 crop circle in Britain shows a coded image representing the first 10 digits of π. The Website 'The Pi-Search Page' finds a person's birthday and other well-known numbers in the digits of π. Several people have endeavored to memorize the value of π with increasing precision, leading to records of over 100,000 digits.

We believe that the study and discoveries of π will never end; there will be books, research articles, new record-setting calculations of the digits, clubs and computer programs dedicated to π. In what follows, we shall discuss the growth and the computation of π chronologically. For our ready reference, we also give some digits of π,

$$\pi = 3.14159265358979323846264338327950288419716939937510$$

$$58209749445923078164062862089986280348253421170679.$$

About 3200 BC. The meaning of the word *sulv* is to measure, and geometry in ancient India came to be known by the name *sulba* or *sulva*. *Sulbasutras* means 'rule of chords', which is another name for geometry. The Sulbasutras are part of the larger corpus of texts called the Shrautasutras, considered to be appendices to the *Vedas*, which give rules for constructing altars. If the ritual sacrifice was to be successful, then the altar had to conform to very precise measurements, so mathematical accuracy was seen to be of the utmost importance. The sulbas contain a large number of geometric constructions for squares, rectangles, parallelograms and trapezia. Sulbas also contain remarkable approximations

$$\sqrt{2} \simeq 1 + \frac{1}{3} + \frac{1}{3 \cdot 4} - \frac{1}{3 \cdot 4 \cdot 34},$$

which gives $\sqrt{2} = 1.4142156\ldots$, and

$$\pi \simeq 18(3 - 2\sqrt{2}) = \left(\frac{6}{2 + \sqrt{2}} \right)^2,$$

which gives $\pi = 3.088311\ldots$.

About 2742 BC. Aryabhatta was born in 2765 BC in Patliputra in Magadha, modern Patna in Bihar (India). He was teaching astronomy and mathematics when he was 23 years of age in 2742 BC. His astronomical knowledge was so advanced that he could claim that the Earth rotated on its own axis, the Earth moves round the Sun and the Moon rotates

Agarwal et al. *Advances in Difference Equations* 2013, **2013**:100
http://www.advancesindifferenceequations.com/content/2013/1/100

round the Earth; incredibly he believed that the orbits of the planets are ellipses. He talks about the position of the planets in relation to its movement around the Sun. He refers to the light of the planets and the Moon as reflection from the Sun. He explains the eclipse of the Moon and the Sun, day and night, the length of the year exactly as 365 days. He calculated the circumference of the Earth as 24,835 miles, which is close to modern day calculation of 24,900 miles. In his *Aryabhattiyam*, which consists of the 108 verses and 13 introductory verses, and is divided into four padas or chapters (written in the very terse style typical of sutra literature, in which each line is an aid to memory for a complex system), Aryabhatta included 33 verses giving 66 mathematical rules *ganita* on pure mathematics. He described various original ways to perform different mathematical operations, including square and cube roots and solving quadratic equations. He provided elegant results for the summation of series of squares and cubes. He made use of decimals, the zero (sunya) and the place value system. To find an approximate value of π, Aryabhatta gives the following prescription: Add 4 to 100, multiply by 8 and add to 62,000. This is 'approximately' the circumference of a circle whose diameter is 20,000. This means $\pi = 62,832/20,000 = 3.1416$. It is important to note that Aryabhatta used the word *asanna* (approaching), to mean that not only is this an approximation of π, but that the value is *incommensurable* or *irrational*, *i.e.*, it cannot be expressed as a ratio of two integers.

About 2600 BC. Great pyramid at Gizeh was built around 2600 BC in Egypt. It is one of the most massive buildings ever erected. It has at least twice the volume and thirty times the mass (the resistance an object offers to a change in its speed or direction of motion) of the Empire Sate Building in New York, and built from individual stones weighing up to 70 tons each. From the dimensions of the Great Pyramid, it is possible to derive the value of π, namely, π = half the perimeter of the base of the pyramid, divided by its height = 3 + 1/7 $\simeq$ 3.14285....

About 2000 BC. In a tablet found in 1936 in Susa (Iraq), Babylonians used the value

$$\frac{3}{\pi} = \frac{57}{60} + \frac{36}{(60)^2},$$

which yields π = 3 1/8 = 3.125. They were also satisfied with π = 3.

About 2000 BC. Ahmes (around 1680-1620 BC) (more accurately Ahmose) was an Egyptian scribe. A surviving work of Ahmes is part of the *Rhind Mathematical Papyrus, 1650 BC* (named after the Scottish Egyptologist Alexander Henry Rhind who went to Thebes for health reasons, became interested in excavating and purchased the papyrus in Egypt in 1858) located in the British Museum since 1863. When new, this papyrus was about 18 feet long and 13 inches high. Ahmes states that he copied the papyrus from a now-lost Middle Kingdom original, dating around 2000 BC. This curious document entitled *directions for knowing all dark things*, deciphered by Eisenlohr in 1877, is a collection of problems in geometry and arithmetic, algebra, weights and measures, business and recreational diversions. The 87 problems are presented with solutions, but often with no hint as to how the solution was obtained. In problem no. 50, Ahmes states that a circular field with a diameter of 9 units in area is the same as a square with sides of 8 units, *i.e.*, $\pi(9/2)^2 = 8^2$, and hence the Egyptian value of π is

$$\pi = 4 \times \left(\frac{8}{9}\right)^2 = 3.16049\ldots,$$

Agarwal et al. *Advances in Difference Equations* 2013, **2013**:100
http://www.advancesindifferenceequations.com/content/2013/1/100

which is only very slightly worse than the Babylonians value, and in contrast to the latter, an overestimation. We have no idea how this very satisfactory result was obtained (probably empirically), although various justifications are available. Maya value of π was as good as that of the Egyptians.

About 1200 BC. The earliest Chinese mathematicians, from the time of Chou-Kong used the approximation $\pi = 3$. Some of those who used this approximation were mathematicians of considerable attainments in other respects. According to the Chinese mythology, 3 is used because it is the number of the Heavens and the circle.

About 950 BC. In the *Old Testament* (I Kings vii.23, and 2 Chronicles iv.2), we find the following verse: 'Also, he made a molten sea of ten cubits from brim to brim, round in compass, and five cubits the height thereof; and a line of thirty cubits did compass it round about'. Hence the biblical value of π is 30/10 = 3. The Jewish Talmud, which is essentially a commentary on the Old Testament, was published about 500 AD. This shows that the Jews did not pay much attention to geometry. However, debates have raged on for centuries about this verse. According to some, it was just a simple approximation, while others say that '...the diameter perhaps was measured from outside, while the circumference was measured from inside'.

About 900 BC. *Shatapatha Brahmana* (Priest manual of 100 paths) is one of the prose texts describing the Vedic ritual. It survives in two recensions, Madhyandina and Kanva, with the former having the eponymous 100 brahmanas in 14 books, and the latter 104 brahmanas in 17 books. In these books, π is approximated by $339/108 = 3.138888\ldots$.

About 440 BC. Anaxagoras of Clazomanae (500-428 BC) came to Athens from near Smyrna, where he taught the results of the Ionian philosophy. He neglected his possessions in order to devote himself to science, and in reply to the question, what was the object of being born, he remarked: 'The investigation of the Sun, Moon and heaven'. He was the first to explain that the Moon shines due to reflected light from the Sun, which explains the Moon's phases. He also said that the Moon had mountains and he believed that it was inhabited. Anaxagoras gave some scientific accounts of eclipses, meteors, rainbows, and the Sun, which he asserted was larger than the Peloponnesus: this opinion, and various other physical phenomena, which he tried to explain which were supposed to have been direct action of the Gods, led him to a prosecution for impiety. While in prison he wrote a treatise on the quadrature of the circle. (The general problem of squaring a figure came to be known as the *quadrature problem*.) Since that time, hundreds of mathematicians tried to find a way to draw a square with equal area to a given circle; some maintained that they have found methods to solve the problem, while others argued that it is impossible. We will see that the problem was finally laid to rest in the nineteenth century.

About 430 BC. Hippocrates of Chios was born about 470 BC, and began life as a merchant. About 430 BC he came to Athens from Chios and opened a school of geometry, and began teaching, thus became one of the few individuals ever to enter the teaching profession for its financial rewards. He established the formula πr^2 for the area of a circle in terms of its radius. It means that a certain number π exists, and is the same for all circles, although his method does not give the actual numerical value of π. In trying to square the circle (unsuccessfully), Hippocrates discovered that two moon-shaped figures (lunes, bounded by pair of circular arcs) could be drawn whose areas were together equal to that of a right-angled triangle. Hippocrates gave the first example of constructing a rectilinear area equal to an area bounded by one or more curves.

Agarwal et al. *Advances in Difference Equations* 2013, **2013**:100
http://www.advancesindifferenceequations.com/content/2013/1/100

About 430 BC. Antiphon of Rhamnos (around 480-411 BC) was a sophist who attempted to find the area of a circle by considering it as the limit of an inscribed regular polygon with an infinite number of sides. Thus, he provided preliminary concept of infinitesimal calculus.

About 420 BC. Bryson of Heraclea was born around 450 BC. He was a student of Socrates. Bryson considered the circle squaring problem by comparing the circle to polygons inscribed within it. He wrongly assumed that the area of a circle was the arithmetical mean between circumscribed and inscribed polygons.

About 420 BC. Hippias of Elis was born about 460 BC. He was a Greek Sophist, a younger contemporary of Socrates. He is described as an expert arithmetician, but he is best known to us through his invention of a curve called the *quadratrix* ($x = y \cot(\pi y/2)$), by means of which an angle can be trisected, or indeed divided in any given ratio. It is not known whether Hippias realized that by means of his curve the circle could be squared; perhaps he realized but could not prove it. He lectured widely on mathematics and as well on poetry, grammar, history, politics, archeology and astronomy. Hippias was also a prolific writer, producing elegies, tragedies and technical treatises in prose. His work on Homer was considered excellent.

414 BC. Aristophanes (446-386 BC) in his play *The Birds* makes fun of circle squarers.

Around 375 BC. Plato of Athens (around 427-347 BC) was one of the greatest Greek philosophers, mathematicians, mechanician, a pupil of Socrates for eight years, and teacher of Aristotle. He is famous for 'Plato's Academy'. 'Let no man ignorant of mathematics enter here' is supposed to have been inscribed over the doors of the Academy. He is supposedly obtained for his day a fairly accurate value for $\pi = \sqrt{2} + \sqrt{3} = 3.146 \ldots$.

About 370 BC. Eudoxus of Cnidus (around 400-347 BC) was the most celebrated mathematician. He developed the theory of proportion, partly to place the doctrine of incommensurables (irrationals) upon a thoroughly sound basis. Specially, he showed that the area of a circle is proportional to its diameter squared. Eudoxus established fully the *method of exhaustions* of Antiphon by considering both the inscribed and circumscribed polygons. He also considered certain curves other than the circle. He explained the apparent motions of the planets as seen from the earth. Eudoxus also wrote a treatise on practical astronomy, in which he supposed a number of moving spheres to which the Sun, Moon and stars were attached, and which by their rotation produced the effects observed. In all, he required 27 spheres.

About 350 BC. Dinostratus (around 390-320 BC) was a Greek mathematician. He used Hippias quadratrix to square the circle. For this, he proved *Dinostratus' theorem*. Hippias quadratrix later became known as the *Dinostratus quadratrix* also. However, his demonstration was not accepted by the Greeks as it violated the foundational principles of their mathematics, namely, using only ruler and compass.

About 240 BC. Archimedes of Syracuse (287-212 BC) ranks with Newton and Gauss as one of the three greatest mathematicians who ever lived, and he is certainly the greatest mathematician of antiquity. Galileo called him 'divine Archimedes, superhuman Archimedes'; Sir William Rowan Hamilton (1805-1865) remarked 'who would not rather have the fame of Archimedes than that of his conqueror Marcellus'?; Alfred North Whitehead (1861-1947) commented 'no Roman ever died in contemplation over a geometrical diagram'; Godfrey Harold Hardy (1877-1947) said 'Archimedes will be remembered when Aeschylus is forgotten, because languages die and mathematical ideas do not'; and

Agarwal et al. *Advances in Difference Equations* 2013, **2013**:100
http://www.advancesindifferenceequations.com/content/2013/1/100

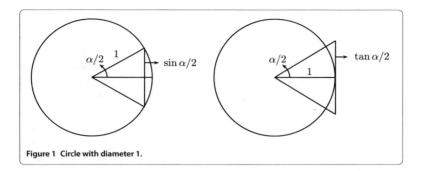

Figure 1 Circle with diameter 1.

Voltaire remarked 'there was more imagination in the head of Archimedes than in that of Homer'. His mathematical work is so modern in spirit and technique that it is barely distinguishable from that of a seventeenth-century mathematician. Among his mathematical achievements, Archimedes developed a general *method of exhaustion* for finding areas bounded by parabolas and spirals, and volumes of cylinders, parabolas, segments of spheres, and specially to approximate π, which he called as *the parameter to diameter*. His approach to approximate π is based on the following fact: the circumference of a circle lies between the perimeters of the inscribed and circumscribed regular polygons (equilateral and equiangular) of n sides, and as n increases, the deviation of the circumference from the two perimeters becomes smaller. Because of this fact, many mathematicians claim that it is more correct to say that a circle has an infinite number of corners than to view a circle as being cornerless. If a_n and b_n denote the perimeters of the inscribed and circumscribed regular polygons of n sides, and C the circumference of the circle, then it is clear that $\{a_n\}$ is an increasing sequence bounded above by C, and $\{b_n\}$ is a decreasing sequence bounded below by C. Both of these sequences converge to the same limit C. To simplify matters, suppose we choose a circle with the diameter 1, then from Figure 1 it immediately follows that

$$a_n = n \sin \frac{\pi}{n} \quad \text{and} \quad b_n = n \tan \frac{\pi}{n}. \tag{1}$$

It is clear that $\lim_{n \to \infty} a_n = \pi = \lim_{n \to \infty} b_n$. Further, b_{2n} is the harmonic mean of a_n and b_n, and a_{2n} is the geometric mean of a_n and b_{2n}, i.e.,

$$b_{2n} = \frac{2a_n b_n}{a_n + b_n} \quad \text{and} \quad a_{2n} = \sqrt{a_n b_{2n}}. \tag{2}$$

From (1) for the hexagon, *i.e.*, $n = 6$ it follows that $a_6 = 3$, $b_6 = 2\sqrt{3}$. Then Archimedes successively took polygons of sides 12, 24, 48 and 96, used the recursive relations (2), and the inequality

$$\frac{265}{153} < \sqrt{3} < \frac{1,351}{780},$$

which he probably found by what is now called Heron's method, to obtain the bounds

$$3.140845\ldots = 3\frac{10}{71} < \pi < 3\frac{1}{7} = 3.142857\ldots. \tag{3}$$

Agarwal et al. *Advances in Difference Equations* 2013, **2013**:100
http://www.advancesindifferenceequations.com/content/2013/1/100

It is interesting to note that during Archimedes time algebraic and trigonometric no-tations, and our present decimal system were not available, and hence he had to de-rive recurrence relations (2) geometrically, and certainly for him the computation of a_{96} and b_{96} must have been a formidable task. The approximation 22/7 is often called the *Archimedean value* of π, and it is good for most purposes. If we take the average of the bounds given in (3), we obtain $\pi = 3.141851\ldots$. The above method of computing π by using regular inscribed and circumscribed polygons is known as the *classical method* of computing π. It follows that an inscribed regular polygon of 2^n sides takes up more than $1 - 1/2^{n-1}$ of the area of a circle. Heron of Alexandria (about 75 AD) in his *Metrica*, which had been lost for centuries until a fragment was discovered in 1894, followed by a complete copy in 1896, refers to an Archimedes work, where he gives the bounds

$$3.14163\ldots = \frac{211,875}{67,441} < \pi < \frac{197,888}{62,351} = 3.173774\ldots.$$

Clearly, in the above right inequality, there is a mistake as it is worse than the upper bound 22/7 found by Archimedes earlier. Heron adds 'Since these numbers are inconvenient for measurements, they are reduced to the ratio of the smaller numbers, namely, 22/7.' Archimedes' polygonal method remained unsurpassed for 18 centuries. Archimedes also showed that a curve discovered by Conon of Samos (around 280-220 BC) could, like Hip-pias' quadratrix, be used to square the circle. The curve is today called the *Archimedean Spiral*.

About 123 BC. Daivajna Varahamihira (working 123 BC) was an astronomer, math-ematician and astrologer. His picture may be found in the Indian Parliament along with Aryabhata. He was one of the nine jewels (Navaratnas) of the court of legendary king Vikramaditya I (101-18 BC). In 123 BC, Varahamihira wrote *Pancha-Siddhanta* (The Five Astronomical Canons), in which he codified the five existing Siddhantas, namely, Paulisa Siddhanta, Romaka Siddhanta, Vasishtha Siddhanta, Surya Siddhanta and Paitamaha Siddhanta. He also made some important mathematical discoveries such as giving certain trigonometric formulae; developing new interpolation methods to produce sine tables; constructing a table for the binomial coefficients; and examining the pandiag-onal magic square of order four. In his work, he approximated π as $\sqrt{10}$.

15 BC. Marcus Vitruvius Pollio (about 80-5 BC), a Roman writer, architect and engineer, in his multi-volume work *De Architectura* (On Architecture) used the value $\pi = 3\ 1/8 = 3.125$, which is the same as Babylonians had used 2,000 years earlier. He was the first to describe direct measurement of distances by the revolution of a wheel.

About 10 BC. Liu Xin (Liu Hsin) (about 50 BC-23 AD) was an astronomer, historian and editor during the Xin Dynasty (9-23 AD). Liu created a new astronomical system, called *Triple Concordance*. He was the first to give a more accurate calculation of π as 3.1547, the exact method he used to reach this figure is unknown. This was first mentioned in the *Sui shu* (387-388). He also found the approximations 3.1590, 3.1497 and 3.1679.

Around 5 AD. Liu Xin (50 BC-AD 23) was a Chinese astronomer, historian and editor during the Xin Dynasty (9-23 AD). He was the son of Confucian scholar Liu Xiang (77-6 BC). Liu created a catalog of 1,080 stars, where he used the scale of 6 magnitudes. He was the first in China to give a more accurate calculation of π as 3.1457. The method he used to reach this figure is unknown.

Agarwal et al. *Advances in Difference Equations* 2013, **2013**:100
http://www.advancesindifferenceequations.com/content/2013/1/100

10 AD. Brahmagupta (born 30 BC) wrote two treatises on mathematics and astronomy: the *Brahmasphutasiddhanta* (The Correctly Established Doctrine of Brahma) but often translated as (The Opening of the Universe), and the *Khandakhadyaka* (Edible Bite) which mostly expands the work of Aryabhata. As a mathematician he is considered as the father of arithmetic, algebra, and numerical analysis. Most importantly, in Brahmasphutasiddhanta he treated zero as a number in its own right, stated rules for arithmetic on negative numbers and zero, and attempted to define division by zero, particularly he wrongly believed that 0/0 was equal to 0. He used a geometric construction for squaring the circle, which amounts to $\pi = \sqrt{10}$.

125. Zhang Heng (78-139 AD) was an astronomer, mathematician, inventor, geographer, cartographer, artist, poet, statesman and literary scholar. He proposed a theory of the universe that compared it to an egg. 'The sky is like a hen's egg and is as round as a crossbow pellet. The Earth is like the yolk of the egg, lying alone at the center. The sky is large and the Earth is small'. According to him the universe originated from chaos. He said that the Sun, Moon and planets were on the inside of the sphere and moved at different rates. He demonstrated that the Moon did not have independent light, but that it merely reflected the light from the sun. He is most famous in the West for his rotating celestial globe, and inventing in 132 the first seismograph for measuring earthquakes. He proposed $\sqrt{10}$ (about 3.1623) for π. He also compared the celestial circle to the width (*i.e.*, diameter) of the earth in the proportion of 736 to 232, which gives π as 3.1724.

150. Claudius Ptolemaeus (around 90-168 AD) known in English as Ptolemy, was a mathematician, geographer, astrologer, poet of a single epigram in the Greek Anthology, and most importantly astronomer. He made a map of the ancient world in which he employed a coordinate system very similar to the latitude and longitude of today. One of his most important achievements was his geometric calculations of semichords. Ptolemy in his famous *Syntaxis mathematica* (more popularly known by its Arabian title of the Almagest), the greatest ancient Greek work on astronomy, obtained, using chords of a circle and an inscribed 360-gon, an approximate value of π in sexagesimal notation, as 3 8' 30", which is the same as 377/120 = 3.141666.... Eutocius of Ascalon (about 480-540) refers to a book Quick delivery by Apollonius of Perga (around 262-200 BC), who earned the title 'The Great Geometer', in which Apollonius obtained an approximation for π, which was better than known to Archimedes, perhaps the same as 377/120.

250. Wang Fan (228-266) was a mathematician and astronomer. He calculated the distance from the Sun to the Earth, but his geometric model was not correct. He has been credited with the rational approximation 142/45 for π, yielding $\pi = 3.155$.

263. Liu Hui (around 220-280) wrote two works. The first one was an extremely important commentary on the *Jiuzhang suanshu*, more commonly called *Nine Chapters on the Mathematical Art*, which came into being in the Eastern Han Dynasty, and believed to have been originally written around 1000 BC. (It should be noted that very little is known about the mathematics of ancient China. In 213 BC, the emperor Shi Huang of the Chin dynasty had all of the manuscript of the kingdom burned.) The other was a much shorter work called *Haidao suanjing* or *Sea Island Mathematical Manual*. In Jiuzhang suanshu, Liu Hui used a variation of the Archimedean inscribed regular polygon with 192 sides to approximate π as 3.141014 and suggested 157/50 = 3.14 as a practical approximation.

About 330. Pappus of Alexandria (around 290-350) was born in Alexandria, Egypt, and either he was a Greek or a Hellenized Egyptian. The written records suggest that, Pappus

Agarwal et al. *Advances in Difference Equations* 2013, **2013**:100
http://www.advancesindifferenceequations.com/content/2013/1/100

lived in Alexandria during the reign of Diocletian (284-305). His major work is *Synagoge* or the Mathematical Collection, which is a compendium of mathematics of which eight volumes have survived. Pappus' Book IV contains various theorems on circles, study of various curves, and an account of the three classical problems of antiquity (the squaring of the circle, the duplication of a cube, and the trisection of an angle). For squaring the circle, he used Dinostratus quadratrix and his proof is a *reductio ad absurdum*. Pappus is remembered for Pappus's centroid theorem, Pappus's chain, Pappus's harmonic theorem, Pappus's hexagon theorem, Pappus's trisection method, and for the focus and directrix of an ellipse.

400. He Chengtian (370-447) gave the approximate value of π as 111,035/35,329 = 3.142885....

475. Tsu Ch'ung-chih (Zu Chongzhi) (429-500) created various formulas that have been used throughout history. With his son he used a variation of Archimedes method to find $3.1415926 < \pi < 3.1415927$. He also obtained a remarkable rational approximation 355/113, which yields π correct to six decimal digits. In Chinese this fraction is known as Milü. To compute this accuracy for π, he must have taken an inscribed regular 6×2^{12}-gon and performed lengthy calculations. Note that $\pi = 355/113$ can be obtained from the values of Ptolemy and Archimedes:

$$\frac{355}{113} = \frac{377 - 22}{120 - 7}.$$

He declared that 22/7 is an inaccurate value whereas 355/113 is the accurate value of π. We also note that $\pi = 355/113$ can be obtained from the values of Liu Hui and Archimedes. In fact, by using the method of averaging, we have

$$\frac{157 + (9 \times 22)}{50 + (9 \times 7)} = \frac{355}{113}.$$

486. Bhaskara II or Bhaskaracharya (working 486) wrote *Siddhanta Siromani* (crown of treatises), which consists of four parts, namely, *Leelavati Bijaganitam, Grahaganitam* and *Goladhyaya*. The first two exclusively deal with mathematics and the last two with astronomy. His popular text *Leelavati* was written in 486 AD in the name of his daughter. His contributions to mathematics include: a proof of the Pythagorean theorem, solutions of quadratic, cubic, and quartic indeterminate equations, solutions of indeterminate quadratic equations, integer solutions of linear and quadratic indeterminate equations, a cyclic Chakravala method for solving indeterminate equations, solutions of the Pell's equation and solutions of Diophantine equations of the second order. He solved quadratic equations with more than one unknown, and found negative and irrational solutions, provided preliminary concept of infinitesimal calculus, along with notable contributions toward integral calculus, conceived differential calculus, after discovering the derivative and differential coefficient, stated Rolle's theorem, calculated the derivatives of trigonometric functions and formulae and developed spherical trigonometry. He conceived the modern mathematical convention that when a finite number is divided by zero, the result is infinity. He speculated the nature of the number 1/0 by stating that it is 'like the Infinite, Invariable God who suffers no change when old worlds are destroyed or new ones created, when innumerable species of creatures are born or as many perish'. He gave several

Agarwal et al. *Advances in Difference Equations* 2013, **2013**:100
http://www.advancesindifferenceequations.com/content/2013/1/100

approximations for π. According to him 3,927/1,250 is an accurate value, 22/7 is an inaccurate value, and $\sqrt{10}$ is for ordinary work. The first value may have been taken from Aryabhatta. This approximation has also been credited to Liu Hui and Zu Chongzhi. He also gave the value 754/240 = 3.1416, which is of uncertain origin; however, it is the same as that by Ptolemy.

510. Anicius Manlius Severinus Boethius (around 475-526) introduced the public use of sun-dials, water-clocks, *etc.* His integrity and attempts to protect the provincials from the plunder of the public officials brought on him the hatred of the Court. King Theodoric sentenced him to death while absent from Rome, seized at Ticinum (now Pavia), and in the baptistery of the church there tortured by drawing a cord round his head till the eyes were forced out of the sockets, and finally beaten to death with clubs on October 23, 526. His *Geometry* consists of the enunciations (only) of the first book of Euclid, and of a few selected propositions in the third and fourth books, but with numerous practical applications to finding areas, *etc.* According to him, the circle had been squared in the period since Aristotle's time, but noted that the proof was too long.

800. Abu Jafar Mohammed Ibn Musa al-Khwarizmi (around 780-850) 'Mohammed the father of Jafar and the son of Musa' was a scholar in the academy *Bait al-Hikma* (House of Wisdom) founded by Caliph al-Mamun (786-833). His task (along with several other scholars) was to translate the Greek and Sanskrit scientific manuscripts. They also studied, and wrote on algebra, geometry and astronomy. There al-Khwarizmi encountered the Hindu place-value system based on the numerals $0, 1, 2, 3, 4, 5, 6, 7, 8, 9$, including the first use of zero as a place holder in positional base notation, and he wrote a treatise around 820 AD, on what we call *Hindu-Arabic numerals*. The Arabic text is lost but a Latin translation, *Algoritmi de numero Indorum* (that is, al-Khwarizmi on the Hindu Art of Reckoning), a name given to the work by Baldassarre Boncompagni in 1857, much changed from al-Khwarizmi's original text (of which even the title is unknown) is known. The French Minorite friar Alexander de Villa Dei, who taught in Paris around 1240, mentions the name of an Indian king named Algor as the inventor of the new 'art', which itself is called the *algorismus*. Thus, the word 'algorithm' was tortuously derived from al-Khwarizmi (Alchwarizmi, al-Karismi, Algoritmi, Algorismi, Algorithm), and has remained in use to this day in the sense of an arithmetic operation. This Latin translation was crucial in the introduction of Hindu-Arabic numerals to medieval Europe. Al-Khwarizmi used $\pi = 22/7$ in algebra, $\pi = \sqrt{10}$ in geometry, and $\pi = 62{,}832/20{,}000 = 3.1416$ in astronomy.

850. Mahavira (817-875) in his work *Ganita Sara Samgraha* summarized and extended the works of Aryabhatta, Bhaskara, Brahmagupta and Bhaskaracharya. This treatise contains: a naming scheme for numbers from 10 up to 10^{24}, formulas for obtaining cubes of sums; techniques for least common denominators (LCM), techniques for combinations $^{n}C_r = n(n-1)(n-2)\cdots(n-r+1)/r!$, techniques for solving linear, quadratic as well higher order equations, arithmetic and geometric series, and techniques for calculating areas and volumes. He was the first person to mention that no real square roots of negative numbers can exist. According to Mahavira whatever is there in all the three worlds, which are possessed of moving and non-moving beings, all that indeed cannot exist without mathematics. He used the approximate value of π as $\sqrt{10}$. He also mentions that the approximate volume of a sphere with diameter d is $(9/2)(d/2)^3$, *i.e.*, $\pi = 3.375$, and exact volume is $(9/10)(9/2)(d/2)^3$, *i.e.*, $\pi = 3.0375$.

Agarwal et al. *Advances in Difference Equations* 2013, **2013**:100
http://www.advancesindifferenceequations.com/content/2013/1/100

About 1040. Franco von Lüttich (around 1015-1083) claimed to have contributed the only important work in the Christian era on squaring the circle. His works are published in six books, but only preserved in fragments.

1220. Fibonacci (Leonardo of Pisa) (around 1170-1250) after the Dark Ages is considered the first to revive mathematics in Europe. He wrote *Liber Abbaci* (Book of the Abacus) in 1202. In this book, he quotes that 'The nine Indian numerals are... with these nine and with the sign 0 which in Arabic is sifr, any desired number can be written'. His *Practica geometria*, a collection of useful theorems from geometry and (what would eventually be named) trigonometry appeared in 1220, which was followed five years later by *Liber quadratorum*, a work on indeterminate analysis. A problem in Liber Abbaci led to the introduction of the Fibonacci sequence for which he is best remembered today; however, this sequence earlier appeared in the works of Pingala (about 500 BC) and Virahanka (about 600 AD). In Practica geometriae, Fibonacci used a 96-sided polygon, to obtain the approximate value of π as 864/275 = 3.141818....

1260. Johannes Campanus (around 1220-1296) was chaplain to three popes, Pope Urban IV, Pope Nicholas IV and Pope Boniface VIII. He was one of the four greatest contemporary mathematicians. Campanus wrote a Latin edition of Euclid's Elements in 15 books around 1260. He used the value of π as 22/7.

About 1300. Zhao Youqin (born 1271) used a regular polygon of 4×2^{12} sides to derive $\pi = 3.1415926$.

About 1360. Albert of Saxony (around 1320-1390) was a German philosopher known for his contributions to logic and physics. He wrote a long treatise *De quadratura circuli* (Question on the Squaring of the Circle) consisting mostly philosophy. He said 'following the statement of many philosophers, the ratio of circumference to diameter is exactly 22/7; of this, there is proof, but a very difficult one'.

1400. Madhava of Sangamagramma's (1340-1425) work has come to light only very recently. Although there is some evidence of mathematical activities in Kerala (India) prior to Madhava, *e.g.*, the text *Sadratnamala* (about 1300), he is considered the founder of the Kerala school of astronomy and mathematics. Madhava was the first to have invented the ideas underlying infinite series expansions of functions, power series, trigonometric series of sine, cosine, tangent and arctangent, which is

$$\tan^{-1} x = x - \frac{x^3}{3} + \frac{x^5}{5} - \frac{x^7}{7} + \cdots + (-1)^{n-1}\frac{x^{2n-1}}{2n-1} + \cdots. \tag{4}$$

This series is valid for $-1 < x < 1$, and also for $x = 1$. He also gave rational approximations of infinite series, tests of convergence of infinite series, estimate of an error term, early forms of differentiation and integration and the analysis of infinite continued fractions. He fully understood the limit nature of the infinite series. Madhava discovered the solutions of transcendental (transcends the power of algebra) equations by iteration, and found the approximation of transcendental numbers by continued fractions. He also gave many methods for calculating the circumference of a circle. The value of π correct to 13 decimal places is attributed to Madhava. However, the text *Sadratnamala*, usually considered as prior to Madhava, while some researchers have claimed that it was compiled by Madhava, gives the astonishingly accurate value of π correct to 17 decimal places.

1429. Jemshid al-Kashi (around 1380-1429), astronomer royal to Ulugh Beg of Samarkand, wrote several important books *Sullam al-sama* (The Stairway of Heaven),

Agarwal et al. *Advances in Difference Equations* 2013, **2013**:100
http://www.advancesindifferenceequations.com/content/2013/1/100

Mukhtasar dar 'ilm-i hay'at (Compendium of the Science of Astronomy), *Khaqani Zij* on astronomical tables, *Risala dar sharh-i alat-i rasd* (Treatise on the Explanation of Observational Instruments), *Nuzha al-hadaiq fi kayfiyya san'a al-ala almusamma bi tabaq al-manatiq* (The Method of Construction of the Instrument Called Plate of Heavens), *Risala al-muhitiyya* (Treatise on the Circumference), *The Key to Arithmetic*, and *The Treatise on the Chord and Sine*. In these works al-Kashi showed a great venality in numerical work. In 1424, he calculated π to 14 decimal places, and later in 1429 to 16 decimal places. For this, he used classical polygon method of 6×2^{27} sides.

1460. George Pürbach (1423-1461) whose real surname is unknown, was born in Pürbach, a town upon the confines of Bavaria and Austria. He studied under Nicholas de Cusa, and one of his most famous pupils is Regiomontanus. Pürbach wrote a work on planetary motions which was published in 1460; an arithmetic, published in 1511; and a table of eclipses, published in 1514. He calculated tables of sines for every minute of arc for a radius of 600,000 units. This table was published in 1541. He approximated π by the rational 62,832/20,000, which is exactly the same as given by Aryabhatta.

1464. Nicholas of Cusa (1401-1464) is often referred to as Nicolaus Cusanus and Nicholas of Kues (Cusa was a Latin place-name for a city on the Mosel). He was a German cardinal of the Roman Catholic Church, a philosopher, jurist, mathematician and an astronomer. Most of his mathematical ideas can be found in his essays, *De Docta Ignorantia* (Of Learned Ignorance), *De Visione Dei* (Vision of God) and *On Conjectures*. He made important contributions to the field of mathematics by developing the concepts of the infinitesimal and of relative motion. He gave the approximations of π as $(3/4)(\sqrt{3} + \sqrt{6})$ and $24\sqrt{21}/35 = 3.142337\ldots$. Nicholas thought this to be the exact value. Nicholas said, if we can approach the Divine only through symbols, then it is most suitable that we use mathematical symbols, for these have an indestructible certainty. He also said that no perfect circle can exist in the universe. In accordance with his wishes, his heart is within the chapel altar at the Cusanusstift in Kues.

1464. Johann Regiomontanus (Johannes Müller) (1436-1476) is considered as one of the most prominent mathematicians of his generation. He was the first to study Greek mathematical works in order to make himself acquainted with the methods of reasoning and results used there. He also well read the works of the Arab mathematicians. In most of this study, he compiled in his *De Triangulis*, which was completed in 1464, however, was published only in 1533. Regiomontanus used algebra to find solutions of geometrical problems. He criticized Nicholas of Cusa's approximations and methods to approximate the value of π and gave the approximation 3.14343.

About 1500. Nilakanthan Somayaji's (around 1444-1544) most notable work *Tantrasangraha* elaborates and extends the contributions of Madhava. He was also the author of *Aryabhatiya-Bhashya*, a commentary of the Aryabhatiya. Of great significance in Nilakanthan's work includes the inductive mathematical proofs, a derivation and proof of the arctangent trigonometric function, improvements and proofs of other infinite series expansions by Madhava, and in Sanskrit poetry the series

$$\frac{\pi}{4} = 1 - \frac{1}{3} + \frac{1}{5} - \frac{1}{7} + \frac{1}{9} - \frac{1}{11} + \cdots, \tag{5}$$

which follows from Madhava's series (4) when $x = 1$. In the literature (5) is known as Gregory-Leibniz series. He also gave sophisticated explanations of the irrationality of

Agarwal et al. *Advances in Difference Equations* 2013, **2013**:100
http://www.advancesindifferenceequations.com/content/2013/1/100

π, the correct formulation for the equation of the center of the planets, and a helio-centric model of the solar system. If s_n denotes the nth partial sum of (5), then $s_1 = 1$, $s_{10} = 0.76045\ldots, s_{100} = 0.78289\ldots, s_{1,000} = 0.78514\ldots, s_{10,000} = 0.78537\ldots$ and Roy North showed that $4s_{500,000} = 3.14159\underline{0}65358979324\underline{04}62643383\underline{26}9502884197$ (where under-lined digits are incorrect) indicating an annoyingly slow convergence of the partial sums. Since this is an alternating series, the error committed by stopping at the nth term does not exceed $1/(2n+1)$ in absolute value. Thus, to compute $\pi/4$ to eight decimals from (5) would require $n > 10^8$ terms. Hence, although it is only of theoretical interest, the expres-sions on the right are arithmetical, while π arises from geometry. We also note that the series (5) can be written as

$$\frac{\pi}{4} = 1 - 2\left(\frac{1}{3\cdot 5} + \frac{1}{7\cdot 9} + \frac{1}{11\cdot 13} + \cdots\right).$$

The following expansion of π is also due to Nilakanthan

$$\pi = 3 + \frac{4}{2\cdot 3\cdot 4} - \frac{4}{4\cdot 5\cdot 6} + \frac{4}{6\cdot 7\cdot 8} - \frac{4}{8\cdot 9\cdot 10} + \cdots.$$

This series converges faster than (5).

Before 1510. Leonardo da Vinci (1452-1519) was an Italian painter, sculptor, architect, musician, scientist, mathematician, engineer, inventor, anatomist, geologist, cartographer, botanist and writer. He briefly worked on squaring the circle, or approximating π.

1525. Michael Stifel (1486-1567) served in several different Churches at different posi-tions; however, every time due to bad circumstances had to resign and flee. He made the error of predicting the end of the world on 3 October 1533, and other time used a clever rearrangement of the letters LEO DECIMVS to 'prove' that Leo X was 666, the number of the beast given in the Book of Revelation. He was forced to take refuge in a prison af-ter ruining the lives of many believing peasants who had abandoned work and property to accompany him to heaven. In the later part of his life, he lectured on mathematics and the-ology. He invented logarithms independently of Napier using a totally different approach. His most famous work is *Arithmetica integra* which was published in 1544. This work contains binomial coefficients, multiplication by juxtaposition, the term 'exponent', and the notation $+$, $-$ and $\sqrt{\ }$, and the opinion that the quadrature of π is impossible. Ac-cording to him 'the quadrature of the circle is obtained when the diagonal of the square contains 10 parts of which the diameter of the circle contains 8'. Thus, $\pi \simeq 3\,1/8$.

1525. Albrecht Dürer (1471-1528) was a famous artist and mathematician. His book *Underweysung der Messung mit dem Zirckel und Richtscheyt* provides measurement of lines, areas and solids by means of compass and ruler, particularly there is a discussion of squaring the circle.

1544. Oronce Fiñe (1494-1555) was a prolific author of mathematical books. He was imprisoned in 1524, probably for practicing judicial astrology. He approximated π as $3\,11/63 = 3.174603\ldots$. Later, he gave $3\,2/15 = 3.133333\ldots$ and, in 1556, $3\,11/78 = 3.141025\ldots$.

1559. Johannes Buteo (1492-1572), a French scholar published a book *De quadratura circuli*, which seems to be the first book that accounts the history of π and related prob-lems.

Agarwal et al. *Advances in Difference Equations* 2013, **2013**:100
http://www.advancesindifferenceequations.com/content/2013/1/100

1573. Valentin Otho (around 1550-1603) was a German mathematician and astronomer. In 1573, he came to Wittenberg and proposed to Johannes Praetorius the Tsu Ch'ung-chih approximate value of π as 355/113.

1580. Tycho Brahe was an astronomer and an alchemist and was known for his most accurate astronomical and planetary observations of his time. His data was used by his assistant, Kepler, to derive the laws of planetary motion. He observed a new star in 1572 and a comet in 1577. In 1566, when he was just 20, he lost his nose partially in a duel with another student in Wittenberg and wore throughout his life a metal insert over his nose. His approximation to π is $88/\sqrt{785} = 3.140854\ldots$.

1583. Simon Duchesne finds $\pi = (39/22)^2 = 3.142561\ldots$.

About 1584. Zhu Zaiyu (1536-1611), a noted musician, mathematician and astronomer-calendarist, Prince of the Ming Dynasty, obtained the twelfth root of two. He also gave the approximate value of π as $\sqrt{2}/0.45 = 3.142696\ldots$. Around the same time Xing Yunlu adopted π as 3.1126 and 3.12132034, while Chen Jinmo and Fang Yizhi, respectively, took as 3.1525 and 52/17.

1584. Simon van der Eycke (Netherland) published an incorrect proof of the quadrature of the circle. He approximated π as $1,521/484 = 3.142561\ldots$. In 1585, he gave the value 3.1416055.

1585. Adriaen Anthoniszoon (1529-1609) was a mathematician and fortification engineer. He rediscovered the Tsu Ch'ung-chih approximation 355/113 to π. This was apparently lucky incident, since all he showed was that $377/120 > \pi > 333/106$. He then averaged the numerators and the denominators to obtain the 'exact' value of π.

1593. Francois Viéte (1540-1603) is frequently called by his semi-Latin name of Vieta. In relation to the three famous problems of antiquity, he showed that the trisection of an angle and the duplication of a cube problems depend upon the solution of cubic equations. He has been called the father of modern algebra and the foremost mathematician of the sixteenth century. In his 1593 book, *Supplementum geometriae*, he showed $3.1415926535 < \pi < 3.1415926537$, *i.e.*, gave the value of π correct to 9 places. For this, he used the classical polygon of $6 \times 2^{16} = 393,216$ sides. He also represented π as an infinite product

$$\frac{2}{\pi} = \cos\frac{\pi}{4}\cos\frac{\pi}{8}\cos\frac{\pi}{16}\cos\frac{\pi}{32}\cdots = \frac{\sqrt{2}}{2}\frac{\sqrt{(2+\sqrt{2})}}{2}\frac{\sqrt{(2+\sqrt{(2+\sqrt{2})})}}{2}\cdots. \tag{6}$$

For this, we note that

$$\sin x = \cos\frac{x}{2}\cdot 2\sin\frac{x}{2} = \cos\frac{x}{2}\cos\frac{x}{2^2}\cdot 2^2\sin\frac{x}{2^2} = \cdots = \left(\prod_{k=1}^{n}\cos\frac{x}{2^k}\right)2^k\sin\frac{x}{2^k}$$

and hence

$$\frac{\sin x}{x} = \left(\prod_{k=1}^{n}\cos\frac{x}{2^k}\right)\frac{\sin x/2^k}{x/2^k},$$

which as $k \to \infty$, and then $x = \pi/2$ gives

$$\frac{2}{\pi} = \cos\frac{\pi}{4}\cos\frac{\pi}{8}\cos\frac{\pi}{16}\cos\frac{\pi}{32}\cdots.$$

Agarwal et al. *Advances in Difference Equations* 2013, **2013**:100
http://www.advancesindifferenceequations.com/content/2013/1/100

Finally, note that

$$\cos\frac{\pi}{4} = \sqrt{\frac{1}{2}\left(1+\cos\frac{\pi}{2}\right)} = \frac{\sqrt{2}}{2},$$

$$\cos\frac{\pi}{8} = \sqrt{\frac{1}{2}\left(1+\cos\frac{\pi}{4}\right)} = \sqrt{\frac{1}{2}\left(1+\frac{\sqrt{2}}{2}\right)} = \frac{1}{2}\sqrt{2+\sqrt{2}},\dots.$$

The above formula (6) is one of the milestones in the history of π. The convergence of Vieta's formula was proved by Ferdinand Rudio (1856-1929) in 1891. It is clear that Vieta's formula cannot be used for the numerical computation of π. In fact, the square roots are much too cumbersome, and the convergence is rather slow. It is clear that if we define $a_1 = \sqrt{1/2}$ and $a_{n+1} = \sqrt{(1+a_n)/2}$, then (6) is the same as $a_1 a_2 a_3 \cdots = 2/\pi$.

1593. Adrianus van Roomen (1561-1615), more commonly referred to as Adrianus Romanus, successively professor of medicine and mathematics in Louvain, professor of mathematics at Würzburg, and royal mathematician (astrologer) in Poland, proposed a challenge to all contemporary mathematicians, to solve a certain 45th degree equation. The Dutch ambassador presented van Roomen's book to King Henry IV with the comment that at present there is no mathematician in France capable of solving this equation. The King summoned and showed the equation to Vieta, who immediately found one solution to the equation, and then the next day presented 22 more. However, negative roots escaped him. In return, Vieta challenged van Roomen to solve the problem of Apollonius, to construct a circle tangent to three given circles, but he was unable to obtain a solution using Euclidean geometry. When van Roomen was shown proposer's elegant solution, he immediately traveled to France to meet Vieta, and a warm friendship developed. The same year Rooman used the classical method with 2^{30} sides, to approximate π to 15 correct decimal places.

1594. Joseph Justus Scaliger (1540-1609) was a religious leader and scholar. He is known for ancient Greek, Roman, Persian, Babylonian, Jewish and Egyptian history. In his work, *Cyclometrica elementa duo* he claimed that π is equal to $\sqrt{10}$.

1596. Ludolph van Ceulen (1539-1610) was a German who emigrated to the Netherlands. He taught Fencing and Mathematics in Delft until 1594, when he moved to Leiden and opened a Fencing School. In 1600, he was appointed to the Engineering School at Leiden, where he spent the remainder of his life teaching Mathematics, Surveying and Fortification. He wrote several books, including *Van den Circkel* (On The Circle, 1596), in which he published his geometric findings, and the approximate value of π correct to 20 decimal places. For this, he reports that he used classical method with 60×2^{33}, *i.e.*, 515,396,075,520 sides. This book ends with 'Whoever wants to, can come closer.'

1610. Ludolph van Ceulen (1539-1610) in his work *De Arithmetische en Geometrische fondamenten*, which was published posthumously by his wife in 1615, computed π correct to 35 decimal places by using classical method with 2^{62} sides. This computational feat was considered so extraordinary that his widow had all 35 digits of *die Ludolphsche Zahl* (the Ludolphine number) was engraved on his tombstone in St. Peter's churchyard in Leiden. The tombstone was later lost but was restored in 2000. This was one of the last major attempts to evaluate π by the classical method; thereafter, the techniques of calculus were employed.

Agarwal et al. *Advances in Difference Equations* 2013, **2013**:100
http://www.advancesindifferenceequations.com/content/2013/1/100

Page 17 of 59

1621. Willebrord Snell (Snellius) (1580-1626) was a Dutch astronomer and mathematician. At the age of 12, he is said to have been acquainted with the standard mathematical works, while at the age of 22, he succeeded his father as Professor of Mathematics at Leiden. His fame rests mainly on his discovery in 1621 of the law of refraction, which played a significant role in the development of both calculus and the wave theory of light. However, it is now known that this law was first discovered by Ibn Sahl (940-1000) in 984. Snell cleverly combined Archimedean method with trigonometry, and showed that for each pair of bounds on π given by the classical method, considerably closer bounds can be obtained. By his method, he was able to approximate π to seven places by using just 96 sides, and to van Ceulen's 35 decimal places by using polygons having only 2^{30} sides. The classical method with such polygons yields only two and fifteen decimal places.

1627. Yoshida Mitsuyoshi (1598-1672) was working during Edo period. His 1627 work named as Jinkoki deals with the subject of soroban arithmetic, including square and cube root operations. In this work, he used 3.16 for π.

1630. Christoph (Christophorus) Grienberger (1561-1636) was an Austrian Jesuit astronomer. The crater Gruemberger on the Moon is named after him. He used Snell's refinement to compute π to 39 decimal places. This was the last major attempt to compute π by the Archimedes method.

1635. *Celiang quanyi* (Complete Explanation of Methods of Planimetry and Stereometry) gives without proof the following bounds $3.14159265358979323846 < \pi < 3.14159265358979323847$, *i.e.*, π correct to 19 digits.

1647. William Oughtred (1575-1660), an English mathematician offered free mathematical tuition to pupils, which included even Wallis. His textbook, *Clavis Mathematicae* (The Key to Mathematics) on arithmetic published in 1631 was used by Wallis and Newton amongst others. In this work, he introduced the $\times$ symbol for multiplication, and the proportion sign (double colon ::). He designated the ratio of the circumference of a circle to its diameter by π/δ. His notation was used by Isaac Barrow (1630-1677) a few years later, and David Gregory (1659-1708). Before him, mathematicians described π in round-about ways such as '*quantitas, in quam cum multipliectur diameter, proveniet circumferential*', which means 'the quantity which, when the diameter is multiplied by it, yields the circumference'.

1647. Grégoire de Saint-Vincent (1584-1667), a Jesuit, was a mathematician who discovered that the area under the hyperbola ($xy = k$) is the same over $[a, b]$ as over $[c, d]$ when $a/b = c/d$. This discovery played an important role in the development of the theory of logarithms and an eventual recognition of the natural logarithm. In 1668, Nicolaus Mercator (Kauffmann) (1620-1687) wrote a treatise entitled *Logarithmo-technica*, and discovered the series

$$\ln(1 + x) = x - \frac{1}{2}x^2 + \frac{1}{3}x^3 - \frac{1}{4}x^4 + \cdots;$$

(7)

however, the same series was independently discovered earlier by Saint-Vincent. In his book, *Opus geometricum quadraturae circuli et sectionum coni* he proposed at least four methods of squaring the circle, but none of them were implemented. The fallacy in his quadrature was pointed out by Huygens.

1650. René Descartes (1596-1650) was a thoughtful child who asked so many questions that his father called him 'my little philosopher'. In 1638, he published his *Discourse*

Agarwal et al. *Advances in Difference Equations* 2013, **2013**:100
http://www.advancesindifferenceequations.com/content/2013/1/100

on Method, which contained important mathematical work, and three essays, Meteors, Dioptrics and Geometry, produced an immense sensation and his name became known throughout Europe. The rectangular coordinate system is credited to Descartes. He is regarded as a genius of the first magnitude. He was one of the most important and influential thinkers in human history and is sometimes called the founder of modern philosophy. After his death, a novel geometric approach to approximate π was found in his papers. His method consisted of doubling the number of sides of regular polygons while keeping the perimeter constant. In modern terms, Descartes' method can be summarized as

$$\pi = \lim_{k \to \infty} 2^k \tan\left(\frac{\pi}{2^k}\right).$$

If we let $a_k = 2^k \tan(\pi/2^k)$, $k \geq 2$, then in view of $\tan 2\theta = 2 \tan \theta / (1 - \tan^2 \theta)$, $x_k = 1/a_k$ satisfies the relation

$$x_{k+1}(x_{k+1} - x_k) = 2^{-2k-2},$$

and hence

$$x_{k+1} = \frac{1}{2}\left(x_k + \left(x_k^2 + 2^{-2k}\right)^{1/2}\right), \quad k \geq 2, \qquad x_2 = 1/4.$$

The sequence $\{x_k\}$ generated by the above recurrence relation converges to $1/\pi$.

1650. John Wallis (1616-1703) in 1649 was appointed as Savilian professor of geometry at the University of Oxford, which he continued for over 50 years until his death. He was the most influential English mathematician before Newton. In his most famous work, *Arithmetica infinitorum*, which he published in 1656, he established the formula

$$\pi = 2 \cdot \frac{2}{1} \cdot \frac{2}{3} \cdot \frac{4}{3} \cdot \frac{4}{5} \cdot \frac{6}{5} \cdot \frac{6}{7} \cdot \frac{8}{7} \cdot \frac{8}{9} \cdots. \tag{8}$$

This formula is a great milestone in the history of π. Like Viéte's formula (6), Wallis had found π in the form of an infinite product, but he was the first in history whose infinite sequence involved only rational operations. In his *Opera Mathematica I* (1695), Wallis introduced the term continued fraction. He rejected as absurd the now usual idea of a negative number as being less than nothing, but accepted the view that it is something greater than infinity, specially showed that $-1 > \infty$. He had great ability to do mental calculations. He slept badly and often did mental calculations as he lay awake in his bed. On 22 December 1669, he when in bed, occupied himself in finding the integral part of the square root of 3×10^{40}; and several hours afterward wrote down the result from memory. Two months later, he was challenged to extract the square root of a number of 53 digits; this he performed mentally, and a month later he dictated the answer which he had not meantime committed to writing. Wallis' life was embittered by quarrels with his contemporaries including Huygens, Descartes, and the political philosopher Hobbes, which continued for over 20 years, ending only with Hobbes' death. Hobbes called Arithmetica infinitorum 'a scab of symbols', and claimed to have squared the circle. It seems that to some, individual's quarrels give strength, encouragement and mental satisfaction. To derive (8), we note that

Agarwal et al. *Advances in Difference Equations* 2013, **2013**:100
http://www.advancesindifferenceequations.com/content/2013/1/100

$I_n = \int_0^{\pi/2} \sin^n x \, dx$ satisfies the recurrence relation

$$I_n = \frac{n-1}{n} I_{n-2}. \tag{9}$$

Thus, in view of $I_0 = \pi/2$ and $I_1 = 1$, we have

$$I_{2m} = \frac{2m-1}{2m} \cdot \frac{2m-3}{2m-2} \cdots \frac{5}{6} \cdot \frac{3}{4} \cdot \frac{1}{2} \cdot \frac{\pi}{2}$$

and

$$I_{2m+1} = \frac{2m}{2m+1} \cdot \frac{2m-2}{2m-1} \cdots \frac{6}{7} \cdot \frac{4}{5} \cdot \frac{2}{3}.$$

From these relations, a termwise division leads to

$$\frac{\pi}{2} = \left(\frac{2 \cdot 4 \cdot 6 \cdots 2m}{3 \cdot 5 \cdots (2m-1)} \right)^2 \frac{1}{2m+1} \frac{I_{2m}}{I_{2m+1}}.$$

Now, it suffices to show that

$$\lim_{m \to \infty} \frac{I_{2m}}{I_{2m+1}} = 1. \tag{10}$$

We know that for all $x \in (0, \pi/2)$ the inequalities $\sin^{2m-1} x > \sin^{2m} x > \sin^{2m+1} x$ hold. Thus, an integration from 0 to $\pi/2$ gives $I_{2m-1} \geq I_{2m} \geq I_{2m+1}$, and hence

$$\frac{I_{2m-1}}{I_{2m+1}} \geq \frac{I_{2m}}{I_{2m+1}} \geq 1. \tag{11}$$

Further, from (9), we have

$$\frac{I_{2m-1}}{I_{2m+1}} = \frac{2m+1}{2m},$$

thus, it follows that

$$\lim_{m \to \infty} \frac{I_{2m-1}}{I_{2m+1}} = \lim_{m \to \infty} \frac{2m+1}{2m} = 1. \tag{12}$$

Finally, a combination of (11) and (12) immediately gives (10). If we define $a_n = 1 - 1/(2n)^2$, then (8) is equivalent to $a_1 a_2 a_3 \cdots = 2/\pi$. We also note that

$$\frac{1}{a_1 a_2 \cdots a_n} = \frac{\pi}{2} + O\left(\frac{1}{n}\right).$$

1650. William Brouncker, 2nd Viscount Brouncker (1620-1684) was one of the founders and the second President of the Royal Society. His mathematical contributions are: reproduction of Brahmagupta's solution of a certain indeterminate equation, calculations of the lengths of the parabola and cycloid, quadrature of the hyperbola which required approximation of the natural logarithm function by infinite series and the study of generalized

Agarwal et al. *Advances in Difference Equations* 2013, **2013**:100
http://www.advancesindifferenceequations.com/content/2013/1/100

continued fractions. He undertook some calculations to verify formula (8), and showed that $3.141592653569\ldots < \pi < 3.141592653696\ldots$, which is very satisfactory. He also converted Wallis' result (8) into the continued fraction

$$\pi = \cfrac{4}{1 + \cfrac{1}{2 + \cfrac{9}{2 + \cfrac{25}{2 + \cdots}}}}. \tag{13}$$

Neither of the expressions (8), and (13); however, later has served for an extensive calculation of π.

Another continued fraction representation of π which follows from the series (5) is

$$\pi = \cfrac{4}{1 + \cfrac{1^2}{3 + \cfrac{2^2}{5 + \cfrac{3^2}{7 + \cdots}}}}.$$

1654. Christiaan Huygens (1629-1695) is famous for his invention of the pendulum clock, which was a breakthrough in timekeeping. He formulated the second law of motion of Newton in a quadratic form, and derived the now well-known formula for the centripetal force, exerted by an object describing a circular motion. Huygens was the first to derive the formula for the period of an ideal mathematical pendulum (with massless rod or cord), $T = 2\pi\sqrt{\ell/g}$. For the computation of π, he gave the correct proof of Snell's refinement, and using an inscribed polygon of only 60 sides obtained the bounds $3.1415926533 < \pi < 3.1415926538$, for the same accuracy the classical method requires almost 400,000 sides.

1663. Muramatsu Shigekiyo (1608-1695) published Sanso, or Stack of Mathematics, in which he used classical polygon method of 2^{15} sides to obtain $\pi = 3.14195264877$.

1665. Sir Isaac Newton (1642-1727), hailed as one of the greatest scientist-mathematicians of the English-speaking world, had the following more modest view of his own monumental achievements: '... to myself I seem to have been only like a boy playing on the seashore, and diverting myself in now and then finding a smoother pebble or a prettier shell than ordinary, whilst the great ocean of truth lay all undiscovered before me'. As he examined these shells, he discovered to his amazement more and more of the intricacies and beauties that lay in them, which otherwise would remain locked to the outside world. At the age of 26, he succeeded Barrow as Lucasian professor of mathematics at Cambridge. About him, Aldous Huxley (1894-1963) had said 'If we evolved a race of Isaac Newtons, that would not be progress. For the price Newton had to pay for being a supreme intellect was that he was incapable of friendship, love, fatherhood and many other desirable things. As a man he was a failure; as a monster he was superb'. Newton made some of the greatest discoveries the world ever knew at that time. Newton discovered: 1. The nature of colors. 2. The law of gravitation and the laws of mechanics. 3. The fluxional calculus. Most of the history books say that to compute π Newton used the series

$$\sin^{-1} x = x + \frac{1 \cdot x^3}{2 \cdot 3} + \frac{1 \cdot 3 \cdot x^5}{2 \cdot 4 \cdot 5} + \frac{1 \cdot 3 \cdot 5 \cdot x^7}{2 \cdot 4 \cdot 6 \cdot 7} + \cdots,$$

Agarwal et al. *Advances in Difference Equations* 2013, **2013**:100
http://www.advancesindifferenceequations.com/content/2013/1/100

Page 21 of 59

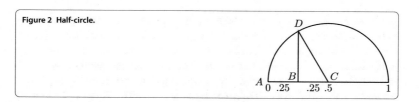

Figure 2 Half-circle.

which for $x = 1/2$ gives

$$\frac{\pi}{6} = \sin^{-1}\left(\frac{1}{2}\right) = \left(\frac{1}{2} + \frac{1}{2 \cdot 3 \cdot 2^3} + \frac{1 \cdot 3}{2 \cdot 4 \cdot 5 \cdot 2^5} + \cdots \right);$$

however, he actually used twenty-two terms to obtain 16 decimal places of the following series

$$\pi = \frac{3\sqrt{3}}{4} + 24\left(\frac{1}{3 \cdot 2^2} - \frac{1}{5 \cdot 2^5} - \frac{1}{28 \cdot 2^7} - \frac{1}{72 \cdot 2^9} - \cdots \right). \tag{14}$$

Later, he wrote 'I am ashamed to tell you to how many figures I carried these computations, having no other business at the time'. His result was not published until 1737 (posthumously).

Using analysis and geometry, the series (14) can be obtained as follows: From Figure 2, the equation of the upper half circle is $y = x^{1/2}(1 - x)^{1/2}$. Thus, binomial expansion gives

$$y = x^{1/2} - \frac{1}{2}x^{3/2} - \frac{1}{8}x^{5/2} - \frac{1}{16}x^{7/2} - \frac{5}{128}x^{9/2} - \frac{7}{256}x^{11/2} - \cdots.$$

Thus, the area of the sector ABD is (integrating the above series from 0 to 1/4)

$$\triangle ABD = \frac{1}{3 \cdot 2^2} - \frac{1}{5 \cdot 2^5} - \frac{1}{28 \cdot 2^7} - \frac{1}{72 \cdot 2^9} - \cdots. \tag{15}$$

Also, from geometry the area of the sector ABD is

$$\triangle ABD = \triangle ABCD - \triangle BCD$$

$$= \frac{1}{3}\left(\frac{1}{2}\pi\left(\frac{1}{2}\right)^2\right) - \frac{1}{2}\left(\frac{1}{4}\right)\sqrt{\left(\frac{1}{2}\right)^2 - \left(\frac{1}{4}\right)^2} = \frac{\pi}{24} - \frac{\sqrt{3}}{32}. \tag{16}$$

Equating (15) and (16), we immediately get (14).

1666. Thomas Hobbes of Malmesbury (1588-1679) was an English philosopher, best known today for his work on political philosophy. He also contributed in several other diverse fields, including history, geometry, the physics of gases, theology, ethics and general philosophy. He approximated π by 3 1/5 = 3.2, which was refuted by Huygens and Wallis. In 1678, he also gave the approximation $\sqrt{10}$.

1671. James Gregory (1638-1675) published two books *Vera circuli et hyperbolae quadratura* in 1667, and *Geometriae pars universalis* in 1668. In the first book particularly, he showed that the area of a circle can be obtained in the form of an infinite convergent series only, and hence inferred that the quadrature of the circle was impossible.

Agarwal et al. *Advances in Difference Equations* 2013, **2013**:100
http://www.advancesindifferenceequations.com/content/2013/1/100

In the second book, he attempted to write calculus systematically, which perhaps made the basis of Newton's fluxions. This book also contains series expansions of $\sin(x)$, $\cos(x)$, $\arcsin(x)$ and $\arccos(x)$; however, as we have seen earlier these expansions were known to Madhava. Gregory anticipated Newton in discovering both the interpolation formula and the general binomial theorem as early as 1670. In early 1671, he discovered Taylor's theorem (published by Taylor in 1715); however, he did not publish. Later in 1671, he rediscovered Nilakanthan's arctangent series (5). In his *Vera circuli et hyperbolae quadratura* of 1667, Gregory tried to show that π was a transcendental number, but his attempt, though very interesting, was not successful. Huygens made detailed and rather biased criticisms of it.

1672. Pietro Mengoli (1626-1686) studied at the University of Bologna, and became a professor there in 1647 for the next 39 years of his life. Besides proposing Basel problem, he proved that the harmonic series does not converge, established that the alternating harmonic series is equal to the natural logarithm of 2, published on the problem of squaring the circle, and provided a proof that Wallis' product (8) for π is correct.

1674. Gottfried Wilhelm von Leibniz (1646-1716) was a universal genius who won recognition in many fields - law, philosophy, religion, literature, politics, geology, metaphysics, alchemy, history and mathematics. He shares credit with Newton in developing calculus independently. He popularized and gave several mathematical symbols. Leibniz tried to reunite the Protestant and Catholic churches. He in binary arithmetic saw the image of Creation. He imagined that Unity represented God, and Zero the void; that the Supreme Being drew all beings from the void, just as unity and zero express all numbers in the binary system of numeration. He communicated his idea to the Jesuit Grimaldi, who was the President of the Chinese tribunal for mathematics in the hopes that it would help convert to Christianity the Emperor of China, who was said to be very fond of the Sciences. Later Leibniz became an expert in the Sanskrit language and the culture of China. For calculating π, he developed a method without any reference to a circle. In 1674, he also rediscovered Nilakanthan's arctangent series (5), whose beauty he described by saying that Lord loves odd numbers. Leibniz even invented a calculating machine that could perform the four operations and extract roots.

1684. Isomura Yoshinori (1640-1710) employed a 2^{17}-sided inscribed polygon to obtain 3.141592664 for π, but for some reason he wrote only $\pi = 3.1416$.

1685. Father Adam Adamad Kochansky (1631-1700) was librarian of the Polish King John III. He was the first to utilize a steel spring for suspension of the pendulum of a clock. He used a new approximate geometric construction for π to obtain

$$\pi \simeq \sqrt{\frac{40}{3} - 2\sqrt{3}} = 3.141533\ldots.$$

His method was later quoted in several geometrical textbooks.

1690. Takebe Katahiro (1664-1739) also known as Takebe Kenko played a critical role in the development of a crude version of the calculus. He also created charts for trigonometric functions. He used polygon (just 1,024 sides) approximation and a numerical method which is essentially equivalent to the Romberg algorithm (rediscovered by Sigmund Romberg, 1887-1951) to compute π to 41 digits. In 1722, Takebe obtained power series expansion of $(\sin^{-1} x)^2$, 15 years earlier than Euler. Around 1729, essentially the same

Agarwal et al. *Advances in Difference Equations* 2013, **2013**:100
http://www.advancesindifferenceequations.com/content/2013/1/100

series was rediscovered by Oyama Shokei who used it to find the expansion

$$\pi^2 = 8\left(1 + \sum_{n=1}^{\infty} \frac{2^{n+1}(n!)^2}{(2n+2)!}\right) = 8\left(1 + \frac{1}{2}\cdot\frac{1}{3} + \frac{1}{3}\cdot\frac{1\cdot2}{3\cdot5} + \frac{1}{4}\cdot\frac{1\cdot2\cdot3}{3\cdot5\cdot7} + \cdots\right).$$

The above expansion of π^2 was also given by Yamaji Nushizumi (1704-1772) around 1765.

1699. Abraham Sharp (1653-1742) was a mathematician and astronomer. In 1688, he joined the Greenwich Royal Observatory and did notable work, improving instruments and showing great skill as a calculator. He also worked on geometry and improved logarithmic tables. In the supervision of Edmund Halley (1656-1742), he realized that by putting $x = 1$ in (4) (see (5)) we lose the benefit of the powers $x^3, x^5, x^7, \ldots$, which tend to increase the rapidity of convergence for smaller values of x. He substituted $x = 1/\sqrt{3}$ in (4), to obtain

$$\frac{\pi}{6} = \frac{1}{\sqrt{3}}\left(1 - \frac{1}{3\cdot3} + \frac{1}{3^2\cdot5} - \frac{1}{3^3\cdot7} + \cdots\right). \tag{17}$$

Sharp used (17) to calculated π to 72 decimal places out of which 71 digits are correct. In (17), the 10th term is $1/(\sqrt{3}\cdot19\cdot3^9)$, which is less than 0.00005, and hence we have at least 4 places correct after just 9 terms. It is believed that Madhava of Sangamagramma used the same series in the fourteenth century to compute the value of π correct to 11 decimal places.

1700. Seki Takakazu also known as Seki Kowa (1642-1708) is generally regarded as the greatest Japanese mathematician. He was a prolific writer, and a number of his publications are either transcripts of mathematics from Chinese into Japanese, or commentaries on certain works of well-known Chinese mathematicians. His interests in mathematics ranged recreational mathematics, magic squares and magic circles, solutions of higher-order and indeterminate equations, conditions for the existence of positive and negative roots of polynomials, and continued fractions. He discovered determinants ten years before Leibniz, and the Bernoulli numbers a year before Bernoulli. He used polygon of 2^{17} sides and Richardson extrapolation (rediscovered by Alexander Craig Aitken, 1895-1967) to compute π to 10 digits. Some authors believe that he also used the formula

$$\pi = \lim_{n\to\infty} \frac{4}{n^2}\sum_{j=0}^{n}\sqrt{n^2 - j^2}.$$

About 1700. Oliver de Serres believed that by weighing a circle and a triangle equal to the equilateral triangle inscribed he had found that the circle was exactly double of the triangle, not being aware that this double is exactly the hexagon inscribed in the same circle. Thus, according to him $\pi = 3$.

1706. William Jones (1675-1749), an obscure English writer, represented the ratio of the circumference of a circle to its diameter by π in his *Synopsis Palmariorum Matheseos* (New Introduction to the Mathematics). He used the letter π as an abbreviation for the Greek word perimetros (periphery) (of a circle with unit diameter). In his book, he published the value of π correct to 100 decimal places.

1706. John Machin (1680-1752) was a professor of astronomy at Gresham College, London. He also served as secretary of the Royal Society during 1718-1747. Machin is best

Agarwal et al. *Advances in Difference Equations* 2013, **2013**:100
http://www.advancesindifferenceequations.com/content/2013/1/100

remembered for computing the value of π to 100 decimal places by using the formula

$$\frac{\pi}{4} = 4\tan^{-1}\left(\frac{1}{5}\right) - \tan^{-1}\left(\frac{1}{239}\right), \tag{18}$$

which in view of (4) is the same as

$$\frac{\pi}{4} = 4\left(\frac{1}{5} - \frac{1}{3 \cdot 5^3} + \frac{1}{5 \cdot 5^5} - \cdots\right) - \left(\frac{1}{239} - \frac{1}{3 \cdot 239^3} + \frac{1}{5 \cdot 239^5} - \cdots\right). \tag{19}$$

To establish (18), we let $\tan\theta = 1/5$, so that

$$\tan 2\theta = \frac{2\tan\theta}{1 - \tan^2\theta} = \frac{5}{12} \quad \text{and} \quad \tan 4\theta = \frac{2\tan 2\theta}{1 - \tan^2 2\theta} = \frac{120}{119}.$$

Thus, it follows that

$$\tan\left(4\theta - \frac{\pi}{4}\right) = \frac{\tan 4\theta - 1}{1 + \tan 4\theta} = \frac{1}{239}$$

and hence

$$\tan^{-1}\left(\frac{1}{239}\right) = 4\theta - \frac{\pi}{4} = 4\tan^{-1}\left(\frac{1}{5}\right) - \frac{\pi}{4}.$$

The proof of (18) also follows by comparing the angles in the identity (the idea originally goes back to Caspar Wessel (1745-1818) who presented his work in 1797 to the Royal Danish Academy of Sciences)

$$(5 + i)^4(-239 + i) = -114{,}244(1 + i),$$

i.e.,

$$4\tan^{-1}\left(\frac{1}{5}\right) + \pi - \tan^{-1}\left(\frac{1}{239}\right) = \pi + \tan^{-1} 1.$$

The series (19) certainly converges significantly faster than (5) and (17). In fact, taking six terms of the first series and two terms of the second and paying attention to the remainders and round-off errors, we get the inequalities $3.141592629 < \pi < 3.141592668$. Thus, the value of π correct to seven decimals is 3.1415926.

Several other Machin-type formulas are known, *e.g.,*

$$\frac{\pi}{4} = 2\tan^{-1}\left(\frac{1}{2}\right) - \tan^{-1}\left(\frac{1}{7}\right)$$

$$= 5\tan^{-1}\left(\frac{1}{5}\right) - 3\tan^{-1}\left(\frac{1}{18}\right) - 2\tan^{-1}\left(\frac{1}{57}\right)$$

$$= 17\tan^{-1}\left(\frac{1}{22}\right) + 3\tan^{-1}\left(\frac{1}{172}\right) - 2\tan^{-1}\left(\frac{1}{682}\right) - 7\tan^{-1}\left(\frac{1}{5{,}357}\right).$$

For a long list of such type of formulas with a discussion of their relative merits in computational work, see Lehmer (1938).

Agarwal et al. *Advances in Difference Equations* 2013, **2013**:100
http://www.advancesindifferenceequations.com/content/2013/1/100

1713. Chapter 15 of *Shu li jing yun* (Collected Basic Principles of Mathematics), which was commissioned by the Emperor Kang Xi and edited by Mei Gucheng and He Guozong, gives $\pi = 3.14159265$, which is correct to eight decimal places.

1719. Thomas Fantet de Lagny (1660-1734) was a French mathematician who is well known for his contributions to computational mathematics. He used the series (17) to determine the value of π up to 127 decimal places; however, only 112 are correct.

1728. Alexander Pope (1688-1744) was an English poet. He is the third-most frequently cited writer in The Oxford Dictionary of Quotations, after Shakespeare and Tennyson. In his *Dunciad* it is mentioned that 'The mad Mathesis, now, running round the circle, finds it square'. This explains the wild and fruitless attempts of squaring the circle.

1728. Sieur Malthulon (France) offered solutions to squaring the circle and to perpetual motion. He offered 1,000 crowns reward in legal form to anyone proving him wrong. Nicoli, who proved him wrong, collected the reward and abandoned it to the Hotel Dieu of Lyons. Later, the courts gave the money to the poor.

1730. Toshikiyo Kamata (1678-1747) used both the circumscribed and inscribed polygons and gave the bounds $3.14159265358979323846265341667 < \pi < 3.14159265358979323846264336658$.

1730. Abraham De Moivre (1667-1754) was an intimate personal friend of Newton, and was elected an FRS of London in 1697. In 1710, he was appointed to the Commission set up by the Royal Society to solve the Newton-Leibniz dispute concerning which of them invented calculus first. He is best known for his memoir *Doctrine of Chances*: A method of calculating the probabilities of events in play, which was first printed in 1618 and dedicated to Newton. In 1722, he published his famous theorem $(\cos x + i \sin x)^n = \cos nx + \sin nx$. In his *Miscellanea Analytica* published in 1730, appears the formula for very large n,

$$n! \simeq (2\pi n)^{1/2} e^{-n} n^n,$$

which is known today as *Stirling's formula*. In 1733, De Moivre used this formula to derive the normal curve as an approximation to the binomial.

1735. Leonhard Euler (1707-1783) was probably the most prolific mathematician who ever lived. He was born in Basel (Switzerland), and had the good fortune to be tutored one day a week in mathematics by a distinguished mathematician, Johann Bernoulli (1667-1748). Euler's energy and capacity for work were virtually boundless. His collected works form about 60 to 80 quarto-sized volumes and it is believed that much of his work has been lost. What is particularly astonishing is that Euler became virtually sightless in his right eye during the mid-1730s, and was blind for the last 17 years of his life, and this was one of the most productive periods. In 1644, Mengoli asked for the precise summation of the infinite series $\sum_{n=1}^{\infty} n^{-2}$. The series is approximately equal to $1.644934\ldots$. In the literature, this problem has been referred after Basel, hometown of Euler as well as of the Bernoulli family who unsuccessfully attacked the problem. *Basel problem* appears in number theory, *e.g.*, if two positive integers are selected at random and independently of each other, then the probability that they are relatively prime is $(\sum_{n=1}^{\infty} n^{-2})^{-1}$ (R. Chartres, 1904). An integer that is not divisible by the square of any prime number is said to be square free. The probability that a randomly selected integer is square free is also $(\sum_{n=1}^{\infty} n^{-2})^{-1}$. Euler considered the function $\sin x/x$, $x \neq 0$ which has the roots at $\pm n\pi$, $n = 1, 2, \ldots$. Thus, it

Agarwal et al. *Advances in Difference Equations* 2013, **2013**:100
http://www.advancesindifferenceequations.com/content/2013/1/100

follows that

$$\frac{\sin x}{x} = 1 - \frac{x^2}{3!} + \frac{x^4}{5!} - \frac{x^6}{7!} + \cdots$$

$$= \left(1 - \frac{x^2}{1^2\pi^2}\right)\left(1 - \frac{x^2}{2^2\pi^2}\right)\left(1 - \frac{x^2}{3^2\pi^2}\right)\cdots. \tag{20}$$

Thus, on equating the coefficients of x^2, we get

$$\frac{1}{6} = \frac{1}{1^2\pi^2} + \frac{1}{2^2\pi^2} + \frac{1}{3^2\pi^2} + \cdots,$$

which is the same as

$$\frac{1}{1^2} + \frac{1}{2^2} + \frac{1}{3^2} + \cdots = \frac{\pi^2}{6}. \tag{21}$$

The above proof of Euler is based on manipulations that were not justified at the time, and it was not until 1741 that he was able to produce a truly rigorous proof. It is interesting to note that (20) with $x = \pi/2$ immediately gives Wallis' formula (8). Today, several different proofs of (21) are known in the literature. Euler also established the following series:

$$3\sum_{m=1}^{\infty} \frac{1}{m^2\binom{2m}{m}} = \frac{\pi^2}{6},$$

$$\frac{1}{1^2} + \frac{1}{3^2} + \frac{1}{5^2} + \cdots = \frac{\pi^2}{8},$$

$$\frac{1}{1^2} - \frac{1}{2^2} + \frac{1}{3^2} - \frac{1}{4^2} + \cdots = \frac{\pi^2}{12}.$$

Later, Euler generalized the Basel problem considerably, in fact, for all positive integers $k = 1, 2, \ldots,$ he established the series

$$\frac{1}{1^{2k}} + \frac{1}{2^{2k}} + \frac{1}{3^{2k}} + \cdots = \frac{(2\pi)^{2k}(-1)^{n+1}B_{2k}}{2(2n)!},$$

where B_{2k} are Bernoulli numbers: $B_2 = 1/6, B_4 = -1/30, B_6 = 1/42, B_8 = -1/30, B_{10} = 5/66, \ldots.$ In particular, he established

$$\frac{1}{1^{26}} + \frac{1}{2^{26}} + \frac{1}{3^{26}} + \cdots = \frac{2^{24} \times 76,977,927}{27!}\pi^{26}.$$

Euler's ideas were taken up years later by George Friedrich Bernhard Riemann (1826-1866) in his seminal 1859 paper, *On the Number of Primes Less Than a Given Magnitude*, in which he defined his zeta function

$$\zeta(s) = 1 + \frac{1}{2^s} + \frac{1}{3^s} + \frac{1}{4^s} + \cdots, \quad s = \sigma + it,$$

and proved its basic properties, and in a sovereign way simply stated a number of others without proof. After his death, many of the finest mathematicians of the world have exerted their strongest efforts and created new branches of analysis in attempts to prove

Agarwal et al. *Advances in Difference Equations* 2013, **2013**:100
http://www.advancesindifferenceequations.com/content/2013/1/100

Page 27 of 59

these statements. Since then with one exception, every statement has been settled in the sense Riemann expected. This exception is the famous *Riemann hypothesis*: that all the zeros of $\zeta(s)$ in the strip $0 \leq \sigma \leq 1$ lie on the central line $\sigma = 1/2$. It stands today as the most important unsolved problem of mathematics, and perhaps the most difficult problem that the mind of man has ever conceived.

1737. The letter π was first used by Euler in 1737 in his *Variae observationes circa series infinitas*. Until that time, he had been using the letters p (1734), or c (1736). In 1742, Christian Goldbach (1690-1764) also used π. After the publication of Euler's treatise: *Introductio in Analysin Infinitorum* (1748), π became a standard symbol, as was the case with other notations he adopted. In 1737, Euler also showed that both e and e^2 are irrational and gave several continued fractions for e. In another paper, *De variis modis circuli quadraturam numeris proxime exprimendi* of 1737, Euler derived the formulas

$$\tan^{-1}\left(\frac{1}{p}\right) = \tan^{-1}\left(\frac{1}{p+q}\right) + \tan^{-1}\left(\frac{q}{p^2 + pq + 1}\right) \tag{22}$$

and

$$\tan^{-1}\left(\frac{x}{y}\right) = \tan^{-1}\left(\frac{ax - y}{ay + x}\right) + \tan^{-1}\left(\frac{b - a}{ab + 1}\right) + \tan^{-1}\left(\frac{c - b}{cb + 1}\right) + \cdots$$

and these give rise to any amount of relations for π; for example, if $x = 1 = y$, and the odd numbers are substituted for $a, b, c, \ldots$, we obtain

$$\frac{\pi}{4} = \tan^{-1}\left(\frac{1}{2}\right) + \tan^{-1}\left(\frac{1}{8}\right) + \tan^{-1}\left(\frac{1}{18}\right) + \cdots.$$

The proof of (22) immediately follows by comparing the angles in the identity

$$(p + q + i)(p^2 + pq + 1 + iq) = [(p + q)^2 + 1](p + i).$$

1739. Matsunaga Yoshisuke (died in 1844) was a prolific writer. In modern terms, he used the hypergeometric series

$$F(a, b, c, x) = 1 + \frac{abx}{1!c} + \frac{a(a + 1)b(b + 1)x^2}{2!c(c + 1)} + \frac{a(a + 1)(a + 2)b(b + 1)(b + 2)x^3}{3!c(c + 1)(c + 2)} + \cdots$$

for $a = 1/2$, $b = 1/2$, $c = 3/2$, and $x = 1/4$, *i.e.*, the series

$$\pi = 3F\left(\frac{1}{2}, \frac{1}{2}, \frac{3}{2}, \frac{1}{4}\right) = 3\left(1 + \frac{1^2}{4 \cdot 2 \cdot 3} + \frac{3^2}{4^2 \cdot 2 \cdot 3 \cdot 4 \cdot 5} + \frac{3^2 \cdot 5^2}{4^3 \cdot 2 \cdot 3 \cdot 4 \cdot 5 \cdot 6 \cdot 7} + \cdots\right)$$

to compute π correct to 50 digits. He also gave the following series:

$$\pi^2 = 9\left(1 + \frac{1^2}{3 \cdot 4} + \frac{1^2 \cdot 2^2}{3 \cdot 4 \cdot 5 \cdot 6} + \frac{1^2 \cdot 2^2 \cdot 3^2}{3 \cdot 4 \cdot 5 \cdot 6 \cdot 7 \cdot 8} + \cdots\right).$$

1748. The following expansion of π is due to Euler

$$\pi = 1 + \frac{1}{2} + \frac{1}{3} + \frac{1}{4} - \frac{1}{5} + \frac{1}{6} + \frac{1}{7} + \frac{1}{8} + \frac{1}{9} - \frac{1}{10} + \frac{1}{11} + \frac{1}{12} - \frac{1}{13} + \cdots,$$

Agarwal et al. *Advances in Difference Equations* 2013, **2013**:100
http://www.advancesindifferenceequations.com/content/2013/1/100

where the signs are determined following the rule: If the denominator is a prime of the form $4m - 1$, the sign is positive; if the denominator is 2 or a prime of the form $4m + 1$, the sign is negative; for composite numbers, the sign is equal to the product of signs of its factors. The following curious infinite product was also given by Euler:

$$\frac{\pi}{4} = \frac{3}{4} \cdot \frac{5}{4} \cdot \frac{7}{8} \cdot \frac{11}{12} \cdot \frac{13}{12} \cdot \frac{17}{16} \cdot \frac{19}{20} \cdot \frac{23}{24} \cdot \frac{29}{28} \cdot \frac{31}{32} \cdots,$$

where the numerators are the odd primes and each denominator is the multiple of four nearest to the numerator.

1750. Henry Sullamar, a real Bedlamite, found the quadrature of the circle in the number 666 inscribed on the forehead of the beast in the Revelations. He published periodically every two or three years some pamphlet in which he endeavored to prop his discovery.

1753. M. de Causans of the Guards cut a circular piece of turf, squared it and from the result deduced original sin and the Trinity. He found that the circle was equal to the square in which it is inscribed, *i.e.*, $\pi = 4$. He offered a reward for the detection of any error, and actually deposited 10,000 francs as earnest of 300,000. But the courts did not allow any one to recover.

1754. Jean Étienne Montucla (1725-1799) was an early French historian of mathematics. He published an anonymous treatise entitled *Histoire des récherches sur la quadrature du cercle*, and in 1758 the first part of his great work *Histoire des mathématiques*.

1755. Euler in his treatise *De relatione inter ternas pluresve quantitates instituenda*, which was published ten years later, wrote 'It appears to be fairly certain that the periphery of a circle constitutes such a peculiar kind of transcendental quantities that it can in no way be compared with other quantities, either roots or other transcendentals'. This conjecture haunted mathematicians for 107 years. The following expansion is due to Euler:

$$\tan^{-1} x = \frac{y}{x} \left(1 + \frac{2}{3} y + \frac{2 \cdot 4}{3 \cdot 5} y^2 + \frac{2 \cdot 4 \cdot 6}{3 \cdot 5 \cdot 7} y^3 + \cdots \right), \tag{23}$$

where $y = x^2/(1 + x^2)$. It converges rapidly.

1760. Georges Louis Leclerc (Comte of Buffon 1707-1788) was a naturalist, mathematician, cosmologist and encyclopedic author. Suppose a number of parallel lines, distance a apart, are ruled on a horizontal plane, and suppose a homogeneous uniform rod of length $\ell < a$ is dropped at random onto the plane. Buffon showed that the probability that the rod will fall across one of the lines in the plane is given by $p = (2\ell/\pi a)$. In the literature, this problem is known as Buffon's needle problem. This was the earliest problem in geometric probability to be solved. By actually performing this experiment, a large number of times and noting the number of successful cases, we can compute an approximation for π.

1761. Johann Heinrich Lambert (1728-1777) was the first to introduce hyperbolic functions into trigonometry. He wrote landmark books on geometry, the theory of cartography, and perspective in art. He is also credited for expressing Newton's second law of motion in the notation of the differential calculus. Lambert used the properties of continued fractions to show that π is irrational. He published a more general result in 1768. Lambert also showed that the functions e^x and $\tan x$ cannot assume rational values if x is

Agarwal et al. *Advances in Difference Equations* 2013, **2013**:100
http://www.advancesindifferenceequations.com/content/2013/1/100

a non-zero rational number. He also gave an interesting continued fraction for π,

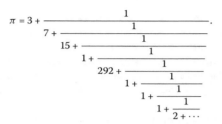

$$\pi = 3 + \cfrac{1}{7 + \cfrac{1}{15 + \cfrac{1}{1 + \cfrac{1}{292 + \cfrac{1}{1 + \cfrac{1}{1 + \cfrac{1}{1 + \cfrac{1}{2 + \cdots}}}}}}}}.$$

Some *inverse convergents* of this continued fraction are as follows:

$$\frac{3}{1}, \quad \frac{22}{7}, \quad \frac{333}{106}, \quad \frac{355}{113}, \quad \frac{103{,}993}{33{,}102}, \quad \frac{104{,}348}{33{,}215}, \quad \frac{208{,}341}{66{,}317},$$

$$\frac{312{,}689}{99{,}532}, \quad \frac{833{,}719}{265{,}381}, \quad \frac{1{,}146{,}408}{364{,}913}, \quad \frac{4{,}272{,}943}{1{,}360{,}120}.$$

1766. Arima Yoriyuki (1714-1783) was a Japanese mathematician of the Edo period. He found the following rational approximation of π, which is correct to 29 digits

$$\pi = \frac{428{,}224{,}593{,}349{,}304}{136{,}308{,}121{,}570{,}117}.$$

1775. The French Academy of Sciences passed a resolution henceforth not to examine any more solutions of the problem of squaring the circle. In fact, it became necessary to protect its officials against the waste of time and energy involved in examining the efforts of circle squarers. A few years later, the Royal Society in London also banned consideration of any further proofs of squaring the circle. This decision of the Royal Society was described by Augustus De Morgan (1806-1871) about 100 years later as the official blow to circle-squarers.

1776. Charles Hutton (1737-1823) was an English mathematician. He wrote several mathematical texts. In 1774, he was elected a fellow of the Royal Society of London. He suggested Machin's stratagem in the form

$$\pi = 20 \tan^{-1}\left(\frac{1}{7}\right) + 8 \tan^{-1}\left(\frac{3}{79}\right); \tag{24}$$

however, he did not carry computations far enough. Euler also developed the formula (24).

1778. M. de Vausenville, one of the deluded individuals, brought an action against the French Academy of Sciences to recover a reward to which he felt himself entitled. It ought to be needless to say that there was no reward offered for squaring the circle.

1779. Euler used his expansion (23) to evaluate right terms of (24), to calculate π to 20 decimal places in one hour!

About 1785. Franz Xaver Freiherr von Zach (1754-1832) discovered a manuscript by an unknown author in the Radcliffe Library, Oxford, which gives the correct value of π to 152 decimal places. Zach was elected a member of the Royal Swedish Academy of Sciences in 1794, a Fellow of the Royal Society in 1804, and an honorary member of the Hungarian

Agarwal et al. *Advances in Difference Equations* 2013, **2013**:100
http://www.advancesindifferenceequations.com/content/2013/1/100

Academy of Sciences in 1832. Asteroid 999 Zachia and the crater Zach on the Moon are named after him.

1789. Baron Jurij Bartolomej Vega (Georg Vega 1754-1802) was a Slovene mathematician, physicist and artillery officer. He wrote six scientific papers. The record of de Lagny of 127 digits seems to have stood until 1789, when Vega, using a new series for the arctangent discovered by Euler in 1755, calculated 140 decimal places (126 correct). Vega's result showed that de Lagny's string of digits had a 7 instead of an 8 in the 113th decimal place. His article was not published until six years later, in 1795 (136 correct). Vega retained his record for 52 years until 1841.

1794. Adrien-Marie Legendre (1752-1833) is remembered for Legendre functions, law of quadratic reciprocity for residues, standardizing weights and measures to the metric system, supervising the major task of producing logarithmic and trigonometric tables, least squares method of fitting a curve to the data available, proof of Fermat's last theorem for the exponent $n = 5$, Gauss-Legendre algorithm, Legendre's constant, Legendre's equation, Legendre polynomials, Legendre's conjecture, and Legendre transformation. The Legendre crater on the Moon is named after him. Legendre, in his *Elements de Géometrie* (1794) used a slightly modified version of Lambert's argument to prove the irrationality of π more rigorously, and also gave a proof that π^2 is irrational. He writes: 'It is probable that the number π is not even contained among the algebraic irrationalities, *i.e.*, that it cannot be the root of an algebraic equation with a finite number of terms, whose coefficients are rational. But, it seems to be very difficult to prove this strictly'.

1795. Ajima Naonobu (1732-1798), also known as Ajima Chokuyen, was a Japanese mathematician of the Edo period. The series he developed can be simplified as

$$\frac{\pi}{2} = F\left(1, 1, \frac{3}{2}, \frac{1}{2}\right) = \left(1 + \frac{1!}{3} + \frac{2!}{3 \cdot 5} + \frac{3!}{3 \cdot 5 \cdot 7} + \frac{4!}{3 \cdot 5 \cdot 7 \cdot 9} \cdots\right)$$
$$= \sum_{i=0}^{\infty} \frac{i!}{(2i+1)!!} = \sum_{i=0}^{\infty} \frac{(i!)^2 2^i}{(2i+1)!}.$$

It is interesting to note that the above series follows from (5) by using an acceleration technique known in the literature as Euler's transform. It can also be derived from the Wallis product formula (8).

1795. Jean-Charles Callet (174-1799) in his tables gave 154 (152 correct) decimal digits of π.

1797. Lorenzo Mascheroni (1750-1800) was educated with the aim of becoming a priest and he was ordained at the age of 17. In 1790, he calculated Euler's constant to 32 (19 correct) decimal places. Lorenzo dedicated his book, *Geometria del compasso*, to Napoleon Bonaparte. In this work, he proved that all Euclidean constructions can be made with compasses alone, so a straight edge in not needed. However, it was proved earlier in 1672 by the Danish mathematician Georg Mohr (1640-1697). He claimed that compasses are more accurate then those of a ruler.

1800. Karl Friedrich Gauss (1777-1855) was one of the greatest mathematicians of all time. Alexander von Humboldt (1769-1859), the famous traveler and amateur of the sciences, asked Pierre Simon de Laplace (1749-1827) who was the greatest mathematician in Germany, Laplace replied Johann Friedrich 'Pfaff' (1765-1825). 'But what about Gauss' the astonished von Humboldt asked, as he was backing Gauss for the position of director at

Agarwal et al. *Advances in Difference Equations* 2013, **2013**:100
http://www.advancesindifferenceequations.com/content/2013/1/100

the Göttingen observatory. 'Oh', said Laplace, 'Gauss is the greatest mathematician in the world'. Gauss suggested to his teacher Pfaff to study the sequences $\{x_n\}$ and $\{y_n\}$ generated by the recurrence relations

$$x_{n+1} = \frac{1}{2}(x_n + y_n), \qquad y_{n+1} = \sqrt{x_{n+1}y_n}, \quad n \geq 0. \tag{25}$$

In his reply, Pfaff showed that for any positive numbers x_0 and y_0 these sequences converge monotonically to a common limit given by

$$B(x_0, y_0) = \begin{cases} (y_0^2 - x_0^2)^{1/2} / \cos^{-1}(x_0/y_0), & 0 \leq x_0 < y_0, \\ (x_0^2 - y_0^2)^{1/2} / \cosh^{-1}(x_0/y_0), & 0 < y_0 < x_0. \end{cases} \tag{26}$$

Pfaff's letter was unpublished. In 1881, Carl Wilhelm Borchardt (1817-1880) work was published in which he rediscovered this result which now bears his name. For this, it suffices to note that:

1. $\{x_n\}$ and $\{y_n\}$ converge monotonically to the same limit.
2. The ratio $r_n = x_n/y_n$ satisfies $r_{n+1}^2 = (1 + r_n)/2$.
3. If $x_0 < y_0$, let $\theta = \cos^{-1} r_0$. Then, $s_n = 2^n \cos^{-1} r_n = \theta$ and $c_n = 4^n(x_n^2 - y_n^2) = (x_0^2 - y_0^2)$ are independent of n.
4. $\lim_{n \to \infty} y_n = \lim_{n \to \infty} \frac{2^{-n}|c_n|^{1/2}}{\sin^{-1}(2^{-n}|c_n|^{1/2}/y_n)} = \lim_{n \to \infty} \frac{|c_n|^{1/2}}{s_n} = \frac{(y_0^2 - x_0^2)^{1/2}}{\theta}$.
   If $y_0 < x_0$, we let $\theta = \cosh^{-1} r_0$, and follow similarly.
   Now we let $x_n = 1/a_n, y_n = 1/b_n$, then (25) and (26) take the form

$$a_{n+1} = \frac{2a_n b_n}{a_n + b_n}, \qquad b_{n+1} = \sqrt{a_{n+1}b_n}, \quad n \geq 0 \tag{27}$$

and

$$A(a_0, b_0) = \frac{1}{B(1/a_0, 1/b_0)} = \begin{cases} a_0 b_0 (a_0^2 - b_0^2)^{-1/2} \cos^{-1}(b_0/a_0), & a_0 > b_0 \geq 0, \\ a_0 b_0 (b_0^2 - a_0^2)^{-1/2} \cosh^{-1}(b_0/a_0), & b_0 > a_0 > 0. \end{cases} \tag{28}$$

Clearly, the recurrence relations (2) are different from (27). In fact, (2) minimize the count of arithmetic operations. In particular, if we let $a_0 = 2\sqrt{3}$, $b_0 = 3$, then (27) in view of (28) converges to π.

In what follows, we let the constant $c = 4^n(x_n^2 - y_n^2) = (x_0^2 - y_0^2)$, then we can uncouple (25) and (27), respectively, to obtain

$$x_{n+1} = \frac{x_n}{2} + \left(\left(\frac{x_n}{2}\right)^2 - 4^{-n-1}c \right)^{1/2}, \qquad y_{n+1}^2 = \frac{y_n^2}{2}\left(1 + \left(1 + 4^{-n}cy_n^{-2}\right)^{1/2}\right) \tag{29}$$

and

$$a_{n+1} = \frac{2^{2n+1}}{ca_n}\left(1 - \left(1 - 4^{-n}ca_n^2\right)^{1/2}\right), \qquad b_{n+1}^2 = \frac{2^{2n+1}}{c}\left(\left(1 + 4^{-n}cb_n^2\right)^{1/2} - 1\right). \tag{30}$$

From (29) and (30), several known and new recurrence relations can be obtained.

Gauss also developed the Machin-type formula

$$\frac{\pi}{4} = 12\tan^{-1}\left(\frac{1}{18}\right) + 8\tan^{-1}\left(\frac{1}{57}\right) - 5\tan^{-1}\left(\frac{1}{239}\right). \tag{31}$$

Agarwal et al. *Advances in Difference Equations* 2013, **2013**:100
http://www.advancesindifferenceequations.com/content/2013/1/100

He also estimated the value of π by using lattice theory and considering a lattice inside a large circle, but he did not pursue it further.

1810. Sakabe Kohan (1759-1824) developed the series

$$\frac{\pi}{4} = 1 - \frac{1}{5} - \frac{1 \cdot 4}{5 \cdot 7 \cdot 9} - \frac{(1 \cdot 3)(4 \cdot 6)}{5 \cdot 7 \cdot 9 \cdot 11 \cdot 13} - \frac{(1 \cdot 3 \cdot 5)(4 \cdot 6 \cdot 8)}{5 \cdot 7 \cdot 9 \cdot 11 \cdot 13 \cdot 15 \cdot 17} - \cdots .$$

1818. Wada Yenzo Nei (known as Wada Yasushi, 1787-1840) developed over one hundred infinite series expressing directly or indirectly π. One of his series can be written as

$$\pi = 2F\left(\frac{1}{2}, \frac{1}{2}, \frac{3}{2}, 1\right) = 2\left(1 + \frac{1^2}{3!} + \frac{1^2 \cdot 3^2}{5!} + \frac{1^2 \cdot 3^2 \cdot 5^2}{7!} + \cdots\right).$$

1825. Malacarne of Italy published a geometric construction in Géométrique (Paris), which leads to the value of π less than 3.

1828. C.G. Specht of Berlin published a geometric construction in Crelle's Journal, Volume 3, p.83, which leads to $\pi = 13\sqrt{146}/50 = 3.1415919\ldots$.

1832. Karl Heinrich Schellbach (1809-1890) began with the relation

$$\frac{\pi i}{2} = \ln(i) = \ln\left(\frac{1+i}{1-i}\right) = \ln(1+i) - \ln(1-i),$$

which is due to Giulio Carlo Fagnano dei Toschi (1682-1766), and used the logarithm expansion (7), to obtain

$$\frac{\pi i}{2} = \left(i + \frac{1}{2} - \frac{1}{3}i - \frac{1}{4} + \frac{1}{5}i + \frac{1}{6} - \cdots\right) - \left(-i + \frac{1}{2} + \frac{1}{3}i - \frac{1}{4} - \frac{1}{5}i + \frac{1}{6} + \cdots\right)$$

$$= 2i - \frac{2}{3}i + \frac{2}{5}i - \cdots,$$

which immediately gives Nilakanthan series (5). He also considered the relation

$$\frac{\pi i}{2} = \ln(i) = \ln\left(\frac{(2+i)(3+i)}{(2-i)(3-i)}\right)$$

$$= \left\{\ln\left(1 + \frac{1}{2}i\right) - \ln\left(1 - \frac{1}{2}i\right)\right\} + \left\{\ln\left(1 + \frac{1}{3}i\right) - \ln\left(1 - \frac{1}{3}i\right)\right\}$$

and used the expansion (7), to obtain, compare to (5), a fast converging expansion

$$\frac{\pi}{4} = \left(\frac{1}{2} + \frac{1}{3}\right) - \frac{1}{3}\left(\frac{1}{2^3} + \frac{1}{3^3}\right) + \frac{1}{5}\left(\frac{1}{2^5} + \frac{1}{3^5}\right) - \cdots .$$

1833. William Baddeley in his work *Mechanical quadrature of the circle, London Mechanics' Magazine, August, 1833* writes 'From a piece of carefully rolled sheet brass was cut out a circle 1 9/10 inches in diameter, and a square 1 7/10 inches in diameter. On weighing them, they were found to be exactly the same weight, which proves that, as each are of the same thickness, the surfaces must also be precisely similar. The rule, therefore, is that the square is to the circle as 17 to 19'. We believe for the square it must be the side (not the diameter). Then it follows that $\pi = 1{,}156/361 = 3.202216\ldots$.

Agarwal et al. *Advances in Difference Equations* 2013, **2013**:100
http://www.advancesindifferenceequations.com/content/2013/1/100

Page 33 of 59

1836. Joseph LaComme 'at a time when he could neither read nor write being desirous to ascertain what quantity of stones would be required to prove a circular reservoir he had constructed, consulted a mathematics professor. He was told that it was impossible to determine the full amount, as no one had yet found the exact relation between the circumference of a circle and its diameter. The well-sinker thereupon, full of self-confidence in his powers, applied himself to the celebrated problem and discovered the solution, which lead to $\pi = 25/8$ by mechanical process. He then taught himself to read and write, and managed to acquire some knowledge of arithmetic by which he verified his mechanical solution'. Joseph was honored for his profound discovery with several medals of the first class, bestowed by Parisian societies.

1841. William Rutherford (1798-1871) was an English mathematician. He calculated π to 208 places of which 152 were later found to be correct. For this, he employed Euler's formula

$$\frac{\pi}{4} = 4\tan^{-1}\left(\frac{1}{5}\right) - \tan^{-1}\left(\frac{1}{70}\right) + \tan^{-1}\left(\frac{1}{99}\right)$$

and Madhava's series expansion (4).

1844. Johann Martin Zacharias Dase (1824-1861) was a calculating prodigy. At the age of 15, he gave exhibitions in Germany, Austria and England. His extraordinary calculating powers were timed by renowned mathematicians including Gauss. He multiplied $79{,}532{,}853 \times 93{,}758{,}479$ in 54 seconds; two 20-digit numbers in 6 minutes; two 40-digit numbers in 40 minutes; and two 100-digit numbers in 8 hours 45 minutes. In 1840, he made acquaintance with Viennese mathematician L.K. Schulz von Strasznicky (1803-1852) who suggested him to apply his powers to scientific purposes. When he was 20, Strasznicky taught him the use of the formula

$$\frac{\pi}{4} = \tan^{-1}\left(\frac{1}{2}\right) + \tan^{-1}\left(\frac{1}{5}\right) + \tan^{-1}\left(\frac{1}{8}\right),$$

and asked him to calculate π. In two months, he carried the approximation to 205 places of decimals, of which 200 are correct. He next calculated a 7-digit logarithm table of the first 1,005,000 numbers; he did this in his off-time from 1844 to 1847, when occupied by the Prussian survey. His next contribution of two years was the compilation of hyperbolic table in his spare time which was published by the Austrian Government in 1857. Next, he offered to make a table of integer factors of all numbers from 7,000,000 to 10,000,000; for this, on the recommendation of Gauss the Hamburg Academy of Sciences agreed to assist him financially, but Dase died shortly thereafter in Hamburg. He also had an uncanny sense of quantity. That is, he could just tell, without counting, how many sheep were in a field, or words in a sentence, and so forth, up to about 30.

1844. Hiromu Hasegawa (1810-1887) and his father Hiroshi Hasegawa (1782-1838) published many Wasan books. Hiromu developed the series

$$\frac{\pi}{4} = 1 - \frac{1}{2\cdot 3} - \frac{1}{5\cdot 8} - \frac{1}{7\cdot 16} - \frac{5}{9\cdot 128} - \cdots .$$

This series can be written as

$$\pi = \frac{4}{3}F\left(\frac{1}{2},\frac{3}{2},\frac{5}{2},1\right).$$

Agarwal et al. *Advances in Difference Equations* 2013, **2013**:100
http://www.advancesindifferenceequations.com/content/2013/1/100

1847. Thomas Clausen (1801-1885) wrote over 150 papers on pure mathematics, applied mathematics, astronomy and geophysics. He used the formula

$$\frac{\pi}{4} = 2\tan^{-1}\left(\frac{1}{3}\right) + \tan^{-1}\left(\frac{1}{7}\right)$$

to calculate π to 250 decimal places, but only 248 are correct. In 1854, he factored the sixth Fermat number as $2^{64} + 1 = 67{,}280{,}421{,}310{,}721 \times 274{,}177$. Clausen also gave a new method of factorising numbers.

1849. Jacob de Gelder (1765-1848) a mathematical ideologist proposed a geometric construction which gives π correct to 6 decimal places. His method is based on the fact that $\pi = 355/113 = 3 + 4^2/(7^2 + 8^2)$. Gelder's result was published in 1849.

1851. Joseph Liouville (1809-1882) was a highly respected professor at the Collége de France in Paris, and the founder and for 39 years the editor of the *Journal des Mathématiques Pures et Appliquées*. His ingenious theory of fractional differentiation answered the long-standing question of what reasonable meaning can be assigned to the symbol $d^n y/dx^n$ when n is not a positive integer. In 1844, Liouville showed that e is not a root of any quadratic equation with integral coefficients. This led him to conjecture that e is transcendental. In 1851, Liouville showed, by using continued fractions, that there are an infinite number of transcendental numbers, a result which had previously been suspected but had not been proved. He produced the first examples of real numbers that are provably transcendent. One of these is

$$\sum_{n=1}^{\infty} \frac{1}{10^{n!}} = \frac{1}{10^1} + \frac{1}{10^2} + \frac{1}{10^6} + \cdots = 0.11000100\ldots.$$

His methods led to extensive further research.

1853. Lehmann correctly calculated 261 decimal places of π. For this, he used Euler's formula

$$\frac{\pi}{4} = \tan^{-1}\left(\frac{1}{2}\right) + \tan^{-1}\left(\frac{1}{3}\right). \tag{32}$$

1853. Rutherford obtained 440 correct decimal places.

1853. William Shanks (1812-1882) was a British amateur mathematician. He used Machin formula (18) to calculate π to 607 decimal places. He was assisted by Rutherford in checking first 440 digits.

1853-1855. Richter in 1853 published 333 digits (330 correct), and in 1855 (after his death in 1854) 500 decimal places.

1860. James Smith published the value of π as 3 1/8 and argued that it is exact and correct. He attempted to bring it before the British Association for the Advancement of Science. Interestingly, even De Morgan and Hamilton could not convince him for his mistake.

1861. Philip H. Vanderweyde published an essay discussing the subject π. He also used several constructions, resulting $\pi = 3.1415926535\ldots$.

1862. Lawrence Sluter Benson published about 20 pamphlets on the area of the circle, three volumes on philosophic essays, and one on geometry *The Elements of Euclid and Legendre*. He demonstrated that the area of the circle is equal to $3R^2$, or the

Agarwal et al. *Advances in Difference Equations* 2013, **2013**:100
http://www.advancesindifferenceequations.com/content/2013/1/100

arithmetical square between the inscribed and circumscribed squares. According to him $\sqrt{12} = 3.4641016\ldots$ is the ratio between the diameter of a circle and the perimeter of its equivalent square. The ratio between the diameter and circumference, he believed, is not a function of the area of the circle. He accepted the value of $\pi = 3.141592\ldots$.

1863. S.M. Drach proved that the circumference of a circle can be obtained as follows: From thrice diameter, deduct 8/1,000 and 7/1,000,000 of a diameter, and add 5% to the result, *i.e.*,

$$2\pi = 6 - \frac{16}{1,000} - \frac{14}{1,000,000} + \frac{5}{100}\left(6 - \frac{16}{1,000} - \frac{14}{1,000,000}\right),$$

which gives $\pi = 3.14159265$.

1868. Cyrus Pitt Grosvenor (1792-1879) was an American anti-slavery Baptist minister. In his retirement, he worked on the problem of squaring the circle. He described his method in a pamphlet titled *The circle squared*, New York: Square the diameter of the circle; multiply the square by 2; extract the square root of the product; from the root subtract the diameter of the circle; square the remainder; multiply this square by four fifths; subtract the square from the diameter of the circle, *i.e.*,

$$\frac{\pi D^2}{4} = D^2 - \frac{5}{4}\left(\sqrt{2D^2} - D\right)^2 = D^2\left[1 - \frac{5}{4}(\sqrt{2} - 1)^2\right] = D^2(0.785533906\ldots),$$

which gives $\pi = 3.142135\ldots$.

1872. Augustus De Morgan (1806-1871) was born in Madura (India), but his family moved to England when he was seven months old. He lost the sight of his right eye shortly after birth. He was an extremely prolific writer. He wrote more than 1,000 articles for more than 15 periodicals. De Morgan also wrote textbooks on many subjects, including logic, probability, calculus and algebra. In 1866, he was a co-founder of the London Mathematical Society and became its first President. His book *A Budget of Paradoxes* of 512 pages, which was edited and published by his wife in 1872, is an entertaining text. This book contains the names of 75 writers on π. In this work, De Morgan reviewed the works of 42 of these writers, bringing the subject down to 1870. He once remarked that it is easier to square the circle then to get round a mathematician. He was the first to point out that in the decimal expansion of π one should expect each of the 10 digits appear uniformly, *i.e.*, roughly one out of ten digits should be a 4, *etc.*

1872. Asaph Hall (1829-1907) was an astronomer. He published the results of an experiment in random sampling that Hall had convinced his friend, Captain O.C. Fox, to perform when Fox was recovering from a wound received at the Second Battle of Bull Run. The experiment was based on Buffon's needle problem. After throwing his needles eleven hundred times, Fox was able to derive $\pi \simeq 3.14$. This work is considered as a very early documentation use of random sampling (which Nicholas Constantine Metropolis (1915-1999) named as the Monte Carlo method during the Manhattan Project of World War II).

1873. Charles Hermite (1822-1901) in 1870 was appointed to a professorship at the Sorbonne, where he trained a whole generation of well-known French mathematicians. He was strongly attracted to number theory and analysis, and his favorite subject was elliptic functions, where these two fields touch in many remarkable ways. His proof of the transcendence of e was high point in his career.

Agarwal et al. *Advances in Difference Equations* 2013, **2013**:100
http://www.advancesindifferenceequations.com/content/2013/1/100

1873-1874. William Shanks again used Machin formula (18) to calculate π to 707 decimal places (published in the Proceedings of the Royal Society, London), but only 527 decimal places are correct. For this, he used mechanical desk calculator and worked for almost 15 years. For a long time, this remained the most fabulous piece of calculation ever performed. In the Palais de la Découverte (a science museum in Paris), there is a circular room known as the 'pi room'. On its wall are inscribed these 707 digits of π. The digits are large wooden characters attached to the dome-like ceiling. Shanks also calculated e and the Euler-Mascheroni constant γ to many decimal places. He published a table of primes up to 60,000 and found the natural logarithms of 2, 3, 5 and 10 to 137 places.

1874. Tseng Chi-Hung (died in 1877) finds 100 digits of π in a month. For this, he used the formula (32).

1874. John A. Parker in his book *The Quadrature of the Circle. Containing Demonstrations of the Errors of Geometers in Finding Approximations in Use* published by John Wiley & Sons, New York claims that $\pi = 20,612/6,561$ exactly. He exclaims, 'all the serial and algebraic formula in the world, or even geometrical demonstration, if it be subjected to any error whatever, cannot overthrow the ratio of circumference to diameter which I have established'. He praises Metius (lived in the sixteenth century) for using the ratio 355/113. His book also contains practical questions on the quadrature applied to the astronomical circles.

1876. Alick Carrick proposed in his book, *The Secret of the Circle, its Area Ascertained*, the value of π as 3 1/7.

1879. Pliny Earle Chase (1820-1886) was a scientist, mathematician, and educator who mainly contributed to the fields of astronomy, electromagnetism and cryptography. In his pamphlet, *Approximate Quadrature of the Circle*, he used a geometric construction to obtain $\pi = 3.14158499\ldots$.

1882. Carl Louis Ferdinand von Lindemann (1852-1939) worked on non-Euclidean geometry. He followed the method of Hermite to show that π is also transcendental. His result showed at last that the age-old problem of squaring the circle by a ruler-and-compass construction is impossible. Lindemann's paper runs to 13 pages of tough mathematics. Karl Wilhelm Weierstrass (1815-1897), the apostle of mathematical rigor, simplified the proof of Lindemann's theorem somewhat in 1885, and it was further simplified in later years by renowned mathematicians (Stieltjes, Hurwitz, Hilbert, and others). The interested reader is referred to the comparatively easy version given by Hobson. Nonetheless, there are still some amateur mathematicians who do not understand the significance of this result, and futilely look for techniques to square the circle. Next, Lindemann spent several years to provide the proof of Fermat's Last Theorem, which is unfortunately wrong. He also worked on projective geometry, Abelian functions and developed a method of solving equations of any degree using transcendental functions.

1888. Sylvester Clark Gould (1840-1909) was the editor of *Notes and Queries*, Manchester, New Hampshire. He compiled the bibliography entitled *What is the Value of Pi*. It contains 100 titles and gives the result of 63 authors. In this work the diagram 16 claims that $\pi = 3\ 3{,}949/27{,}889$ exactly.

1892. A writer announced in the *New York Tribune* the rediscovery of a long-lost secret that gives 3.2 as the exact value of π. This announcement caused considerable discussion, and even near the beginning of the twentieth century 3.2 had its advocates as against the value 22/7.

Agarwal et al. *Advances in Difference Equations* 2013, **2013**:100
http://www.advancesindifferenceequations.com/content/2013/1/100

1896. Fredrik Carl Mülertz Störmer (1874-1957) was a mathematician and physicist, known for his work in number theory. He gave the following Machin-like formulas for calculating π

$$\frac{\pi}{4} = 44\tan^{-1}\left(\frac{1}{57}\right) + 7\tan^{-1}\left(\frac{1}{239}\right) - 12\tan^{-1}\left(\frac{1}{682}\right) + 24\tan^{-1}\left(\frac{1}{12{,}943}\right) \qquad (33)$$

and

$$\frac{\pi}{4} = 6\tan^{-1}\left(\frac{1}{8}\right) + 2\tan^{-1}\left(\frac{1}{57}\right) + \tan^{-1}\left(\frac{1}{239}\right). \qquad (34)$$

1897. In the State of Indiana, the House of Representatives unanimously passed the Bill No. 246 (known as the 'π bill') introducing a new mathematical truth 'Be it enacted by the General Assembly of the State of Indiana: It has been found that a circular area is to the square on a line equal to the quadrant of the circumference, as the area of an equilateral rectangle is to the square on one side...' ($\pi = 3.2$). The author of the bill was a physician, Edwin J. Goodman (1825-1902), M.D., of Solitude, Posey County, Indiana, and it was introduced in the Indiana House on 18 January 1897, by Mr. Taylor I. Record, representative from Posey County. Edwin offered this contribution as a free gift for the sole use of the State of Indiana (the others would evidently have to pay royalties). Copies of the bill are preserved in the Archives Division of the Indiana State Library. The bill was sent to the Senate for approval. Fortunately, during the House's debate on the bill, Purdue University Mathematics Professor Clarence Abiathar Waldo (1852-1926) was present. When Professor Waldo informed the Indiana Senate of the 'merits' of the bill, the Senate, after some ridicule at the expense of their colleagues, indefinitely postponed voting on the bill and let it die.

1900. H.S. Uhler used Machin's formula (18) to compute π to 282 decimal places.

1901. Mario Lazzarini an Italian mathematician performed the Buffon's needle experiment. Tossing a needle 3,408 times, he obtained the well-known estimate 355/113 for π. Although it is an impressive observation, but suspiciously good. In fact, statisticians Sir Maurice George Kendall (1907-1983) and Patrick Alfred Pierce Moran (1917-1988) FRS have commented that one can do better to cut out a large circle and use a tape to measure to find its circumference and diameter. On the same theme of phoney experiments, Gridgeman, in 1960, pours scorn on Lazzerini and others, created some amusement by using a needle of carefully chosen length $k = 0.7857$, throwing it twice, and hitting a line once. His estimate for π was thus given by $2 \times 0.7857/\pi = 1/2$ from which he got the highly creditable value of $\pi = 3.1428$. Of course, he was not being serious.

1902. Duarte used Machin's formula (18) to compute π to 200 decimal places.

1906. Various mnemonic devices have been given for remembering the decimal digits of π. The most common type of mnemonic is the word-length mnemonic in which the number of letters in each word corresponds to a digit, for example, *How I wish I could calculate pi* (by C. Heckman), *May I have a large container of coffee* (by Martin Gardner), and *How I want a drink, alcoholic of course, after the heavy lectures involving quantum mechanics* (by Sir James Jeans), respectively, give π to seven, eight, and fifteen decimal places. Adam C. Orr in *Literary Digest*, vol. 32 (1906), p.84 published the following poem which gives π to 30 decimal places:

Agarwal et al. *Advances in Difference Equations* 2013, **2013**:100
http://www.advancesindifferenceequations.com/content/2013/1/100

Page 38 of 59

> Now I, even I, would celebrate
> In rhymes inapt, the great
> Immortal Syracusan, rivaled nevermore
> Who in his wondrous lore,
> Passed on before
> Left men his guidance,
> How to circles mensurate.

Several other such poems not only in English, but almost in every language including Albanian, Bulgarian, Czech, Dutch, French, German, Italian, Latin, Polish, Portuguese, Romanian, Spanish and Swedish are known. However, there is a problem with this type of mnemonic, namely, how to represent the digit zero. Fortunately, a zero does not occur in π until the thirty-second place. Several people have come up with ingenious methods of overcoming this, most commonly using a ten-letter word to represent zero. In other cases, a certain piece of punctuation is used to indicate a naught. Michael Keith (with such similar understanding) in his work *Circle digits*: *a self-referential story, Mathematical Intelligencer*, vol. 8 (1986), 56-57, wrote an interesting story which gives first 402 decimals of π.

1913. Ernest William Hobson (1856-1933) was Sadleirian Professor at the University of Cambridge from 1910 to 1931. His 1907 work on real analysis was very influential in England. In his book, *Squaring the circle*: *A History of the Problem*, he used a geometrical construction to obtain $\pi = 3.14164079\ldots$.

1914. Srinivasa Ramanujan (1887-1920) was a famous mathematical prodigy. He collaborated with Hardy for five years, proving significant theorems about the number of partitions of integers. Ramanujan also made important contributions to number theory and also worked on continued fractions, infinite series and elliptic functions. In 1918, he became the youngest Fellow of the Royal Society. According to Hardy, 'the limitations of Ramanujan's knowledge were as startling as its profundity'. *Here was a man who could workout modular equations and theorems of complex multiplication, to orders unheard of, whose mastery of continued fractions was, beyond that of any mathematician in the world, who had found for himself the functional equation of the zeta-function, and the dominant terms of the many of the most famous problems in the analytic theory of numbers; and he had never heard of a doubly periodic function or of Cauchy's theorem, and had indeed but the vaguest idea of what a function of a complex variable was.* Ramanujan considered mathematics and religion to be linked. He said, 'an equation for me has no meaning unless it expresses a thought of God'. He was endowed with an astounding memory and remembered the idiosyncrasies of the first 10,000 integers to such an extent that each number became like a personal friend to him. *Once Hardy went to see Ramanujan when he was in a nursing home and remarked that he had traveled in a taxi with a rather dull number, viz 1,729, Ramanujan exclaimed, 'No, Hardy, 1,729 is a very interesting number. It is the smallest number that can be expressed as the sum of two cubes viz* $1,729 = 1^3 + 12^3 = 9^3 + 10^3$, *and the next such number is very large'*. His life can be summed up in his own words, 'I really love my subject'. His 1914 paper on 'Modulus functions and approximation to π' contains several new innovative empirical formulas and geometrical constructions for approximating π. One of the remarkable formulas for its elegance and inherent mathematical depth

Agarwal et al. *Advances in Difference Equations* 2013, **2013**:100
http://www.advancesindifferenceequations.com/content/2013/1/100

is

$$\frac{1}{\pi} = \frac{\sqrt{8}}{9{,}801} \sum_{m=0}^{\infty} \frac{(4m)!}{(m!)^4} \frac{(1{,}103 + 26{,}390m)}{396^{4m}}. \tag{35}$$

It has been used to compute π to a level of accuracy, never attained earlier. Each additional term of the series adds roughly 8 digits. He also developed the series

$$\frac{1}{\pi} = \sum_{m=0}^{\infty} \binom{2m}{m}^3 \frac{42m + 5}{2^{12m+4}} \quad \text{and} \quad \frac{2}{\pi} = \sum_{m=0}^{\infty} \frac{(-1)^m (4m+1)[(2m-1)!!]^3}{((2k)!!)^3}.$$

The first series has the property that it can be used to compute the second block of k (binary) digits in the decimal expansion of π without calculating the first k digits. The following mysterious approximation which approximates π to 18 correct decimal places is also due to Ramanujan

$$\pi \simeq \frac{12}{\sqrt{190}} \ln\big((2\sqrt{2} + \sqrt{10})(3 + \sqrt{10})\big).$$

1914. T.M.P. Hughes in his work *A triangle that gives the area and circumference of any circle, and the diameter of a circle equal in area to any given square, Nature* 93, 110, doi:10.1038/093110a0 uses a geometric construction to obtain $\pi = 3.14159292035\ldots$.

1928. In March 1928, the University of Minnesota was notified that Gottfried Lenzer (a native of Germany who lived in St. Paul for many years) had bequeathed to the university a series of 60 drawings from 1911-1927 and explanatory notes concerning the three classical problems of antiquity. He used a geometrical construction for squaring the circle to obtain $\pi = 3.1378\ldots$.

1934. Alexander Osipovich Gelfond (1906-1968) was a Soviet mathematician. He proved that e^{π} (Gelfond's constant) is transcendental, but nothing yet is known about the nature of any of the numbers $\pi + e$, πe or π^e.

1934. Helen Abbot Merrill (1864-1949) earned her Ph.D. from Yale in 1903 on the thesis *On Solutions of Differential Equations which Possess an Oscillation Theorem*. She served as an associate editor of *The American Mathematical Monthly* during 1916-1919, and was a vice-president from 1920 to 1921 of the Mathematical Association of America. Her book *Mathematical Excursions: Side Trips Along Paths not Generally Traveled in Elementary Courses in Mathematics, Bruce Humphries, Inc., Boston*, 1934 was a text for the general public. In this book, a geometric construction is given (perhaps by an earlier author) which leads to $\pi = 3.141591953\ldots$.

1934. Edmund Georg Hermann (Yehezkel) Landau (1877-1938) was a child prodigy. In 1903, he gave a simpler proof of the *prime number theorem*. His masterpiece of 1909 was a treatise *Handbuch der Lehre von der Verteilung der Primzahlen* a two volume work giving the first systematic presentation of analytic number theory. Landau wrote over 250 papers on number theory, which had a major influence on the development of the subject. Despite his outstanding talents as both a teacher and researcher, Landau annoyed many of his colleagues at Göttingen. He started criticizing privately, and often publicly, their results. Landau in his work defined $\pi/2$ as the value of x between 1 and 2 for which $\cos x$ vanishes. One cannot believe this definition was used, at least as an excuse, for a racial attack on

Agarwal et al. *Advances in Difference Equations* 2013, **2013**:100
http://www.advancesindifferenceequations.com/content/2013/1/100

Landau. This unleashed an academic dispute which was to end in Landau's dismissal from his chair at Göttingen. Ludwig Georg Elias Moses Bieberbach (1886-1982) famous for his conjecture, explained the reasons for Landau's dismissal: 'Thus the valiant rejection by the Göttingen student body which a great mathematician, Edmund Landau, has experienced is due in the final analysis to the fact that the un-German style of this man in his research and teaching is unbearable to German feelings. A people who have perceived how members of another race are working to impose ideas foreign to its own must refuse teachers of an alien culture'. Hardy replied immediately to Bieberbach about the consequences of this un-German definition of π: 'There are many of us, many Englishmen and many Germans, who said things during the War which we scarcely meant and are sorry to remember now. Anxiety for one's own position, dread of falling behind the rising torrent of folly, determination at all cost not to be outdone, may be natural if not particularly heroic excuses. Professor Bieberbach's reputation excludes such explanations of his utterances, and I find myself driven to the more uncharitable conclusion that he really believes them true.'

1934. A Cleveland businessman Carl Theodore Heisel published a book *Mathematical and Geometrical Demonstrations* in which he announced the grand discovery that π was exactly equal to 256/81, a value that the Egyptians had used some 4,000 years ago. Substituting this value for calculations of areas and circumferences of circles with diameters $1, 2, \ldots$ up to 9, he obtained numbers which showed consistency of circumference and area, 'thereby furnishing incontrovertible proof of the exact truth' of his ratio. He also rejected decimal fractions as inexact (whereas ratios of integers as exact and scientific), and extracted roots of negative numbers thus: $\sqrt{-a} = \sqrt{a}-$, $\sqrt{a-2} = -a$. He published this book on his own expense and distributed to colleges and public libraries throughout the United States without charge.

1934. Miff Butler claimed discovery of a new relationship between π and e. He stated his work to be the first basic mathematical principle ever developed in USA. He convinced his congressman to read it into the Congressional Record on 5 June 1940.

1940. H.S. Uhler used Machin's formula (18) to compute π to 333 decimal places.

1945-1947. D.F. Ferguson of England used the formula

$$\frac{\pi}{4} = 3\tan^{-1}\left(\frac{1}{4}\right) + \tan^{-1}\left(\frac{1}{20}\right) + \tan^{-1}\left(\frac{1}{1,985}\right)$$

to find that his value disagreed with that of William Shanks in the 528th place. In 1946, he approximated π to 620 decimal places, and in January 1947 to 710 decimal places. In the same month William Shanks used Machin's formula (18) to compute 808-place value of π, but Ferguson soon found an error in the 723rd place. For all the calculations, he used desk calculator.

1947. Ivan Morton Niven (1915-1999) gave an elementary proof that π is irrational.

1949. Ferguson and John William Wrench, Jr. (1911-2009) using a desk calculator, computed 1,120 decimal digits of π. This record was broken only by the electronic computers.

September 1949. John Wrench and L.R. Smith (also attributed to George Reitwiesner *et al.*) were the first to use an electronic computer Electronic Numerical Integrator and Computer (ENIAC) at the Army Ballistic Research Laboratories in Aberdeen, Maryland, to calculate π to 2,037 decimal places. For this, they programed Machin's formula (18). It took 70 hours, a pitifully long time by today's standards. In this project, John Louis von Neumann (1903-1957), one of the most versatile and smartest mathematicians of the

Agarwal et al. *Advances in Difference Equations* 2013, **2013**:100
http://www.advancesindifferenceequations.com/content/2013/1/100

twentieth century, also took part. In 1965, The ENIAC became obsolete, and it was dismembered and moved to the Smithsonian Institution as a museum piece.

1951. Konrad Knopp gave the following two expansions of π:

$$\frac{\pi}{4} = \sum_{k=1}^{\infty} \tan^{-1}\left(\frac{1}{k^2 + k + 1}\right) \quad \text{and} \quad \frac{\pi^2}{16} = \sum_{k=0}^{\infty} \frac{(-1)^k}{k+1}\left(1 + \frac{1}{3} + \cdots + \frac{1}{2k+1}\right).$$

1953. Kurt Mahler (1903-1988) showed that π is not a Liouville number: A real number x is called a Liouville number if for every positive integer n, there exist integers p and q with $q > 1$ and such that

$$0 < \left|x - \frac{p}{q}\right| < \frac{1}{q^n}.$$

A Liouville number can thus be approximated 'quite closely' by a sequence of rational numbers. In 1844, Liouville showed that all Liouville numbers are transcendental.

1954. S.C. Nicholson and J. Jeenel programmed NORC (Naval Ordnance Research Calculator) at Dahlgren, Virginia to compute π to 3,092 decimals. For this, they used Machin's formula (18). The run took only 13 minutes.

1956. John Gurland established that for all positive integers n,

$$\frac{4n+3}{(2n+1)^2}\left(\frac{(2n)!!}{(2n-1)!!}\right)^2 < \pi < \frac{4}{4n+1}\left(\frac{(2n)!!}{(2n-1)!!}\right)^2. \tag{36}$$

March 1957. G.E. Felton used the Ferranti Pegasus computer to find 10,021 decimal places of π in 33 hours. The program was based on Klingenstierna's formula

$$\pi = 32\arctan\left(\frac{1}{10}\right) - 4\arctan\left(\frac{1}{239}\right) - 16\arctan\left(\frac{1}{515}\right). \tag{37}$$

However, a subsequent check revealed that a machine error had occurred, so that 'only' 7,480 decimal places were correct. The run was therefore repeated in May 1958, but the correction was not published.

January 1958. Francois Genuys programmed an IBM 704 at the Paris Data Processing Center. He used Machin's type formula (18). It yielded 10,000 decimal places of π in 1 hour and 40 minutes.

July 1959. Genuys programmed an IBM 704 at the Commissariat á l'Energie Atomique in Paris to compute π to 16,167 decimal places. He used Machin's type formula (18). It took 4 hours and 30 minutes.

July 1961. Daniel Shanks (1917-1996) and William Shanks used Störmer's formula (34) on an IBM 7090 (at the IBM Data Processing Center, New York) to compute π to 100,265 digits, of which the first 100,000 digits were published by photographically reproducing the print-out with 5,000 digits per page. The time required for this computation was 8 hours and 43 minutes. They also checked the calculations by using Gauss' formula (31), which required 4 hours and 22 minutes.

1961. Machin's formula (18) was also the basis of a program run on an IBM 7090 at the London Data Center in July 1961, which resulted in 20,000 decimal places and required only 39 minutes running time.

Agarwal et al. *Advances in Difference Equations* 2013, **2013**:100
http://www.advancesindifferenceequations.com/content/2013/1/100

February 1966. Jean Guilloud and J. Filliatre used an IBM 7030 at the Commissariat á l'Energie Atomique in Paris to obtain an approximation of π extending to 250,000 decimal places on a STRETCH computer. For this, they used Störmer's and Gauss' formulas (34) and (31). It took almost 28 hours.

February 1967. Guilloud and M. Dichampt used CDC (Control Data Corporation) 6600 in Paris to approximate π to 500,000 decimal places. For this, they used Störmer's and Gauss' formulas (34) and (31). The computer that churned out half a million digits needed only 26 hours and 40 minutes (plus 1 hour and 30 minutes to convert that final result from binary to decimal notation).

1968. In the Putnam Competition, the first problem was

$$\pi = \frac{22}{7} - \int_0^1 \frac{x^4(1-x)^4}{1+x^2}\,dx.$$

This integral was known to Mahler in the mid-1960s, and has later appeared in several exams. It is also discussed by Borwein, Bailey, and Girgensohn in their book on p.3.

1971. K.Y. Choong, D.E. Daykin and C.R. Rathbone used 100,000 digits of Daniel Shanks and William Shanks (1961) to generate the first 21,230 partial quotients of the continued fraction expansion of π.

1974. Ralph William Gosper, Jr. (born 1943), known as Bill Gosper, is a mathematician and programmer. He is best known for the symbolic computation, continued fraction representations of real numbers, Gosper's algorithm, and Gosper curve. He used a refinement of Euler transform on (5) to obtain the series

$$\pi = 3 + \frac{1}{60}8 + \frac{1}{60}\frac{2 \cdot 3}{7 \cdot 8 \cdot 3}13 + \frac{1}{60}\frac{2 \cdot 3}{7 \cdot 8 \cdot 3}\frac{3 \cdot 5}{10 \cdot 11 \cdot 3}18$$
$$+ \frac{1}{60}\frac{2 \cdot 3}{7 \cdot 8 \cdot 3}\frac{3 \cdot 5}{10 \cdot 11 \cdot 3}\frac{4 \cdot 7}{13 \cdot 14 \cdot 3}23 + \cdots.$$

1974. Guilloud with Martine Bouyer (Paris) used formulas (34) and (31) on a CDC 7600 to compute π to 1,000,250 digits. The run time required for this computation was 23 hours and 18 minutes, of which 1 hour 7 minutes was used to convert the final result from binary to decimal. Results of statistical tests, which generally support the conjecture that π is *simply normal* (in 1909, Félix Édouard Justin Émil Borel (1871-1956) defined: A real number a is simply normal in base b if in its representation in base b all digits occur, in an asymptotic sense, equally often) were also performed.

1974. Louis Comtet developed the following Euler's type expansion of π:

$$\frac{\pi^4}{90} = \frac{36}{17}\sum_{m=1}^{\infty}\frac{1}{m^4\binom{2m}{m}}.$$

1976. Richard Brent and Eugene Salamin independently discovered an algorithm which is based on an arithmetic-geometric mean and modifies slightly Gauss-Legendre algorithm. Set $a_0 = 1$, $b_0 = 1/\sqrt{2}$ and $s_0 = 1/2$. For $k = 1, 2, 3, \ldots$ compute

$$a_k = \frac{a_{k-1} + b_{k-1}}{2},$$
$$b_k = \sqrt{a_{k-1}b_{k-1}},$$

Agarwal et al. *Advances in Difference Equations* 2013, **2013**:100
http://www.advancesindifferenceequations.com/content/2013/1/100

$$c_k = a_k^2 - b_k^2, \tag{38}$$

$$s_k = s_{k-1} - 2^k c_k,$$

$$p_k = \frac{2a_k^2}{s_k}.$$

Then p_k converges quadratically to π, *i.e.*, each iteration doubles the number of accurate digits. In fact, successive iterations must produce 1, 4, 9, 20, 42, 85, 173, 347 and 697 correct digits of π. The twenty-fifth iteration must produce 45 million correct decimal digits of π.

1981. Kazunori Miyoshi and Kazuhika Nakayama of the University of Tsukuba, Japan calculated π to 2,000,038 significant figures in 137.30 hours on a FACOM M-200 computer. They used Klingenstierna's formula (37) and checked their result with Machin's formula (18).

1981. Guilloud computed 2,000,050 decimal digits of π.

1981. Rajan Srinivasan Mahadevan (born 1957) recited from memory the first 31,811 digits of π. This secured him a place in the 1984 Guinness Book of World Records, and he has been featured on Larry King Live and Reader's Digest.

1982. Kikuo Takano (1927-2006) was a Japanese poet and mathematician. He developed the following Machin-like formula for calculating π:

$$\frac{\pi}{4} = 12\tan^{-1}\left(\frac{1}{49}\right) + 32\tan^{-1}\left(\frac{1}{57}\right) - 5\tan^{-1}\left(\frac{1}{239}\right) + 12\tan^{-1}\left(\frac{1}{110,443}\right). \tag{39}$$

1982. Yoshiaki Tamura on MELCOM 900II computed 2,097,144 decimal places of π. For this, he used the Salamin-Brent algorithm (38).

1982. Yoshiaki Tamura and Yasumasa Kanada (born 1948, life-long 'pi digit-hunter', set the record 11 of the past 21 times) on HITAC M-280H computed 4,194,288 decimal places of π. For this, they used the Salamin-Brent algorithm (38).

1982. Yoshiaki Tamura and Yasumasa Kanada on HITAC M-280H computed 8,388,576 decimal places of π. For this, they used the Salamin-Brent algorithm (38).

October 1983. Yasumasa Kanada, Yoshiaki Tamura, Sayaka Yoshino and Yasunori Ushiro on HITAC S-810/20 computed 10,013,395 decimal places of π. For this, they used the Salamin-Brent algorithm (38). In this work to gather evidence that π is simply normal, they also performed statistical analysis. It showed expected behavior. In the first ten million digits, the frequencies for each ten digits are 999,440; 999,333; 1,000,306; 999,964; 1,001,093; 1,000,466; 999,337; 1,000,207; 999,814; and 1,000,040. Further, the rate at which the relative frequencies approach 1/10 agrees with theory. As an example, for the digit 7 relative frequencies in the first 10^i, $i = 0, 1, 2, 3, 4, 5, 6, 7$ digits are $0, 0.08, 0.095, 0.097, 0.10025, 0.0998, 0.1000207$, which seem to be approaching 1/10 at rate predicted by the probability theory for random digits, *i.e.*, a speed approximately proportional to $1/\sqrt{n}$. But this is far from a formal proof of simple normalcy perhaps for a proof the current mathematics is not sufficiently developed. In spite of the fact that the digits of π pass statistical tests for randomness, π contains some sequences of digits that, to some, may appear non-random, such as Feynman point, which is a sequence of six consecutive 9s that begins at the 762nd decimal place. A number is said to be *normal* if all blocks of digits of the same length occur with equal frequency. Mathematicians expect π to be normal, so that every pattern possible eventually will occur in the digits of π.

Agarwal et al. *Advances in Difference Equations* 2013, **2013**:100
http://www.advancesindifferenceequations.com/content/2013/1/100

1983. Yasumasa Kanada, Sayaka Yoshino and Yoshiaki Tamura on HITAC M-280H computed 16,777,206 decimal places of π. For this, they used the Salamin-Brent algorithm (38).

1984. Jonathan Borwein and Peter Borwein gave the following algorithm. Set $x_0 = \sqrt{2}$, $y_0 = 0$ and $\alpha_0 = 2 + \sqrt{2}$. Iterate

$$x_{k+1} = (\sqrt{x_k} + 1/\sqrt{x_k})/2,$$

$$y_{k+1} = \sqrt{x_k}\left(\frac{1 + y_k}{y_k + x_k}\right), \tag{40}$$

$$\alpha_{k+1} = \alpha_k y_{k+1}\left(\frac{1 + x_{k+1}}{1 + y_{k+1}}\right).$$

Then α_k converges to π quartically. The algorithm is not self-correcting; each iteration must be performed with the desired number of correct digits of π.

1984. Morris Newman and Daniel Shanks proved the following: Set

$$a = \frac{1{,}071}{2} + 92\sqrt{34} + \frac{3}{2}\sqrt{255{,}349 + 43{,}792\sqrt{34}},$$

$$b = \frac{1{,}533}{2} + 133\sqrt{34} + \frac{1}{2}\sqrt{4{,}817{,}509 + 826{,}196\sqrt{34}},$$

$$c = 429 + 304\sqrt{2} + 2\sqrt{92{,}218 + 65{,}208\sqrt{2}},$$

$$d = \frac{627}{2} + 221\sqrt{2} + \frac{1}{2}\sqrt{783{,}853 + 554{,}268\sqrt{2}},$$

then

$$\left|\pi - \frac{6}{\sqrt{3{,}502}}\ln(2abcd)\right| < 7.4 \times 10^{-82}.$$

1985. Gosper used Symbolics 3,670 and Ramanujan's formula (35) to compute π to 17,526,200 decimal digits.

1985. Jonathan Borwein and Peter Borwein gave the following algorithm. Set $a_0 = 6 - 4\sqrt{2}$ and $y_0 = \sqrt{2} - 1$. Iterate

$$y_{k+1} = \frac{1 - (1 - y_k^4)^{1/4}}{1 + (1 - y_k^4)^{1/4}}, \tag{41}$$

$$a_{k+1} = a_k(1 + y_{k+1})^4 - 2^{2k+3}y_{k+1}\left(1 + y_{k+1} + y_{k+1}^2\right).$$

Then a_k converges quartically to $1/\pi$, *i.e.*, each iteration approximately quadruples the number of correct digits.

1985. The following is not an identity, but is correct to over 42 billion digits

$$\left(\frac{1}{10^5}\sum_{n=-\infty}^{\infty} e^{-n^2/10^{10}}\right)^2 \simeq \pi.$$

1985. Carl Sagan in his novel deals with the theme of contact between humanity and a more technologically advanced, extraterrestrial life form. He suggests that the creator of the universe buried a message deep within the digits of π.

Agarwal et al. *Advances in Difference Equations* 2013, **2013**:100
http://www.advancesindifferenceequations.com/content/2013/1/100

January 1986. David H. Bailey used Borweins' algorithms (40) and (41) on CRAY-2 to compute 29,360,111 decimal places of π.

September 1986. Yasumasa Kanada and Yoshiaki Tamura on HITAC S-810/20 computes 33,554,414 decimal places of π. For this, they used algorithms (38) and (41).

October 1986. Yasumasa Kanada and Yoshiaki Tamura on HITAC S-810/20 computed 67,108,839 decimal places of π. For this, they used algorithm (38).

January 1987. Yasumasa Kanada, Yoshiaki Tamura, Yoshinobu Kubo and others on NEC SX-2 computed 134,214,700 decimal places of π. For this they used algorithms (38) and (41).

1987. Jonathan Borwein and Peter Borwein gave the following algorithm. Set $x_0 = 2^{1/2}$, $y_1 = 2^{1/4}$ and $p_0 = 2 + 2^{1/2}$. Iterate

$$x_k = \frac{1}{2}\left(x_{k-1}^{1/2} + x_{k-1}^{-1/2}\right),$$

$$y_k = \frac{y_{k-1}x_{k-1}^{1/2} + x_{k-1}^{-1/2}}{y_{k-1} + 1},$$

$$p_k = p_{k-1}\frac{x_k + 1}{y_k + 1}.$$

Then p_k decreases monotonically to π and $|p_k - \pi| \leq 10^{-2^{k+1}}$ for $k \geq 4$.

1987. Hideaki Tomoyori (born 1932) recited π from memory to 40,000 places taking 17 hours 21 minutes, including breaks totaling 4 hours 15 minutes, at Tsukuba University Club House.

January 1988. Yasumasa Kanada on HITAC S-820/80 computed 201,326,551 decimal places of π. For this, he used algorithms (38) and (41).

1988. Jonathan Borwein and Peter Borwein developed the series

$$\frac{1}{\pi} = 12\sum_{n=0}^{\infty}\frac{(-1)^n(6n)!}{(n!)^3(3n)!}\frac{(A + Bn)}{C^{n+1/2}},$$

where

$$A = 212{,}175{,}710{,}912\sqrt{61} + 1{,}657{,}145{,}277{,}365,$$

$$B = 13{,}773{,}980{,}892{,}672\sqrt{61} + 107{,}578{,}229{,}802{,}750,$$

$$C = \left[5{,}280(236{,}674 + 30{,}303\sqrt{61})\right]^3.$$

Each additional term of the series adds roughly 31 digits.

1988. Dario Castellanos gave the following approximation:

$$\pi \simeq \left(\frac{77{,}729}{254}\right)^{1/5} = 3.1415926541\ldots.$$

May 1989. David Volfovich Chudnovsky (born 1947) and Gregory Volfovich Chudnovsky (born 1952) have published hundreds of research papers and books on number theory and mathematical physics. Gregory solved Hilbert's tenth problem at the age of 17. They on CRAY-2 and IBM 3090/VF computed 480,000,000 decimal places of π.

Agarwal et al. *Advances in Difference Equations* 2013, **2013**:100
http://www.advancesindifferenceequations.com/content/2013/1/100

June 1989. David and Gregory Chudnovsky on IBM 3090 computed 535,339,270 decimal places of π.

July 1989. Yasumasa Kanada and Yoshiaki Tamura on HITAC S-820/80 computed 536,870,898 decimal places of π. For this, they used algorithm (38).

August 1989. David and Gregory Chudnovsky developed the following rapidly convergent generalized hypergeometric series:

$$\frac{1}{\pi} = 12 \sum_{n=0}^{\infty} (-1)^n \frac{(6n)!}{(n!)^3(3n)!} \frac{13{,}591{,}409 + 545{,}140{,}134n}{(640{,}320^3)^{n+1/2}}. \tag{42}$$

Each additional term of the series adds roughly 15 digits. This series is an improved version to that of Ramanujan's (35). It was used by the Chudnovsky brothers to calculate more than one billion (to be exact 1,011,196,691) digits on IBM 3090.

November 1989. Yasumasa Kanada and Yoshiaki Tamura on HITAC S-820/80 computed 1,073,740,799 decimal places of π. For this, they used algorithms (38) and (41).

August 1991. David and Gregory Chudnovsky used a home made parallel computer (they called it m zero, where m stands for machine, and zero for the success) to obtain 2,260,000,000 decimal places of π. For this they used series (42).

1991. David Boll discovered an occurrence of π in the Mandelbrot set fractal.

1991. Jonathan Borwein and Peter Borwein improved on the Salamin-Brent algorithm (38). Set $a_0 = 1/3$ and $s_0 = (\sqrt{3}-1)/2$. Iterate

$$r_{k+1} = \frac{3}{1 + 2(1 - s_k^3)^{1/3}},$$

$$s_{k+1} = \frac{r_{k+1} - 1}{2},$$

$$a_{k+1} = r_{k+1}^2 a_k - 3^k (r_{k+1}^2 - 1).$$

Then $1/a_k$ converges cubically to π, *i.e.*, each iteration approximately triples the number of correct digits.

Among the several other known iterative schemes, we list the following two which are easy to implement on a computer: Set $a_0 = 1/2$ and $s_0 = 5(\sqrt{5} - 2)$. Iterate

$$x_{n+1} = 5/s_n - 1,$$

$$y_{n+1} = (x_{n+1} - 1)^2 + 7,$$

$$z_{n+1} = \left(\frac{1}{2} x_{n+1} \left(y_{n+1} + \sqrt{y_{n+1}^2 - 4x_{n+1}^3} \right) \right)^{1/5},$$

$$a_{n+1} = s_n^2 a_n - 5^n \left(\frac{s_n^2 - 5}{2} + \sqrt{s_n(s_n^2 - 2s_n + 5)} \right),$$

$$s_{n+1} = \frac{25}{(z_{n+1} + x_{n+1}/z_{n+1} + 1)^2 s_n}.$$

Then a_k converges quintically to $1/\pi$, *i.e.*, each iteration approximately quintuples the number of correct digits, and $0 < a_n - 1/\pi < 16 \cdot 5^n \cdot e^{-5^n \pi}$.

Agarwal et al. *Advances in Difference Equations* 2013, **2013**:100
http://www.advancesindifferenceequations.com/content/2013/1/100

Set $a_0 = 1/3$, $r_0 = (\sqrt{3} - 1)/2$ and $s_0 = (1 - r_0^3)^{1/3}$. Iterate

$$t_{n+1} = 1 + 2r_n,$$

$$u_{n+1} = \left(9r_n\left(1 + r_n + r_n^2\right)\right)^{1/3},$$

$$v_{n+1} = t_{n+1}^2 + t_{n+1}u_{n+1} + u_{n+1}^2,$$

$$w_{n+1} = \frac{27(1 + s_n + s_n^2)}{v_{n+1}},$$

$$a_{n+1} = w_{n+1}a_n + 3^{2n-1}(1 - w_{n+1}),$$

$$s_{n+1} = \frac{(1 - r_n)^3}{(t_{n+1} + 2u_{n+1})v_{n+1}},$$

$$r_{n+1} = \left(1 - s_{n+1}^3\right)^{1/3}.$$

Then a_k converges nonically to $1/\pi$, *i.e.*, each iteration approximately multiplies the number of correct digits by nine.

1993. Jonathan Borwein and Peter Borwein developed the series

$$\frac{\sqrt{-C^3}}{\pi} = \sum_{m=0}^{\infty} \frac{(6m)!}{(3m)!(m!)^3} \frac{A + mB}{C^{3m}},$$

where

$$A = 63{,}365{,}028{,}312{,}971{,}999{,}585{,}426{,}220$$
$$+ 28{,}337{,}702{,}140{,}800{,}842{,}046{,}825{,}600\sqrt{5}$$
$$+ 384\sqrt{5}(10{,}891{,}728{,}551{,}171{,}178{,}200{,}467{,}436{,}212{,}395{,}209{,}160{,}385{,}656{,}017}$$
$$+ 4{,}870{,}929{,}086{,}578{,}810{,}225{,}077{,}338{,}534{,}541{,}688{,}721{,}351{,}255{,}040\sqrt{5})^{1/2},$$

$$B = 7{,}849{,}910{,}453{,}496{,}627{,}210{,}289{,}749{,}000$$
$$+ 3{,}510{,}586{,}678{,}260{,}932{,}028{,}965{,}606{,}400\sqrt{5}$$
$$+ 2{,}515{,}968\sqrt{3{,}110}(6{,}260{,}208{,}323{,}789{,}001{,}636{,}993{,}322{,}654{,}444{,}020{,}882{,}161}$$
$$+ 2{,}799{,}650{,}273{,}060{,}444{,}296{,}577{,}206{,}890{,}718{,}825{,}190{,}235\sqrt{5})^{1/2}$$

and

$$C = -214{,}772{,}995{,}063{,}512{,}240 - 96{,}049{,}403{,}338{,}648{,}032\sqrt{5}$$
$$- 1{,}296\sqrt{5}(10{,}985{,}234{,}579{,}463{,}550{,}323{,}713{,}318{,}473}$$
$$+ 4{,}912{,}746{,}253{,}692{,}362{,}754{,}607{,}395{,}912\sqrt{5})^{1/2}.$$

Each additional term of the series adds approximately 50 digits. However, computation of this series on a computer does not seem to be easy.

May 1994. David and Gregory Chudnovsky used a home made parallel computer *m* zero to obtain 4,044,000,000 decimal places of π. For this they used series (42).

Agarwal et al. *Advances in Difference Equations* 2013, **2013**:100
http://www.advancesindifferenceequations.com/content/2013/1/100

June 1995. Yasumasa Kanada and Daisuke Takahashi on HITAC S-3800/480 (dual CPU) computed 3,221,225,466 decimal places of π. For this, they used algorithms (38) and (41).

August 1995. Yasumasa Kanada and Daisuke Takahashi on HITAC S-3800/480 (dual CPU) computed 4,294,967,286 decimal places of π. For this, they used algorithms (38) and (41).

October 1995. Yasumasa Kanada and Daisuke Takahashi on HITAC S-3800/480 (dual CPU) computed 6,442,450,938 decimal places of π. For this, they used algorithms (38) and (41).

1995. David Bailey, Peter Borwein and Simon Plouffe developed the following formula (known as BBP formula) to compute the nth hexadecimal digit (base 16) of π without having the previous $n-1$ digits

$$\pi = \sum_{m=0}^{\infty} \frac{1}{16^m}\left(\frac{4}{8m+1} - \frac{2}{8m+4} - \frac{1}{8m+5} - \frac{1}{8m+6}\right). \tag{43}$$

To show the validity of (43), for any $k < 8$, we have

$$\int_0^{1/\sqrt{2}} \frac{x^{k-1}}{1-x^8}\,dx = \int_0^{1/\sqrt{2}} \sum_{m=0}^{\infty} x^{k-1+8m}\,dx = \frac{1}{2^{k/2}}\sum_{m=0}^{\infty}\frac{1}{16^m(8m+k)},$$

therefore

$$\sum_{m=0}^{\infty} \frac{1}{16^m}\left(\frac{4}{8m+1} - \frac{2}{8m+4} - \frac{1}{8m+5} - \frac{1}{8m+6}\right)$$

$$= \int_0^{1/\sqrt{2}} \frac{4\sqrt{2} - 8x^3 - 4\sqrt{2}x^4 - 8x^5}{1-x^8}\,dx. \tag{44}$$

Substituting $u = \sqrt{2}x$ in equation (44), we obtain

$$\int_0^1 \frac{16u-16}{u^4 - 2u^3 + 4u - 4}\,du = \int_0^1 \frac{4u}{u^2-2}\,du - \int_0^1 \frac{4u-8}{u^2-2u+2}\,du = \pi.$$

The discovery of this formula came as a surprise. For centuries, it had been assumed that there was no way to compute the nth digit of π without calculating all of the preceding $n-1$ digits. Since this discovery, many such formulas for other irrational numbers have been discovered. Such formulas have been called as *spigot algorithms* because, like water dripping from a spigot, they produce digits that are not reused after they are calculated.

1996. Simon Plouffe discovered an algorithm for the computation of π in any base. Later he expressed regrets for having shared credit for his discovery of this formula with Bailey and Borwein.

March 1996. David and Gregory Chudnovsky used a home made parallel computer m zero to obtain 8,000,000,000 decimal places of π. For this, they used series (42). They said 'we are looking for the appearance of some rules that will distinguish the digits of π from other numbers, *i.e.*, if someone were to give you a million digits from somewhere in π, could you tell it was from π? The digits of π form the most nearly perfect random sequence of digits that has ever been discovered. However, each digit appears to be orderly. If a single

Agarwal et al. *Advances in Difference Equations* 2013, **2013**:100
http://www.advancesindifferenceequations.com/content/2013/1/100

digit in π were to be changed anywhere between here and infinity, the resulting number would no longer be π, it would be garbage. Around the three-hundred-millionth decimal place of π, the digits go 88888888-eight eights pop up in a row. Does this mean anything? It appears to be random noise. Later, ten sixes erupt: 6666666666. What does this mean? Apparently nothing, only more noise. Somewhere past the half-million mark appears the string 123456789. It is an accident, as it were. We do not have a good, clear, crystallized idea of randomness. It cannot be that π is truly random. Actually, a truly random sequence of numbers has not yet been discovered.'

1996. Gosper posted the following fascinating formula

$$\lim_{n\to\infty} \prod_{m=n}^{2n} \frac{\pi}{2\tan^{-1} m} = 4^{1/\pi} = 1.554682\ldots.$$

April 1997. Yasumasa Kanada and Daisuke Takahashi on HITACHI SR2201 (1,024 CPU) computed 17,179,869,142 decimal places of π. For this, they used algorithms (38) and (41).

July 1997. Yasumasa Kanada and Daisuke Takahashi on HITACHI SR2201 (1,024 CPU) computed 51,539,600,000 decimal places of π. The computation tool just over 29 hours, at an average rate of nearly 500,000 digits per second. For this, they used algorithms (38) and (41).

1997. Fabrice Bellard developed the following formula:

$$\pi = \frac{1}{2^6} \sum_{m=0}^{\infty} \frac{(-1)^m}{2^{10m}}$$

$$\times \left(-\frac{2^5}{4m+1} - \frac{1}{4m+3} + \frac{2^8}{10m+1} - \frac{2^6}{10m+3} - \frac{2^2}{10m+5} - \frac{2^2}{10m+7} + \frac{1}{10m+9} \right),$$

which can used to compute the nth digit of π in base 2. It is about 43% faster then (43). The following exotic formula is also due to him:

$$\pi = \frac{1}{740,025} \left[\sum_{m=1}^{\infty} \frac{3P(m)}{\binom{7m}{2m} 2^{m-1}} - 20,379,280 \right],$$

where

$$P(m) = -885,673,181m^5 + 3,125,347,237m^4 - 2,942,969,225m^3$$

$$+ 1,031,962,795m^2 - 196,882,274m + 10,996,648.$$

April 1999. Yasumasa Kanada and Daisuke Takahashi on HITACHI SR8000 (64 of 128 nodes) computed 68,719,470,000 decimal places of π. For this, they used algorithms (38) and (41).

September 1999. Yasumasa Kanada and Daisuke Takahashi on HITACHI SR8000/MPP (128 nodes) computed 206,158,430,000, *i.e.*, 206 billion decimal places of π. For this, they used algorithms (38) and (41).

Agarwal et al. *Advances in Difference Equations* 2013, **2013**:100
http://www.advancesindifferenceequations.com/content/2013/1/100

1999. Leo Jerome Lange developed the following continued fraction of π:

$$\pi = 3 + \cfrac{1^2}{6 + \cfrac{3^2}{6 + \cfrac{5^2}{6 + \cfrac{7^2}{6 + \cdots}}}}.$$

2000. J. Munkhammar gave the following formula which is related to Viéte's (6):

$$\pi = \lim_{n \to \infty} 2^{n+1} \sqrt{2 - \frac{\sqrt{2 + \sqrt{2 + \sqrt{2 + \cdots + \sqrt{2}}}}}{n}},$$

which as a recurrence relation can be written as $\pi = \lim_{n \to \infty} 2^{n+1} a_n$, where $a_0 = \sqrt{2}$, and

$$a_n = \sqrt{\left(\frac{1}{2}a_{n-1}\right)^2 + \left[1 - \sqrt{1 - \left(\frac{1}{2}a_{n-1}\right)^2}\right]^2}.$$

Another closely related formula is

$$\pi = 2 \lim_{n \to \infty} \sum_{m=1}^{n} \sqrt{\left[\sqrt{1 - \left(\frac{m-1}{n}\right)^2} - \sqrt{1 - \left(\frac{m}{n}\right)^2}\right]^2 + \frac{1}{n^2}}.$$

2001. Robert Palais believes that the notation π is wrongly used right from the beginning. According to him, some suitable symbol (now popular as tau τ) must have been used for 2π. He justifies his claim by giving several formulas where τ appears naturally rather than just π. For some people, June 28, is *Tau's Day* and they celebrate.

November 2002. Yasumasa Kanada used Machin-like formulas (33) and (39) to compute the value of π to 1,241,100,000,000 decimal places. The calculation took more than 600 hours on 64 nodes of a HITACHI SR8000/MPP supercomputer. The work was done at the Department of Information Science at the University of Tokyo. For this, he used arctangent formulas (33) and (39).

2004. Daniel Tammet, at age 25, recited 22,514 decimal places of π, scoring the European record. For an audience at the Museum of the History of Science in Oxford, he said these numbers aloud for 5 hours and 9 minutes. Unfortunately, he made his first mistake at position 2,965 and did not correct this error immediately and without outside help, but only after he was told that there was a mistake.

2005. Stephen K. Lucas found that

$$\pi = \frac{355}{113} - \int_0^1 \frac{x^8(1-x)^8(25+816x^2)}{3,164(1+x^2)} \, dx.$$

Several other integral formulas of this type are known, here we give the following:

$$\pi = \frac{741,269,838,109}{235,953,517,800} - \int_0^1 \frac{x^{16}(1-x)^{16}}{64(1+x^2)} \, dx,$$

Agarwal et al. *Advances in Difference Equations* 2013, **2013**:100
http://www.advancesindifferenceequations.com/content/2013/1/100

which gives $3.14159265358955 < \pi$. If we substitute $x = 1$ in the above integral and note that

$$\int_0^1 \frac{1}{128} x^{16}(1-x)^{16}\, dx = \frac{1}{2{,}538{,}963{,}567{,}360}$$

then it follows that $\pi < 3.14159265358996$.

November 2005. Chao Lu, a chemistry student, at age 23, broke the Guiness record by reciting π from memory to 67,890 places. For this, he practiced for 4 years. The attempt lasted 24 hrs 4 min and was recorded on 26 video tapes. The attempt was witnessed by 8 officials, math professors and 20 students.

2005. Kate Bush in the song π (in her album Aerial) sings the number to its 137th decimal place (though she omits the 79th to 100th decimal places).

October 2006. Akira Haraguchi a retired engineer from Chiba recited π from memory to 54,000 digits in September 2004, 68,000 digits in December 2004, 83,431 digits in July 2005, and 100,000 digits in October 2006. He accomplished the last recitation in 16 1/2-hours in Tokyo. He says memorization of the digits of π is 'the religion of the universe'.

2006. Simon Plouffe found the following curious formula:

$$\pi = 72 \sum_{k=1}^{\infty} \frac{1}{k(e^{k\pi}-1)} - 96 \sum_{k=1}^{\infty} \frac{1}{k(e^{2k\pi}-1)} + 24 \sum_{k=1}^{\infty} \frac{1}{k(e^{4k\pi}-1)}.$$

2008. In Midnight (tenth episode of the fourth series of British science fiction television series Doctor Who), the character, the businesswoman, Sky Silvestry mimics the speech of The Doctor by repeating the square root of π to 30 decimal places 1.772453850905516027298167483334.

2008. Syamal K. Sen and Ravi P. Agarwal suggested four Matlab based procedures, *viz*, (i) Exhaustive search, (ii) Principal convergents of continued fraction based procedure, (iii) Best rounding procedure for decimal (rational) approximation, and (iv) Continued fraction based algorithm with intermediate convergents. While the first procedure is exponential-time, the remaining three are polynomial-time. Roughly speaking, they have demonstrated that the absolute best k-digit rational approximation of π will be as good as $2k$-digit decimal approximation of π. The absolute best k-digit rational approximation is most desired for error-free computation involving π/any other irrational number.

2008. Syamal K. Sen, Ravi P. Agarwal and Ghoolam A. Shaykhian have demonstrated through numerical experiment using Matlab that π has always scored over ϕ (golden ratio), as a source of uniformly distributed random numbers, statistically in one-dimensional Monte Carlo (M.C.) integration; whether π fares better than ϕ for double, triple and higher dimensional M.C. integration or not deserves exploration.

2009. Syamal K. Sen, Ravi P. Agarwal and Ghoolam A. Shaykhian compared the four procedures they proposed in (2008) for computing best k-digit rational approximations of irrational numbers in terms of quality (error) and cost (complexity). They have stressed on the fact that ultra-high-speed computing along with abundance of unused computing power allows employing an exponential-time algorithm for most real-world problems. This obviates the need for acquiring and employing the mathematical knowledge involving principal and intermediate convergents computed using a polynomial-time algorithm for practical problems. Since π is the most used irrational number in the physical world, the

Agarwal et al. *Advances in Difference Equations* 2013, **2013**:100
http://www.advancesindifferenceequations.com/content/2013/1/100

simple concise Matlab program would do the job wherever π/any other irrational number is involved.

2009. Syamal K. Sen, Ravi P. Agarwal and Raffela Pavani have provided, using Matlab, the best possible rational bounds bracketing π/any irrational number with absolute error and the time complexity involved. Any better bounds are impossible. In these rational bounds, either the lower bound or the upper bound will always be the absolute best rational approximation. The absolute error computed provides the overall error bounds in an error-free computational environment involving π/any other irrational number.

2009. Tue N. Vu has given Machin-type formula (http://seriesmathstudy.com/sms/machintypetv): For each positive integer n,

$$\frac{\pi}{4} = \tan^{-1}\left(\frac{1}{4+2n}\right) + \tan^{-1}\left(\frac{1}{5+2n}\right) + \sum_{k=0}^{n}\left[\tan^{-1}\left(\frac{1}{2(2+k)^2}\right) + \tan^{-1}\left(\frac{2}{(3+2k)^2}\right)\right].$$

2009. Cetin Hakimoglu-Brown developed the following expansion:

$$\pi = \frac{\sqrt{3}}{6^5}\sum_{k=0}^{\infty}\frac{((4k)!)^2(6k)!}{9^{k+1}(12k)!(2k)!}\left(\frac{127{,}169}{12k+1} - \frac{1{,}070}{12k+5} - \frac{131}{12k+7} + \frac{2}{12k+11}\right),$$

which can be written as

$$\pi = \frac{\sqrt{3}}{1{,}155}\sum_{k=0}^{\infty}\frac{(4k)!(671{,}840k^3 + 1{,}289{,}936k^2 + 782{,}458k + 150{,}835)}{(72)^{4k+1}(13/12)_k(17/12)_k(19/12)_k(23/12)_k},$$

where $(x)_k = x(x+1)(x+2)\cdots(x+k-1)$ is the Pochhammer notation. He also gave the expansion

$$\pi = \frac{1}{2^{10}\sqrt{3}}\sum_{k=0}^{\infty}\frac{1}{\binom{8k}{4k}9^k}\left(\frac{5{,}717}{8k+1} - \frac{413}{8k+3} - \frac{45}{8k+5} + \frac{5}{8k+7}\right).$$

August 2009. Daisuke Takahashi *et al.* used a massive parallel computer called the T2K Tsukuba System to compute π to 2,576,980,377,524 decimal places in 73 hours 36 minutes. For this, they used algorithms (38) and (41).

December 2009. Fabrice Bellard used Chudnovsky brothers series (42) to compute 2,699,999,990,000, *i.e.*, 2.7 trillion decimal places of π in 131 days. For this, he used a single desktop PC, costing less than \$3,000.

August 2010. Shigeru Kondo and Alexander J. Yee used Chudnovsky brothers series (42) to compute 5,000,000,000,000, *i.e.*, 5 trillion decimal places of π in 90 days. For this, they used a server-class machine running dual Intel Xeons, equipped with 96 GB of RAM.

2010. Michael Keith used 10,000 digits of π to establish a new form of constrained writing, where the word lengths are required to represent the digits of π. His book contains a collection of poetry, short stories, a play, a movie script, crossword puzzles and other surprises.

2011. Syamal K. Sen and Ravi P. Agarwal in their monograph systematically organized their work of 2008 and 2009 on π and other irrational numbers. They also included several examples to illustrate the importance of their findings.

Agarwal et al. *Advances in Difference Equations* 2013, **2013**:100
http://www.advancesindifferenceequations.com/content/2013/1/100

2011. During the auction for Nortel's portfolio of valuable technology patents, Google made a series of strange bids based on mathematical and scientific constants, including π.

October 2011. Shigeru Kondo and Alexander J. Yee used Chudnovsky brothers series (42) to compute 10,000,000,000,050, *i.e.*, 10 trillion decimal places of π in 371 days.

2011. Cristinel Mortici improved Gurland's bounds (36) to $\alpha_n < \pi < \beta_n$, $n \geq 1$ where

$$\alpha_n = \left(\frac{n + \frac{1}{4}}{n^2 + \frac{1}{2}n + \frac{3}{32}} + \frac{9}{2{,}048n^5} - \frac{45}{8{,}192n^6} \right) \left(\frac{(2n)!!}{(2n-1)!!} \right)^2$$

and

$$\beta_n = \left(\frac{n + \frac{1}{4}}{n^2 + \frac{1}{2}n + \frac{3}{32}} + \frac{9}{2{,}048n^5} \right) \left(\frac{(2n)!!}{(2n-1)!!} \right)^2 .$$

It follows that

$$\alpha_n = \pi + O\left(\frac{1}{n^6} \right) \quad \text{and} \quad \beta_n = \pi + O\left(\frac{1}{n^5} \right).$$

2012. Long Lin has improved Mortici's bounds to $\lambda_n < \pi < \mu_n$, $n \geq 1$ where

$$\lambda_n = \left(1 + \frac{1}{4n} - \frac{3}{32n^2} + \frac{3}{128n^3} + \frac{3}{2{,}048n^4} - \frac{33}{8{,}192n^5} - \frac{39}{65{,}536n^6} \right) \frac{2}{2n+1} \left(\frac{(2n)!!}{(2n-1)!!} \right)^2$$

and

$$\mu_n = \left(1 + \frac{1}{4n} - \frac{3}{32n^2} + \frac{3}{128n^3} + \frac{3}{2{,}048n^4} \right) \frac{2}{2n+1} \left(\frac{(2n)!!}{(2n-1)!!} \right)^2 .$$

It follows that

$$\lambda_n = \pi + O\left(\frac{1}{n^7} \right) \quad \text{and} \quad \mu_n = \pi + O\left(\frac{1}{n^5} \right).$$

He has also obtained the higher order bounds $\delta_n < \pi < \omega_n$, $n \geq 1$ where

$$\delta_n = \frac{1}{n} \exp\left\{ -\frac{1}{4n} + \frac{1}{96n^3} - \frac{1}{320n^5} + \frac{17}{7{,}168n^7} - \frac{31}{9{,}216n^9} \right\} \left(\frac{(2n)!!}{(2n-1)!!} \right)^2$$

and

$$\mu_n = \frac{1}{n} \exp\left\{ -\frac{1}{4n} + \frac{1}{96n^3} - \frac{1}{320n^5} + \frac{17}{7{,}168n^7} \right\} \left(\frac{(2n)!!}{(2n-1)!!} \right)^2 .$$

It follows that

$$\delta_n = \pi + O\left(\frac{1}{n^{11}} \right) \quad \text{and} \quad \mu_n = \pi + O\left(\frac{1}{n^9} \right).$$

Agarwal et al. *Advances in Difference Equations* 2013, **2013**:100
http://www.advancesindifferenceequations.com/content/2013/1/100

Conclusions

No number system can capture π exactly. We are deeply and almost completely involved in the conventional decimal number system in representing any real quantity. This is not the only number system for the representation. There are other number systems such as binary, octal, hexadecimal, binary-coded decimal, negative radix, p-adic and modular number systems. If the circumference of a circle is exactly represented, then its diameter will not have exact representation and *vice versa*.

Reading the mathematicians in pre-computer days. An important focus of this paper is that the reader besides, however, knowing the usual chronology of the events in the life of π, could get a feel and also read how the mind of a mathematician has been working when he ponders over π either independently without much knowledge/concern of what has been done in the past or with considerable knowledge of the work done by his predecessors. Hyper-computers (10^{18} flops) of 2012 were completely non-existence and even beyond the imagination of all the mathematicians/scientists until almost the mid-twentieth century. Also, publication machinery was too poor until the beginning of the twentieth century. Consequently, all the work on π that has been carried out during thousands of years prior to the twentieth century was not a monotonic improvement in the π value as well as in the exploration of its wonderful character. Many have worked on π stand-alone while others have contributed with some prior knowledge of the earlier work. All of them were severely handicapped due to the non-existence of today's ultra-high speed computers. They entirely depended on their ingenuity and on whatever negligible computing device they had. It is really interesting under this environment to read these scientists/mathematicians and realize how fortunate we are in the gigantic computer age. All that has been done during the last 20 years (1990-2010) amounts to much more than what has been achieved during the past several millennia.

Matlab is well-suited to check/evaluate merits of all past π formulas. Widely used user-friendly Matlab that needs no formal programming knowledge along with the vpa (variable precision arithmetic) and *format long g* commands can be used to easily and readily check all that has been done during the past several thousand years and possibly appreciate the inherent intellectual import of the bygone scientists (having practically no computing device) and their expected pitfalls, bias and incorrect beliefs.

Checking exactness of billions of digits of π is difficult. Are all the billions of digit of π computed 100% error-free? We are familiar with the age-old proverb that 'To err is human (living being)'. Maybe a new proverb 'Not to err is computer (non-living being)' can be taken as true in the modern computer age. Here, 'err' means mistake. The arithmetic operations, particularly subtraction operations of two nearly equal numbers, involved in a formula could be sometimes error introducer. However, different computers with different formulas used to compute π would help verification and obviate possible error in computation.

Computing nth decimal digit exactly always without preceding digits seems yet an open computational problem. While probabilistically one may determine the nth digit of π without computing the preceding $n-1$ digits, obtaining nth digit exactly (correctly) always for any n does not seem to be possible without a large precision. It seems yet an open computational problem that needs exploration. Thus, formulas such as (43) seem more of theoretical/academic interest than of practical usage as of now.

Agarwal et al. *Advances in Difference Equations* 2013, **2013**:100
http://www.advancesindifferenceequations.com/content/2013/1/100

PI for testing performance and stability of a computer. Super PI is a computer program that calculates π to a specified number of digits after decimal point up to a maximum of 32 million digits. It uses the Gauss-Legendre algorithm and is a Windows port of the program used by Yasumasa Kanada in 1995 to compute π to 2^{32} digits. Super PI is used by many overclockers to test the performance and stability of their computers. Overclocking is the process of making a computer run faster than the clock frequency specified by the manufacturer by modifying system parameters.

Competing interests
The authors declare that they have no competing interests.

Authors' contributions
All authors contributed equally and read and approved the final manuscript.

Author details
[1]Department of Mathematics, Texas A&M University-Kingsville, Kingsville, TX, 78363, USA. [2]1540 Ravena Street, Bethlehem, PA 18015, USA. [3]GVP-Prof. V. Lakshmikantham Institute for Advanced Studies, #1-83-21/3, Sector 8, MVP Colony, Visakhapatnam, A.P., India.

Acknowledgements
Dedicated to V Lakshmikantham (1924-2012).

Received: 22 January 2013 Accepted: 25 March 2013 Published: 11 April 2013

References
1. Adamchik, V, Wagon, S: A simple formula for π. Am. Math. Mon. **104**, 852-855 (1997)
2. Adamchik, V, Wagon, S: Pi: a 2000-year search changes direction. Educ. Res. **5**, 11-19 (1996)
3. Ahmad, A: On the π of Aryabhata I. Ganita Bharati **3**, 83-85 (1981)
4. Akira, H: History of π. Kyoiku Tosho, Osaka (1980)
5. Al-Kashi, J: Treatise on the Circumference of the Circle (1424)
6. Almkvist, G: Many correct digits of π, revisited. Am. Math. Mon. **104**, 351-353 (1997)
7. Anderson, DV: A polynomial for π. Math. Gaz. **55**, 67-68 (1971)
8. Anonymous: Cyclometry and Circle-Squaring in a Nutshell. Simpkin, Marshall & Co., Stationer's Hall Court, London (1871)
9. Arndt, J: Cryptic Pi related formulas. http://www.jjj.de/hfloat/pise.dvi
10. Arndt, J, Haenel, C: π-Unleashed. Springer, Berlin (2000)
11. Assmus, EF: Pi. Am. Math. Mon. **92**, 213-214 (1985)
12. Backhouse, N: Note 79.36. Pancake functions and approximations to π. Math. Gaz. **79**, 371-374 (1995)
13. Badger, L: Lazzarini's lucky approximation of π. Math. Mag. **67**, 83-91 (1994)
14. Bai, S: An exploration of Liu Xin's value of π from Wang Mang's measuring vessel. Sugaku-shi Kenkyu **116**, 24-31 (1988)
15. Bailey, DH: Numerical results on the transcendence of constants involving π, e, and Euler's constant. Math. Comput. **50**, 275-281 (1988)
16. Bailey, DH: The computation of π to 29,360,000 decimal digits using Borweins' quartically convergent algorithm. Math. Comput. **50**, 283-296 (1988)
17. Bailey, DH, Borwein, JM, Borwein, PB, Plouffe, S: The quest for pi. Math. Intell. **19**, 50-57 (1997)
18. Bailey, DH, Borwein, PB, Plouffe, S: On the rapid computation of various polylogarithmic constants. Math. Comput. **66**, 903-913 (1997)
19. Beck, G, Trott, M: Calculating Pi from antiquity to modern times. http://library.wolfram.com/infocenter/Demos/107/
20. Beckmann, P: A History of π. St Martin's, New York (1971)
21. Bellard, F: Fabrice Bellard's, Pi page. http://bellard.org/pi/
22. Berggren, L, Borwein, JM, Borwein, PB: Pi: A Source Book, 3rd edn. Springer, New York (2004)
23. Beukers, F: A rational approximation to π. Nieuw Arch. Wiskd. **5**, 372-379 (2000)
24. Blatner, D: The Joy of Π. Penguin, Toronto (1997)
25. Bokhari, N: Piece of Pi. Dandy Lion, San Luis Obispo (2001)
26. Boll, D: Pi and the Mandelbrot set. http://www.pi314.net/eng/mandelbrot.php
27. Borwein, JM, Bailey, DH, Girgensohn, R: Experimentation in Mathematics: Computational Paths to Discovery. AK Peters, Wellesley (2004)
28. Borwein, JM, Borwein, PB: A very rapidly convergent product expansion for π. BIT Numer. Math. **23**, 538-540 (1983)
29. Borwein, JM, Borwein, PB: Cubic and higher order algorithms for π. Can. Math. Bull. **27**, 436-443 (1984)
30. Borwein, JM, Borwein, PB: Explicit algebraic nth order approximations to π. In: Singh, SP, Burry, JHW, Watson, B (eds.) Approximation Theory and Spline Functions, pp. 247-256. Reidel, Dordrecht (1984)
31. Borwein, JM, Borwein, PB: The arithmetic-geometric mean and fast computation of elementary functions. SIAM Rev. **26**, 351-365 (1984)
32. Borwein, JM, Borwein, PB: An explicit cubic iteration for π. BIT Numer. Math. **26**, 123-126 (1986)
33. Borwein, JM, Borwein, PB: More quadratically converging algorithms for π. Math. Comput. **46**, 247-253 (1986)
34. Borwein, JM, Borwein, PB: Pi and the AGM - A Study in Analytic Number Theory and Computational Complexity. Wiley-Interscience, New York (1987)

Agarwal et al. *Advances in Difference Equations* 2013, **2013**:100
http://www.advancesindifferenceequations.com/content/2013/1/100

35. Borwein, JM, Borwein, PB: Explicit Ramanujan-type approximations to π of high order. Proc. Indian Acad. Sci. Math. Sci. **97**, 53-59 (1987)
36. Borwein, JM, Borwein, PB: Ramanujan's rational and algebraic series for $1/\pi$. J. Indian Math. Soc. **51**, 147-160 (1987)
37. Borwein, JM, Borwein, PB: Ramanujan and π. Sci. Am. **258**, 112-117 (1988)
38. Borwein, JM, Borwein, PB: More Ramanujan-type series for $1/\pi$. In: Ramanujan Revisited, pp. 359-374. Academic Press, Boston (1988)
39. Borwein, JM, Borwein, PB: Approximating π with Ramanujan's modular equations. Rocky Mt. J. Math. **19**, 93-102 (1989)
40. Borwein, JM, Borwein, PB: Class number three Ramanujan type series for $1/\pi$. J. Comput. Appl. Math. **46**, 281-290 (1993)
41. Borwein, JM, Borwein, PB, Bailey, DH: Ramanujan, modular equations, and approximations to π, or how to compute one billion digits of π. Am. Math. Mon. **96**, 201-219 (1989)
42. Borwein, JM, Borwein, PB, Dilcher, K: Pi, Euler numbers, and asymptotic expansions. Am. Math. Mon. **96**, 681-687 (1989)
43. Borwein, JM, Borwein, PB, Garvan, F: Hypergeometric analogues of the arithmetic-geometric mean iteration. Constr. Approx. **9**, 509-523 (1993)
44. Borwein, PM: The amazing number II. Nieuw Arch. Wiskd. **1**, 254-258 (2000)
45. Brent, RP: The complexity of multiple-precision arithmetic. In: Andressen, RS, Brent, RP (eds.) Complexity of Computational Problem Solving. University of Queensland Press, Brisbane (1976)
46. Brent, RP: Fast multiple-precision evaluation of elementary functions. J. ACM **23**, 242-251 (1976)
47. Breuer, S, Zwas, G: Mathematical-educational aspects of the computation of π. Int. J. Math. Educ. Sci. Technol. **15**, 231-244 (1984)
48. Brown, CH: An algorithm for the derivation of rapidly converging infinite series for universal mathematical constants. Preprint (2009)
49. Bruins, EM: With roots towards Aryabhata's π-value. Ganita Bharati **5**, 1-7 (1983)
50. Carlson, BC: Algorithms involving arithmetic and geometric means. Am. Math. Mon. **78**, 496-505 (1971)
51. Castellanos, D: The ubiquitous pi, part I. Math. Mag. **61**, 67-98 (1988)
52. Castellanos, D: The ubiquitous pi, part II. Math. Mag. **61**, 148-163 (1988)
53. Chan, J: As easy as Pi. Math Horizons, Winter 1993, 18-19
54. Choong, KY, Daykin, DE, Rathbone, CR: Rational approximations to π. Math. Comput. **25**, 387-392 (1971)
55. Choong, KY, Daykin, DE, Rathbone, CR: Regular continued fractions for π and γ. Math. Comput. **25**, 403 (1971)
56. Chudnovsky, DV, Chudnovsky, GV: Approximations and complex multiplication according to Ramanujan. In: Ramanujan Revisited, pp. 375-396 & 468-472. Academic Press, Boston, (1988)
57. Chudnovsky, DV, Chudnovsky, GV: The computation of classical constants. Proc. Natl. Acad. Sci. USA **86**, 8178-8182 (1989)
58. Cohen, GL, Shannon, AG: John Ward's method for the calculation of π. Hist. Math. **8**, 133-144 (1981)
59. Colzani, L: La quadratura del cerchio e dell'iperbole (The squaring of the circle and hyperbola). Universitá degli studi di Milano-Bicocca, Matematica, Milano, Italy
60. Cox, DA: The arithmetic-geometric mean of Gauss. Enseign. Math. **30**, 275-330 (1984)
61. Dahse, Z: Der Kreis-Umfang für den Durchmesser 1 auf 200 Decimalstellen berechnet. J. Reine Angew. Math. **27**, 198 (1944)
62. Dalzell, DP: On 22/7. J. Lond. Math. Soc. **19**, 133-134 (1944)
63. Dalzell, DP: On 22/7 and 355/113. Eureka Archimed. J. **34**, 10-13 (1971)
64. Datta, B: Hindu values of π. J. Asiat. Soc. Bengal **22**, 25-42 (1926)
65. Davis, PJ: The Lore of Large Numbers. New Mathematical Library, vol. 6. Math. Assoc. of America, Washington (1961)
66. Delahaye, JP: Le Fascinant Nombre π. Bibliothéque Pour la Science, Belin (1997)
67. Dixon, R: The story of pi (π). In: Mathographics. Dover, New York (1991)
68. Engels, H: Quadrature of the circle in ancient Egypt. Hist. Math. **4**, 137-140 (1977)
69. Eymard, P, Lafon, JP: The Number Pi. Am. Math. Soc., Providence (1999) (Translated by S.S. Wilson)
70. Ferguson, DF: Evaluation of π. Are Shanks' figures correct? Math. Gaz. **30**, 89-90 (1946)
71. Ferguson, DF: Value of π. Nature **17**, 342 (1946)
72. Finch, SR: Mathematical Constants. Cambridge University Press, Cambridge (2003)
73. Frisby, E: On the calculation of *pi*. Messenger Math. **2**, 114 (1872)
74. Fox, L, Hayes, L: A further helping of π. Math. Gaz. **59**, 38-40 (1975)
75. Fuller, R: Circle and Square. Springfield Printing and Binding Co., Springfield (1908)
76. Genuys, F: Dix milles décimales de π. Chiffres **1**, 17-22 (1958)
77. Goggins, JR: Formula for $\pi/4$. Math. Gaz. **57**, 134 (1973)
78. Goldsmith, C: Calculation of ln 2 and π. Math. Gaz. **55**, 434-436 (1971)
79. Goodrich, LC: Measurements of the circle in ancient China. Isis **39**, 64-65 (1948)
80. Gosper, RW: Acceleration of series. Memo no. 304., M.I.T., Artificial Intelligence Laboratory, Cambridge, Mass. (1974)
81. Gosper, RW: math-fun@cs.arizona.edu posting, Sept. (1996)
82. Gosper, RW: A product, math-fun@cs.arizona.edu posting, Sept. 27 (1996)
83. Gould, SC: What is the value of Pi. Notes and Queries, Manchester, N.H. (1888)
84. Gourdon, X, Sebah, P: Collection of series for π. http://numbers.computation.free.fr/Constants/Pi/piSeries.html
85. Greenblatt, MH: The 'legal' value of π and some related mathematical anomalies. Am. Sci. **53**, 427A-432A (1965)
86. Gregory, RT, Krishnamurthy, EV: Methods and Applications of Error-Free Computation. Springer, New York (1984)
87. Gridgeman, NT: Geometric probability and the number π. Scr. Math. **25**, 183-195 (1960)
88. Guilloud, J, Bouyer, M: Un Million de Décimales de π. Commissariat á l'Energie Atomique, Paris (1974)
89. Gupta, RC: Aryabhata I's value of π. Math. Educ. **7**, 17-20 (1973)
90. Gupta, RC: Madhava's and other medieval Indian values of π. Math. Educ. **9**, 45-48 (1975)
91. Gupta, RC: Some ancient values of pi and their use in India. Math. Educ. **9**, 1-5 (1975)
92. Gupta, RC: Lindemann's discovery of the transcendence of π: a centenary tribute. Ganita Bharati **4**, 102-108 (1982)

Agarwal et al. *Advances in Difference Equations* 2013, **2013**:100
http://www.advancesindifferenceequations.com/content/2013/1/100

93. Gupta, RC: New Indian values of π from the 'Manava'sulba sutra'. Centaurus **31**, 114-125 (1988)
94. Gupta, RC: On the values of π from the bible. Ganita Bharati **10**, 51-58 (1988)
95. Gupta, RC: The value of π in the 'Mahabharata'. Ganita Bharati **12**, 45-47 (1990)
96. Gurland, J: On Wallis' formula. Am. Math. Mon. **63**, 643-645 (1956)
97. Hall, A: On an experimental determination of pi. Messenger Math. **2**, 113-114 (1873)
98. Hata, M: Improvement in the irrationality measures of π and π^2. Proc. Jpn. Acad., Ser. A, Math. Sci. **68**, 283-286 (1992)
99. Hata, M: Rational approximations to π and some other numbers. Acta Arith. **63**, 335-349 (1993)
100. Hayashi, T: The value of π used by the Japanese mathematicians of the 17th and 18th centuries. In: Bibliotheca Mathematica, vol. 3, pp. 273-275 (1902)
101. Hayashi, T, Kusuba, T, Yano, M: Indian values for π derived from Aryabhata's value. Hist. Sci. **37**, 1-16 (1989)
102. Hermann, E: Quadrature of the circle in ancient Egypt. Hist. Math. **4**, 137-140 (1977)
103. Hobson, EW: Squaring the Circle: A History of the Problem. Cambridge University Press, Cambridge (1913)
104. Huygens, C: De circuli magnitudine inventa. Christiani Hugenii Opera Varia, vol. I, pp. 384-388. Leiden (1724)
105. Huylebrouck, D: Van Ceulen's tombstone. Math. Intell. **4**, 60-61 (1995)
106. Hwang, CL: More Machin-type identities. Math. Gaz. **81**, 120-121 (1997)
107. Jami, C: Une histoire chinoise du nombre π. Arch. Hist. Exact Sci. **38**, 39-50 (1988)
108. Jha, P: Aryabhata I and the value of π. Math. Educ. **16**, 54-59 (1982)
109. Jha, SK, Jha, M: A study of the value of π known to ancient Hindu and Jaina mathematicians. J. Bihar Math. Soc. **13**, 38-44 (1990)
110. Jones, W: Synopsis palmiorum matheseos, London, 263 (1706)
111. Jörg, A, Haenel, C: Pi Unleashed, 2nd edn. Springer, Berlin (2000) (Translated by C. Lischka and D. Lischka)
112. Kanada, Y: Vectorization of multiple-precision arithmetic program and 201,326,000 decimal digits of π calculation. In: Supercomputing: Science and Applications, vol. 2, pp. 117-128 (1988)
113. Kanada, Y, Tamura, Y, Yoshino, S, Ushiro, Y: Calculation of π to 10,013,395 decimal places based on the Gauss-Legendre algorithm and Gauss arctangent relation. Technical report 84-01, Computer Center, University of Tokyo (1983)
114. Keith, M: Not a Wake: A Dream Embodying (pi)'s Digits Fully for 10,000 Decimals. Vinculum Press, Baton Rouge (2010) (Diana Keith (Illustrator))
115. Knopp, K: Theory and Application of Infinite Series. Blackie, London (1951)
116. Kochansky, AA: Observationes Cyclometricae ad facilitandam Praxin accomodatae. Acta Erud. **4**, 394-398 (1685)
117. Krishnamurhty, EV: Complementary two-way algorithms for negative radix conversions. IEEE Trans. Comput. **20**, 543-550 (1971)
118. Kulkarni, RP: The value of π known to Sulbasutrakaras. Indian J. Hist. Sci. **13**, 32-41 (1978)
119. Laczkovich, M: On Lambert's proof of the irrationality of π. Am. Math. Mon. **104**, 439-443 (1997)
120. de Lagny, F: Mémoire sur la quadrature du cercle et sur la mesure de tout arc, tout secteur et tout segment donné. In: Histoire de L'Académie Royale des Sciences. Académie des sciences, Paris (1719)
121. Lakshmikantham, V, Leela, S, Vasundhara Devi, J: The Origin and History of Mathematics. Cambridge Scientific Publishers, Cambridge (2005)
122. Lambert, JH: Mémoire sur quelques propriétés remarquables des quantités transcendantes circulaires et logarithmiques. In: Mémoires de l'Académie des Sciences de Berlin, vol. 17, pp. 265-322 (1761)
123. Lange, LJ: An elegant continued fraction for π. Am. Math. Mon. **106**, 456-458 (1999)
124. Lay-Yong, L, Tian-Se, A: Circle measurements in ancient China. Hist. Math. **13**, 325-340 (1986)
125. Lazzarini, M: Un' applicazione del calcolo della probabilitá alla ricerca sperimentale di un valore approssimato di π. Period. Mat. **4**, 140-143 (1901)
126. Legendre, AM: Eléments de Géométrie. Didot, Paris (1794)
127. Lehmer, DH: On arctangent relations for π. Am. Math. Mon. **45**, 657-664 (1938)
128. Lin, L: Further refinements of Gurland's formula for π. J. Inequal. Appl. **2013**, 48 (2013). doi:10.1186/1029-242X-2013-48
129. Lindemann, F: Über die Zahl π. Math. Ann. **20**, 213-225 (1882)
130. Le Lionnais, F: Les Nombres Remarquables. Hermann, Paris (1983)
131. Lucas, SK: Integral proofs that $355/113 > \pi$. Aust. Math. Soc. Gaz. **32**, 263-266 (2005)
132. Mao, Y: A short history of π in China. Kexue **3**, 411-423 (1917)
133. Maor, E: The history of π on the pocket calculator. J. Coll. Sci. Teach. Nov., 97-99 (1976)
134. Matar, KM, Rajagopal, C: On the Hindu quadrature of the circle. J. Bombay Branch R. Asiat. Soc. **20**, 77-82 (1944)
135. Mikami, Y: The Development of Mathematics in China and Japan. Chelsea, New York (1913)
136. Miel, G: An algorithm for the calculation of π. Am. Math. Mon. **86**, 694-697 (1979)
137. Miel, G: Of calculations past and present: the Archimedean algorithm. Am. Math. Mon. **90**, 17-35 (1983)
138. Moakes, AJ: The calculation of π. Math. Gaz. **54**, 261-264 (1970)
139. Moakes, AJ: A further note on machine computation for π. Math. Gaz. **55**, 306-310 (1971)
140. Mortici, C: Refinement of Gurland's formula for pi. Comput. Math. Appl. **62**, 2616-2620 (2011)
141. Myers, WA: The Quadrature of the Circle, the Square Root of Two, and the Right-Angled Triangle. Wilstach, Baldwin & Co. Printers, Cincinnati (1873)
142. Nagell, T: Irrationality of the numbers e and π. In: Introduction to Number Theory, pp. 38-40. Wiley, New York (1951)
143. Nakamura, K: On the sprout and setback of the concept of mathematical 'proof' in the Edo period in Japan: regarding the method of calculating number π. Hist. Sci. **3**, 185-199 (1994)
144. Nanjundiah, TS: On Huygens' approximation to π. Math. Mag. **44**, 221-223 (1971)
145. Newman, M, Shanks, D: On a sequence arising in series for π. Math. Comput. **42**, 199-217 (1984)
146. Nicholson, SC, Jeenel, J: Some comments on a NORC computation of π. Math. Tables Other Aids Comput. **9**, 162-164 (1955)
147. Niven, IM: A simple proof that π is irrational. Bull. Am. Math. Soc. **53**, 507 (1947)
148. Niven, IM: Irrational Numbers. Wiley, New York (1956)

Agarwal et al. *Advances in Difference Equations* 2013, **2013**:100
http://www.advancesindifferenceequations.com/content/2013/1/100

149. Palais, R: *pi* is wrong. Math. Intell. **23**, 7-8 (2001)
150. Parker, JA: The Quadrature of the Circle: Setting Forth the Secrete Teaching of the Bible. Kessinger Publ., Whitefish (2010)
151. Pereira da Silva, C: A brief history of the number π. Bol. Soc. Parana. Mat. **7**, 1-8 (1986)
152. Plouffe, S: Identities inspired from Ramanujan notebooks (Part 2). Apr. 2006. http://www.lacim.uqam.ca/~plouffe/inspired2.pdf
153. Posamentier, AS, Lehmann, I: Pi: A Biography of the World's Most Mysterious Number. Prometheus Books, New York (2004)
154. Preston, R: The mountains of π. The New Yorker, March 2, 36-67 (1992)
155. Puritz, CW: An elementary method of calculating π. Math. Gaz. **58**, 102-108 (1974)
156. Qian, B: A study of π found in Chinese mathematical books. Kexue **8**, 114-129 and 254-265 (1923)
157. Rabinowitz, S, Wagon, S: A spigot algorithm for the digits of π. Am. Math. Mon. **102**, 195-203 (1995)
158. Rajagopal, CT, Vedamurti Aiyar, TV: A Hindu approximation to π. Scr. Math. **18**, 25-30 (1952)
159. Ramanujan, S: Modular equations and approximations to π. Q. J. Pure Appl. Math. **45**(1914), 350-372 (1913-1914)
160. Reitwiesner, G: An ENIAC determination of π and e to more than 2,000 decimal places. Math. Tables Other Aids Comput. **4**, 11-15 (1950)
161. Roy, R: The discovery of the series formula for π by Leibniz, Gregory, and Nilakantha. Math. Mag. **63**, 291-306 (1990)
162. Rutherford, W: Computation of the ratio of the diameter of a circle to its circumference to 208 places of figures. Philos. Trans. R. Soc. Lond. **131**, 281-283 (1841)
163. Sagan, C: Contact. Simon & Schuster, New York (1985)
164. Salamin, E: Computation of π using arithmetic-geometric mean. Math. Comput. **30**, 565-570 (1976)
165. Salikhov, V: On the irrationality measure of π. Russ. Math. Surv. **53**, 570-572 (2008)
166. Schepler, HC: The chronology of PI. Math. Magazine, January-February 1950: 165-170; March-April 1950: 216-228; May-June 1950: 279-283
167. Schröder, EM: Zur irrationalität von π^2 und π. Mitt. Math. Ges. Hamb. **13**, 249 (1993)
168. Schubert, H: Squaring of the circle. Smithsonian Institution Annual Report (1890)
169. Sen, SK, Agarwal, RP: Best k-digit rational approximation of irrational numbers: pre-computer versus computer era. Appl. Math. Comput. **199**, 770-786 (2008)
170. Sen, SK, Agarwal, RP: π, e, ϕ with MATLAB: Random and Rational Sequences with Scope in Supercomputing Era. Cambridge Scientific Publishers, Cambridge (2011)
171. Sen, SK, Agarwal, RP, Shaykhianb, GA: Golden ratio versus pi as random sequence sources for Monte Carlo integration. Math. Comput. Model. **48**, 161-178 (2008)
172. Sen, SK, Agarwal, RP, Shaykhian, GA: Best k-digit rational approximations-true versus convergent, decimal-based ones: quality, cost, scope. Adv. Stud. Contemp. Math. **19**, 59-96 (2009)
173. Sen, SK, Agarwal, RP, Pavani, R: Best k-digit rational bounds for irrational numbers: pre- and super-computer era. Math. Comput. Model. **49**, 1465-1482 (2009)
174. Shanks, D: Dihedral quartic approximations and series for π. J. Number Theory **14**, 397-423 (1982)
175. Shanks, D, Wrench, JW Jr.: Calculation of π to 100,000 decimals. Math. Comput. **16**, 76-99 (1962)
176. Shanks, W: Contributions to Mathematics Comprising Chiefly the Rectification of the Circle to 607 Places of Decimals. Bell, London (1853)
177. Shanks, W: On the extension of the numerical value of π. Proc. R. Soc. Lond. **21**, 315-319 (1873)
178. Singmaster, D: The legal values of π. Math. Intell. **7**, 69-72 (1985)
179. Smith, DE: History and transcendence of pi. In: Young, WJA (ed.) Monograms on Modern Mathematics. Longmans, Green, New York (1911)
180. Smith, DE: The history and transcendence of π. In: Young, JWA (ed.) Monographs on Topics of Modern Mathematics Relevant to the Elementary Field, chapter 9, pp. 388-416. Dover, New York (1955)
181. Smith, DE, Mikami, Y: A History of Japanese Mathematics. Open-Court, Chicago (1914)
182. van Roijen Snell, W: Cyclometricus. Leiden (1621)
183. Sondow, J: A faster product for π and a new integral for ln(π/2). Am. Math. Mon. **112**, 729-734 (2005)
184. Stern, MD: A remarkable approximation to π. Math. Gaz. **69**, 218-219 (1985)
185. Stevens, J: Zur irrationalität von π. Mitt. Math. Ges. Hamb. **18**, 151-158 (1999)
186. Störmer, C: Sur l'application de la théorie des nombres entiers complexes á la solution en nombres rationnels $x_1, x_2, \ldots, x_n, c_1, c_2, \ldots, c_n, k$ de l'équation $c_1 \arctan x_1 + c_2 \arctan x_2 + \cdots + c_n \arctan x_n = kp/4$. Arch. Math. Naturvidensk. **19**, 75-85 (1896)
187. Takahasi, D, Kanada, Y: Calculation of π to 51.5 billion decimal digits on distributed memory and parallel processors. Trans. Inf. Process. Soc. Jpn. **39**(7) (1998)
188. Tamura, Y, Kanada, Y: Calculation of π to 4,194,293 decimals based on Gauss-Legendre algorithm. Technical report 83-01, Computer Center, University of Tokyo (1982)
189. Todd, J: A problem on arctangent relations. Am. Math. Mon. **56**, 517-528 (1949)
190. Trier, PE: Pi revisited. Bull. - Inst. Math. Appl. **25**, 74-77 (1989)
191. Tweddle, I: John Machin and Robert Simson on inverse-tangent series for π. Arch. Hist. Exact Sci. **42**, 1-14 (1991)
192. Uhler, HS: Recalculation and extension of the modulus and of the logarithms of 2, 3, 5, 7 and 17. Proc. Natl. Acad. Sci. USA **26**, 205-212 (1940)
193. Vega, G: Thesaurus Logarithmorum Completus. Leipzig (1794)
194. Viéta, F: Uriorum de rebus mathematicis responsorum. Liber VII (1593)
195. Volkov, A: Calculations of π in ancient China: from Liu Hui to Zu Chongzhi. Hist. Sci. **4**, 139-157 (1994)
196. Volkov, A: Supplementary data on the values of π in the history of Chinese mathematics. Philos. Hist. Sci. Taiwan. J. **3**, 95-120 (1994)
197. Volkov, A: Zhao Youqin and his calculation of π. Hist. Math. **24**, 301-331 (1997)
198. Wagon, S: Is π normal. Math. Intell. **7**, 65-67 (1985)
199. Wells, D: The Penguin Dictionary of Curious and Interesting Numbers. Penguin, Middlesex (1986)
200. Wrench, JW Jr.: The evolution of extended decimal approximations to π. Math. Teach. **53**, 644-650 (1960)

Agarwal et al. *Advances in Difference Equations* 2013, **2013**:100
http://www.advancesindifferenceequations.com/content/2013/1/100

201. Wrench, JW Jr., Smith, LB: Values of the terms of the Gregory series for arccot 5 and arccot 239 to 1,150 and 1,120 decimal places, respectively. Math. Tables Other Aids Comput. **4**, 160-161 (1950)
202. Yeo, A: The Pleasures of π, e and Other Interesting Numbers. World Scientific, Singapore (2006)
203. Zebrowski, E: A History of the Circle: Mathematical Reasoning and the Physical Universe. Rutgers University Press, Pisacataway (1999)
204. Zha, Y-L: Research on Tsu Ch'ung-Chih's approximate method for π. In: Science and Technology in Chinese Civilization, pp. 77-85. World Scientific, Teaneck (1987)
205. http://mathworld.wolfram.com/PiFormulas.html
206. http://en.wikipedia.org/wiki/Pi
207. http://en.wikipedia.org/wiki/Pi_approximations
208. en.wikipedia.org/wiki/Negative_base
209. en.wikipedia.org/wiki/Super_Pi
210. en.wikipedia.org/wiki/Overclocking

doi:10.1186/1687-1847-2013-100
Cite this article as: Agarwal et al.: **Birth, growth and computation of pi to ten trillion digits.** *Advances in Difference Equations* 2013 **2013**:100.

23. Pi day is upon us again and we still do not know if pi is normal (2014)

Paper 23: David H. Bailey and Jonathan Borwein, "Pi day is upon us again and we still do not know if pi is normal," *American Mathematical Monthly*, vol. 121 (2014), p. 191–206. Copyright 2014 Mathematical Association of America. All Rights Reserved.

Synopsis:

This paper, which appeared, appropriately enough, on Pi day (March 14, or 3/14 in North American notation) of 2014, discusses the enduring appeal of π in both the popular media and also in serious state-of-the-art mathematical research.

The article provides a brief review of the origins of decimal arithmetic, the original formulas found to compute π, recent formulas by Ramanujan, Salamin, Brent and others, the BBP formula for computing binary digits of π beginning at an arbitrary starting position, and new computer-based techniques for analyzing the digits of π, such as the graphical techniques discussed in paper #21 of this collection, recent results on normality and, in conclusion, a list of unanswered questions regarding π.

Keywords: Algorithms, General Audience, Normality

© Springer International Publishing Switzerland 2016
D.H. Bailey, J.M. Borwein, *Pi: The Next Generation*,
DOI 10.1007/978-3-319-32377-0_23

Pi Day Is Upon Us Again and We Still Do Not Know if Pi Is Normal

David H. Bailey and Jonathan Borwein

Abstract. The digits of π have intrigued both the public and research mathematicians from the beginning of time. This article briefly reviews the history of this venerable constant, and then describes some recent research on the question of whether π is normal, or, in other words, whether its digits are statistically random in a specific sense.

1. PI AND ITS DAY IN MODERN POPULAR CULTURE. The number π, unique among the pantheon of mathematical constants, captures the fascination both of the public and of professional mathematicians. Algebraic constants such as $\sqrt{2}$ are easier to explain and to calculate to high accuracy (e.g., using a simple Newton iteration scheme). The constant e is pervasive in physics and chemistry, and even appears in financial mathematics. Logarithms are ubiquitous in the social sciences. But none of these other constants has ever gained much traction in the popular culture.

In contrast, we see π at every turn. In an early scene of Ang Lee's 2012 movie adaptation of Yann Martel's award-winning book *The Life of Pi*, the title character Piscine ("Pi") Molitor writes hundreds of digits of the decimal expansion of π on a blackboard to impress his teachers and schoolmates, who chant along with every digit.[1] This has even led to humorous take-offs such as a 2013 Scott Hilburn cartoon entitled "Wife of Pi," which depicts a 4 figure seated next to a π figure, telling their marriage counselor "He's irrational and he goes on and on." [**22**].

This attention comes to a head on March 14 of each year with the celebration of "Pi Day," when in the United States, with its taste for placing the day after the month, 3/14 corresponds to the best-known decimal approximation of Pi (with 3/14/15 promising a gala event in 2015). Pi Day was originally founded in 1988, the brainchild of Larry Shaw of San Francisco's Exploratorium (a science museum), which in turn was founded by Frank Oppenheimer (the younger physicist brother of Robert Oppenheimer) after he was blacklisted by the U.S. Government during the McCarthy era.

Originally a light-hearted gag where folks walked around the Exploratorium in funny hats with pies and the like, by the turn of the century Pi Day was a major educational event in North American schools, garnering plenty of press.[2] In 2009, the U.S. House of Representatives made Pi Day celebrations official by passing a resolution designating March 14 as "National Pi Day," and encouraging "schools and educators to observe the day with appropriate activities that teach students about Pi and engage them about the study of mathematics." [**23**].[3]

As a striking example, the March 14, 2007 *New York Times* crossword puzzle featured clues, where, in numerous locations, π (standing for PI) must be entered at the

http://dx.doi.org/10.4169/amer.math.monthly.121.03.191

MSC: Primary 01A99, Secondary 11Z05

[1]Good scholarship requires us to say that in the book Pi contents himself with drawing a circle of unit diameter.

[2]Try `www.google.com/trends?q=Pi+` to see the seasonal interest in 'Pi'.

[3]This seems to be the first legislation on Pi to have been adopted by a government, though in the late 19th century Indiana came embarrassingly close to legislating its value, see [**12**, Singmaster, Entry 27] and [**14**]. This MONTHLY played an odd role in that affair.

intersection of two words. For example, 33 across "Vice president after Hubert" (answer: SπRO) intersects with 34 down "Stove feature" (answer: πLOT). Indeed, 28 down, with clue "March 14, to mathematicians," was, appropriately enough, PIDAY, while PIPPIN is now a four-letter word. The puzzle and its solution are reprinted with permission in [**15**, pp. 312–313].

π **Mania in popular culture.** Many instances are given in [**14**]. They include the following:

1. On September 12, **2012**, five aircraft armed with dot-matrix-style skywriting technology wrote 1000 digits of π in the sky above the San Francisco Bay area as a spectacular and costly piece of *piformance* art.

2. On March 14, **2012**, U.S. District Court Judge Michael H. Simon dismissed a copyright infringement suit relating to the lyrics of a song by ruling that "Pi is a non-copyrightable fact."

3. On the September 20, **2005** edition of the North American TV quiz show *Jeopardy!*, in the category "By the numbers," the clue was "'How I want a drink, alcoholic of course' is often used to memorize this." (Answer: What is Pi?).

4. On August 18, **2005**, Google offered 14,159,265 "new slices of rich technology" in their initial public stock offering. On January 29, **2013** they offered a π million dollar prize for successful hacking of the Chrome Operating System on a specific Android phone.

5. In the first **1999** *Matrix* movie, the lead character Neo has only 314 seconds to enter the Source. *Time* noted the similarity to the digits of π.

6. The **1998** thriller "Pi" received an award for screenplay at the Sundance film festival. When the authors were sent advance access to its website, they diagnosed it a fine hoax.

7. The May 6, **1993** edition of *The Simpsons* had Apu declaring "I can recite pi to 40,000 places. The last digit is 1." This digit was supplied to the screenwriters by one of the present authors.

8. In Carl Sagan's **1986** book *Contact*, the lead character (played by Jodie Foster in the movie) searched for patterns in the digits of π, and after her mysterious experience sought confirmation in the base-11 expansion of π.

With regards to item #3 above, there are many such "pi-mnemonics" or "piems" (i.e., phrases or verse whose letter count, ignoring punctuation, gives the digits of π) in the popular press [**12, 14**]. Another is "Sir, I bear a rhyme excelling / In mystic force and magic spelling / Celestial sprites elucidate / All my own striving can't relate." [**13**, p. 106]. Some are very long [**12**, Keith, Entry 59, pp. 560–561].

Sometimes the attention given to π is annoying, such as when on August 14th, 2012, the U.S. Census Office announced the population of the country had passed *exactly* 314,159,265. Such precision was, of course, completely unwarranted. Sometimes the attention is breathtakingly pleasurable.[4,5]

Poems versus piems. While piems are fun they are usually doggerel. To redress this, we include examples of excellent π poetry and song.[6] In Figure 1 we present the much anthologised poem "PI," by Polish poet Wislawa Szymborska (1923–2012) who won

---

[4]See the 2013 movie at http://www.youtube.com/watch?v=Vp9zLbIE8zo.

[5]A comprehensive Pi Day presentation is lodged at http://www.carma.newcastle.edu.au/jon/piday.pdf.

[6]See also [**12**, Irving Kaplansky's "A song about Pi."].

the 1996 Novel prize for literature [**29**, p. 174]. In Figure 2 we present the lyrics of "Pi" by the influential British singer songwriter Kate Bush [**18**]. The *Observer* review of her 2005 collection *Aerial*, on which the song appears, wrote that it is

a sentimental ode to a mathematician, audacious in both subject matter and treatment. The chorus is the number sung to many, many decimal places.[7]

The admirable number pi:
three point one four one.
All the following digits are also just a start,
five nine two because it never ends.
It can't be grasped, six five three five, at a glance,
eight nine, by calculation,
seven nine, through imagination,
or even three two three eight in jest, or by comparison
four six to anything
two six four three in the world.
The longest snake on earth ends at thirty-odd feet.
Same goes for fairy tale snakes, though they make it a little longer.
The caravan of digits that is pi
does not stop at the edge of the page,
but runs off the table and into the air,
over the wall, a leaf, a bird's nest, the clouds, straight into the sky,
through all the bloatedness and bottomlessness.
Oh how short, all but mouse-like is the comet's tail!
How frail is a ray of starlight, bending in any old space!
Meanwhile two three fifteen three hundred nineteen
my phone number your shirt size
the year nineteen hundred and seventy-three sixth floor
number of inhabitants sixty-five cents
hip measurement two fingers a charade and a code,
in which we find how blithe the trostle sings!
and please remain calm,
and heaven and earth shall pass away,
but not pi, that won't happen,
it still has an okay five,
and quite a fine eight,
and all but final seven,
prodding and prodding a plodding eternity
to last.

Figure 1. "PI," by Wislawa Szymborska

2. PRE-DIGITAL HISTORY.

π is arguably the only mathematical topic from very early history that is still being researched today. The Babylonians used the approximation $\pi \approx 3$. The Egyptian Rhind Papyrus, dated roughly 1650 BCE, suggests $\pi = 32/18 = 3.16049\ldots$. Early Indian mathematicians believed $\pi = \sqrt{10} = 3.162277\ldots$. Archimedes, in the first mathematically rigorous calculation, employed a clever iterative construction of inscribed and circumscribed polygons to establish that

$$3\ 10/71 = 3.14084\ldots < \pi < 3\ 1/7 = 3.14285\ldots$$

---

[7]She sings over 150 digits but errs after 50 places. The correct digits occurred with the published lyrics.

Sweet and gentle sensitive man
With an obsessive nature and deep fascination
For numbers
And a complete infatuation with the calculation
Of PI

Oh he love, he love, he love
He does love his numbers
And they run, they run, they run him
In a great big circle
In a circle of infinity

3.1415926535 897932
3846 264 338 3279

Oh he love, he love, he love
He does love his numbers
And they run, they run, they run him
In a great big circle
In a circle of infinity
But he must, he must, he must
Put a number to it

50288419 716939937510
582319749 44 59230781
6406286208 821 4808651 32

Oh he love, he love, he love
He does love his numbers
And they run, they run, they run him
In a great big circle
In a circle of infinity

82306647 0938446095 505 8223...

Figure 2. "Pi," by Kate Bush

This amazing work, done without trigonometry or floating point arithmetic, is charmingly described by George Phillips [**12**, Entry 4].

Life after modern arithmetic. The advent of modern positional, zero-based decimal arithmetic, most likely discovered in India prior to the fifth century [**4, 27**], significantly reduced computational effort. Even though the Indo-Arabic system, as it is now known, was introduced to Europeans first by Gerbert of Aurillac (c. 946–1003, who became Pope Sylvester II in 999) in the 10th century, and again, in greater detail and more successfully, by Fibonacci in the early 13th century, Europe was slow to adopt it, hampering progress in both science and commerce. In the 16th century, prior to the widespread adoption of decimal arithmetic, a wealthy German merchant was advised, regarding his son's college plans,

> If you only want him to be able to cope with addition and subtraction, then any French or German university will do. But if you are intent on your son going on to multiplication and division—assuming that he has sufficient gifts—then you will have to send him to Italy. [**24**, p. 577]

Life after calculus. Armed with decimal arithmetic and modern calculus, 17th-, 18th-, and 19th-century mathematicians computed π with aplomb. Newton recorded

16 digits in 1665, but later admitted, "I am ashamed to tell you how many figures I carried these computations, having no other business at the time." In 1844 Dase, under the guidance of Strassnitzky, computed 212 digits correctly in his head [14]. These efforts culminated with William Shanks (1812–1882), who employed John *Machin's formula*

$$\frac{\pi}{4} = 4\arctan\left(\frac{1}{5}\right) - \arctan\left(\frac{1}{239}\right), \tag{1}$$

where $\arctan x = x - x^3/3 + x^5/5 - x^7/7 + x^9/9 - \cdots$, to compute 707 digits in 1874. His 1853 work to 607 places was funded by 30 subscriptions from such notables as Rutherford, De Morgan (two copies), Herschel (Master of the Mint and son of the astronomer) and Airy.[8]

Alas, only 527 digits were correct (as Ferguson found nearly a century later in 1946 using a calculator), confirming the suspicions of De Morgan at the time, who asserted that there were too many sevens in Shanks' published result (although the statistical deviation was not as convincing as De Morgan thought [26]). A brief summary of this history is shown in Table 1. We note that Sharp was a cleric, Ferguson was a school teacher, and Dase a "kopfrechnenner." Many original documents relating to this history can be found in [12].

Table 1. Brief chronicle of pre-20th-century π calculations

Archimedes	250? BCE	3	3.1418 (ave.)
Liu Hui	263	5	3.14159
Tsu Ch'ung Chi	480?	7	3.1415926
Al-Kashi	1429	14	
Romanus	1593	15	
Van Ceulen	1615	39	(35 correct)
Newton	1665	16	
Sharp	1699	71	
Machin	1706	100	
De Lagny	1719	127	(112 correct)
Vega	1794	140	
Rutherford	1824	208	(152 correct)
Strassnitzky and Dase	1844	200	
Rutherford	1853	440	
Shanks	1853	607	(527 correct)
Shanks	1873	707	(527 correct)

Mathematics of Pi. Alongside these numerical developments, the mathematics behind π enjoyed comparable advances. In 1761, using improper continued fractions, Lambert [12, Entry 20] proved that π is irrational, thus establishing that the digits of π never repeat. Then in 1882, Lindemann [12, Entry 22] proved that e^α is transcendental for every nonzero algebraic number α, which immediately implied that π is transcendental (since $e^{i\pi} = -1$). This result settled in decisive terms the 2000-year-old question of whether a square could be constructed with the same area as a circle, using compass and straightedge (it cannot, because if it could then π would be a geometrically *constructible number* and hence algebraic).

---

[8]He had originally intended to present only about 500 places, and evidently added the additional digits while finishing the galleys a few months later [12, Entry 20]. Errors introduced in a rush to publish are not new.

3. THE TWENTIETH CENTURY AND BEYOND. With the development of computer technology in the 1950s and 1960s, π was computed to thousands of digits, facilitated in part by new algorithms for performing high-precision arithmetic, notably the usage of fast Fourier transforms to dramatically accelerate multiplication.

Ramanujan-type series for $1/\pi$. Even more importantly, computations of π began to employ some entirely new mathematics, such as Ramanujan's 1914 formula

$$\frac{1}{\pi} = \frac{2\sqrt{2}}{9801} \sum_{k=0}^{\infty} \frac{(4k)!\,(1103 + 26390k)}{(k!)^4\,396^{4k}}, \tag{2}$$

each term of which produces an additional eight correct digits in the result [16]. David and Gregory Chudnovsky employed the variant

$$\frac{1}{\pi} = 12 \sum_{k=0}^{\infty} \frac{(-1)^k (6k)!\,(13591409 + 545140134k)}{(3k)!\,(k!)^3\,640320^{3k+3/2}}, \tag{3}$$

each term of which adds 14 correct digits. Both of these formulas rely on rather deep number theory [14] and related modular-function theory [16].

Reduced complexity algorithms [17] for $1/\pi$. Another key development in the mid 1970s was the *Salamin–Brent algorithm* [12, Entries 46 and 47] for π: Set $a_0 = 1, b_0 = 1/\sqrt{2}$, and $s_0 = 1/2$. Then for $k \geq 1$, iterate

$$a_k = \frac{a_{k-1} + b_{k-1}}{2} \qquad b_k = \sqrt{a_{k-1} b_{k-1}}$$

$$c_k = a_k^2 - b_k^2 \qquad s_k = s_{k-1} - 2^k c_k \qquad p_k = \frac{2a_k^2}{s_k}. \tag{4}$$

The value of p_k converges *quadratically* to π—each iteration approximately *doubles* the number of correct digits.

A related algorithm, inspired by a 1914 Ramanujan paper, was found in 1986 by one of us and Peter Borwein [16]: Set $a_0 = 6 - 4\sqrt{2}$ and $y_0 = \sqrt{2} - 1$. Then for $k \geq 0$, iterate

$$y_{k+1} = \frac{1 - (1 - y_k^4)^{1/4}}{1 + (1 + y_k^4)^{1/4}} \tag{5}$$

$$a_{k+1} = a_k (1 + y_{k+1})^4 - 2^{2k+3} y_{k+1} (1 + y_{k+1} + y_{k+1}^2).$$

Then a_k converges *quartically* to $1/\pi$—each iteration approximately quadruples the number of correct digits. Just twenty-one iterations suffices to produce an algebraic number that agrees with π to more than six trillion digits (provided all iterations are performed with this precision).

With discoveries such as these, combined with prodigious improvements in computer hardware (thanks to *Moore's Law*) and clever use of parallelism, π was computed to millions, then billions, and, in 2011, to 10 *trillion* decimal digits. A brief chronicle of π computer-age computations is shown in Table 2.[9]

---

[9]It is probably unnecessary to note that the Shanks of this table is not the Shanks of Table 1.

Table 2. Brief chronicle of computer-age π calculations

Ferguson	1945	620
Smith and Wrench	1949	1,120
Reitwiesner et al. (ENIAC)	1949	2,037
Guilloud	1959	16,167
Shanks and Wrench	1961	100,265
Guilloud and Bouyer	1973	1,001,250
Kanada, Yoshino and Tamura	1982	16,777,206
Gosper	1985	17,526,200
Bailey	Jan. 1986	29,360,111
Kanada and Tamura	Jan. 1988	201,326,551
Kanada and Tamura	Nov. 1989	1,073,741,799
David and Gregory Chudnovsky	Aug. 1991	2,260,000,000
Kanada and Takahashi	Apr. 1999	51,539,600,000
Kanada and Takahashi	Sep. 1999	206,158,430,000
Kanada and 9 others	Nov. 2002	1,241,100,000,000
Bellard	Dec. 2009	2,699,999,990,000
Kondo and Yee	Aug. 2010	5,000,000,000,000
Kondo and Yee	Oct. 2011	10,000,000,000,000

4. COMPUTING DIGITS OF π AT AN ARBITRARY STARTING POSITION.

A recent reminder of the folly of thinking that π is fully understood was the 1996 discovery of a simple scheme for computing binary or hexadecimal digits of π, beginning at an arbitrary starting position, without needing to compute any of the preceding digits. This scheme is based on the following formula, which was *discovered by a computer program* implementing Ferguson's "PSLQ" algorithm [9, 20]:

$$\pi = \sum_{k=0}^{\infty} \frac{1}{16^k} \left(\frac{4}{8k+1} - \frac{2}{8k+4} - \frac{1}{8k+5} - \frac{1}{8k+6} \right). \tag{6}$$

The proof of this formula (now known as the "BBP" formula for π) is a relatively simple exercise in calculus. It is perhaps puzzling that it had not been discovered centuries before. But then no one was looking for such a formula.

How bits are extracted. The scheme to compute digits of π beginning at an arbitrary starting point is best illustrated by considering the similar (and very well known) formula for $\log 2$:

$$\log 2 = \sum_{k=1}^{\infty} \frac{1}{k 2^k}. \tag{7}$$

Note that the binary expansion of $\log 2$ beginning at position $d + 1$ is merely the fractional part of $2^d \log 2$, so that we can write (where $\{\cdot\}$ denotes fractional part):

$$\{2^d \log 2\} = \left\{ \left\{ \sum_{k=1}^{d} \frac{2^{d-k} \bmod k}{k} \right\} + \left\{ \sum_{k=d+1}^{\infty} \frac{2^{d-k}}{k} \right\} \right\}. \tag{8}$$

Now note that the numerators of the first summation can be computed very rapidly by means of the binary algorithm for exponentiation, namely the observation, for example, that $3^{17} \bmod 10 = ((((3^2 \bmod 10)^2 \bmod 10)^2 \bmod 10)^2 \bmod 10) \cdot 3 \bmod 10$.

This same approach can be used to compute binary or hexadecimal digits of π using (6).

This scheme has been implemented to compute hexadecimal digits of π beginning at stratospherically high positions. In July 2010, for example, Tsz-Wo Sze of Yahoo! Cloud Computing computed base-16 digits of π beginning at position 2.5×10^{14}. Then on March 14 (Pi Day), 2013, Ed Karrels of Santa Clara University computed 26 base-16 digits beginning at position one quadrillion [25]. His result: 8353CB3F7F0C9ACCFA9AA215F2.

Beyond utility. Certainly, there is no need for computing π to millions or billions of digits in practical scientific or engineering work. A value of π to 40 digits is more than enough to compute the circumference of the Milky Way galaxy to an error less than the size of a proton. There are certain scientific calculations that require intermediate calculations to be performed to higher than standard 16-digit precision (typically 32 or 64 digits may be required) [3], and certain computations in the field of experimental mathematics have required as high as 50,000 digits [6], but we are not aware of any "practical" applications beyond this level.

Computations of digits of π are, however, excellent tests of computer integrity— if even a single error occurs during a large computation, almost certainly the final result will be in error, resulting in disagreement with a check calculation done with a different algorithm. For example, in 1986, a pair of π-calculating programs using (4) and (5) detected some obscure hardware problems in one of the original Cray-2 supercomputers.[10] Also, some early research into efficient implementations of the fast Fourier transform on modern computer architectures had their origins in efforts to accelerate computations of π [2].

5. NEW TECHNIQUES TO EXPLORE NORMALITY AND RELATED PROPERTIES. Given an integer $b \geq 2$, a real number α is said to be *b-normal* or *normal base b* if every m-long string of base-b digits appears in the base-b expansion of α with limiting frequency $1/b^m$. It is easy to show via measure theory that almost all real numbers are b-normal for every $b \geq 2$ (a condition known as *absolute normality*), but establishing normality for specific numbers has proven to be very difficult.

In particular, no one has been able to establish that π is b-normal for any integer b, much less for all bases simultaneously. It is a premier example of an intriguing mathematical question that has occurred to countless schoolchildren as well as professional mathematicians through the ages, but which has defied definitive answer to the present day. A proof for any specific base would not only be of great interest worldwide, but would also have potential practical application as a provably effective pseudorandom number generator. This ignorance extends to other classical constants of mathematics, including e, $\log 2$, $\sqrt{2}$, and γ (Euler's constant). Borel conjectured that all irrational algebraic numbers are absolutely normal, but this has not been proven in even a single instance, to any base.

Two examples where normality has been established are Champernowne's number $C_{10} = 0.12345678910111213\ldots$ (constructed by concatenating the positive integers), which is provably 10-normal, and Stoneham's number $\alpha_{2,3} = \sum_{k \geq 0} 1/(3^k 2^{3^k})$, which is provably 2-normal—see below [10, 11, 28]. One relatively weak result for algebraic numbers is that the number of 1-bits in the binary expansion of a degree-D algebraic number α must exceed $Cn^{1/D}$ for all sufficiently large n, for a positive number C that

---

[10]Cray's own tests did not find these errors. After that, these π algorithms were included in Cray's test suite in the factory.

depends on α [**8**]. Thus, for example, the number of 1-bits in the first n bits of the binary expansion of $\sqrt{2}$ must exceed $\sqrt{n}$.

In spite of these intriguing developments, it is clear that more powerful techniques must be brought to bear on the question of normality, either for π or other well-known constants of mathematics, before significant progress can be achieved. Along this line, modern computer technology suggests several avenues of research.

Statistical analysis. One approach is simply to perform large-scale statistical analyses on the digits of π, as has been done, to some degree, on nearly all computations since ENIAC. In [**7**], for example, the authors empirically tested the normality of its first roughly four trillion hexadecimal (base-16) digits using a Poisson process model, and concluded that, according to this test, it is "extraordinarily unlikely" that π is not 16-normal (of course, this result does not pretend to be a proof).

Graphical representations. Another fruitful approach is to display the digits of π or other constants graphically, cast as a random walk [**1**]. For example, Figure 3 shows a walk based on one million base-4 pseudorandom digits, where at each step the graph moves one unit east, north, west, or south, depending on the whether the pseudorandom iterate at that position is 0, 1, 2, or 3. The color indicates the path followed by the walk—shifted up the spectrum (red-orange-yellow-green-cyan-blue-purple-red) following an HSV scheme with S and V equal to one. The HSV (hue, saturation, and value) model is a cylindrical-coordinate representation that yields a rainbow-like range of colors.

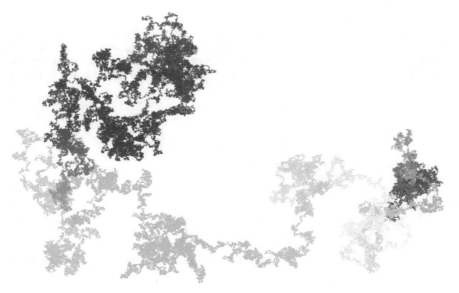

Figure 3. A uniform pseudorandom walk

Figure 4 shows a walk on the first 100 billion base-4 digits of π. This may be viewed dynamically in more detail online at `http://gigapan.org/gigapans/106803`, where the full-sized image has a resolution of 372,224×290,218 pixels (108.03 billion pixels in total). This is one of the largest mathematical images ever produced and, needless to say, its production was by no means easy [**1**].

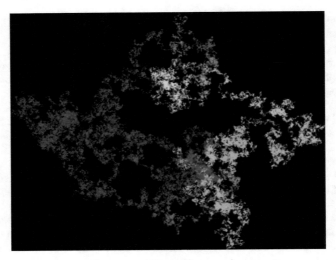

Figure 4. A walk on the first 100 billion base-4 digits of π

Although no clear inference regarding the normality of π can be drawn from these figures, it is plausible that π is 4-normal (and thus 2-normal), since the overall appearance of its graph is similar to that of the graph of the pseudorandomly generated base-4 digits.

The Champernowne numbers. We should emphasize what a poor surrogate for randomness the notion of normality actually is. The base-b *Champernowne* number, C_b, is formed by concatenating the natural numbers base b as a floating-point number in that base. It was the first type of number proven to be normal and fails stronger normality tests [**1**]. Thus,

$$C_b := \sum_{k=1}^{\infty} \frac{\sum_{m=b^{k-1}}^{b^k-1} m b^{-k\left[m-(b^{k-1}-1)\right]}}{b \sum_{m=0}^{k-1} m(b-1)b^{m-1}}$$

$$C_{10} = 0.123456789101112\ldots$$

$$C_4 = 0.1231011121320212223\ldots_4. \tag{9}$$

In Figure 5 we show how far from random a walk on a normal number may be—pictorially or by many quantitative measures—as illustrated by C_4.[11]

Stoneham numbers. This same tool can be employed to study the digits of Stoneham's constant, namely

$$\alpha_{2,3} = \sum_{k=0}^{\infty} \frac{1}{3^k 2^{3^k}}. \tag{10}$$

This constant is one of the few that is provably 2-normal (and thus 2^n-normal, for every positive integer n) [**10, 11, 28**]. What's more, it is provably *not* 6-normal, so that it is

---

[11]The subscript four denotes a base-four representation.

Figure 5. A walk on Champernowne's base-4 number

an explicit example of the fact that normality in one base does not imply normality in another base [**5**]. For other number bases, including base 3, its normality is not yet known one way or the other.

Figures 6, 7, and 8 show walks generated from the base-3, base-4, and base-6 digit expansions, respectively, of $\alpha_{2,3}$. The base-4 digits are graphed using the same scheme mentioned above, with each step moving east, north, west, or south, according to whether the digit is 0, 1, 2, or 3. The base-3 graph is generated by moving unit distance at an angle 0, $\pi/3$, or $2\pi/3$, respectively, for 0, 1, or 2. Similarly, the base-6 graph is generated by moving unit distance at angle $k\pi/6$ for $k = 0, 1, \ldots, 5$.

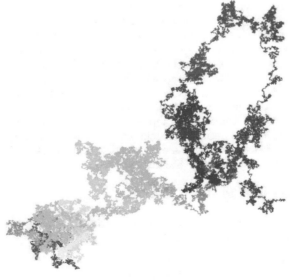

Figure 6. A walk on the base-3 digits of Stoneham's constant ($\alpha_{2,3}$)

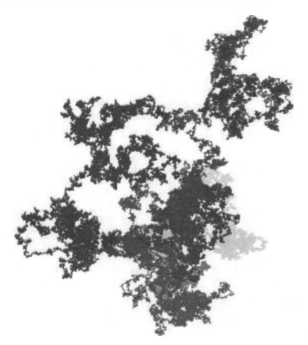

Figure 7. A walk on the provably normal base-4 digits of Stoneham's constant ($\alpha_{2,3}$)

Figure 8. A walk on the abnormal base-6 digits of Stoneham's constant ($\alpha_{2,3}$)

From these three figures, it is clear that while the base-3 and base-4 graphs appear to be plausibly random (since they are similar in overall structure to Figures 3 and 4), the base-6 walk is vastly different, mostly a horizontal line. Indeed, we discovered the fact that $\alpha_{2,3}$ fails to be 6-normal by a similar empirical analysis of the base-6 digits—there are long stretches of zeroes in the base-6 expansion [5]. Results of this type are given in [1] for numerous other constants besides π, both "man-made" and "natural."

Such results certainly do not constitute formal proofs, but they do suggest, often in dramatic form, as we have seen, that certain constants are not normal, and can further be used to bound statistical measures of randomness. For example, remarkable structure was uncovered in the normal Stoneham numbers [1]. Moreover, many related quantitative measures of random walks were examined, as were other graphical representations. Much related information, including animations, is stored at http://carma.newcastle.edu.au/walks/.

6. OTHER UNANSWERED QUESTIONS.

Mathematical questions. There are, of course, numerous other unanswered mathematical questions that can be posed about π.

1. Are the continued fraction terms of π bounded or unbounded? The continued fraction expansion provides information regarding how accurately π can be written as a fraction.

2. Is the limiting fraction of zeroes in the binary expansion of π precisely $1/2$? Is the limiting fraction of zeroes in the decimal expansion precisely $1/10$? We do not know the answers to these questions even for simple algebraic constants such as $\sqrt{2}$, much less π.

3. Are there infinitely many ones in the ternary expansion of π? Are there infinitely many sevens in the decimal expansion of π? Sadly, we cannot definitively answer such basic questions one way or the other.

Meta-mathematical questions. For that matter, there are numerous historical questions that are worth asking, if only rhetorically.

1. Why was π not known more accurately in ancient times? It could have been known to at least two-digit accuracy by making careful measurements with a rope.

2. Why did Archimedes, in spite of his astonishing brilliance in geometry and calculus, fail to grasp the notion of positional, zero-based decimal arithmetic? This would have greatly facilitated his computations (and likely would have changed history as well).

3. Why did Indian mathematicians fail to extend their system of decimal arithmetic for integers to decimal fractions? Decimal fraction notation was first developed in the Arabic world in the 12th century. They managed by scaling their results, but missed the obvious.

4. Why did Gauss and Ramanujan fail to exploit their respective identities for π? After all, the Salamin-Brent quadratically convergent algorithm for π is derived directly from some identities of Gauss, and other algorithms for π follow from (then largely unproven) formulas of Ramanujan. For that matter, why was the notion of an algorithm, fundamental in our computer age, so foreign to their way of thinking?

5. Why did centuries of mathematicians fail to find the BBP formula for π, namely formula (6), not to mention the associated "trick" for computing digits at an arbitrary starting position? After all, as mentioned above, it can be proven in just a few steps with freshman-level calculus.

In any event, it is clear that modern computer technology has changed the game for π. Modern systems are literally billions of times faster and more capacious than their predecessors when the present authors began their careers, and advances in software (such as fast Fourier transforms for high-precision numerical computation and symbolic computing facilities for algebraic manipulations) have improved computational productivity just as much as hardware improvements.

And computers are no longer merely passive creatures. A computer program discovered the BBP formula for π, as well as similar formulas for numerous other constants. Other formulas for π have been discovered by computer in a similar way, using high-precision implementations of the PSLQ algorithm or related integer relation algorithms.

Two unproven facts. In some of these cases, such as the following two formulas, proofs remain elusive:

$$\frac{4}{\pi^3} \overset{?}{=} \sum_{k=0}^{\infty} \frac{r^7(k)\left(1 + 14k + 76k^2 + 168k^3\right)}{8^{2k+1}}, \tag{11}$$

$$\frac{2048}{\pi^4} \overset{?}{=} \sum_{k-0}^{\infty} \frac{\left(\frac{1}{4}\right)_k \left(\frac{1}{2}\right)_k^7 \left(\frac{3}{4}\right)_k}{(1)_k^9 \, 2^{12k}} \left(21 + 466k + 4340k^2 + 20632k^3 + 43680k^4\right), \tag{12}$$

where, in the first (due to Gourevich in 2001), $r(k) = 1/2 \cdot 3/4 \cdots (2k-1)/(2k)$, and, in the second (due to Cullen in 2010), the notation $(x)_n = x(x+1)(x+1) \cdots (x+n-1)$ is the *Pochhammer symbol*.

7. CONCLUSION. The mathematical constant π has intrigued both the public and professional mathematicians for millennia. Countless facts have been discovered about π and published in the mathematical literature. But, as we have seen, much misunderstanding abounds. We must also warn the innocent reader to beware of mathematical terrorists masquerading as nice people, in their evil attempt to replace π by $\tau = 2\pi$ (which is pointless in any event, since the binary expansion of τ is the same as π, except for a shift of the decimal point).[12]

Yet there are still very basic questions that remain unanswered, among them whether (and why) π is normal to any base. Indeed, why do the basic algorithms of arithmetic, implemented to compute constants such as π, produce such random-looking results? And can we reliably exploit these randomness-producing features for benefit, say, as commercial-quality pseudorandom number generators?

Other challenges remain as well. But the advent of the computer might at last give humankind the power to answer some of them. Will computers one day be smarter than human mathematicians? Probably not any time soon, but for now they are remarkably pleasant research assistants.

ACKNOWLEDGMENTS. David H. Bailey was supported in part by the Director, Office of Computational and Technology Research, Division of Mathematical, Information, and Computational Sciences of the U.S. Department of Energy, under contract number DE-AC02-05CH11231.

REFERENCES

1. F. J. Aragón Artacho, D. H. Bailey, J. M. Borwein, P. B. Borwein, Walking on real numbers, *Mathematical Intelligencer* **35** (2013) 42–60.
2. D. H. Bailey, FFTs in external or hierarchical memory, *Journal of Supercomputing* **4** (1990) 23–35.
3. D. H. Bailey, R. Barrio, J. M. Borwein, High precision computation: Mathematical physics and dynamics, *Applied Mathematics and Computation* **218** (2012) 10106–10121, available at http://dx.doi.org/10.1016/j.amc.2012.03.087.
4. D. H. Bailey, J. M. Borwein, Ancient Indian square roots: An exercise in forensic paleo-mathematics, *Amer. Math. Monthly* **119** (2012) 646–657.
5. ———, Nonnormality of Stoneham constants, *Ramanujan Journal* **29** (2012) 409–422, available at http://dx.doi.org/10.1007/s11139-012-9417-3.
6. D. H. Bailey, J. M. Borwein, R. E. Crandall, J. Zucker, Lattice sums arising from the Poisson equation, *Journal of Physics A: Mathematical and Theoretical* **46** (2013) 115–201.
7. D. H. Bailey, J. M. Borwein, C. S. Calude, M. J. Dinneen, M. Dumitrescu, A. Yee, An empirical approach to the normality of pi, *Experimental Mathematics* **21** no. 4 (2012) 375–384.

---

[12]See http://tauday.com/ and www.pbs.org/newshour/rundown/2013/03/for-the-love-of-pi-and-the-tao-of-tau.html.

8. D. H. Bailey, J. M. Borwein, R. E. Crandall, C. Pomerance, On the binary expansions of algebraic numbers, *Journal of Number Theory Bordeaux* **16** (2004) 487–518.

9. D. H. Bailey, P. B. Borwein, S. Plouffe, On the rapid computation of various polylogarithmic constants, *Mathematics of Computation* **66** (1997) 903–913, available at http://dx.doi.org/10.1090/S0025-5718-97-00856-9.

10. D. H. Bailey, R. E. Crandall, Random generators and normal numbers, *Experimental Mathematics* **11** (2002) 527–546, available at http://dx.doi.org/10.1080/10586458.2002.10504704.

11. D. H. Bailey, M. Misiurewicz, A strong hot spot theorem, *Proceedings of the American Mathematical Society* **134** (2006) 2495–2501, available at http://dx.doi.org/10.1090/S0002-9939-06-08551-0.

12. L. Berggren, J. M. Borwein, P. B. Borwein, *A Sourcebook on Pi*. Third edition. Springer, New York, 2004.

13. B. Bolt, *More Mathematical Activities: A Further Resource Book for Teachers*. Cambridge University Press, Cambridge, 1985.

14. J. M. Borwein, The Life of Pi, extended and updated version of "La vita di pi greco," volume 2 of *Mathematics and Culture*, *La matematica: Problemi e teoremi*, Guilio Einaudi Editori, Turino, Italian, 2008. (French, in press). 532–561, in *From Alexandria Through Baghdad: Surveys and Studies in the Ancient Greek and Medieval Islamic Mathematical Sciences in Honor of J. L. Berggren*. Edited by N. Sidol and G. Van Brummelen. Springer, New York, 2014.

15. J. M. Borwein, D. H. Bailey, *Mathematics by Experiment: Plausible Reasoning in the 21st Century*. Second expanded edition, A. K. Peters, Natick, MA, 2008.

16. J. M. Borwein, P. B. Borwein, D. H. Bailey, Ramanujan, modular equations, and approximations to pi, or how to compute one billion digits of pi, *Amer. Math. Monthly* **96** (1989) 201–219.

17. R. P. Brent, Fast multiple-precision evaluation of elementary functions, *Journal of the ACM* **23** (1976) 242–251, available at http://dx.doi.org/10.1145/321941.321944.

18. Kate Bush, Aerial (Audio CD), Sony, 2005.

19. R. Corliss, Unlocking the Matrix, *Time*, 12 May 2003, available at http://www.time.com/time/magazine/article/0,9171,1004809,00.html.

20. H. R. P. Ferguson, D. H. Bailey, S. Arno, Analysis of PSLQ, an integer relation finding algorithm, *Mathematics of Computation* **68** (1999) 351–369, available at http://dx.doi.org/10.1090/S0025-5718-99-00995-3.

21. The Works of Archimedes, *Great Books of the Western World*, Edited by R. M. Hutchins. Translated by T. L. Heath. Vol. 11, Encyclopedia Britannica, 1952. 447–451.

22. S. Hilburn, Wife of Pi, cartoon, 8 Feb 2013, available at http://www.gocomics.com/theargylesweater/2013/02/08.

23. House Resolution 224 (111th): Supporting the designation of Pi Day, and for other purposes, 9 Mar 2009, available at http://www.govtrack.us/congress/bills/111/hres224.

24. G. Ifrah, *The Universal History of Numbers: From Prehistory to the Invention of the Computer*. Translated by David Bellos, E. F. Harding, Sophie Wood and Ian Monk, John Wiley and Sons, New York, 2000.

25. E. Karrels, Computing digits of pi with CUDA, 14 Mar 2013, available at http://www.karrels.org/pi.

26. G. Marsaglia, On the randomness of pi and other decimal expansions, 2005, available at http://interstat.statjournals.net/YEAR/2005/articles/0510005.pdf.

27. K. Plofker, *Mathematics in India*. Princeton University Press, Princeton, NJ, 2009.

28. R. Stoneham, On absolute (j, ε)-normality in the rational fractions with applications to normal numbers, *Acta Arithmetica* **22** (1973) 277–286.

29. W. Symborska, PI, *Poems, New and Collected, 1957–1997*. Houghton Mifflin Harcourt, Boston, 2000.

DAVID H. BAILEY. In June 2013, David H. Bailey retired from the Lawrence Berkeley National Laboratory (LBNL), after 15 years of service, although he still holds a Research Fellow appointment at the University of California, Davis. Prior to coming to LBNL in 1998, he was a computer scientist for 15 years at the NASA Ames Research Center. In the field of high-performance scientific computing, Bailey has published research in numerical algorithms, parallel computing, high-precision computation and supercomputer performance. He was the first recipient of the Sidney Fernbach Award from the IEEE Computer Society, and more recently the Gordon Bell Prize from the Association for Computing Machinery.

Bailey is also active in computational and experimental mathematics, applying techniques from high performance computing to problems in research mathematics. His best-known paper in this area (co-authored with Peter Borwein and Simon Plouffe) describes what is now known as the "BBP" formula for pi. In two more recent papers, Bailey and the late Richard Crandall demonstrated a connection between these formulas and a fundamental question about the digit randomness of pi. Bailey has received the Chauvenet Prize and the Merten Hesse Prize from the Mathematical Association of America, in each case jointly with Jonathan Borwein and Peter Borwein.

Lawrence Berkeley National Laboratory, Berkeley, CA 94720
DHBailey@lbl.gov

JONATHAN MICHAEL BORWEIN is a Laureate Professor of mathematics at the University of Newcastle, NSW Australia. His research interests include functional analysis, optimization, high performance computing, number theory, and he is one of the leading advocates of experimental mathematics. He has authored over a dozen books and more than 400 articles. With more than 5500 ISI citations, he is an ISI "highly cited mathematician."

Prof. Borwein received his D.Phil. from Oxford University in 1974 as a Rhodes Scholar. He has worked at Dalhousie University (1974–1991, 2004–2009), Carnegie Mellon University (1980–1982), the University of Waterloo (1991–1993), and Simon Fraser University (1993–2003). He is a fellow of the Royal Society of Canada, the American Association for the Advancement of Science, the Australian Academy of Science and a foreign member of the Bulgarian Academy of Sciences. Jon was a Governor at large of the Mathematical Association of America (2004–2007), is past president of the Canadian Mathematical Society (2000–2002) and past chair of NATO's scientific programs. He currently chairs the Scientific Advisory Committee of the Australian Mathematical Sciences Institute (AMSI). In 1993 he shared the Chauvenet Prize with David Bailey and his brother Peter. Among many other achievements, he is known for co-authoring the Borwein–Preiss variational principle, and Borwein's algorithm for pi.

Centre for Computer Assisted Research Mathematics and its Applications (CARMA),
University of Newcastle, Callaghan, NSW 2308, Australia
jonathan.borwein@newcastle.edu.au

Haiku 2:57

I will come to bed
when all three numbers in the
clock are prime . . . again.

—Submitted by Terry Trowbridge

---

24. The Life of π (2014)

Paper 24: Jonathan M. Borwein, "The life of Pi: From Archimedes to ENIAC and beyond," extended and updated version of "La vita di pi greco," volume 2 of *Mathematics and Culture, La mathematica: Problemi e teoremi*, Guilio Einaudi Editori, Turino, Italian, 2008 (French, in press). Pages 523–561 of *From Alexandria, through Baghdad: Surveys and Studies in the Ancient Greek and Medieval Islamic Mathematical Sciences in Honor of J. L. Berggren*, Sidoli, Nathan; Van Brummelen, Glen (eds.), Springer-Verlag, Berlin, 2014. With permission of Springer.

Synopsis:

This paper presents the panorama of π through the ages, with a brief summary of π from Archimedes and others of antiquity to Renaissance times and finally to the computer age. Significant mathematical detail is provided, including a reprise of Ivan Niven's 1947 elegant proof that π is irrational, numerous formulas that have been used to compute π, quadratically convergent algorithms for π, the BBP formula and algorithm for π and various curiosities, such as the fact that

$$\int_0^t \frac{(1-x)^4 x^4}{1+x^2}\,\mathrm{d}x = \frac{t^7}{7} - \frac{2t^6}{3} + t^5 - \frac{4t^3}{3} + 4t - 4\arctan t,$$

from which one can prove the age-old approximation

$$0 < \int_0^1 \frac{(1-x)^4 x^4}{1+x^2}\,\mathrm{d}x = \frac{22}{7}\pi.$$

Keywords: Algorithms, Computation, General Audience, History, Irrationality

The Life of Pi: From Archimedes to ENIAC and Beyond [1]

JONATHAN M. BORWEIN, FRSC

Prepared for *Berggren Festschrift*   Draft VIII. 19/06/2012

Laureate Professor & Director CARMA
Research currently supported by the Australian Research Council.
Email: jonathan.borwein@newcastle.edu.au

1 Preamble: π and Popular Culture

The desire to understand π, the challenge, and originally the need, to calculate ever more accurate values of π, the ratio of the circumference of a circle to its diameter, has challenged mathematicians–great and less great—for many centuries. It has also, especially recently, provided compelling examples of computational mathematics. π, uniquely in mathematics, is pervasive in popular culture and the popular imagination.

I shall intersperse this largely chronological account of π's mathematical status with examples of its ubiquity. More details will be found in the selected references at the end of the chapter—especially in *Pi: a Source Book* [9]. In [9] all material not otherwise referenced may be followed up upon, as may much other material, both serious and fanciful. Other interesting material is to be found in [21], which includes attractive discussions of topics such as continued fractions and elliptic integrals.

Fascination with π is evidenced by the many recent popular books, television shows, and movies—even perfume—that have mentioned π. In the 1967 *Star Trek* episode "Wolf in the Fold," Kirk asks "*Aren t there some mathematical problems that simply can t be solved?*" And Spock 'fries the brains' of a rogue computer by telling it: "*Compute to the last digit the value of Pi.*" (Figure 1 illustrates how much more is now known, see also http://carma.newcastle.edu.au/piwalk.shtml.) The May 6, 1993 episode of The Simpsons has the character Apu boast "*I can recite pi to 40,000 places. The last digit is one.*"

In November 1996, MSNBC aired a Thanksgiving Day segment about π, including that scene from Star Trek and interviews with the present author and several other mathematicians at Simon Fraser University. The 1997 movie *Contact*, starring Jodie Foster, was based on the 1986 novel by noted astronomer Carl Sagan. In the book, the lead character searched for patterns in the digits of π, and after her mysterious experience found confirmation—that the universe had meaning— in the base-11 expansion of π. The 1997 book *The Joy of Pi* [11] has sold many thousands of copies and continues to sell well. The 1998 movie entitled *Pi* began with decimal digits of π displayed on the screen. And in the 2003 movie *Matrix Reloaded*, the Key Maker warns that a door will be accessible for exactly 314 seconds, a number that *Time* speculated was a reference to π.

As a striking example, imagine the following excerpt from Yann Mandel's 2002 Booker Prize winning novel *Life of Pi* being written about any other transcendental number:

> "My name is
> Piscine Molitor Patel
> known to all as Pi Patel.

For good measure I added

$$\pi = 3.14$$

and I then drew a large circle which I sliced in two with a diameter, to evoke that basic lesson of geometry."

Equally, National Public Radio reported on April 12, 2003 that novelty automatic teller machine withdrawal slips, showing a balance of $314,159.26$, were hot in New York City. One could jot a note on the back and,

---

[1]This paper is an updated and revised version of [13] and is made with permission of the editor.

apparently innocently, let the intended target be impressed by one's healthy saving account. Scott Simon, the host, noted the close resemblance to π. Correspondingly, according to the *New York Times* of August 18 2005, Google offered exactly "$14,159,265$ New Slices of Rich Technology" as the number of shares in its then new stock offering. Likewise, March 14 in North America has become π *Day*, since in the USA the month is written before the day ('314'). In schools throughout North America, it has become a reason for mathematics projects, especially focussing on π.

In another sign of true legitimacy, on March 14, 2007 the *New York Times* published a crossword in which to solve the puzzle, one had first to note that the clue for 28 DOWN was "March 14, to Mathematicians," to which the answer is PIDAY. Moreover, roughly a dozen other characters in the puzzle are PI—for example, the clue for 5 DOWN was "More pleased" with the six character answer HAPπER. The puzzle is reproduced in [14].

Other and more recent examples—including the US Congressional House Resolution 224 designating of National Pi Day in 2009—may be found at `http://www.carma.newcastle.edu.au/jon/piday.pdf` which is annually updated.

It is hard to imagine e, γ or $\log 2$ playing the same role. A corresponding scientific example [4, p. 11] is

> *A coded message, for example, might represent gibberish to one person and valuable information to another. Consider the number 14159265... Depending on your prior knowledge, or lack thereof, it is either a meaningless random sequence of digits, or else the fractional part of pi, an important piece of scienti c information.*

Again, a scientist can use π confident that it is part of shared societal knowledge—even if when pressed his definition might not be as good as Pi's.

For those who know *The Hitchhiker's Guide to the Galaxy*, it is amusing that 042 occurs at the digits ending at the fifty-billionth decimal place in each of π and $1/\pi$—thereby providing an excellent answer to the ultimate question of 'life, the universe and everything', which was "*What is forty two?*" A more intellectual offering is "The Deconstruction of Pi" given by Umberto Eco on page three of his 1988 book *Foucault s Pendulum*, [9, p. 658]. The title says it all.

Pi.   Our central character
$$\pi = 3.14159265358979323\ldots$$
is traditionally defined in terms of the area or perimeter of a unit circle, but see Figure 4 where the subtlety of showing the two are the same is illustrated. The notation of π itself was introduced by William Jones in 1737, replacing 'p' and the like, and was popularized by Leonhard Euler who is responsible for much modern nomenclature. A more formal modern definition of π uses the first positive zero of the sine function defined as a power series. The first thousand decimal digits of π are recorded in Figure 2.

Despite continuing rumours to the contrary, π is not equal to $22/7$ (see End Note 1). Of course $22/7$ is one of the early continued fraction approximations to π. The first six convergents are

$$3, \frac{22}{7}, \frac{333}{106}, \frac{355}{113}, \frac{103993}{33102}, \frac{104348}{33215}.$$

The convergents are necessarily good rational approximations to π. The sixth differs from π by only $3.31\,10^{-10}$. The corresponding simple continued fraction starts

$$\pi = [3, 7, 15, 1, 292, 1, 1, 1, 2, 1, 3, 1, 14, 2, 1, 1, 2, 2, 2, 2, 1, 84, 2, 1, 1, \ldots],$$

using the standard concise notation. This continued fraction is still very poorly understood. Compare that for e which starts

$$e = [2, 1, 2, 1, 1, 4, 1, 1, 6, 1, 1, 8, 1, 1, 10, 1, 1, 12, 1, 1, 14, 1, 1, 16, 1, 1, 18, \ldots].$$

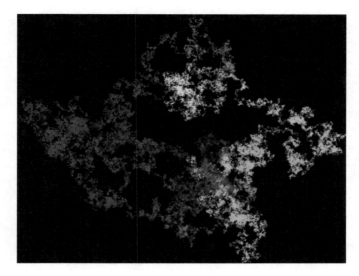

Figure 1: A 100 billion step planar walk on the binary digits of π. Colors change from red and orange to violet and indigo as we proceed (snapshot taken from image in [1])

3 . 1415926535897932384626433832795028841971693993751058209749445923078164062862089986280348253421170679
8214808651328230664709384460955058223172535940812848111745028410270193852110555964462294895493038196
4428810975665933446128475648233786783165271201909145648566923460348610454326648213393607260249141273
7245870066063155881748815209209628292540917153643678925903600113305305488204665213841469519415116094
3305727036575959195309218611738193261179310511854807446237996274956735188575272489122793818301194912
9833673362440656643086021394946395224737190702179860943702770539217176293176752384674818467669405132
0005681271452635608277857713427577896091736371787214684409012249534301465495858537105079227968925892 35
4201995611212902196086403441815981362977477130996051870721134999999837297804995105973173281609631859
5024459455346908302642522308253344685035261931188171010003137838752886587533208381420617177669147303
5982534904287554687311595628638823537875937519577818577805321712268066130019278766111959092164201989 3

Figure 2: 1,001 **Decimal digits of π with the '3' included**

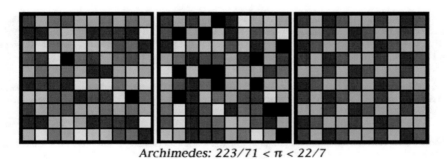

Archimedes: 223/71 < π < 22/7

Figure 3: A pictorial proof of Archimedes' inequalities

3

MEASUREMENT OF A CIRCLE.

Proposition 1.

The area of any circle is equal to a right-angled triangle in which one of the sides about the right angle is equal to the radius, and the other to the circumference, of the circle.

Let $ABCD$ be the given circle, K the triangle described.

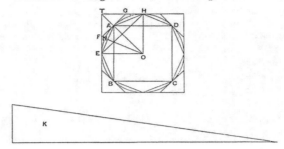

Figure 4: Construction showing uniqueness of π, taken from Archimedes' *Measurement of a Circle*

A proof of this observation shows that e is not a quadratic irrational since such numbers have eventually periodic continued fractions.

Archimedes' famous computation discussed below is:

(1)
$$3\frac{10}{71} < \pi < 3\frac{10}{70}.$$

Figure 3 shows this estimate graphically, with the digits shaded modulo ten; one sees structure in 22/7, less obviously in 223/71, and presumptively not in π.

2   The Childhood of π

Four thousand years ago, the Babylonians used, among other values, the approximation $3\frac{1}{8} = 3.125$. Then, or earlier, according to ancient papyri, Egyptians assumed a circle with diameter nine has the same area as a square of side eight, which implies $\pi = 256/81 = 3.1604\ldots$. Some have argued that the ancient Hebrews were satisfied with $\pi = 3$:

> *Also, he made a molten sea of ten cubits from brim to brim, round in compass, and   ve cubits the height thereof; and a line of thirty cubits did compass it round about.* (I Kings 7:23; see also 2 Chronicles 4:2)

One should know that the cubit was a personal not universal measurement. In Judaism's further defense, several millennia later, the great Rabbi Moses ben Maimon Maimonedes (1135–1204), also known as the RaM-BaM, is translated by Langermann, in "The 'true perplexity' [9, p. 753] as fairly clearly asserting the irrational nature of π:

4

"You ought to know that the ratio of the diameter of the circle to its circumference is unknown, nor will it ever be possible to express it precisely. This is not due to any shortcoming of knowledge on our part, as the ignorant think. Rather, this matter is unknown due to its nature, and its discovery will never be attained." (Maimonedes)

In each of these three cases the interest of the civilization in π was primarily in the practical needs of engineering, astronomy, water management and the like. With the Greeks, as with the Hindus, interest was centrally metaphysical and geometric.

Archimedes and π. Around 250 BCE, Archimedes of Syracuse (287–212 BCE) is thought to be the first to show that the "two possible Pi's" are the same. Clearly for a circle of radius r and diameter d, $Area = \pi_1 r^2$ while $Perimeter = \pi_2 d$, but that $\pi_1 = \pi_2$ is not obvious, and is often overlooked. Figure 4. reproduces his proof (construction) showing the coincidence of the two constants.

Archimedes' Method. The first rigorous mathematical calculation of π was also due to Archimedes, who used a brilliant scheme based on *d*oubling inscribed and circumscribed polygons

$$6 \mapsto 12 \mapsto 24 \mapsto 48 \mapsto 96$$

and computing the perimeters to obtain the bounds $3\frac{10}{71} < \pi < 3\frac{1}{7}$, that we have recapitulated above. The case of 6-gons and 12-gons is shown in Figure 5; for $n = 48$ one already 'sees' near-circles. Arguably no computational mathematics approached this level of rigour again until the 19th century. Phillips [9, pp. 15-19] calls Archimedes the 'first numerical analyst'.

Archimedes' scheme constitutes the first true algorithm for π, in that it is capable of producing an arbitrarily accurate value for π. It also represents the birth of numerical and error analysis—all without positional notation or modern trigonometry. As discovered severally in the 19th century, this scheme can be stated as a simple, numerically stable, recursion, as follows [15].

Archimedean Mean Iteration (Pfaff-Borchardt-Schwab). Set $a_0 = 2\sqrt{3}$ and $b_0 = 3$—the values for circumscribed and inscribed 6-gons. Then define

$$(2) \qquad a_{n+1} = \frac{2a_n b_n}{a_n + b_n} \quad (H) \qquad b_{n+1} = \sqrt{a_{n+1} b_n} \quad (G).$$

This converges to π, with the error decreasing by a factor of four with each iteration. In this case the error is easy to estimate, the limit somewhat less accessible but still reasonably easy [14, 15].

Variations of Archimedes' geometrical scheme were the basis for all high-accuracy calculations of π for the next 1800 years—well beyond its 'best before' date. For example, in fifth century CE China, Tsu Chung-Chih used a variation of this method to get π correct to seven digits. A millennium later, al-Kāshī in Samarkand *"who could calculate as eagles can y"* obtained 2π in *sexagecimal*:

$$2\pi \approx 6 + \frac{16}{60^1} + \frac{59}{60^2} + \frac{28}{60^3} + \frac{01}{60^4} + \frac{34}{60^5} + \frac{51}{60^6} + \frac{46}{60^7} + \frac{14}{60^8} + \frac{50}{60^9},$$

good to 16 decimal places (using $3 \cdot 2^{28}$-gons). This is a personal favourite; reentering it in my computer centuries later and getting the predicted answer gave me horripilation ('goose-bumps').

3   Pre-calculus Era π Calculations

In Figures 6, 8, and 11 we chronicle the main computational records during the indicated period, only commenting on signal entries.

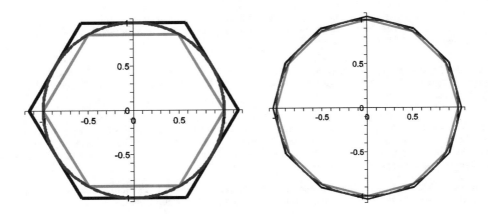

Figure 5: Archimedes' method of computing π with 6- and 12-gons

Little progress was made in Europe during the 'dark ages', but a significant advance arose in India (450 CE): *modern positional, zero-based decimal arithmetic*—the "Indo-Arabic" system. This greatly enhanced arithmetic in general, and computing π in particular. The Indo-Arabic system arrived with the Moors in Europe around 1000 CE. Resistance ranged from accountants who feared losing their livelihood to clerics who saw the system as 'diabolical'—they incorrectly assumed its origin was Islamic. European commerce resisted into the 18th century, and even in scientific circles usage was limited until the 17th century. This is, of course a greatly simplified version of extraordinary events, for a recent article on the matter see [6].

The prior difficulty of doing arithmetic is indicated by college placement advice given a wealthy German merchant in the 16th century:

> *A wealthy (15th Century) German merchant, seeking to provide his son with a good business edu-cation, consulted a learned man as to which European institution o ered the best training. If you only want him to be able to cope with addition and subtraction, the expert replied, then any French or German university will do. But if you are intent on your son going on to multiplication and division assuming that he has su cient gifts then you will have to send him to Italy. " (George Ifrah, [14])*

Discussions about Roman arithmetic continue. Claude Shannon (1916–2001) had a mechanical calculator wryly called *Throback 1* built to compute in Roman, at Bell Labs in 1953 to show that it was practicable, if a tad messy, to compute using Roman numerals!

Ludolph van Ceulen (1540–1610). The last great Archimedean calculation, performed by van Ceulen using 2^{62}-gons—to 39 places with 35 correct—was published posthumously. The number is still called Ludolph's number in parts of Europe and was inscribed on his head-stone. This head-stone disappeared centuries ago but was rebuilt, in part from surviving descriptions, recently as shown in Figure 7. It was reconsecrated on July 5th 2000 with Dutch royalty in attendance. Ludolph van Ceulen, a very serious mathematician, was also the discoverer of the double angle formula for the cosine.

6

Name	Year	Digits
Babylonians	2000? BCE	1
Egyptians	2000? BCE	1
Hebrews (1 Kings 7:23)	550? BCE	1
Archimedes	250? BCE	3
Ptolemy	150	3
Liu Hui	263	5
Tsu Ch'ung Chi	480?	7
Al-Kashi	1429	14
Romanus	1593	15
van Ceulen (**Ludolph's number**[*])	1615	35

Figure 6: Pre-calculus π Calculations

4   Pi's Adolescence

François Viéte (1540–1603). The dawn of modern mathematics appears in *Viête s* or *Viêta s product* (1579)

$$\frac{2}{\pi} = \frac{\sqrt{2}}{2} \frac{\sqrt{2+\sqrt{2}}}{2} \frac{\sqrt{2+\sqrt{2+\sqrt{2}}}}{2} \cdots$$

considered to be the first truly infinite product; and in the *rst in nite continued fraction* for $2/\pi$ given by Lord Brouncker (1620–1684), first President of the Royal Society of London:

$$\frac{2}{\pi} = \cfrac{1}{1 + \cfrac{9}{2 + \cfrac{25}{2 + \cfrac{49}{2 + \cdots}}}}.$$

This was based on the following brilliantly 'interpolated' product of John Wallis[2] (1616–1703):

$$(3) \qquad \prod_{k=1}^{\infty} \frac{4k^2 - 1}{4k^2} = \frac{2}{\pi},$$

which led to the discovery of the Gamma function (see End Note 2) and a great deal more. Variations on these formulas of Víete and Wallis continue to be published.

A flavour of Viéte's writings can be gleaned in this quotation from his work, first given in English in [9, p. 759]. What we now take for granted was reason for much passionate argument.

> *"Arithmetic is absolutely as much science as geometry [is]. Rational magnitudes are conveniently designated by rational numbers, and irrational [magnitudes] by irrational [numbers]. If someone measures magnitudes with numbers and by his calculation get them di erent from what they really are, it is not the reckoning s fault but the reckoner s.*
>
> *Rather, says Proclus,* ARITHMETIC IS MORE EXACT THAN GEOMETRY.[3] *To an accurate calculator, if the diameter is set to one unit, the circumference of the inscribed dodecagon will be the side of the binomial [i.e. square root of the di erence] $72 - \sqrt{3888}$. Whosoever declares any other result, will be mistaken, either the geometer in his measurements or the calculator in his numbers."* (Viéte)

---

[2] One of the few mathematicians whom Newton admitted respecting, and also a calculating prodigy!
[3] The capitalized phrase was written in Greek.

7

Figure 7: Ludolph's rebuilt tombstone in Leiden

This fluent rendition is due to Marinus Taisbak, and the full text is worth reading. It certainly underlines how influential an algebraist and geometer Viéte was—as an early proponent of methods we now take for granted. Viéte, who was the first to introduce literals ('x' and 'y') into algebra, nonetheless rejected the use of negative numbers.

Leonard Euler (1707–1783). Not surprisingly the great Euler 'master of us all' [20] made many contributions to the literature on π. Equation (3) may be derived from Euler's product formula for π, given below in (4), with $x = 1/2$, or by repeatedly integrating $\int_0^{\pi/2} \sin^{2n}(t)\,dt$ by parts. One may divine (4) as Euler did by *considering* $\sin(\pi x)$ *as an infinite polynomial* and obtaining a product in terms of the roots—which are $0, \{1/n^2 : n = \pm 1, \pm 2, \cdots\}$. It is thus plausible that

$$(4) \qquad\qquad \frac{\sin(\pi x)}{x} \;=\; c \prod_{n=1}^{\infty} \left(1 - \frac{x^2}{n^2}\right).$$

Euler in 1735, full well knowing that the whole argument was heuristic, argued that, as with a polynomial, c was the value at zero, 1, and the coefficient of x^2 in the Taylor series must be the sum of the roots. Hence, he was able to pick off coefficients to evaluate the *zeta-function* at two:

$$\zeta(m) := \sum_{n \geq 1} \frac{1}{n^m} \text{ and marvellously } \zeta(2) = \frac{\pi^2}{6}.$$

The explicit formula for $\zeta(2)$ solved the so called *Basel problem* posed in 1644. This method also leads to the evaluation of $\zeta(2n)$ as a rational multiple of π^{2n}:

$$\zeta(2) = \frac{\pi^2}{6},\ \zeta(4) = \frac{\pi^4}{90},\ \zeta(6) = \frac{\pi^6}{945},\ \zeta(8) = \frac{\pi^8}{9450}, \cdots$$

Indeed, it produces:

$$\zeta(2m) = (-1)^{m-1} \frac{(2\pi)^{2m}}{2\,(2m)!}\, B_{2m},$$

in terms of the *Bernoulli numbers*, B_n, where $t/(\exp(t) - 1) = \sum_{n \geq 0} B_n t^n/n!$, gives a generating function for the B_n which are perforce rational; see also [28].

Much less is known about odd integer values of ζ, though they are almost certainly not rational multiples of powers of π. More than two centuries later, in 1976 Roger Apéry, [9, p. 439], [15], showed $\zeta(3)$ to be irrational, and we now also can prove that *at least one of* $\zeta(5), \zeta(7), \zeta(9)$ or $\zeta(11)$ is irrational, but we cannot guarantee which one. All positive integer values of ζ are strongly believed to be irrational. Though it is not relevant to our story, Euler's work on the zeta-function also lead ultimately to the celebrated Riemann hypothesis [14].

5 Pi's Adult Life with Calculus

In the later 17th century, Newton and Leibniz founded the calculus, and this powerful tool was quickly exploited to find new formulae for π. One early calculus-based formula comes from the integral:

$$\tan^{-1} x \;=\; \int_0^x \frac{dt}{1+t^2} \;=\; \int_0^x (1 - t^2 + t^4 - t^6 + \cdots)\, dt$$

$$=\; x - \frac{x^3}{3} + \frac{x^5}{5} - \frac{x^7}{7} + \frac{x^9}{9} - \cdots$$

Substituting $x = 1$ *formally* proves the well-known *Gregory-Leibniz formula* (1671–74)

$$\text{(5)} \qquad \frac{\pi}{4} \;=\; 1 - \frac{1}{3} + \frac{1}{5} - \frac{1}{7} + \frac{1}{9} - \frac{1}{11} + \cdots$$

James Gregory (1638–75) was the greatest of a large Scottish mathematical family. The point, $x = 1$, however, is on the boundary of the interval of convergence of the series. Justifying substitution requires a careful error estimate for the remainder or Lebesgue's monotone convergence theorem, but most introductory texts ignore the issue, as they do the issue of Figure 4. The arctan integral and series was known centuries earlier to the Kerala school: identified with Madhava (c. 1350 – c. 1425) of Sangamagrama near Kerala, India, who may well have computed 13 digits of π by methods similar to those described in the next section.

A Curious Anomaly in the Gregory Series. In 1988, it was observed that Gregory's series for π,

$$\text{(6)} \qquad \pi \;=\; 4 \sum_{k=1}^{\infty} \frac{(-1)^{k+1}}{2k-1} \;=\; 4 \left(1 - \frac{1}{3} + \frac{1}{5} - \frac{1}{7} + \frac{1}{9} - \frac{1}{11} + \cdots \right)$$

when truncated to 5,000,000 terms, differs strangely from the true value of π:

```
3.14159245358979323846464338327950278419716939938730582097494182
230781640...
3.14159265358979323846264338327950288419716939937510582097494459
230781640...
         2           -2           10           -122           2770
```

Values differ as expected from truncating an alternating series, in the seventh place—a "4" which should be a "6." But the next 13 digits are correct, and after another blip, for 12 digits. Of the first 46 digits, only four differ from the corresponding digits of π. Further, the "error" digits seemingly occur with a period of 14, as shown above. Such anomalous behavior begs for explanation. A great place to start is by using Neil Sloane's Internet-based integer sequence recognition tool, available at `www.research.att.com/~njas/sequences`. This tool has no difficulty recognizing the sequence of errors as twice *Euler numbers*. Even Euler numbers are generated by $\sec x = \sum_{k=0}^{\infty} (-1)^k E_{2k} x^{2k}/(2k)!$. The first few are $1, -1, 5, -61, 1385, -50521, 2702765$. This discovery led to the following asymptotic expansion:

$$\text{(7)} \qquad \frac{\pi}{2} - 2 \sum_{k=1}^{N/2} \frac{(-1)^{k+1}}{2k-1} \;\approx\; \sum_{m=0}^{\infty} \frac{E_{2m}}{N^{2m+1}}.$$

9

Name	Year	Correct Digits
Sharp (and Halley)	1699	71
Machin	1706	100
Strassnitzky and Dase	1844	200
Rutherford	1853	440
Shanks	1874	(707) 527
Ferguson (**Calculator**)	1947	808
Reitwiesner et al. (**ENIAC**)	1949	2,037
Genuys	1958	10,000
Shanks and Wrench	1961	100,265
Guilloud and Bouyer	1973	1,001,250

Figure 8: Calculus π Calculations

Now the genesis of the anomaly is clear: by chance the series had been truncated at 5,000,000 terms—exactly one-half of a fairly large power of ten. Indeed, setting $N = 10,000,000$ in Equation (7) shows that the first hundred or so digits of the truncated series value are small perturbations of the correct decimal expansion for π. And the asymptotic expansions show up on the computer screen, as we observed above.

On a hexadecimal computer with $N = 16^7$ the corresponding strings and hex-errors are:

3.243F6A8885A308D313198A2E03707344A4093822299F31D0082EFA98EC4E6C8
9452821E...
3.243F6A6885A308D31319AA2E03707344A3693822299F31D7A82EFA98EC4DBF6
9452821E...
        2   -2   A   -7A   2AD2

with the first being the correct value of π. (In hexadecimal or *hex* one uses 'A,B, ..., F' to write 10 through 15 as single 'hex-digits'.) Similar phenomena occur for other constants. (See [9].) Also, knowing the errors means we can correct them and use (7) to make Gregory's formula computationally tractable, notwithstanding the following discussion of complexity!

6   Calculus Era π Calculations

Used naively, the beautiful formula (5) is computationally useless—so slow that hundreds of terms are needed to compute two digits. Sharp, under the direction of Halley[4], (see Figure 8) actually used $\tan^{-1}(1/\sqrt{3})$ which is geometrically convergent. Moreover, Euler's (1738) trigonometric identity

$$(8) \qquad \tan^{-1}(1) = \tan^{-1}\left(\frac{1}{2}\right) + \tan^{-1}\left(\frac{1}{3}\right)$$

produces the geometrically convergent rational series

$$(9) \qquad \frac{\pi}{4} = \frac{1}{2} - \frac{1}{3 \cdot 2^3} + \frac{1}{5 \cdot 2^5} - \frac{1}{7 \cdot 2^7} + \cdots + \frac{1}{3} - \frac{1}{3 \cdot 3^3} + \frac{1}{5 \cdot 3^5} - \frac{1}{7 \cdot 3^7} + \cdots$$

An even faster formula, found earlier by John Machin, lies similarly in the identity

$$(10) \qquad \frac{\pi}{4} = 4\tan^{-1}\left(\frac{1}{5}\right) - \tan^{-1}\left(\frac{1}{239}\right).$$

---

[4]The astronomer and mathematician who largely built the Greenwich Observatory and after whom the comet is named.

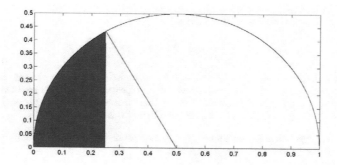

Figure 9: Newton's method of computing Pi: *"I am ashamed to tell you to how many gures I carried these computations, having no other business at the time."* Issac Newton, 1666.

This was used in numerous computations of π, given in Figure 8, starting in 1706 and culminating with Shanks' famous computation of π to 707 decimal digits accuracy in 1873 (although it was *found in 1945 to be wrong* after the 527-th decimal place, by Ferguson, during the last adding machine-assisted pre-computer computations[5]).

Newton's arcsin computation. Newton discovered a different more effective—actually a disguised arcsin—formula. He considering the area A of the left-most region shown in Figure 9. Now, A is the integral

(11)
$$A \;=\; \int_0^{1/4} \sqrt{x - x^2}\, dx.$$

Also, A is the area of the circular sector, $\pi/24$, less the area of the triangle, $\sqrt{3}/32$. Newton used his newly developed *binomial theorem* in (11):

$$A \;=\; \int_0^{\frac{1}{4}} x^{1/2}(1-x)^{1/2}\, dx = \int_0^{\frac{1}{4}} x^{1/2}\left(1 - \frac{x}{2} - \frac{x^2}{8} - \frac{x^3}{16} - \frac{5x^4}{128} - \cdots\right) dx$$
$$= \int_0^{\frac{1}{4}} \left(x^{1/2} - \frac{x^{3/2}}{2} - \frac{x^{5/2}}{8} - \frac{x^{7/2}}{16} - \frac{5x^{9/2}}{128}\cdots\right) dx$$

Integrate term-by-term and combining the above produces

$$\pi = \frac{3\sqrt{3}}{4} + 24\left(\frac{1}{3\cdot 8} - \frac{1}{5\cdot 32} - \frac{1}{7\cdot 128} - \frac{1}{9\cdot 512}\cdots\right).$$

Newton used this formula to compute 15 digits of π. As noted, he later 'apologized' for "having no other business at the time." (1665-1666 was the year of the great plague which closed Cambridge, and of the great fire of London of September 1666.). It was also directly after Newton's *Annus mirabilis* that led ultimately to his *Principia*.) A standard chronology ([26] and [9, p. 294]) says *Newton signi cantly never gave a value for π."* *Caveat emptor*, all users of secondary sources.

The Viennese *computer*. Until quite recently—around 1950—a computer was a person. As a teenager, this computer, one Johan Zacharias Dase (1824–1861), would demonstrate his extraordinary computational skill by, for example, multiplying

$$79532853 \times 93758479 = 7456879327810587$$

---
[5]This must be some sort a record for the length of time needed to detect a mathematical error.

in 54 seconds; two 20-digit numbers in six minutes; two 40-digit numbers in 40 minutes; two 100-digit numbers in 8 hours and 45 minutes. In 1844, after being shown

$$\frac{\pi}{4} = \tan^{-1}\left(\frac{1}{2}\right) + \tan^{-1}\left(\frac{1}{5}\right) + \tan^{-1}\left(\frac{1}{8}\right)$$

he calculated π to 200 places *in his head* in two months, completing correctly—to my mind—the greatest mental computation ever. Dase later calculated a seven-digit logarithm table, and extended a table of integer factorizations to 10,000,000. On Gauss's recommendation Dase was hired to assist this project, but Dase died not long afterwards in 1861 by which time Gauss himself already was dead.

An amusing *Machin-type identity*—meaning an equation that expresses π as a linear combination of arctan's—due to the Oxford logician Charles Dodgson is

$$\tan^{-1}\left(\frac{1}{p}\right) = \tan^{-1}\left(\frac{1}{p+q}\right) + \tan^{-1}\left(\frac{1}{p+r}\right),$$

valid whenever $1 + p^2$ factors as qr. Dodgson is much better known as Lewis Carroll, the author of *Alice in Wonderland*.

7   The Irrationality and Transcendence of π

One motivation for computations of π was very much in the spirit of modern experimental mathematics: to see if the decimal expansion of π repeats, which would mean that π is the ratio of two integers (i.e., rational), or to recognize π as *algebraic*—the root of a polynomial with integer coefficients—and later to look at digit distribution. The question of the *rationality of* π was settled in the late 1700s, when Lambert and Legendre proved (using continued fractions) that the constant is irrational.

The question of whether π was algebraic was settled in 1882, when Lindemann proved that π *is transcendental*. Lindemann's proof also settled, once and for all, the ancient Greek question of whether the circle could be squared with straight-edge and compass. It cannot be, because numbers that are the lengths of lines that can be constructed using ruler and compasses (often called *constructible numbers*) are necessarily algebraic, and squaring the circle is equivalent to constructing the value π. The classical Athenian playwright Aristophanes already 'knew' this and perhaps derided 'circle-squarers' ($\tau\varepsilon\tau\rho\alpha\gamma\omega\sigma\iota\varepsilon\iota\nu$) in his play *The Birds* of 414 BCE. Likewise, the French Academy had stopped accepting proofs of the three great constructions of antiquity—squaring the circle, doubling the cube and trisecting the angle—centuries before it was proven impossible.

We next give, *in extenso*, Ivan Niven's 1947 short proof of the irrationality of π. It well illustrates the ingredients of more difficult later proofs of irrationality of other constants, and indeed of Lindemann's proof of the transcendence of π building on Hermite's 1873 proof of the transcendence of e.

8   A Proof that π is Irrational

Proof. *Let* $\pi = a/b$, *the quotient of positive integers. We de ne the polynomials*

$$f(x) = \frac{x^n(a - bx)^n}{n!}$$

$$F(x) = f(x) - f^{(2)}(x) + f^{(4)}(x) - \cdots + (-1)^n f^{(2n)}(x)$$

the positive integer being speci ed later. Since $n!f(x)$ *has integral coe cients and terms in* x *of degree not less than* n, $f(x)$ *and its derivatives* $f^{(j)}(x)$ *have integral values for* $x = 0$; *also for* $x = \pi = a/b$, *since*

12

$f(x) = f(a/b - x)$. *By elementary calculus we have*

$$\frac{d}{dx}\{F'(x)\sin x - F(x)\cos x\} = F''(x)\sin x + F(x)\sin x = f(x)\sin x$$

and

$$\int_0^\pi f(x)\sin x\,dx = [F'(x)\sin x - F(x)\cos x]_0^\pi$$

(12)
$$= F(\pi) + F(0).$$

Now $F(\pi) + F(0)$ is an integer, since $f^{(j)}(0)$ and $f^{(j)}(\pi)$ are integers. But for $0 < x < \pi$,

$$0 < f(x)\sin x < \frac{\pi^n a^n}{n!},$$

so that the integral in (12) is positive but arbitrarily small for n su ciently large. Thus (12) is false, and so is our assumption that π is rational. **QED**

Irrationality measures. We end this section by touching on the matter of *measures of irrationality*. The infimum $\mu(\alpha)$ of those $\mu > 0$ for which

$$\left|\alpha - \frac{p}{q}\right| \geq \frac{1}{q^\mu}$$

for all integers p, q with sufficiently large q, is called the *Liouville-Roth constant* for α and we say that we have an irrationality measure for α if $\mu(\alpha) < \infty$.

Irrationality measures are difficult. Roth's theorem, [15], implies that $\mu(\alpha) = 2$ for all algebraic irrationals, as is the case for almost all reals. Clearly, $\mu(\alpha) = 1$ for rational α and $\mu(\alpha) = \infty$ if and only if α is a so-called *Liouville number* such as $\sum 1/10^{n!}$. It is known that $\mu(e) = 2$ while in 1993 Hata showed that $\mu(\pi) \leq 8.02$. Similarly, it is known that $\mu(\zeta(2)) \leq 5.45, \mu(\zeta(3)) \leq 4.8$ and $\mu(\log 2) \leq 3.9$.

A consequence of the existence of an irrationality measure μ for π, is the ability to estimate quantities such as $\limsup |\sin(n)|^{1/n} = 1$ for integer n, since for large integer m and n with $m/n \to \pi$, we have eventually

$$|\sin(n)| = |\sin(m\pi) - \sin(n)| \geq \frac{1}{2}|m\pi - n| \geq \frac{1}{2\,m^{\mu-1}}.$$

Related matters are discussed at more length in [2].

9 π in the Digital Age

With the substantial development of computer technology in the 1950s, π was computed to thousands and then millions of digits. These computations were greatly facilitated by the discovery soon after of advanced algorithms for the underlying high-precision arithmetic operations. For example, in 1965 it was found that the newly-discovered *fast Fourier transform* (FFT) [15, 14] could be used to perform high-precision multiplications much more rapidly than conventional schemes. Such methods (e.g., for $\div, \sqrt{x}$ see [15, 16, 14]) dramatically lowered the time required for computing π and other constants to high precision. We are now able to compute algebraic values of algebraic functions essentially as fast as we can multiply, $O_B(M(N))$, where $M(N)$ is the cost of multiplication and O_B counts 'bits' or 'flops'. To convert this into practice: a state-of-the-art processor in 2010, such as the latest AMD Opteron, which runs at 2.4 GHz and has four floating-point cores, each of

13

which can do two 64-bit floating-point operations per second, can produce a total of 9.6 billion floating-point operations per second.

In spite of these advances, into the 1970s all computer evaluations of π still employed classical formulae, usually of Machin-type, see Figure 8. We will see below methods that can compute N digits of π with time complexity $O_B(M(N)) \log O_B(M(N))$. Proving that the log term is unavoidable, as seems likely, would yield an algorithmic proof—quite different from current proofs—that π is not algebraic. Such a proof would be a significant contribution to number theory and complexity theory.

Electronic Numerical Integrator and Calculator. The first computer calculation of π was performed on ENIAC—a behemoth with a tiny brain from today's vantage point. The ENIAC was built in Aberdeen, Maryland by the US Army:

> **Size/weight.** ENIAC had 18,000 vacuum tubes, 6,000 switches, 10,000 capacitors, 70,000 resistors, 1,500 relays, was 10 feet tall, occupied 1,800 square feet and weighed 30 tons.
> **Speed/memory.** A, now aged, 1.5GHz Pentium does 3 million adds/sec. ENIAC did 5,000, three orders faster than earlier machines. The first stored-memory computer, ENIAC stored 200 digits.
> **Input/output.** Data flowed from one accumulator to the next, and after each accumulator finished a calculation, it communicated its results to the next in line. The accumulators were connected to each other manually. The 1949 computation of π to 2,037 places on ENIAC took 70 hours in which output had to be constantly reintroduced as input.

A fascinating description of the ENIAC's technological and commercial travails is to be found in [25]. Note that ENIAC as a child of the 1940's was called a 'calculator' not a computer.

Ballantine's (1939) Series for π. Another formula of Euler for arccot is

$$x \sum_{n=0}^{\infty} \frac{(n!)^2 \, 4^n}{(2\,n+1)! \, (x^2+1)^{n+1}} \;=\; \arctan\left(\frac{1}{x}\right).$$

This, intriguingly and usefully, allowed Guilloud and Boyer to reexpress the formula, used by them in 1973 to compute a million digits of π, viz, $\pi/4 = 12 \arctan(1/18) + 8 \arctan(1/57) - 5 \arctan(1/239)$ in the efficient form

$$\pi \;=\; 864 \sum_{n=0}^{\infty} \frac{(n!)^2 \, 4^n}{(2\,n+1)! \, 325^{n+1}} + 1824 \sum_{n=0}^{\infty} \frac{(n!)^2 \, 4^n}{(2\,n+1)! \, 3250^{n+1}} - 20 \arctan\left(\frac{1}{239}\right),$$

where the terms of the second series are now just decimal shifts of the first.

Ramanujan-type elliptic series. Truly new types of infinite series formulae, based on elliptic integral approximations, were discovered around 1910 by Srinivasa Ramanujan (1887–1920) (shown in Figure 10) but were not well known (nor fully proven) until quite recently when his writings were fully decoded and widely published. They are based on elliptic functions and are described at length in [9, 15, 14].

G.N. Watson (see [14]) elegantly describes his feelings on viewing formulae of Ramanujan, such as (13) below:

> ... a thrill which is indistinguishable from the thrill which I feel when I enter the Sagrestia Nuova of the Cappella Medici and see before me the austere beauty of the four statues representing Day, Night, Evening, and Dawn which Michelangelo has set over the tomb of Giuliano de Medici and Lorenzo de Medici

One of these series is the remarkable:

$$(13) \qquad\qquad \frac{1}{\pi} \;=\; \frac{2\sqrt{2}}{9801} \sum_{k=0}^{\infty} \frac{(4k)! \, (1103 + 26390k)}{(k!)^4 396^{4k}}.$$

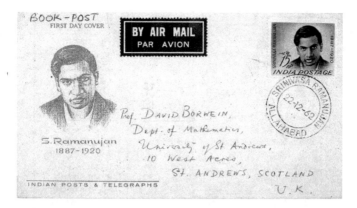

Figure 10: Ramanujan's seventy-fifth birthday stamp (courtesy Naomi Borwein-Yao

Each term of this series produces an additional *eight* correct digits in the result. When Gosper used this formula to compute 17 million digits of π in 1985, and it agreed to many millions of places with the prior estimates, *this concluded the first proof* of (13). As described in [17], this computation can be shown to be exact enough to constitute a bona fide proof ! Actually, Gosper first computed the simple continued fraction for π, hoping to discover some new things in its expansion, but found none. At the time of writing 500 million terms of the continued fraction for π have been computed by Neil Bickford (a teenager) without shedding light on whether the sequence is unbounded (see [1]).

At about the same time, David and Gregory Chudnovsky found the following rational variation of Ramanujan's formula. Amazingly one can show that it exists because $\sqrt{-163}$ corresponds to an imaginary quadratic field with class number one:

$$(14) \qquad \frac{1}{\pi} = 12 \sum_{k=0}^{\infty} \frac{(-1)^k \, (6k)! \, (13591409 + 545140134k)}{(3k)! \, (k!)^3 \, 640320^{3k+3/2}}$$

Each term of this series produces an additional 14 correct digits. The Chudnovskys, shown in Figure 14, implemented this formula using a clever scheme that enabled them to use the results of an initial level of precision to extend the calculation to even higher precision. They used this in several large calculations of π, culminating with a **then record computation** of over four billion decimal digits in 1994. Their remarkable story was told compellingly by Richard Preston in a prizewinning *New Yorker* article "The Mountains of Pi" (March 2, 1992).

While the Ramanujan and Chudnovsky series are in practice considerably more efficient than classical formulae, they share the property that the number of terms needed increases linearly with the number of digits desired: *if you want to compute twice as many digits of π, you must evaluate twice as many terms of the series.*

Relatedly, the Ramanujan-type series

$$(15) \qquad \frac{1}{\pi} = \sum_{n=0}^{\infty} \left(\frac{\binom{2n}{n}}{16^n} \right)^3 \frac{42\,n + 5}{16}$$

allows one to compute the billionth binary digit of $1/\pi$, or the like, *without computing the first half* of the series, and is a foretaste of our later discussion of Borwein-Bailey-Plouffe (or BBP) formulae.

Name	Year	Correct Digits
Miyoshi and Kanada	1981	2,000,036
Kanada-Yoshino-Tamura	1982	16,777,206
Gosper	1985	17,526,200
Bailey	Jan. 1986	29,360,111
Kanada and Tamura	Sep. 1986	33,554,414
Kanada and Tamura	Oct. 1986	67,108,839
Kanada et. al	Jan. 1987	134,217,700
Kanada and Tamura	Jan. 1988	201,326,551
Chudnovskys	May 1989	480,000,000
Kanada and Tamura	Jul. 1989	536,870,898
Kanada and Tamura	Nov. 1989	1,073,741,799
Chudnovskys	Aug. 1991	2,260,000,000
Chudnovskys	May 1994	4,044,000,000
Kanada and Takahashi	Oct. 1995	6,442,450,938
Kanada and Takahashi	Jul. 1997	51,539,600,000
Kanada and Takahashi	Sep. 1999	206,158,430,000
Kanada-Ushiro-Kuroda	Dec. 2002	1,241,100,000,000
Takahashi	Jan. 2009	1,649,000,000,000
Takahashi	April. 2009	2,576,980,377,524
Bellard	Dec. 2009	2,699,999,990,000

Figure 11: Post-calculus π Calculations

10　Reduced Operational Complexity Algorithms

In 1976, Eugene Salamin and Richard Brent independently discovered a *reduced complexity* algorithm for π. It is based on the **arithmetic-geometric mean iteration** (AGM) and some other ideas due to Gauss and Legendre around 1800, although neither Gauss, nor many after him, ever directly saw the connection to effectively computing π.

Quadratic Algorithm (Salamin-Brent).　Set $a_0 = 1, b_0 = 1/\sqrt{2}$ and $s_0 = 1/2$. Calculate

$$(16) \qquad a_k = \frac{a_{k-1} + b_{k-1}}{2} \quad \text{(Arithmetic)} \qquad b_k = \sqrt{a_{k-1}b_{k-1}} \quad \text{(Geometric)}$$

$$(17) \qquad c_k = a_k^2 - b_k^2, \qquad s_k = s_{k-1} - 2^k c_k \quad \text{and compute} \quad p_k = \frac{2a_k^2}{s_k}.$$

Then p_k converges *quadratically* to π. Note the similarity between the arithmetic-geometric mean iteration (16), (which for general initial values converges quickly to a non-elementary limit) and the out-of-kilter harmonic-geometric mean iteration (2) (which in general converges slowly to an elementary limit), and which is an arithmetic-geometric iteration in the reciprocals (see [15]).

Each iteration of the algorithm *doubles* the correct digits. Successive iterations produce $1, 4, 9, 20, 42, 85, 173, 347$ and 697 good decimal digits of π, and takes $\log N$ operations for N digits. Twenty-five iterations computes π to over 45 million decimal digit accuracy. A disadvantage is that each of these iterations must be performed to the precision of the final result. In 1985, my brother Peter and I discovered families of algorithms of this type. For example, here is a genuinely third-order iteration[6]:

---

[6] A fourth-order iteration might be a compound of two second-order ones; this third order one can not be so decomposed.

Cubic Algorithm. Set $a_0 = 1/3$ and $s_0 = (\sqrt{3} - 1)/2$. Iterate

$$r_{k+1} = \frac{3}{1 + 2(1 - s_k^3)^{1/3}}, \qquad s_{k+1} = \frac{r_{k+1} - 1}{2} \quad \text{and} \qquad a_{k+1} = r_{k+1}^2 a_k - 3^k(r_{k+1}^2 - 1).$$

Then $1/a_k$ converges *cubically* to π. Each iteration *triples* the number of correct digits.

Quartic Algorithm. Set $a_0 = 6 - 4\sqrt{2}$ and $y_0 = \sqrt{2} - 1$. Iterate

$$y_{k+1} = \frac{1 - (1 - y_k^4)^{1/4}}{1 + (1 - y_k^4)^{1/4}} \qquad \text{and} \qquad a_{k+1} = a_k(1 + y_{k+1})^4 - 2^{2k+3} y_{k+1}(1 + y_{k+1} + y_{k+1}^2).$$

Then $1/a_k$ converges *quartically* to π. Note that only the power of 2 or 3 used in a_k depends on k.

Let us take an interlude and discuss:

'Piems' or pi-mnemonics. *Piems* are mnemonics in which the length of each word is the corresponding digit of π. Punctuation is ignored. A better piem is both longer and better poetry.

Mnemonics for π

Now I, even I, would celebrate
In rhyme inapt, the great
Immortal Syracusan, rivaled nevermore,
Who in his wondrous lore,
Passed on before
Left men for guidance
How to circles mensurate. *(30)*

How I want a drink, alcoholic of course, after the heavy lectures involving quantum mechanics. *(15)*

See I have a rhyme assisting my feeble brain its tasks ofttimes resisting. *(13)*

There are many more and much longer mnemonics than the famous examples given in the inset box—see [9, p. 405, p.560, p. 659] for a fine selection. Indeed, there is whole cottage industry around the matter, http://www.huffingtonpost.com/jonathan-m-borwein/pi-day_b_1341569.html?ref=science.

Philosophy of mathematics. In 1997 the first occurrence of the sequence 0123456789 was found (later than expected heuristically) in the decimal expansion of π starting at the $17,387,594,880$-th digit after the decimal point. In consequence the status of several famous *intuitionistic examples* due to Brouwer and Heyting has changed. These challenge the *principle of the excluded middle*—either a predicate holds or it does not— and involve classically well-defined objects that for an intuitionist are ill-founded until one can determine when or if the sequence occurred (see [12]).

For example, consider the sequence which is '0' except for a '1' in the first place where 0123456789 first begins to appear in order if it ever occurs. Did it converge when first used by Brouwer as an example? Does it now? Was it then and is it now well defined? Classically it always was and converged to '0'. But, until the computation was done any argument about its convergence relied on the *principle of the excluded middle* which intuitionists reject. Intuitionistically, it converges now. Of course, if we redefine the sequence to have its '1' in the first place that 0123456789101112 first begins, then we are still completely in the dark. But is a sign of how far computation of π has come recently that this was never thought necessary.

手にしているのは π の値が入ったカートリッジテープ

Figure 12: Yasumasa Kanada in his Tokyo office

11   Back to the Future

In December 2002, Kanada computed π to over **1.24 trillion decimal digits**. His team first computed π in hexadecimal (base 16) to 1,030,700,000,000 places, using the following two arctangent relations:

$$\pi = 48 \tan^{-1} \frac{1}{49} + 128 \tan^{-1} \frac{1}{57} - 20 \tan^{-1} \frac{1}{239} + 48 \tan^{-1} \frac{1}{110443}$$

$$\pi = 176 \tan^{-1} \frac{1}{57} + 28 \tan^{-1} \frac{1}{239} - 48 \tan^{-1} \frac{1}{682} + 96 \tan^{-1} \frac{1}{12943}.$$

The first formula was found in 1982 by K. Takano, a high school teacher and song writer. The second formula was found by F. C. W. Störmer in 1896. Kanada verified the results of these two computations agreed, and then converted the hex digit sequence to decimal. The resulting decimal expansion was checked by converting it back to hex. These conversions are themselves non-trivial, requiring massive computation.

This process is quite different from those of the previous quarter century. One reason is that reduced operational complexity algorithms require full-scale multiply, divide and square root operations. These in turn require large-scale *fast Fourier transform* (FFT) operations, which demand huge amounts of memory, and massive all-to-all communication between nodes of a large parallel system. For this latest computation, even the very large system available in Tokyo did not have sufficient memory and network bandwidth to perform these operations at reasonable efficiency levels—at least not for trillion-digit computations. Utilizing arctans again meant using many more arithmetic operations, but no system-scale FFTs, and so the method can be implemented using $\times, \div$ by smallish integer values—additionally, hex is somewhat more efficient!

Kanada and his team evaluated these two formulae using a scheme analogous to that employed by Gosper and by the Chudnovskys in their series computations, in that they were able to avoid explicitly storing the multiprecision numbers involved. This resulted in a scheme that is roughly competitive in *numerical* efficiency with the Salamin-Brent and Borwein quartic algorithms they had previously used, but with a significantly lower total memory requirement. Kanada used a 1 Tbyte main memory system, as with the previous computation, yet got six times as many digits. Hex and decimal evaluations included, it ran 600 hours on a 64-node Hitachi, with the main segment of the program running at a sustained rate of nearly 1 Tflop/sec.

Decimal Digit	Occurrences
0	99999485134
1	99999945664
2	100000480057
3	99999787805
4	100000357857
5	99999671008
6	99999807503
7	99999818723
8	100000791469
9	99999854780
Total	**1000000000000**

Hex Digit	Occurrences
0	62499881108
1	62500212206
2	62499924780
3	62500188844
4	62499807368
5	62500007205
6	62499925426
7	62499878794
8	62500216752
9	62500120671
A	62500266095
B	62499955595
C	62500188610
D	62499613666
E	62499875079
F	62499937801
Total	**1000000000000**

Figure 13: Seemingly random behaviour of single digits of π in base 10 and 16

12   Why π?

What possible motivation lies behind modern computations of π, given that questions such as the irrationality and transcendence of π were settled more than 100 years ago? One motivation is the raw challenge of harnessing the stupendous power of modern computer systems. Programming such calculations are definitely not trivial, especially on large, distributed memory computer systems.

There have been substantial practical spin-offs. For example, some new techniques for performing the fast Fourier transform (FFT), heavily used in modern science and engineering computing, had their roots in attempts to accelerate computations of π. And always the computations help in road-testing computers—often uncovering subtle hardware and software errors.

Beyond practical considerations lies the abiding interest in the fundamental question of the *normality (digit randomness)* of π. Kanada, for example, has performed detailed statistical analysis of his results to see if there are any statistical abnormalities that suggest π is not normal, so far the answer is "no" (see Figures 1 and 13). (Kanada reports that the 10 decimal digits ending in position one trillion are 6680122702, while the 10 hexadecimal digits ending in position one trillion are 3F89341CD5.) Indeed, the first computer computation of π and e on ENIAC, discussed above, was so motivated by John von Neumann. The digits of π have been studied more than any other single constant, in part because of the widespread fascination with and recognition of π. Very recent work, suggesting π may well be normal, can be found in and traced from [1].

Changing world views.   In retrospect, we may wonder why in antiquity π was not *measured* to an accuracy in excess of 22/7? Perhaps it reflects not an inability to do so but a very different mindset to a modern more experimental one. One can certainly find Roman ampitheatres where more accurate measurement than 22/7 would have been helpful. Bailey and I discuss this issue in more detail in [6].

In the same vein, one reason that Gauss and Ramanujan did not further develop the ideas in their identities for π is that an iterative algorithm, as opposed to explicit results, was not as satisfactory for them (especially

Figure 14: The remarkable Chudnovsky brothers π (courtesy D. and G. Chudnovsky)

Ramanujan). Ramanujan much preferred formulae like

$$\pi \approx \frac{3}{\sqrt{67}} \log{(5280)}, \qquad \frac{3}{\sqrt{163}} \log{(640320)} \approx \pi$$

correct to *9 and 15 decimal places*; both of which rely on deep number theory. Contrastingly, Ramanujan in his famous 1914 paper *Modular Equations and Approximations to Pi* [9, p.253] found

$$\left(9^2 + \frac{19^2}{22}\right)^{1/4} = 3.14159265\overline{2}58\cdots$$

"empirically, and it has no connection with the preceding theory." Only the marked digit is wrong.

Discovering the π Iterations.   The genesis of the π algorithms and related material is an illustrative example of experimental mathematics. My brother and I in the early 1980's had a family of quadratic algorithms for π, [15], call them $\mathcal{P}_N$, of the kind we saw above. For $N = 1, 2, 3, 4$ we could prove they were correct but were only conjectured for $N = 5, 7$. In each case the algorithm *appeared* to converge quadratically to π. On closer inspection while the provable cases were correct to $5,000$ digits, the empirical versions agreed with π to roughly 100 places only. Now in many ways to have discovered a "natural" number that agreed with π to that level—and no more—would have been more interesting than the alternative. That seemed unlikely but recoding and rerunning the iterations kept producing identical results.

Two decades ago even moderately high precision calculation was less accessible, and the code was being run remotely over a phone-line in a Berkeley Unix integer package. After about six weeks, it transpired that the package's *square root algorithm was badly awed*, but *only if run with an odd precision of more than sixty digits!* And for idiosyncratic reasons that had only been the case in the two unproven cases. Needless to say, tracing the bug was a salutary and somewhat chastening experience. And it highlights why one checks computations using different sub-routines and methods.

13   How to Compute the *N*-th Digits of π

One might be forgiven for thinking that essentially everything of interest with regards to π has been dealt with. This is suggested in the closing chapters of Beckmann's 1971 book *A History of π*. Ironically, the Salamin–Brent quadratically convergent iteration was discovered only five years later, and the higher-order convergent algorithms followed in the 1980s. Then in 1990, Rabinowitz and Wagon discovered a "spigot" algorithm for π—the digits 'drip out' one by one. This permits successive digits of π (in any desired base) to be computed by a relatively simple recursive algorithm based on *all previously* generated digits.

Even insiders are sometimes surprised by a new discovery. Prior to 1996, most researchers thought if you want to determine the d-th digit of π, you had to generate the (order of) the entire first d digits. This is not true, at least for hex (base 16) or binary (base 2) digits of π. In 1996, Peter Borwein, Plouffe, and Bailey found an algorithm for computing individual hex digits of π. It (1) yields a modest-length hex or binary digit string for π, from an arbitrary position, using no prior bits; (2) is implementable on any modern computer; (3) requires no multiple precision software; (4) requires very little memory; and (5) has a computational cost growing only slightly faster than the digit position. For example, the millionth hexadecimal digit (four millionth binary digit) of π could be found in four seconds on a 2005 Apple computer.

This new algorithm is not fundamentally faster than the best known schemes if used for computing *all* digits of π up to some position, but its storage requirements, elegance and simplicity are of considerable interest, and it is easy to parallelize. It is based on the following at-the-time new formula for π:

$$(18) \qquad \pi = \sum_{i=0}^{\infty} \frac{1}{16^i} \left(\frac{4}{8i+1} - \frac{2}{8i+4} - \frac{1}{8i+5} - \frac{1}{8i+6} \right)$$

which was discovered using *integer relation methods* (see [14]), with a computer search that ran over several months and ultimately produced the (equivalent) relation:

$$(19) \qquad \pi = 4 \cdot {}_2F_1 \left(\begin{matrix} 1, \frac{1}{4} \\ \frac{5}{4} \end{matrix} \middle| -\frac{1}{4} \right) + 2\arctan\left(\frac{1}{2}\right) - \log 5,$$

where the first term is a generalized Gaussian hypergeometric function evaluation. *Maple* and *Mathematica* can both now prove (18). They could not at the time of its discovery. A human proof may be found in [14].

The algorithm in action. In 1997, Fabrice Bellard at INRIA—whom we shall meet again in Section 15—computed 152 binary digits of π starting at the trillionth position. The computation took 12 days on 20 workstations working in parallel over the Internet. Bellard's scheme is based on the following variant of (18):

$$\pi = 4 \sum_{k=0}^{\infty} \frac{(-1)^k}{4^k(2k+1)} - \frac{1}{64} \sum_{k=0}^{\infty} \frac{(-1)^k}{1024^k} \left(\frac{32}{4k+1} + \frac{8}{4k+2} + \frac{1}{4k+3} \right),$$

which permits hex or binary digits of π to be calculated somewhat faster than (18) depending on the implementation. (Most claims of improved speed of algorithms are subject to many caveats.)

In 1998 Colin Percival, then a 17-year-old student at Simon Fraser University (see Figure 15), accessed 25 machines to calculate first the five trillionth hexadecimal digit of π, and then the ten trillionth hex digit. In September 2000, he found the quadrillionth binary digit is **0**, a computation that required *250 CPU-years, using 1734 machines in 56 countries*. We record some of Percival's computational results in Figure 15.

Nor have matters stopped there. As described in [5, 8] in the most recent computation of π using the BBP formula, Tse-Wo Zse of Yahoo! Cloud Computing calculated 256 binary digits of π starting at the *two quadrillionth* bit. He then checked his result using Bellard's variant. In this case, both computations verified that the 24 hex digits beginning immediately after the 500 trillionth hex digit (i.e., after the two quadrillionth binary bit) are: **E6C1294A ED40403F 56D2D764**.

Kanada was able to confirm his 2002 computation in only 21 hours by computing a 20 hex digit string starting at the trillionth digit, and comparing this string to the hex string he had initially obtained in over 600 hours. Their agreement provided enormously strong confirmation. We shall see this use of BBP for verification again when we discuss the most recent record computations of π.

21

Colin Percival (1998)

Position	Hex strings starting at this Position
10^6	26C65E52CB4593
10^7	17AF5863EFED8D
10^8	ECB840E21926EC
10^9	85895585A0428B
10^{10}	921C73C6838FB2
10^{11}	9C381872D27596
1.25×10^{12}	07E45733CC790B
2.5×10^{14}	E6216B069CB6C1

Figure 15: Percival's hexadecimal findings

14   Further BBP Digit Formulae

Motivated as above, constants α of the form

$$(20) \qquad \alpha \;=\; \sum_{k=0}^{\infty} \frac{p(k)}{q(k)2^k},$$

where $p(k)$ and $q(k)$ are integer polynomials, are said to be in the class of *binary (Borwein-Bailey-Plou e) BBP numbers*. I illustrate for $\log 2$ why this permits one to calculate isolated digits in the binary expansion:

$$(21) \qquad \log 2 \;=\; \sum_{k=0}^{\infty} \frac{1}{k2^k}.$$

We wish to compute a few binary digits beginning at position $d+1$. This is equivalent to calculating $\{2^d \log 2\}$, where $\{\cdot\}$ denotes fractional part. We can write

$$(22) \qquad \{2^d \log 2\} \;=\; \left\{ \left\{ \sum_{k=0}^{d} \frac{2^{d-k}}{k} \right\} + \left\{ \sum_{k=d+1}^{\infty} \frac{2^{d-k}}{k} \right\} \right\}$$

$$(23) \qquad \qquad\qquad\; =\; \left\{ \left\{ \sum_{k=0}^{d} \frac{2^{d-k} \bmod k}{k} \right\} + \left\{ \sum_{k=d+1}^{\infty} \frac{2^{d-k}}{k} \right\} \right\}.$$

The key observation is that the numerator of the first sum in (23), $2^{d-k} \bmod k$, can be calculated rapidly by *binary exponentiation*, performed modulo k. That is, it is economically performed by a factorization based on the binary expansion of the exponent. For example,

$$3^{17} = ((((3^2)^2)^2)^2) \cdot 3$$

uses only five multiplications, not the usual 16. It is important to reduce each product modulo k. Thus, 3^{17} mod 10 is done as

$$3^2 = 9; 9^2 = 1; 1^2 = 1; 1^2 = 1; 1 \times 3 = 3.$$

A natural question in light of (18) is whether there is a formula of this type and an associated computational strategy to compute individual *decimal* digits of π. Searches conducted by numerous researchers have been unfruitful and recently D. Borwein (my father), Gallway and I have shown that there are no BBP formulae of the *Machin-type* (as defined in [14]) of (18) for π unless the base is a power of two [14].

22

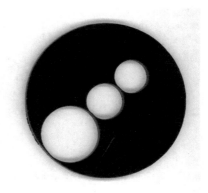

Figure 16: Ferguson's "Eight-Fold Way" and his BBP acrylic circles. These three 'subtractive' acrylic circles (white) and the black circle represent the weights $[4, -2, -2, -1]$ in Equation (18)

Ternary BBP formulae. Yet, BBP formulae exist in other bases for some constants. For example, for π^2 we have both binary and ternary formulae (discovered by Broadhurst):

$$\pi^2 = \frac{9}{8} \sum_{k=0}^{\infty} \frac{1}{64^k} \left(\frac{16}{(6k+1)^2} - \frac{24}{(6k+2)^2} - \frac{8}{(6k+3)^2} - \frac{6}{(6k+4)^2} + \frac{1}{(6k+5)^2} \right).$$

(24)

$$\pi^2 = \frac{2}{27} \sum_{k=0}^{\infty} \frac{1}{729^k} \left(\frac{243}{(12k+1)^2} - \frac{405}{(12k+2)^2} - \frac{81}{(12k+4)^2} - \frac{27}{(12k+5)^2} \right.$$

(25)
$$\left. - \frac{72}{(12k+6)^2} - \frac{9}{(12k+7)^2} - \frac{9}{(12k+8)^2} - \frac{5}{(12k+10)^2} + \frac{1}{(12k+11)^2} \right).$$

These two formulae have recently been used for record digit computations performed on an *IBM Blue Gene* system in conjunction with IBM Australia [8].

Remarkably the volume V_8 in *hyperbolic space* of the *gure-eight knot complement* is well known to be

$$V_8 = 2\sqrt{3} \sum_{n=1}^{\infty} \frac{1}{n\binom{2n}{n}} \sum_{k=n}^{2n-1} \frac{1}{k} = 2.0298832128193072500422405108549\ldots$$

Surprisingly, it is also expressible as

$$V_8 = \frac{\sqrt{3}}{9} \sum_{n=0}^{\infty} \frac{(-1)^n}{27^n} \left\{ \frac{18}{(6n+1)^2} - \frac{18}{(6n+2)^2} - \frac{24}{(6n+3)^2} - \frac{6}{(6n+4)^2} + \frac{2}{(6n+5)^2} \right\},$$

again discovered numerically by Broadhurst, and proved in [14]. A beautiful representation by Helaman Ferguson the mathematical sculptor is given in Figure 16. Ferguson produces art inspired by deep mathematics, but not by a formulaic approach. For instance, his knowledge of hyperbolic geometry allows him to exploit surfaces of negative curvature as shown in his "Eight-Fold Way".

Normality and dynamics. Finally, Bailey and Crandall in 2001 made exciting connections between the existence of a b-ary BBP formula for α and its *normality* base b (uniform distribution of base-b digits)[7]. They

---

[7]See www.sciencenews.org/20010901/bob9.asp.

make a reasonable, hence very hard, conjecture about the *uniform distribution of a related chaotic dynamical system*. This conjecture implies: *Existence of a BBP formula base b for α ensures the normality base b of α.* Illustratively, or $\log 2^8$, the dynamical system, base 2, is to set $x_0 = 0$ and compute

$$x_{n+1} \leftrightarrow 2\left(x_n + \frac{1}{n}\right) \mod 1.$$

15   Pi in the Third Millennium

15.1   Reciprocal series

A few years ago Jesús Guillera found various Ramanujan-like identities for π, using integer relation methods. The three most basic are:

$$(26) \qquad \frac{4}{\pi^2} = \sum_{n=0}^{\infty} (-1)^n r(n)^5 (13 + 180n + 820n^2)\left(\frac{1}{32}\right)^{2n+1}$$

$$(27) \qquad \frac{2}{\pi^2} = \sum_{n=0}^{\infty} (-1)^n r(n)^5 (1 + 8n + 20n^2)\left(\frac{1}{2}\right)^{2n+1}$$

$$(28) \qquad \frac{4}{\pi^3} \overset{?}{=} \sum_{n=0}^{\infty} r(n)^7 (1 + 14n + 76n^2 + 168n^3)\left(\frac{1}{8}\right)^{2n+1},$$

where $r(n) := (1/2 \cdot 3/2 \cdot \cdots \cdot (2n-1)/(2n))/n!$.

Guillera proved (26) and (27) in tandem, using most ingeniously the *Wilf Zeilberger algorithm* for formally proving hypergeometric-like identities [14, 7, 29]. No other proof is known and there seem to be no like formulae for $1/\pi^d$ with $d \geq 4$. The third (28) is certainly true,[9] but has no proof; nor does anyone have an inkling of how to prove it; especially as experiment suggests that it has no 'mate' unlike (26) and (27) [7]. My intuition is that if a proof exists it is more a verification than an explication and so I stopped looking. I am happy just to know the beautiful identity is true. A very nice account of the current state of knowledge for Ramanujan-type series for $1/\pi$ is to be found in [10].

In 2008 Guillera [22] produced another lovely pair of third millennium identities—discovered with integer relation methods and proved with creative telescoping—this time for π^2 rather than its reciprocal. They are

$$(29) \qquad \sum_{n=0}^{\infty} \frac{1}{2^{2n}} \frac{\left(x+\frac{1}{2}\right)_n^3}{(x+1)_n^3}\,(6(n+x)+1) = 8x \sum_{n=0}^{\infty} \frac{\left(\frac{1}{2}\right)_n^2}{(x+1)_n^2},$$

and

$$(30) \qquad \sum_{n=0}^{\infty} \frac{1}{2^{6n}} \frac{\left(x+\frac{1}{2}\right)_n^3}{(x+1)_n^3}\,(42(n+x)+5) = 32x \sum_{n=0}^{\infty} \frac{\left(x+\frac{1}{2}\right)_n^2}{(2x+1)_n^2}.$$

Here $(a)_n = a(a+1)\cdots(a+n-1)$ is the rising factorial. Substituting $x = 1/2$ in (29) and (30), he obtained respectively the formulae

$$\sum_{n=0}^{\infty} \frac{1}{2^{2n}} \frac{(1)_n^3}{\left(\frac{3}{2}\right)_n^3}\,(3n+2) = \frac{\pi^2}{4} \quad \text{and} \quad \sum_{n=0}^{\infty} \frac{1}{2^{6n-2}} \frac{(1)_n^3}{\left(\frac{3}{2}\right)_n^3}\,(21n+13) = \frac{\pi^2}{3}.$$

[8]In this case it is easy to use Weyl's criterion for equidistribution to establish this equivalence without mention of BBP numbers.
[9]Guillera ascribes (28) to Gourevich, who used integer relation methods. I 'rediscovered' (28) using integer relation methods with 30 digits. I then checked it to 500 places in 10 seconds, 1200 in 6.25 minutes, and 1500 in 25 minutes: with a naive command-line instruction in *Maple* on a light laptop.

15.2   Computational records

The last decade has seen the record for computation of π broken in some very interesting ways. We have already described Kanada's 2002 computation in Section 11 and noted that he also took advantage of the BBP formula of Section 13. This stood as a record until 2009 when it was broken three times—twice spectacularly.

Daisuke Takahashi.   The record for computation of π of under 29.37 million decimal digits, by Bailey in 1986 had increased to over 2.649 trillion places by Takahashi in January 2009. Since the same algorithms were used for each computation, it is interesting to review the performance in each case:

In 1986 it took 28 hours to compute 29.36 million digits on 1 cpu of the then new CRAY-2 at NASA Ames using (18). Confirmation using the quadratic algorithm (16) took 40 hours. (The computation uncovered hardware and software errors on the CRAY. Success required developing a speedup of the underlying FFT [14].) In comparison, on 1024 cores of a 2592 core *Appro Xtreme-X3* system 2.649 trillion digits via (16) took 64 hours 14 minutes with 6732 GB of main memory, and (18) took 73 hours 28 minutes with 6348 GB of main memory. (The two computations differed only in the last 139 places.) In April Takahashi upped his record to 2,576,980,377,524 places.

Fabrice Bellard.   Near the end of 2009, Bellard computed nearly 2.7 trillion decimal digits of π (first in binary) using the Chudnovsky series (14). This took 131 days but he only used a single 4-core workstation with a lot of storage and even more human intelligence! For full details of this feat and of Takahashi's most recent computation one can look at

http://en.wikipedia.org/wiki/Chronology_of_computation_of_pi

Nor is that the current end of the matter:

Alexander Yee and Shigeru Kondo.   In August 2010, they announced that they had used the Chudnovsky formula to compute 5 trillion digits of π over a 90-day period, mostly on a two-core Intel Xeon system with 96 Gbyte of memory. They confirmed the result in two ways, using the BBP formula (as discussed above), which required 66 hours, and a variant of the BBP formula due to Bellard, which required 64 hours. Changing from binary to decimal required 8 days. This was upped to *10 trillion digits* in October 2011. Full details are available at http://www.numberworld.org/misc_runs/pi-5t/details.html.

16   ... Life of π

Paul Churchland, writing about the sorry creationist battles of the Kansas school board, [19, Kindle ed, loc 1589] observes that:

> *Even mathematics would not be entirely safe. (Apparently, in the early 1900 s, one legislator in a southern state proposed a bill to rede ne the value of pi as 3.3 exactly, just to tidy things up.)*

As we have seen, the life of π captures a great deal of mathematics—algebraic, geometric and analytic, both pure and applied—along with some history and philosophy. It engages many of the greatest mathematicians and some quite interesting characters along the way. Among the saddest and least-well understood episodes was an abortive 1896 attempt in Indiana to legislate the value(s) of π. The bill, reproduced in [9, p. 231-235], is is accurately described by David Singmaster, [27] and [9, p. 236-239].

At the end of the novel, Piscine (Pi) Molitor writes

I am a person who believes in form, in harmony of order. Where we can, we must give things a meaningful shape. For example I wonder could you tell my jumbled story in exactly one hundred chapters, not one more, not one less? I ll tell you, that s one thing I hate about my nickname, the way that number runs on forever. It s important in life to conclude things properly. Only then can you let go.

We may well not share the sentiment, but we should celebrate that Pi knows π to be irrational.

17   End Notes

1. Why π is not $22/7$.   Today, even the computer algebra systems *Maple* or *Mathematica* 'know' this since

$$(31) \qquad\qquad 0 \;<\; \int_0^1 \frac{(1-x)^4 x^4}{1+x^2}\, dx \;=\; \frac{22}{7} - \pi,$$

though it would be prudent to ask 'why' each can perform the integral and 'whether' to trust it. *Assuming we do trust it*, then the integrand is strictly positive on $(0,1)$, and the answer in (31) is an area and so strictly positive, despite claims that π is $22/7$ ranging over millennia.[10] In this case, requesting the indefinite integral provides immediate reassurance. We obtain

$$\int_0^t \frac{x^4 (1-x)^4}{1+x^2}\, dx \;=\; \frac{1}{7} t^7 - \frac{2}{3} t^6 + t^5 - \frac{4}{3} t^3 + 4\,t - 4\arctan(t),$$

as differentiation easily confirms, and so the Newtonian fundamental theorem of calculus proves (31).

One can take the idea in (31) a bit further, as in [14]. Note that

$$(32) \qquad\qquad \int_0^1 x^4 (1-x)^4\, dx \;=\; \frac{1}{630},$$

and we observe that

$$(33) \qquad\qquad \frac{1}{2} \int_0^1 x^4 (1-x)^4\, dx \;<\; \int_0^1 \frac{(1-x)^4 x^4}{1+x^2}\, dx < \int_0^1 x^4 (1-x)^4\, dx.$$

Combine this with (31) and (32) to derive: $223/71 < 22/7 - 1/630 < \pi < 22/7 - 1/1260 < 22/7$, and so re-obtain Archimedes' famous computation

$$(34) \qquad\qquad 3\frac{10}{71} < \pi < 3\frac{10}{70}.$$

The derivation above was first popularized in *Eureka*, a Cambridge student journal in 1971.[11] A recent study of related approximations is [24]. (See also [14].)

2. More about Gamma.   One may define

$$\Gamma(x) = \int_0^\infty t^{x-1} e^{-t}\, dt$$

---

[10]One may still find adverts in newspapers offering such proofs for sale. A recent and otherwise very nice children's book "Sir Cumference and the the Dragon of Pi (A Math Adventure)" published in (1999) repeats the error, and email often arrives in my in-box offering to show why this and things like this are true.

[11](31) was on a Sydney University examination paper in the early sixties and the earliest source I know of dates from the 1940's [14].

for Re $x > 0$. The starting point is that

(35)
$$x\,\Gamma(x) = \Gamma(x+1), \qquad \Gamma(1) = 1.$$

In particular, for integer n, $\Gamma(n+1) = n!$. Also for $0 < x < 1$

$$\Gamma(x)\,\Gamma(1-x) = \frac{\pi}{\sin(\pi x)},$$

since for $x > 0$ we have

$$\Gamma(x) = \lim_{n\to\infty} \frac{n!\,n^x}{\prod_{k=0}^{n}(x+k)}.$$

This is a nice consequence of the *Bohr-Mollerup theorem* [15, 14] which shows that Γ is the unique log-convex function on the positive half line satisfying (35). Hence, $\Gamma(1/2) = \sqrt{\pi}$ and equivalently we evaluate the *Gaussian integral*

$$\int_{-\infty}^{\infty} e^{-x^2}\,dx = \sqrt{\pi},$$

so central to probability theory. In the same vein, the improper *sinc* function integral evaluates as

$$\int_{-\infty}^{\infty} \frac{\sin(x)}{x}\,dx = \pi.$$

Considerable information about the relationship between Γ and π is to be found in [14, 21].

The Gamma function is as ubiquitous as π. For example, it is shown in [18] that the *expected length*, W_3, of a three-step unit-length random walk in the plane is given by

(36)
$$W_3 = \frac{3}{16}\frac{2^{1/3}}{\pi^4}\Gamma^6\left(\frac{1}{3}\right) + \frac{27}{4}\frac{2^{2/3}}{\pi^4}\Gamma^6\left(\frac{2}{3}\right).$$

We recall that $\Gamma(1/2)^2 = \pi$ and that similar algorithms exist for $\Gamma(1/3), \Gamma(1/4)$, and $\Gamma(1/6)$ [15, 14].

2. More about Complexity Reduction.

To illustrate the stunning complexity reduction in the elliptic algorithms for Pi, let us explicitly provide a *complete set of algebraic equations* approximating π to well over a trillion digits.

> The number π is transcendental and the number $1/a_{20}$ computed next is algebraic; nonetheless they coincide for over 1.5 trillion places.
> Set $a_0 = 6 - 4\sqrt{2}$, $y_0 = \sqrt{2} - 1$ and then solve the system in Figure 17.

This quartic algorithm, with the Salamin–Brent scheme, was first used by Bailey in 1986 [17] and was used repeatedly by Yasumasa Kanada (see Figure 12), in Tokyo in computations of π over 15 years or so, culminating in a 200 billion decimal digit computation in 1999. As recorded in Figure 11, it has been used twice very recently by Takahashi. Only thirty five years earlier in 1963, Dan Shanks—a very knowledgeable participant—was confident that computing a billion digits was forever impossible. Today it is 'reasonably easy' on a modest laptop. A fine self-contained study of this quartic algorithm—along with its cubic confrere also described in Section 10—can be read in [23]. The proofs are nicely refined specializations of those in [16].

3. Following π on the Web.

One can now follow Pi on the web through *Wikipedia, MathWorld* or elsewhere, and indeed one may check the performance of π by looking up 'Pi' at http://www.google.com/trends. This link shows very clear seasonal trends. with a large spike around Pi Day. The final spike (F) is for *Tau Day* (6.28)—a joke that many seem not have realized is a joke.[12]

---

[12] www.washingtonpost.com/blogs/blogpost/post/tau-day-replace-pi-make-music-with-tau/2011/06/28/AG6ub6oH_blog.html

$$y_1 = \frac{1 - \sqrt[4]{1-y_0^4}}{1 + \sqrt[4]{1-y_0^4}}, a_1 = a_0\left(1+y_1\right)^4 - 2^3 y_1\left(1+y_1+y_1{}^2\right)$$

$$y_2 = \frac{1 - \sqrt[4]{1-y_1^4}}{1 + \sqrt[4]{1-y_1^4}}, a_2 = a_1\left(1+y_2\right)^4 - 2^5 y_2\left(1+y_2+y_2{}^2\right)$$

$$y_3 = \frac{1 - \sqrt[4]{1-y_2^4}}{1 + \sqrt[4]{1-y_2^4}}, a_3 = a_2\left(1+y_3\right)^4 - 2^7 y_3\left(1+y_3+y_3{}^2\right)$$

$$y_4 = \frac{1 - \sqrt[4]{1-y_3^4}}{1 + \sqrt[4]{1-y_3^4}}, a_4 = a_3\left(1+y_4\right)^4 - 2^9 y_4\left(1+y_4+y_4{}^2\right)$$

$$y_5 = \frac{1 - \sqrt[4]{1-y_4^4}}{1 + \sqrt[4]{1-y_4^4}}, a_5 = a_4\left(1+y_5\right)^4 - 2^{11} y_5\left(1+y_5+y_5{}^2\right)$$

$$y_6 = \frac{1 - \sqrt[4]{1-y_5^4}}{1 + \sqrt[4]{1-y_5^4}}, a_6 = a_5\left(1+y_6\right)^4 - 2^{13} y_6\left(1+y_6+y_6{}^2\right)$$

$$y_7 = \frac{1 - \sqrt[4]{1-y_6^4}}{1 + \sqrt[4]{1-y_6^4}}, a_7 = a_6\left(1+y_7\right)^4 - 2^{15} y_7\left(1+y_7+y_7{}^2\right)$$

$$y_8 = \frac{1 - \sqrt[4]{1-y_7^4}}{1 + \sqrt[4]{1-y_7^4}}, a_8 = a_7\left(1+y_8\right)^4 - 2^{17} y_8\left(1+y_8+y_8{}^2\right)$$

$$y_9 = \frac{1 - \sqrt[4]{1-y_8^4}}{1 + \sqrt[4]{1-y_8^4}}, a_9 = a_8\left(1+y_9\right)^4 - 2^{19} y_9\left(1+y_9+y_9{}^2\right)$$

$$y_{10} = \frac{1 - \sqrt[4]{1-y_9^4}}{1 + \sqrt[4]{1-y_9^4}}, a_{10} = a_9\left(1+y_{10}\right)^4 - 2^{21} y_{10}\left(1+y_{10}+y_{10}{}^2\right)$$

$$y_{11} = \frac{1 - \sqrt[4]{1-y_{10}^4}}{1 + \sqrt[4]{1-y_{10}^4}}, a_{11} = a_{10}\left(1+y_{11}\right)^4 - 2^{23} y_{11}\left(1+y_{11}+y_{11}{}^2\right)$$

$$y_{12} = \frac{1 - \sqrt[4]{1-y_{11}^4}}{1 + \sqrt[4]{1-y_{11}^4}}, a_{12} = a_{11}\left(1+y_{12}\right)^4 - 2^{25} y_{12}\left(1+y_{12}+y_{12}{}^2\right)$$

$$y_{13} = \frac{1 - \sqrt[4]{1-y_{12}^4}}{1 + \sqrt[4]{1-y_{12}^4}}, a_{13} = a_{12}\left(1+y_{13}\right)^4 - 2^{27} y_{13}\left(1+y_{13}+y_{13}{}^2\right)$$

$$y_{14} = \frac{1 - \sqrt[4]{1-y_{13}^4}}{1 + \sqrt[4]{1-y_{13}^4}}, a_{14} = a_{13}\left(1+y_{14}\right)^4 - 2^{29} y_{14}\left(1+y_{14}+y_{14}{}^2\right)$$

$$y_{15} = \frac{1 - \sqrt[4]{1-y_{14}^4}}{1 + \sqrt[4]{1-y_{14}^4}}, a_{15} = a_{14}\left(1+y_{15}\right)^4 - 2^{31} y_{15}\left(1+y_{15}+y_{15}{}^2\right)$$

$$y_{16} = \frac{1 - \sqrt[4]{1-y_{15}^4}}{1 + \sqrt[4]{1-y_{15}^4}}, a_{16} = a_{15}\left(1+y_{16}\right)^4 - 2^{33} y_{16}\left(1+y_{16}+y_{16}{}^2\right)$$

$$y_{17} = \frac{1 - \sqrt[4]{1-y_{16}^4}}{1 + \sqrt[4]{1-y_{16}^4}}, a_{17} = a_{16}\left(1+y_{17}\right)^4 - 2^{35} y_{17}\left(1+y_{17}+y_{17}{}^2\right)$$

$$y_{18} = \frac{1 - \sqrt[4]{1-y_{17}^4}}{1 + \sqrt[4]{1-y_{17}^4}}, a_{18} = a_{17}\left(1+y_{18}\right)^4 - 2^{37} y_{18}\left(1+y_{18}+y_{18}{}^2\right)$$

$$y_{19} = \frac{1 - \sqrt[4]{1-y_{18}^4}}{1 + \sqrt[4]{1-y_{18}^4}}, a_{19} = a_{18}\left(1+y_{19}\right)^4 - 2^{39} y_{19}\left(1+y_{19}+y_{19}{}^2\right)$$

$$y_{20} = \frac{1 - \sqrt[4]{1-y_{19}^4}}{1 + \sqrt[4]{1-y_{19}^4}}, \mathbf{a_{20}} = a_{19}\left(1+y_{20}\right)^4 - 2^{41} y_{20}\left(1+y_{20}+y_{20}{}^2\right)$$

Figure 17: The system of equations used to compute π to 1.5 trillion places

4. The Difficulty of Popularizing Accurately. Let me finish on this theme. Even after many helpful comments from readers, errors probably remain in my article. So I tell the story below with no particular rancour.

Paul Churchland in [19] offers a fascinating set of essays full of interesting anecdotes—which I have no particular reason to doubt. Nonetheless, the very brief quote at the start of Section 16, regarding the legislation of values of π, contains four inaccuracies. As noted above: (i) the event took place in 1896/7 and (ii) in Indiana (a northern state); (iii) the prospective bill, #246, offered a geometric construction with inconsistent conclusions and certainly offers no one exact value. Finally, (iv) the intent seems to have been pecuniary, not hygienic [27].

As often, this makes me wonder whether mathematics popularization is especially prone to error or if the other disciplines just seem better described because of my relative ignorance. On April 1, 2009, an article entitled "The Changing Value of Pi" appeared in the *New Scientist* with an analysis of how the value of pi has been increasing over time. I hope but am not confident that all readers noted that April 1st is "April Fool's day." (See also entry seven of http://www.museumofhoaxes.com/hoax/aprilfool/.)

Acknowledgements. Thanks are due to many, especially my close collaborators P. Borwein and D. Bailey.

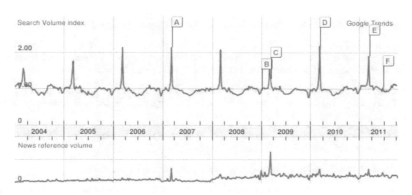

Figure 18: Google's trend line for 'Pi'

References

[1] F. Aragon, D. H. Bailey, J.M. Borwein and P.B. Borwein, "Tools for visualizing real numbers: Planar number walks." Submitted *Mathematical Intelligencer*, June 2012.

[2] F. Amoroso and C. Viola, "Irrational and Transcendental numbers," in volume **2** of *Mathematics and Culture, La matematica: Problemi e teoremi*, Guilio Einaudi Editori, Turino. September 2008.

[3] J. Arndt and C. Haenel, *Pi Unleashed,* Springer-Verlag, New York, 2001.

[4] H.Christian von Baeyer, *Information The New Language of Science*, Weidenfeld and Nicolson, 2003.

[5] D.H. Bailey and J. M. Borwein, "Exploratory Experimentation and Computation." *Notices of the AMS*, **58** (10) (2011), 1410–1419. See also `http://www.ams.orolnotices`.

[6] David H. Bailey and Jonathan M. Borwein, "Ancient Indian Square Roots: An Exercise in Forensic Paleo-Mathematics." *American Mathematical Monthly*. October (2012).

[7] D. Bailey, J. Borwein, N. Calkin, R. Girgensohn, R. Luke, and V. Moll, *Experimental Mathematics in Action*, A K Peters Inc, 2007.

[8] D.H. Bailey, J.M. Borwein, A. Mattingly, and G. Wightwick, "The Computation of Previously Inaccessible Digits of π^2 and Catalan's Constant." *Notices of the AMS*, Accepted, July 2011.

[9] L. Berggren, J.M. Borwein and P.B. Borwein, *Pi: a Source Book,* Springer-Verlag, (2004). Third Edition, 2004.

[10] N.D. Baruah, B.C. Berndt, and H.H. Chan, "Ramanujan's series for $1/\pi$: A survey," *Amer. Math. Monthly* **116** (2009), 567–587.

[11] D. Blatner, *The Joy of Pi*, Walker and Co., New York, 1997.

[12] J.M. Borwein, "Brouwer-Heyting sequences converge," *Mathematical Intelligencer*, **20** (1998), 14–15.

[13] J.M. Borwein, "La vita di pi greco: from Archimedes to ENIAC and beyond," in volume **2** of *Mathematics and Culture, La matematica: Problemi e teoremi*, Guilio Einaudi Editori, Turino. September 2008 (Italian). November 2009 (French).

[14] J.M. Borwein and D.H. Bailey, *Mathematics by Experiment: Plausible Reasoning in the 21st Century,* Second expanded edition, AK Peters Ltd, 2008.

[15] J.M. Borwein and P.B. Borwein, *Pi and the AGM*, John Wiley and Sons, 1987.

[16] J.M. Borwein and P.B. Borwein, "Ramanujan and Pi," *Scienti c American*, February 1988, 112–117. Reprinted in pp. 187-199 of *Ramanujan: Essays and Surveys*, Bruce C. Berndt and Robert A. Rankin Eds., AMS-LMS History of Mathematics, vol. 22, 2001. (Collected in [9].)

[17] J.M. Borwein, P.B. Borwein, and D.A. Bailey, "Ramanujan, modular equations and pi or how to compute a billion digits of pi," *MAA Monthly*, **96** (1989), 201–219. Reprinted in *Organic Mathematics Proceedings*, www.cecm.sfu.ca/organics, 1996. (Collected in [9].)

[18] J. Borwein, D. Nuyens, A. Straub, and J. Wan. "Some Arithmetic Properties of Short Random Walk Integrals." *Ramanujan Journal.* **26** (2011), 109–132.

[19] P. Churchland, *Neurophilosophy at work,* Cambridge University Press, 2007.

[20] William Dunham, *Euler: The Master of Us All* Dolciani Mathematical Expositions, No 22, Mathematical Association of America, 1999.

[21] P. Eymard and J.-P. Lafon, *The Number π*, American Mathematical Society, Providence, 2003.

[22] J. Guillera, "Hypergeometric identities for 10 extended Ramanujan-type series," *Ramanujan J.*, **15** (2008), 219–234.

[23] J. Guillera, "Easy proofs of some Borwein algorithms for π," *American Math. Monthly*, **115** (2008), 850–854.

[24] S.K. Lucas, "Integral approximations to Pi with nonnegative integrands," *American Math. Monthly*, **116** (2009), 166–172.

[25] S. McCartney, *ENIAC: The Triumphs and Tragedies of the World s First Computer*, Walker and Co., New York, 1999.

[26] H. C. Schloper, "The Chronology of Pi," *Mathematics Magazine*, Jan-Feb 1950, 165–170; Mar-Apr 1950, 216–288; and May-Jun 1950, 279–283. (Collected in [9].)

[27] D. Singmaster, "The legal values of Pi," *Mathematical Intelligencer*, **7** (1985), 69–72. (Collected in [9].)

[28] H. Tsumura, "An elementary proof of Euler's formula for $\zeta(2m)$." *American Math. Monthly*, May (2004), 430–431.

[29] W. Zudilin, "Ramanujan-type formulae for $1/\pi$: A second wind." ArXiv:0712.1332v2, 19 May 2008.

There are many other Internet resources on π, a reliable selection is kept at www.experimentalmath.info.

25. I prefer pi: A brief history and anthology of articles in the American Mathematical Monthly (2015)

Paper 25: Jonathan M. Borwein, "I prefer pi: A brief history and anthology of articles in the American Mathematical Monthly," *American Mathematical Monthly*, vol. 122 (2015), p. 195–216. Copyright 2015 Mathematical Association of America. All Rights Reserved.

Synopsis:

This paper, which appeared on Pi Day 2015 (3/14/15 in North American notation), presents a brief summary of papers that have appeared in the *American Mathematical Monthly* on the topic of π. The most frequently cited papers are listed.

The remainder of the article presents a historical overview of π, ranging from a complete analysis of Archimedes' technique for approximating π via inscribed and circumscribed polygons, arctangent-based formulas used in Renaissance times, one of the numerous proofs that π is irrational, the Salamin-Brent quadratically convergent algorithm for π, the Borwein quartic algorithm, spigot algorithms, infinite product formulas, the curious behavior of erroneous digits in the Gregory series for π, and more.

Keywords: Algorithms, Computation, History, Irrationality

© Springer International Publishing Switzerland 2016
D.H. Bailey, J.M. Borwein, *Pi: The Next Generation*,
DOI 10.1007/978-3-319-32377-0_25

I Prefer Pi: A Brief History and Anthology of Articles in the American Mathematical Monthly

Jonathan M. Borwein and Scott T. Chapman

Abstract. In celebration of both a special "big" π Day (3/14/15) and the 2015 centennial of the Mathematical Association of America, we review the illustrious history of the constant π in the pages of the *American Mathematical Monthly*.

1. INTRODUCTION. Once in a century, Pi Day is accurate not just to three digits but to five. The year the MAA was founded (1915) was such a year and so is the MAA's centennial year (2015). To arrive at this auspicious conclusion, we consider the date to be given as month–day–two-digit year.[1] This year, Pi Day turns 26. For a more detailed discussion of Pi and its history, we refer to last year's article [**46**]. We do note that "I prefer pi" is a succinct palindrome.[2]

In honor of this happy coincidence, we have gone back and *selected* roughly 76 representative papers relating to Pi (the constant not the symbol) published in this journal since its inception in 1894 (which predates that of the MAA itself). Those 76 papers listed in three periods (before 1945, 1945–1989, and 1990 on) form the core bibliography of this article. The first author and three undergraduate research students[3] ran a seminar in which they looked at the 76 papers. Here is what they discovered.

Common themes. In each of the three periods, one observes both the commonality of topics and the changing style of presentation. We shall say more about this as we proceed.

- We see authors of varying notoriety. Many are top-tier research mathematicians whose names remain known. Others once famous are unknown. Articles come from small colleges, Big Ten universities, Ivy League schools, and everywhere else. In earlier days, articles came from people at big industrial labs, but nowadays, those labs no longer support research as they used to.
- These papers cover relatively few topics.
  - Every few years a "simple proof" of the irrationality of π is published. Such proofs can be found in [**58, 26, 29, 31, 39, 52, 59, 62, 76*].
  - Many proofs of $\zeta(2) := \sum_{n \geq 1} 1/n^2 = \pi^2/6$ appear, each trying to be a bit more slick or elementary than the last. Of course, whether you prefer your proofs concise and high tech or more leisurely and lower tech is a matter of taste and context. See [**38, *58, 20, 28, 34, 42, 57, 68, 69*].
  - Articles on mathematics outside the European tradition have appeared since the MONTHLY's earliest days. See the papers [**3, 9, 11, 15**].

---

http://dx.doi.org/10.4169/amer.math.monthly.122.03.195
MSC: Primary 01A99, Secondary 11Z05

[1]For advocates of $\tau = 2\pi$, your big day 6/28/31 will come in 2031.

[2]Given by the Professor in Yōko Ozawa, *The Housekeeper and the Professor*, Picador Books, 2003. Kindle location 1095, as is "a nut for a jar of tuna?"

[3]The students are Elliot Catt from Newcastle and Ghislain McKay and Corey Sinnamon from Waterloo.

- In the past 30 years, computer algebra begins to enter the discussions – sometimes in a fundamental way.
- Of course, the compositing style of the MONTHLY has changed several times.
- The process of constructing this selection highlights how much our scholarly life has changed over the past 30 years. Much more can be found and studied easily, but there is even more to find than in previous periods. The ease of finding papers in Google Scholar has the perverse consequence – like Gresham's law in economics – of making less easily accessible material even more likely to be ignored.

While our list is not completely exhaustive, almost every paper listed in the bibliography has been cited in the literature. In fact, several have been highly cited. Some highly used research, such as Ivan Niven's proof of the irrationality of π in 1947 is rarely cited as it has been fully absorbed into the literature [76]. Indeed, a quick look at the AMS's Mathematical Reviews reveals only 15 citations of Niven's paper.

We deem as pi-star (or $\pi^\star$) papers from our MONTHLY bibliography that have been cited in the literature more than 30 times. The existence of JSTOR means that most readers can access all these papers easily, but we have arranged for the $\pi^\star$s to be available free for the next year on our website (www.maa.org/amm_supplements). Here are the $\pi^\star$s with citation numbers according to Google Scholar (as of 1/7/2015). These papers are marked with a $\star$ in the regular bibliography.

1. 133 citations: J. M. Borwein, P. B. Borwein, D. H. Bailey, Ramanujan, modular equations, and approximations to pi or how to compute one billion digits of pi, **96**(1989) 201–219.
2. 119 citations: G. Almkvist, B. Berndt, Gauss, Landen, Ramanujan, the arithmetic-geometric mean, ellipses, π, and the ladies diary, **95**(1988) 585–608.
3. 73 citations: A. Kufner, L. Maligrand, The prehistory of the Hardy inequality, **113**(2006) 715–732.
4. 63 citations: J. M. Borwein, P. B. Borwein, K. Dilcher, Pi, Euler numbers, and asymptotic expansions, **96**(1989) 681–687.
5. 56 citations: N. D. Baruah, B. C. Berndt, H. H. Chan, Ramanujan's series for $1/\pi$: a survey, **116**(2009) 567–587.
6. 40 citations: J. Sondow, Double integrals for Euler's constant and $\ln \pi/4$ and an analog of Hadjicostas's formula, **112**(2005) 61–65.
7. 39 citations: D. H. Lehmer, On arccotangent relations for π, **45**(1938) 657–664.
8. 39 citations: I. Papadimitriou, A simple proof of the formula $\sum_{k=1}^{\infty} 1/k^2 = \pi^2/6$, **80**(1973) 424–425.
9. 36 citations: V. Adamchik, S. Wagon, A simple formula for π, **104**(1997) 852–855.
10. 35 citations: D. Huylebrouck, Similarities in irrationality proofs for π, $\ln 2$, $\zeta(2)$, and $\xi(3)$, **108**(2001) 222–231.
11. 35 citations: L. J. Lange, An elegant continued fraction for π, **106**(1999) 456–458.
12. 33 citations: S. Rabinowitz, S. Wagon, A spigot algorithm for the digits of π, **102**(1995) 195–203.
13. 32 citations: W. S. Brown, Rational exponential expressions and a conjecture concerning π and e, **76**(1969) 28–34.

The remainder of this article. We begin with a very brief history of Pi, both mathematical and algorithmic, which can be followed in more detail in [80] and [46]. We then turn to our three periods and make a very few extra comments about some of the articles. For the most part the title of each article is a pretty good abstract. We then make a few summatory remarks and list a handful of references from outside the MONTHLY, such as David Blattner's *Joy of Pi* [**79**] and Arndt and Haenel's *Pi Unleashed* [**78**].

2. PI: A BRIEF HISTORY. Pi is arguably the most resilient of mathematical objects. It has been studied seriously over many millennia and by every major culture, remaining as intensely examined today as in the Syracuse of Archimedes' time. Its role in popular culture was described in last year's Pi Day article [**46**]. We also recall the recent movies *Life of Pi* ((2012, PG) directed by Ang Lee) and *Pi* ((1998, R) directed by Darren Aronofsky)[4].

From both an analytic and computational viewpoint, it makes sense to begin with Archimedes. Around 250 BCE, Archimedes of Syracuse (287–212 BCE) is thought to have been the first (in *Measurement of the Circle*) to show that the "two possible Pi's" are the same. For a circle of radius r and diameter d, *Area*$= \pi_1 r^2$ while *Perimeter* $= \pi_2 d$ but that $\pi_1 = \pi_2$ is not obvious and is often overlooked; see [**55**].

Archimedes' method. The first rigorous mathematical calculation of π was also due to Archimedes, who used a brilliant scheme based on *doubling inscribed and circumscribed polygons*,

$$6 \mapsto 12 \mapsto 24 \mapsto 48 \mapsto 96,$$

and computing the perimeters to obtain the bounds $3\frac{10}{71} < \pi < 3\frac{10}{70} = \ldots .$[5] The case of 6-gons and 12-gons is shown in Figure 1; for $n = 48$ one already "sees" near-circles. No computational mathematics approached this level of rigor again until the 19th century. Phillips in [**41**] or [**80**, pp. 15-19] calls Archimedes the "first numerical analyst."

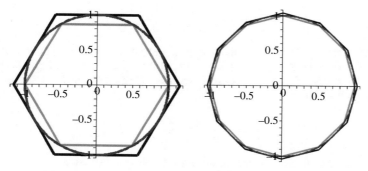

Figure 1. Archimedes' method of computing π with 6- and 12-gons

Archimedes' scheme constitutes the first true algorithm for π in that it can produce an arbitrarily accurate value for π. It also represents the birth of numerical and error analysis – all without positional notation or modern trigonometry. As discovered in the 19th century, this scheme can be stated as a simple, numerically stable, recursion, as follows [**82**].

---

[4]Imagine, an R–rated movie involving Pi!

[5]All rules are meant to be broken. Writing 10/70 without cancellation makes it easier to see that 1/7 is larger than 10/71.

Archimedean mean iteration (Pfaff–Borchardt–Schwab). Set $a_0 = 2\sqrt{3}$ and $b_0 = 3$, which are the values for circumscribed and inscribed 6-gons. If

$$a_{n+1} = \frac{2a_n b_n}{a_n + b_n} \quad (H) \quad \text{and} \quad b_{n+1} = \sqrt{a_{n+1} b_n} \quad (G), \tag{1}$$

then a_n and b_n converge to π, with the error decreasing by a factor of four with each iteration. In this case, the error is easy to estimate—look at $a_{n+1}^2 - b_{n+1}^2$—and the limit is somewhat less accessible but still reasonably easy to determine [**82**].

Variations of Archimedes' geometrical scheme were the basis for all high-accuracy calculations of π over the next 1,800 years—far after its "best before" date. For example, in fifth century China, Tsu Chung-Chih used a variant of this method to obtain π correct to seven digits. A millennium later, al-Kāshī in Samarkand "*who could calculate as eagles can fly*" obtained 2π in *sexadecimal*:

$$2\pi \approx 6 + \frac{16}{60^1} + \frac{59}{60^2} + \frac{28}{60^3} + \frac{01}{60^4} + \frac{34}{60^5} + \frac{51}{60^6} + \frac{46}{60^7} + \frac{14}{60^8} + \frac{50}{60^9},$$

good to 16 decimal places (using $3 \cdot 2^{28}$-gons). This is a personal favorite; reentering it in a computer centuries later and getting the predicted answer gives the authors horripilation ("goose-bumps").

Pi's centrality is emphasised by the many ways it turns up early in new subjects from irrationality theory to probability and harmonic analysis. For instance, Francois Viéta's (1540–1603) formula

$$\frac{2}{\pi} = \frac{\sqrt{2}}{2} \frac{\sqrt{2 + \sqrt{2}}}{2} \frac{\sqrt{2 + \sqrt{2 + \sqrt{2}}}}{2} \cdots \tag{2}$$

and John Wallis' (1616–1703) infinite product [**67, 74, 75**]

$$\frac{\pi}{2} = \frac{2 \cdot 2 \cdot 4 \cdot 4 \cdot 6 \cdot 6 \cdot 8 \cdot 8}{1 \cdot 3 \cdot 3 \cdot 5 \cdot 5 \cdot 7 \cdot 7 \cdot 9} \cdots \tag{3}$$

are counted among the first infinitary objects in mathematics. The latter leads to the gamma function, Stirling's formula, and much more [**64**], including the *first infinite continued fraction*[6] for $2/\pi$ by Lord Brouncker (1620–1684), first president of the Royal Society of London:

$$\frac{2}{\pi} = \frac{1}{1} + \frac{9}{2} + \frac{25}{2} + \frac{49}{2} \cdots. \tag{4}$$

Here, we use the modern concise notation for a continued fraction.

Arctangents and Machin formulas. With the development of calculus, it became possible to extend calculations of π dramatically as shown in Figure 4. Almost all calculations between 1700 and 1980 reduce to exploiting the series for the arctangent (or another inverse trig function) and using identities to require computation only near the center of the interval of convergence. Thus, one starts with

$$\arctan(x) = x - \frac{x^3}{3} + \frac{x^5}{5} - \frac{x^7}{7} + \cdots \quad \text{for } -1 \le x \le 1 \tag{5}$$

---

[6]This was discovered without proof as was (3).

and arctan$(1) = \pi/4$. Substituting $x = 1$ proves the *Gregory–Leibniz formula* (1671–1674)

$$\frac{\pi}{4} = 1 - \frac{1}{3} + \frac{1}{5} - \frac{1}{7} + \frac{1}{9} - \frac{1}{11} + \cdots. \tag{6}$$

James Gregory (1638–1675) was the greatest of a large Scottish mathematical family. The point $x = 1$, however, is on the boundary of the interval of convergence of the series. Justifying substitution requires a careful error estimate for the remainder or Lebesgue's monotone convergence theorem, but most introductory calculus texts ignore the issue. The arctan integral and series were known centuries earlier to the Kerala school, which was identified with Madhava (c. 1350 – c. 1425) of Sangamagrama near Kerala, India. Madhava may well have computed 13 digits of π.

To make (5) computationally feasible, we can use one of many formulas such as:

$$\arctan(1) = 2\arctan\left(\frac{1}{3}\right) + \arctan\left(\frac{1}{7}\right) \quad \text{(Hutton)} \tag{7}$$

$$\arctan(1) = \arctan\left(\frac{1}{2}\right) + \arctan\left(\frac{1}{5}\right) + \arctan\left(\frac{1}{8}\right) \quad \text{(Euler)} \tag{8}$$

$$\arctan(1) = 4\arctan\left(\frac{1}{5}\right) - \arctan\left(\frac{1}{239}\right) \quad \text{(Machin).} \tag{9}$$

All of this, including the efficiency of different *Machin formulas* as they are now called, is lucidly described by the early and distinguished computational number theorist D.H. Lehmer [*13]. See also [2, 5, 49] and [19] by Wrench, who in 1961 with Dan Shanks performed extended computer computation of π using these formulas; see Figure 5.

In [*13] Lehmer gives what he considered to be a best possible self-checking pair of arctan relations for computing π. The pair was

$$\arctan(1) = 8\arctan\left(\frac{1}{10}\right) - \arctan\left(\frac{1}{239}\right) - 4\arctan\left(\frac{1}{515}\right) \tag{10}$$

$$\arctan(1) = 12\arctan\left(\frac{1}{18}\right) + 8\arctan\left(\frac{1}{57}\right) - 5\arctan\left(\frac{1}{239}\right). \tag{11}$$

In [2], Ballantine shows that this pair makes a good choice since the series for arctan$(1/18)$ and arctan$(1/57)$ has terms that differ by a constant factor of "0," a decimal shift. This observation was implemented in both the 1961 and 1973 computations listed in Figure 4.

Mathematical landmarks in the life of Pi. The irrationality of π was first shown by Lambert in 1761 using continued fractions [*63]. This is a good idea since a number α has an eventually repeating nonterminating simple continued fraction if and only if α is a quadratic irrational, as made rigorous in 1794 by Legendre. Legendre conjectured that π is nonalgebraic[7], that is, that π is *transcendental*. Unfortunately, all the pretty continued fractions for π are not simple [*63, 70, 83]. In [*63], Lange examines various proofs of

---

[7]It can be argued that he was anticipated by Maimonides (the Rambam, 1135–1204) [81].

$$\pi = 3 + \frac{1^2}{2} + \frac{3^2}{2} + \frac{5^2}{2} + \frac{7^2}{2} \cdots. \tag{12}$$

Legendre was validated when in 1882 Lindemann proved π transcendental. He did this by extending Hermite's 1873 proof of the transcendence of e. There followed a spate of simplifications by Weierstrass in 1885, Hilbert in 1893, and many others. Oswald Veblen's article [18], written only ten years later, is a lucid description of the topic by one of the leaders of the early 20th century American mathematical community.[8] A 1939 proof of the transcendence of π by Ivan Niven [14] is reproduced exactly in Appendix A since it remains entirely appropriate for a class today.

We next reproduce our personal favorite MONTHLY proof of the irrationality of π. All such proofs eventually arrive at a putative integer that must lie strictly between zero and one.

Theorem 1 (Breusch [26]). *π is irrational.*

Proof. Assume $\pi = a/b$ with a and b integers. Then, with $N = 2a$, $\sin N = 0$, $\cos N = 1$, and $\cos(N/2) = \pm 1$. If m is zero or a positive integer, then

$$A_m(x) \equiv \sum_{k=0}^{\infty} (-1)^k (2k+1)^m \frac{x^{2k+1}}{(2k+1)!} = P_m(x) \cos x + Q_m(x) \sin x$$

where $P_m(x)$ and $Q_m(x)$ are polynomials in x with integral coefficients. (The proof follows by induction on m : $A_{m+1} = x \, dA_m/dx$, and $A_0 = \sin x$.) Thus, $A_m(N)$ is an integer for every positive integer m.

If t is any positive integer, then

$$B_t(N) \equiv \sum_{k=0}^{\infty} (-1)^k \frac{(2k+1-t-1)(2k+1-t-2)\cdots(2k+1-2t)}{(2k+1)!} N^{2k+1}$$

$$= \sum_{k=0}^{\infty} (-1)^k \frac{(2k+1)^t - b_1(2k+1)^{t-1} + \cdots \pm b_t}{(2k+1)!} N^{2k+1}$$

$$= A_t(N) - b_1 A_{t-1}(N) + \cdots \pm b_t A_0(N).$$

Since all the b_i are integers, $B_t(N)$ must be an integer too. Break the sum for $B_t(N)$ into the three pieces

$$\sum_{k=0}^{[(t-1)/2]}, \quad \sum_{k=[(t+1)/2]}^{t-1}, \quad \text{and} \sum_{k=t}^{\infty}.$$

In the first sum, the numerator of each fraction is a product of t consecutive integers; therefore, it is divisible by $t!$ and hence by $(2k+1)!$ since $2k+1 \le t$. Thus, each term of the first sum is an integer. Each term of the second sum is zero. Thus, the third sum must be an integer for every positive integer t.

---

[8]He was also nephew of Thorstein Veblen, one of the founders of sociology and originator of the term "conspicuous consumption."

This third sum is

$$\sum_{k=t}^{\infty}(-1)^k\frac{(2k-t)!}{(2k+1)!(2k-2t)!}N^{2k+1}$$

$$= (-1)^t\frac{t!}{(2t+1)!}N^{2t+1}\left(1 - \frac{(t+1)(t+2)}{(2t+2)(2t+3)}\frac{N^2}{2!}\right.$$

$$\left. + \frac{(t+1)(t+2)(t+3)(t+4)}{(2t+2)(2t+3)(2t+4)(2t+5)}\frac{N^4}{4!} - \cdots\right).$$

Let $S(t)$ stand for the sum in the parenthesis. Certainly

$$|S(t)| < 1 + N + \frac{N^2}{2!} + \cdots = e^N.$$

Thus, the whole expression is absolutely less than

$$\frac{t!}{(2t+1)!}N^{2t+1}e^N < \frac{N^{2t+1}}{t^{t+1}}e^N < (N^2/t)^{t+1}e^N,$$

which is less than 1 for $t > t_0$.

Therefore, $S(t) = 0$ for every integer $t > t_0$. But this is impossible because

$$\lim_{t\to\infty}S(t) = 1 - \frac{1}{2^2}\cdot\frac{N^2}{2!} + \frac{1}{2^4}\cdot\frac{N^4}{4!} - \cdots = \cos(N/2) = \pm 1. \qquad\blacksquare$$

A similar argument shows that the natural logarithm of a rational number must be irrational. From $\log(a/b) = c/d$ would follow that $e^c = a^d/b^d = A/B$. Then

$$B\cdot\sum_{k=0}^{\infty}\frac{(k-t-1)(k-t-2)\cdots(k-2t)}{k!}c^k$$

would have to be an integer for every positive integer t, which leads to a contradiction.

Irrationality measures, denoted $\mu(\alpha)$, as described in [83] seem not to have seen much attention in the MONTHLY. The *irrationality measure* of a real number is the infimum over $\mu > 0$ such that the inequality

$$\left|\alpha - \frac{p}{q}\right| \leq \frac{1}{q^{\mu}}$$

has at most finitely many solutions in $p \in Z$ and $q \in N$. Currently, the best irrationality measure known for π is 7.6063. For π^2, it is 5.095412, and for $\log 2$, it is 3.57455391. For every rational number, the irrationality measure is 1 and the Thue-Siegel-Roth theorem states that if α is a real algebraic irrational then $\mu(\alpha) = 2$. Indeed, almost all real numbers have an irrationality measure of 2, and transcendental numbers have irrationality measure 2 or greater. For example, the transcendental number e has $\mu(e) = 2$ while *Liouville numbers* such as $\sum_{n\geq 0}1/10^{n!}$ are precisely those numbers having infinite irrationality measure. The fact that $\mu(\pi) < \infty$ (equivalently π is not a Liouville number) was first proved by Mahler [85] in 1953.[9] This fact does figure in the solution of many MONTHLY problems over the years; for instance, it lets one estimate how far $\sin(n)$ is from zero.

The *Riemann zeta* function[10] is defined for $s > 1$ by $\zeta(s) = \sum_{n \geq 1} 1/n^s$. The *Basel problem*, first posed by Pietro Mengoli in 1644, which asked for the evaluation of $\zeta(2) = \sum_{n \geq 1} 1/n^2$, was popularized by the Bernoullis, who came from Basel in Switzerland and, hence, the name. In 1735, all even values of ζ were evaluated by Euler. He argued that $\sin(\pi x)$ could be thought of as an infinite polynomial and so

$$\frac{\sin(\pi x)}{x} = \pi \prod_{n=1}^{\infty} \left(1 - \frac{x^2}{n^2} \right), \tag{13}$$

since both sides have the same zeros and value at zero. Comparing the coefficients of the Taylor series of both sides of (13) establishes that $\zeta(2) = \pi^2/6$ and then one recursively can determine a closed form (involving Bernoulli polynomials). In particular, $\zeta(4) = \pi^4/90$, $\zeta(6) = \pi^6/945$, and $\zeta(8) = \pi^8/9450$ and so on. By contrast, $\zeta(3)$ was only proven irrational in the late 1970s, and the status of $\zeta(5)$ is unsettled—although every one who has thought about this *knows* it is irrational. It is a nice exercise to confirm the values of $\zeta(4)$, $\zeta(6)$ from (13). A large number of the papers in this collection center on the Basel problem and its extensions; see [*58, *73, 50, 72]. An especially nice accounting is in [43]. As is discussed in [*24, 46], it is striking how little more is known about the number–theoretic structure of π.

Algorithmic high spots in the life of Pi. In the large, only three methods have been used to make significant computations of π: before 1700 by Archimedes' method, between 1700 and 1980 using calculus methods (usually based on the arctangent's Maclaurin series and Machin formulas), and since 1980 using spectacular series or iterations both based on elliptic integrals and the arithmetic–geometric mean. The progress of this multicentury project is shown in Figures 2, 4, and 5. If plotted on a log linear scale, the records line up well, especially in Figure 5, which neatly tracks Moore's law.

Name	Year	Digits
Babylonians	2000? BCE	1
Egyptians	2000? BCE	1
Hebrews (1 Kings 7:23)	550? BCE	1
Archimedes	250? BCE	3
Ptolemy	150	3
Liu Hui	263	5
Tsu Ch'ung Chi	480?	7
Al-Kashi	1429	14
Romanus	1593	15
van Ceulen (**Ludolph's number**)	1615	35

Figure 2. Pre-calculus π calculations

---

[9]He showed $\mu(\pi) \leq 42$. Douglas Adams would be pleased. The entire Mahler archive is on line at http://carma.newcastle.edu.au/mahler/.

[10]As expressed in Stigler's law of eponymy, discoveries are often named after later researchers, but in Euler's case, he needs no more glory.

Decimal Digit	Occurrences
0	99999485134
1	99999945664
2	100000480057
3	99999787805
4	100000357857
5	99999671008
6	99999807503
7	99999818723
8	100000791469
9	99999854780
Total	**1000000000000**

Hex Digit	Occurrences
0	62499881108
1	62500212206
2	62499924780
3	62500188844
4	62499807368
5	62500007205
6	62499925426
7	62499878794
8	62500216752
9	62500120671
A	62500266095
B	62499955595
C	62500188610
D	62499613666
E	62499875079
F	62499937801
Total	**1000000000000**

Figure 3. Seemingly random behavior of single digits of π in base 10 and 16

Name	Year	Correct Digits
Sharp (and Halley)	1699	71
Machin	1706	100
Strassnitzky and Dase	1844	200
Rutherford	1853	440
Shanks	1874	(707) 527
Ferguson (**Calculator**)	1947	808
Reitwiesner et al. (**ENIAC**)	1949	2,037
Genuys	1958	10,000
Shanks and Wrench	1961	100,265
Guilloud and Bouyer	1973	1,001,250

Figure 4. Calculus π calculations

The "post-calculus" era was made possible by the simultaneous discovery by Eugene Salamin and Richard Brent in 1976 of identities—actually known to Gauss but not recognized for their value [*24, 37, 82]—that lead to the following two illustrative reduced complexity algorithms.

Quadratic algorithm (Salamin–Brent). Set $a_0 = 1$, $b_0 = 1/\sqrt{2}$, and $s_0 = 1/2$. Calculate

$$a_k = \frac{a_{k-1} + b_{k-1}}{2} \quad \text{(Arithmetic)}, \quad b_k = \sqrt{a_{k-1}b_{k-1}} \quad \text{(Geometric)}, \quad (14)$$

$$c_k = a_k^2 - b_k^2, \quad s_k = s_{k-1} - 2^k c_k \quad \text{and compute} \quad p_k = \frac{2a_k^2}{s_k}. \quad (15)$$

Then p_k converges *quadratically* to π. Note the similarity between the arithmetic–geometric mean iteration (14) (which for general initial values converges quickly to a nonelementary limit) and the out-of-kilter harmonic–geometric mean iteration (1) (which in general converges slowly to an elementary limit) and which is an arithmetic–geometric iteration in the reciprocals (see [**82**]).

Name	Year	Correct Digits
Miyoshi and Kanada	1981	2,000,036
Kanada-Yoshino-Tamura	1982	16,777,206
Gosper	1985	17,526,200
Bailey	Jan. 1986	29,360,111
Kanada and Tamura	Sep. 1986	33,554,414
Kanada and Tamura	Oct. 1986	67,108,839
Kanada et. al	Jan. 1987	134,217,700
Kanada and Tamura	Jan. 1988	201,326,551
Chudnovskys	May 1989	480,000,000
Kanada and Tamura	Jul. 1989	536,870,898
Kanada and Tamura	Nov. 1989	1,073,741,799
Chudnovskys	Aug. 1991	2,260,000,000
Chudnovskys	May 1994	4,044,000,000
Kanada and Takahashi	Oct. 1995	6,442,450,938
Kanada and Takahashi	Jul. 1997	51,539,600,000
Kanada and Takahashi	Sep. 1999	206,158,430,000
Kanada-Ushiro-Kuroda	Dec. 2002	1,241,100,000,000
Takahashi	Jan. 2009	1,649,000,000,000
Takahashi	April. 2009	2,576,980,377,524
Bellard	Dec. 2009	2,699,999,990,000
Kondo and Yee	Aug. 2010	5,000,000,000,000
Kondo and Yee	Oct. 2011	10,000,000,000,000
Kondo and Yee	Dec. 2013	12,200,000,000,000

Figure 5. Post-calculus π calculations

Each iteration of the Brent–Salamin algorithm *doubles* the correct digits. Successive iterations produce 1, 4, 9, 20, 42, 85, 173, 347, and 697 good decimal digits of π, and take log N operations to compute N digits. Twenty-five iterations compute π to over 45 million decimal digit accuracy. A disadvantage is that each of these iterations must be performed to the precision of the final result. Likewise, we have the following.

Quartic Algorithm (The Borweins). Set $a_0 = 6 - 4\sqrt{2}$ and $y_0 = \sqrt{2} - 1$. Iterate

$$y_{k+1} = \frac{1 - (1 - y_k^4)^{1/4}}{1 + (1 - y_k^4)^{1/4}} \quad \text{and} \quad a_{k+1} = a_k(1 + y_{k+1})^4 - 2^{2k+3} y_{k+1}(1 + y_{k+1} + y_{k+1}^2).$$

Then $1/a_k$ *converges quartically*[11] to π. Note that only the power of 2 used in a_k depends on k. Twenty-five iterations yield an algebraic number that agrees with π to in excess of a quadrillion digits. This iteration is nicely derived in [**56**].

As charmingly detailed in [*21], see also [*47, 82], Ramanujan discovered that

$$\frac{1}{\pi} = \frac{2\sqrt{2}}{9801} \sum_{k=0}^{\infty} \frac{(4k)! \, (1103 + 26390k)}{(k!)^4 396^{4k}}. \tag{16}$$

Each term of this series produces an additional *eight* correct digits in the result. When Gosper used this formula to compute 17 million digits of π in 1985, it agreed to many millions of places with the prior estimates, *this concluded the first proof* of (16). As described in [*24], this computation can be shown to be exact enough to constitute a bona fide proof! Actually, Gosper first computed the simple continued fraction for π,

---

[11]A fourth-order iteration might be a compound of two second-order ones; this one cannot be so decomposed.

hoping to discover some new things in its expansion, but found none. At the time of this writing, 500 million terms of the continued fraction for π have been computed by Neil Bickford (then a teenager) without shedding light on whether the sequence is unbounded (see [77]).

G. N. Watson, on looking at various of Ramanujan's formulas such as (16), reports the following sensations [86]:

...a thrill which is indistinguishable from the thrill I feel when I enter the Sagrestia Nuovo of the Capella Medici and see before me the austere beauty of the four statues representing 'Day', 'Night', 'Evening', and 'Dawn' which Michelangelo has set over the tomb of Guiliano de'Medici and Lorenzo de'Medici. – G. N. Watson, 1886–1965.

Soon after Gosper did his computation, David and Gregory Chudnovsky found the following even more rapidly convergent variation of Ramanujan's formula. It is a consequence of the fact that $\sqrt{-163}$ corresponds to an imaginary quadratic field with class number one:

$$\frac{1}{\pi} = 12 \sum_{k=0}^{\infty} \frac{(-1)^k (6k)! (13591409 + 545140134k)}{(3k)! (k!)^3 \, 640320^{3k+3/2}}. \tag{17}$$

Each term of this series produces an extraordinary additional 14 correct digits. Note that in both (16) and (17), one computes a rational series and has a single multiplication by a surd to compute at the end.

Some less familiar themes. While most of the articles in our collection fit into one of the big themes (irrationality [57], transcendence, arctangent formulas, Euler's product for $\sin x$, evaluation of $\zeta(2)$, π in other cultures), there are of course some lovely sporadic examples. These include the following.

- *Spigot algorithms,* **which drip off one more digit at a time for π and use only integer arithmetic [*71, 54].** As described in [*44], the first spigot algorithm was discovered for e. While the ideas are simple, the specifics for π need some care; we refer to Rabinowitz and Wagon [*71] for the carefully explained details.
- **Products for $\pi \cdot e$ and π / e [35].** Melzack, then at Bell Labs, proved[12] that

$$\frac{\pi}{2e} = \lim_{N \to \infty} \prod_{n=1}^{2N} \left(1 + \frac{2}{n}\right)^{(-1)^{n+1} n} \tag{18}$$

$$\frac{6}{\pi e} = \lim_{N \to \infty} \prod_{n=2}^{2N+1} \left(1 + \frac{2}{n}\right)^{(-1)^n n}. \tag{19}$$

Melzak begins by showing that $\lim_{n \to \infty} V(C_n)/V(S_n) = \sqrt{2/(\pi e)}$, where S_n is the n-sphere and C_n is the inscribed n-dimensional cylinder of greatest volume. He then proves (18) and (19), saying the proof follows that of the derivation of Wallis' formula, and he *conjectures* that (18) can be used to prove that e/π is irrational. We remind the reader that the transcendentality of e^π follows from the *Gelfond–Schneider* theorem (1934) [82] since $e^{\pi/2} = i^{-i}$, but the statuses of $e + \pi, e/\pi, e \cdot \pi,$ and π^e are unsettled.

---

[12]We correct errors in Melzack's original formulas.

Both (18) and (19) are very slowly convergent. To check (19), one may take logs and expand the series for log then exchange the order of summation to arrive at the more rapidly convergent "zeta"-series

$$\sum_{n=2}^{\infty} \frac{(-2)^n}{n} \left(\alpha (n-1) - 1 \right) = \log \left(\frac{\pi e}{6} \right)$$

where $\alpha(s) := \sum_{k \geq 0} (-1)^k / (k+1)^s$ is the alternating zeta function, which is well defined for $\operatorname{Re} s > 0$.

If we consider the partial products for (18), then we obtain

$$\left(\frac{2}{1} \cdot \frac{2}{3} \cdot \frac{4}{3} \cdot \frac{4}{5} \cdot \frac{6}{5} \cdot \frac{6}{7} \cdot \frac{8}{7} \cdot \frac{8}{9} \cdots \frac{2N}{2N+1} \right) \cdot \left(\frac{2N+1}{2N+2} \right)^{2N}.$$

As $N \to \infty$, the left factor yields Wallis's product for $\pi/2$ and the right factor tends to $1/e$, which confirms (18). A similar partial product can be obtained from (19).

- **A curious predictability in the error in the Gregory–Liebnitz series (6) for $\pi/4$** [*25, 45]. In 1988, it was observed that the series

$$\pi = 4 \sum_{k=1}^{\infty} \frac{(-1)^{k+1}}{2k-1} = 4 \left(1 - \frac{1}{3} + \frac{1}{5} - \frac{1}{7} + \frac{1}{9} - \frac{1}{11} + \cdots \right), \qquad (20)$$

when truncated to 5,000,000 terms, differs strangely from the true value of π:

3.14159245358979323846464338327950278419716939938730582097494182230781640...
3.14159265358979323846426433832795028841971693993751058209749445923078164 0...
    2   -2   10   -122   2770.

Values differ as expected from truncating an alternating series: in the seventh place a "4" that should be a "6." But the next 13 digits are correct and, after another blip, for 12 digits. Of the first 46 digits, only four differ from the corresponding digits of π. Further, the "error" digits seemingly occur with a period of 14. Such anomalous behavior begs for explanation. A great place to start is by using Neil Sloane's Internet-based integer sequence recognition tool, available at www.oeis.org. This tool has no difficulty recognizing the sequence of errors as twice the *Euler numbers*. Even Euler numbers are generated by $\sec x = \sum_{k=0}^{\infty} (-1)^k E_{2k} x^{2k} / (2k)!$. The first few are $1, -1, 5, -61, 1385, -50521, 2702765$. This discovery led to the following *asymptotic expansion*:

$$\frac{\pi}{2} - 2 \sum_{k=1}^{N/2} \frac{(-1)^{k+1}}{2k-1} \approx \sum_{m=0}^{\infty} \frac{E_{2m}}{N^{2m+1}}. \qquad (21)$$

Now the genesis of the anomaly is clear: by chance, the series had been truncated at 5,000,000 terms—exactly one-half of a fairly large power of ten. Indeed, setting $N = 10,000,000$ in equation (21) shows that the first hundred or so digits of the truncated series value are small perturbations of the correct decimal expansion for π.

On a hexadecimal computer with $N = 16^7$, the corresponding strings and hex errors are

© THE MATHEMATICAL ASSOCIATION OF AMERICA

3.243F6A8885A308D313198A2E03707344A4093822299F31D0082EFA98EC4E6C89452821E...
3.243F6A6885A308D31319AA2E03707344A3693822299F31D7A82EFA98EC4DBF69452821E...
        2   -2   A   -7A   2AD2

with the first being the correct value of π. (In hexadecimal or *hex* one uses "A,B, ..., F" to write 10 through 15 as single "hex-digits.") Similar phenomena occur for other constants; see [80]. Also, knowing the errors means we can correct them and use (21) to make Gregory's formula computationally tractable.

- **Hilbert's inequality** [*61, 48] In its simplest incarnation, Hilbert's inequality is

$$\sum_{m,n=1}^{\infty} \frac{a_n b_m}{n+m} \leq \pi \sqrt{\sum_{n=1}^{\infty} a_n^2 \sum_{n=1}^{\infty} b_n^2} \quad \text{(for } a_n, b_m \in \mathbb{R}, \ a_n, b_m > 0) \quad (22)$$

with the assertion that the constant π is best possible. Actually, 2π was the best constant that Hilbert could obtain. Hardy's inequality, which originated in his successful attempt to prove (22) early in the development of the modern theory of inequalities, is well described in [*61]. One could write a nice book on the places in which π or $\zeta(2)$ arise as the best possible constant in an inequality.

- **The distribution of the digits of π** [46]. Single-digit distribution of the first trillion digits base 10 and 16 is shown in Figure 3. All the counts in these figures are consistent with π being random.

3. PI IN THIS MONTHLY: 1894–1944. This period yielded 20 papers for our selection. The July 1894 issue of this MONTHLY contained the most embarrassing article on Pi [10] ever to grace the pages of the MONTHLY. Flagged only by "published by the request of the author," who indicated it was copyrighted in 1889, it is the origin of the famous usually garbled story of the attempt by Indiana in 1897 to legislate the value of π; see [81] and [80, D. Singmaster, The legal values of pi]. It contains a nonsensical geometric construction of π. So π and the MONTHLY got off on a bad footing.

Luckily, the future was brighter. While most early articles would meet today's criteria for publication, this is not true of all. For example, [20] offers a carefully organized list of 68 consequences of Euler's product for sin given in (13) with almost no English. By contrast, [6] is perhaps the first discussion of the efficiency of calculation in the MONTHLY.

REFERENCES FROM 1894 TO 1944

1. R. C. Archibald, Historical notes on the relation $e^{-(\pi/2)} = i^i$, *Amer. Math. Monthly* **28**(1921) 116–121. MR1519723
2. J. P. Ballantine, The best (?) formula for computing π to a thousand places, *Amer. Math. Monthly* **46**(1939), 499–501. MR3168990
3. J. M. Barbour, A sixteenth century Chinese approximation for π, *Amer. Math. Monthly* **40**(1933) 69–73. MR1522708
4. A. A. Bennett, Discussions: Pi and the factors of $x^2 + 1$, *Amer. Math. Monthly* **32**(1925) 375–377. MR1520736
5. A. A. Bennett, Two new arctangent relations for π, *Amer. Math. Monthly* **32**(1925) 253–255. MR1520682
6. C. C. Camp, Discussions: A new calculation of π, *Amer. Math. Monthly* **33**(1926) 472–473. MR1521028
7. J. S. Frame, A series useful in the computation of π, *Amer. Math. Monthly* **42**(1935) 499–501. MR1523462
8. M. G. Gaba, A simple approximation for π, *Amer. Math. Monthly* **45**(1938) 373–375. MR1524313
9. S. Ganguli, The elder Aryabhata's value of π, *Amer. Math. Monthly* **37**(1930) 16–22. MR1521892
10. E. J. Goodwin, Quadrature of the circle, *Amer. Math. Monthly* **1**(1894) 246–248.
11. G. B. Halsted, Pi in Asia, *Amer. Math. Monthly* **15**(1908) 84. MR1517012

12. W. E. Heal, Quadrature of the circle, *Amer. Math. Monthly* **3**(1896) 41–45. MR1514010

*13. D. H. Lehmer, On arccotangent relations for π, *Amer. Math. Monthly* **45**(1938) 657–664. MR1524440

14. I. Niven, The transcendence of π, *Amer. Math. Monthly* **46**(1939) 469–471. MR0000415

15. C. Schoy, Discussions: Al-Biruni's computation of the value of π, *Amer. Math. Monthly* **33**(1926) 323–325. MR1520959

16. D. E. Smith, Historical survey of the attempts at the computation and construction of π, *Amer. Math. Monthly* **2**(1895) 348–351. MR1513968

17. R. S. Underwood, Discussions: Some results involving π, *Amer. Math. Monthly* **31**(1924) 392–394. MR1520517

18. O. Veblen, The transcendence of π and e, *Amer. Math. Monthly* **11**(1904) 219–223. MR1516235

19. J. W. Wrench, On the derivation of arctangent equalities, *Amer. Math. Monthly* **45**(1938) 108–109. MR1524198

20. G. B. Zerr, Summation of series, *Amer. Math. Monthly* **5**(1898) 128–135. MR1514571

4. PI IN THIS MONTHLY: 1945–1989. This second period collects 22 papers. It saw the birth and evolution of the digital computer with many consequences for the computation of π. Even old topics are new when new ideas and tools arise. A charming example is as follows.

Why π is not 22/7. Did you know that

$$0 < \int_0^1 \frac{(1-x)^4 x^4}{1+x^2}\, dx = \frac{22}{7} - \pi ? \tag{23}$$

The integrand is strictly positive on $(0, 1)$, so the integral in (23) is strictly positive—despite claims that π is 22/7 that rage over the millennia.[13] Why is this identity true? We have

$$\int_0^t \frac{x^4 (1-x)^4}{1+x^2}\, dx = \frac{1}{7} t^7 - \frac{2}{3} t^6 + t^5 - \frac{4}{3} t^3 + 4t - 4 \arctan(t),$$

as differentiation easily confirms, and so the Newtonian fundamental theorem of calculus proves (23).

One can take the idea in (23) a bit further. Note that

$$\int_0^1 x^4 (1-x)^4\, dx = \frac{1}{630}, \tag{24}$$

and we observe that

$$\frac{1}{2} \int_0^1 x^4 (1-x)^4\, dx < \int_0^1 \frac{(1-x)^4 x^4}{1+x^2}\, dx < \int_0^1 x^4 (1-x)^4\, dx. \tag{25}$$

Combine this with (23) and (24) to derive

$$\frac{223}{71} < \frac{22}{7} - \frac{1}{630} < \pi < \frac{22}{7} - \frac{1}{1260} < \frac{22}{7},$$

and so we re-obtain Archimedes' famous computation

$$3\frac{10}{71} < \pi < 3\frac{10}{70}. \tag{26}$$

---

[13]One may still find adverts in newspapers offering such proofs for sale. A recent and otherwise very nice children's book "Sir Cumference and the the Dragon of Pi (A Math Adventure)" published in 1999 repeats the error, and email often arrives in our in-boxes offering to show why things like this are true.

This derivation was popularized in *Eureka*, a Cambridge University student journal, in 1971.[14] A recent study of related approximations is made by Lucas [**65**]. It seems largely happenstance that 22/7 is an early continued fraction approximate to π.

Another less standard offering is in [**33**] where Y. V. Matiyasevich shows that

$$\pi = \lim_{m \to \infty} \sqrt{\frac{6 \log \mathrm{fcm}(F_1, \ldots, F_m)}{\log \mathrm{lcm}(u_1, \ldots, u_m)}}. \tag{27}$$

Here, lcm is the least common multiple, fcm is the formal common multiple (the product), and F_n is the n-th Fibonacci number with $F_0 = 0$, $F_1 = 1$, $F_n = F_{n-1} + F_{n-2}$, $n \geq 2$ (without the square root we obtain a formula for $\zeta(2)$).

REFERENCES FROM 1945 TO 1989

*21.   G. Almkvist, B. Berndt, Gauss, Landen, Ramanujan, the arithmetic–geometric mean, ellipses, π, and the ladies diary, *Amer. Math. Monthly* **95**(1988) 585–608. MR0966232

22.   B. H. Arnold, H. Eves, A simple proof that, for odd $p > 1$, arccos $1/p$ and π are incommensurable, *Amer. Math. Monthly* **56**(1949) 20. MR0028343

23.   L. Baxter, Are π, e, and $\sqrt{2}$ equally difficult to compute?, *Amer. Math. Monthly* **88**(1981) 50–51. MR1539586

*24.   J. M. Borwein, P. B. Borwein, D. H. Bailey, Ramanujan, modular equations, and approximations to pi or how to compute one billion digits of pi, **96**(1989) 201–219. MR099186

*25.   J. M. Borwein, P. B. Borwein, K. Dilcher, Pi, Euler numbers, and asymptotic expansions, *Amer. Math. Monthly* **96**(1989) 681–687. MR1019148

26.   R. Breusch, A proof of the irrationality of π, *Amer. Math. Monthly* **61**(1954) 631–632. MR0064087

*27.   W. S. Brown, Rational exponential expressions and a conjecture concerning π and e, *Amer. Math. Monthly* **76**(1969) 28–34. MR0234933

28.   B. R. Choe, An elementary proof of $\sum_{n=1}^{\infty} 1/n^2 = \pi^2/6$, *Amer. Math. Monthly* **94**(1987) 662–663. MR0935853

29.   J. D. Dixon, π is not algebraic of degree one or two, *Amer. Math. Monthly* **69**(1962) 636. MR1531775

30.   J. Gurland, On Wallis' formula, *Amer. Math. Monthly* **63**(1956) 643–645. MR0082117

31.   J. Hancl, A simple proof of the irrationality of π^4, *Amer. Math. Monthly* **93**(1986) 374–375. MR0841114

32.   D. K. Kazarinoff, A simple derivation of the Leibnitz–Gregory series for $\pi/4$, *Amer. Math. Monthly* **62**(1955) 726–727. MR1529178

33.   Y. V. Matiyasevich, A new formula for π, *Amer. Math. Monthly* **93**(1986) 631–635. MR1712797

34.   Y. Matsuoka, An elementary proof of the formula $\sum_{k=1}^{\infty} 1/k^2 = \pi^2/6$, *Amer. Math. Monthly* **68**(1961) 485–487. MR0123858

35.   Z. A. Melzak, Infinite products for $\pi \cdot e$, and π/e, *Amer. Math. Monthly* **68**(1961) 39–41. MR0122920

36.   K. Menger, Methods of presenting e and π, *Amer. Math. Monthly* **52**(1945) 28–33. MR0011319

37.   G. Miel, An algorithm for the calculation of π, *Amer. Math. Monthly* **86**(1979) 694–697. MR0546184

*38.   I. Papadimitriou, A simple proof of the formula $\sum_{k=1}^{\infty} 1/k^2 = \pi^2/6$, *Amer. Math. Monthly* **80**(1973) 424–425. MR0313666

39.   A. E. Parks, π, e, and other irrational numbers, *Amer. Math. Monthly* **93**(1986) 722–723. MR0863976

40.   L. L. Pennisi, Expansions for π and π^2, *Amer. Math. Monthly* **62**(1955) 653–654. MR1529151

41.   G. M. Phillips, Archimedes the numerical analyst, *Amer. Math. Monthly* **88**(1981) 165–169. MR0619562

42.   E. L. Stark, Another proof of the formula $\sum_{k=1}^{\infty} 1/k^2 = \pi^2/6$, *Amer. Math. Monthly* **76**(1969) 552–553. MR1535429

43.   K. Venkatachaliengar, Elementary proofs of the infinite product for sin z and allied formulae, *Amer. Math. Monthly* **69**(1962) 541–545. MR1531736

5. PI IN THIS MONTHLY: 1990–2015.

In the final period, we have collected 32 papers and see no sign that interest in π is lessening. A new topic [***44, 46, 51, 81**] is that of *BBP formulas*, which can compute individual digits of certain constants such as π in base 2 or π^2 in bases 2 and 3 without using the earlier digits. The phenomenon

---

[14]Equation (23) was on a Sydney University examination paper in the early sixties and the earliest source we know of dates from the 1940s [**65**] in an article by Dalzell, who lamentably did not cite himself in [**84**].

is based on the formula

$$\pi = \sum_{i=0}^{\infty} \frac{1}{16^i} \left(\frac{4}{8i+1} - \frac{2}{8i+4} - \frac{1}{8i+5} - \frac{1}{8i+6} \right). \tag{28}$$

On August 27, 2012, Ed Karrel used (28) to extract 25 hex digits of π starting after the 10^{15} position. They are 353CB3F7F0C9ACCFA9AA215F2.[15] In 1990, a billion digits had not yet been computed; see [**80**], and even now, it is inconceivable to compute the full first quadrillion digits in any base.

Over this period, the use of the computer has become more routine even in pure mathematics, and concrete mathematics is back in fashion. In this spirit, we record the following evaluation of $\zeta(2)$, which to our knowledge first appeared as an exercise in [**82**].

Theorem 2 (Sophomore's Dream). *One may square term-wise to obtain*

$$\left(\sum_{n=-\infty}^{\infty} \frac{(-1)^n}{2n+1} \right)^2 = \sum_{n=-\infty}^{\infty} \frac{1}{(2n+1)^2}. \tag{29}$$

In particular $\zeta(2) = \pi^2/6$.

Proof. Let

$$\delta_N := \sum_{n=-N}^{N} \sum_{m=-N}^{N} \frac{(-1)^{m+n}}{(2m+1)(2n+1)} - \sum_{k=-N}^{N} \frac{1}{(2k+1)^2},$$

and note that $\delta_N = \sum_{n=-N}^{N} \frac{(-1)^n}{(2n+1)} \sum_{n \neq m=-N}^{N} \frac{(-1)^m}{m-n}$. We leave it to the reader to show that for large N the inner sum $\epsilon_N(n)$ is of order $1/(N-n+1)$, which goes to zero.

The proof is finished by evaluating the left side of (29) to $\pi^2/4$ using Gregory's formula (6) and then noting that this means $\sum_{n=0}^{\infty} 1/(2n+1)^2 = \pi^2/8$. ∎

Another potent and concrete way to establish an identity is to obtain an appropriate differential equation. For example, consider

$$f(x) := \left(\int_0^x e^{-s^2} \, ds \right)^2 \quad \text{and} \quad g(x) := \int_0^1 \frac{\exp(-x^2(1+t^2))}{1+t^2} \, dt.$$

The derivative of $f+g$ is zero: in *Maple*,

```
f:=x->Int(exp(-s^2),s=0..x)^2;
g:=x->Int(exp(-x^2*(1+t^2))/(1+t^2),t=0..1);
with(student):d:=changevar(s=x*t,diff(f(x),x),t)+diff(g(x),x);
d:=expand(d);
```

---

[15]All processing was done on four NVIDIA GTX 690 graphics cards (GPUs) installed in CUDA; the computation took 37 days. CUDA is a parallel computing platform and programming mode developed by NVIDIA for use in its graphics processing units (GPUs).

shows this. Hence, $f(x) + g(x)$ is constant for $0 \leq x \leq \infty$ and so, after justifying taking the limit at ∞,

$$\left(\int_0^\infty \exp(-t^2) \, dt \right)^2 = f(\infty) = g(0) = \arctan(1) = \frac{\pi}{4}.$$

Thus, we have evaluated the Gaussian integral using only elementary calculus and Gregory's formula (6). The change of variables $t^2 = x$ shows that this evaluation of the normal distribution agrees with $\Gamma(1/2) = \sqrt{\pi}$.

In similar fashion, we may evaluate

$$F(y) := \int_0^\infty \exp(-x^2) \cos(2xy) \, dx$$

by checking that it satisfies the differential equation $F'(y) + 2y \, F(y) = 0$. We obtain

$$F(y) = \frac{\sqrt{\pi}}{2} \exp(-y^2),$$

since we have just evaluated $F(0) = \sqrt{\pi}/2$.

REFERENCES FROM 1990 TO PRESENT

*44. V. Adamchik, S. Wagon, A simple formula for π, Amer. Math. Monthly **104**(1997) 852–855. MR1479991
45. G. Almkvist, Many correct digits of π, revisited, Amer. Math. Monthly **104**(1997) 351–353. MR1450668
46. D. H. Bailey, J. M. Borwein, Pi Day is upon us again and we still do not know if pi is normal, Amer. Math. Monthly **121**(2014) 191–206. MR3168990
*47. N. D. Baruah, B. C. Berndt, H. H. Chan, Ramanujan's series for $1/\pi$: A survey, Amer. Math. Monthly **116**(2009) 567–587. MR2549375
48. J. M. Borwein, Hilbert's inequality and Witten's zeta-function, Amer. Math. Monthly **115**(2008) 125–137. MR2384265
49. J. S. Calcut, Gaussian integers and arctangent identities for π, Amer. Math. Monthly **116**(2009) 515–530. MR2519490
50. S. D. Casey, B. M. Sadler, Pi, the primes, periodicities, and probability, Amer. Math. Monthly **120**(2013) 594–608. MR3096466
51. H. Chan, More formulas for π, Amer. Math. Monthly **113**(2006) 452–455. MR2225478
52. D. Desbrow, On the irrationality of π^2, Amer. Math. Monthly **97**(1990) 903–906. MR1079978
53. F. J. Dyson, N. E. Frankel, M. L. Glasser, Lehmer's interesting series, Amer. Math. Monthly **120**(2013) 116–130. MR3029937
54. J, Gibbons, Unbounded spigot algorithms for the digits of Pi, Amer. Math. Monthly **113**(2006) 318–328. MR2211758
55. L. Gillman, π and the limit of $\sin \alpha / \alpha$, Amer. Math. Monthly **98**(1991) 346–349. MR1541886
56. J. Guillera, Easy proofs of some Borwein algorithms for π, Amer. Math. Monthly **115**(2008) 850–854. MR2463297
57. J. Hofbauer, A simple proof of $1 + 1/2^2 + 1/3^2 + \cdots = \pi^2/6$ and related identities, Amer. Math. Monthly **109**(2002) 196–200. MR1903157
*58. D. Huylebrouck, Similarities in irrationality proofs for π, $\ln 2$, $\zeta(2)$, and $\xi(3)$, Amer. Math. Monthly **108**(2001) 222–231. MR1834702
59. T. W. Jones, Discovering and proving that π is irrational, Amer. Math. Monthly **117**(2010) 553–557. MR2662709
60. J. B. Keller, R. Vakil, π_p, the value of π in ℓ_p, Amer. Math. Monthly **116**(2009) 931–935. MR2589224
*61. A. Kufner, L. Maligrand, The prehistory of the Hardy inequality, Amer. Math. Monthly **113**(2006) 715–732. MR2256532
62. M. Laczkovich, On Lambert's proof of the irrationality of π, Amer. Math. Monthly **104**(1997) 439–443. MR1447977
*63. L. J. Lange, An elegant continued fraction for π, Amer. Math. Monthly **106**(1999) 456–458. MR1699266

64. P. Levrie, W. Daems, Evaluating the probability integral using Wallis's product formula for π, *Amer. Math. Monthly* **116**(2009) 538–541. MR2519493

65. S. Lucas, Approximations to π derived from integrals with nonnegative integrands, *Amer. Math. Monthly* **116**(2009) 166–172. MR2478060

66. D. Manna, V. H. Moll, A simple example of a new class of Landen transformations, *Amer. Math. Monthly* **114**(2007) 232–241. MR2290287

67. S. Miller, A probabilistic proof of Wallis's formula for π, *Amer. Math. Monthly* **115**(2008) 740–745. MR2456095

68. H. B. Muzaffar, A new proof of a classical formula, *Amer. Math. Monthly* **120**(2013) 355–358. MR3035128

69. L. Pace, Probabilistically proving that $\zeta(2) = \pi^2/6$, *Amer. Math. Monthly* **118**(2011) pp. 641–643. MR2826455

70. T. J. Pickett, A. Coleman, Another continued fraction for π, *Amer. Math. Monthly* **115**(2008) 930–933. MR2468553

*71. S. Rabinowitz, S. Wagon, A spigot algorithm for the digits of π, *Amer. Math. Monthly* **102**(1995) 195–203. MR1317842

72. J. Sondow, A faster product for and a new integral for $\ln \pi/2$, *Amer. Math. Monthly* **112**(2005) 729–734. MR216777

*73. J. Sondow, Double integrals for Euler's constant and $\ln \pi/4$ and an analog of Hadjicostas's formula, *Amer. Math. Monthly* **112**(2005) 61–65. MR2110113

74. J. Sondow, New Wallis- and Catalan-type infinite products for π, e, and $\sqrt{2 + \sqrt{2}}$, *Amer. Math. Monthly* **117**(2010) 912–917. MR2759364

75. J. Wästlund, An elementary proof of the Wallis product formula for π, *Amer. Math. Monthly* **114**(2007) 914–917. MR23057

76. L. Zhou, L. Markov, Recurrent proofs of the irrationality of certain trigonometric values, *Amer. Math. Monthly* **117**(2010) 360–362. MR2647819

6. CONCLUDING REMARKS.

It's generally the way with progress that it looks much greater than it really is. – Ludwig Wittgenstein[16]

It is a great strength of mathematics that "old" and "inferior" are not synonyms. As we have seen in this selection, many seeming novelties are actually rediscoveries. That is not at all a bad thing, but it does behoove authors to write "I have not seen this before" or "this is to my knowledge new" rather than unnecessarily claiming ontological or epistemological primacy.

ACKNOWLEDGMENT. The authors thank Ivars Peterson for his help in assembling our bibliography. They also thank three undergraduate research students, Elliot Catt from Newcastle and Ghislain McKay and Corey Sinnamon from Waterloo. Last but not least, they thank two former MONTHLY editors, Roger Horn and Dan Velleman, for extensive comments on an earlier draft of this paper.

OTHER REFERENCES

77. F. Aragon, D. H. Bailey, J. M. Borwein, P. B. Borwein, Walking on real numbers. *Math. Intelligencer.* **35** no. 1 (2013) 42–60. DOI: `http://link.springer.com/article/10.1007%2Fs00283-012-9340-x`.

78. J. Arndt, C. Haenel, *Pi Unleashed*, Springer Science & Business Media, Berlin, 2001.

79. D. Blatner, *The Joy of Pi*, Walker/Bloomsbury, US, 1997. See also `http://www.joyofpi.com/thebook.html`.

80. L. Berggren, J. M. Borwein, P. B. Borwein, *Pi: A Source Book*, Springer-Verlag, (1997), ISBN: 0-387-94924-0. Second edition, 2000, ISBN: 0-387-94946-3. Third edition, incorporating "A Pamphlet on Pi," 2004.

---

[16]From "The Wittgenstein Controversy," by Evelyn Toynton in the *Atlantic Monthly*, June 1997, pp. 28–41.

81. J. M. Borwein, The Life of Pi." Extended and updated version of "La vita di pi greco," volume 2 of *Mathematics and Culture, La matematica: Problemi e teoremi*, Guilio Einaudi Editori, Turino, Italian, 2008.

82. J. M. Borwein, P. B. Borwein, *Pi and the AGM*, John Wiley and Sons, New York, 1987.

83. J. M. Borwein, A. van der Poorten, J. Shallit, W. Zudilin, *Neverending Fractions*, Australia Mathematical Society Lecture Series, Cambridge Univ. Press, Cambridge, UK, 2014.

84. D. Dalzell, On 22/7, *J. London Math. Soc.* **19**(1944) 133–134.

85. K. Mahler, On the approximation of π, *Nederl. Akad. Wetensch. Proc. Ser. A. 56 = Indagationes Math.* **15**(1953) 30–42.

86. G. N. Watson, The final problem: An account of the mock theta functions *J. London Math. Soc.* **11**(1936) 55–80.

JONATHAN M. BORWEIN is Laureate Professor in the School of Mathematical and Physical Sciences and director of the Priority Research Centre in *Computer Assisted Research Mathematics and Its Applications* at the University of Newcastle. An ISI highly cited scientist and former Chauvenet prize winner, he has published widely in various fields of mathematics. His most recent books are *Convex Functions* (with John Vanderwerff, vol 109, Encyclopedia of Mathematics, Cambridge Univ. Press, 2010), *Modern Mathematical Computation with Maple* (with Matt Skerritt, Springer Undergraduate Mathematics and Technology, 2011), *Lattice Sums Then and Now* (with Glasser, McPhedran, Wan and Zucker, vol 150, Encyclopedia of Mathematics, Cambridge Univ. Press, 2013), and *Neverending Fractions* with the late Alf van der Poorten, Jeff Shallit, and Wadim Zudilin (Australian Mathematical Society Lecture Series, Cambridge Univ. Press, 2014).
Division of Mathematics, School of Mathematical and Physical Sciences, University of Newcastle, NSW 2308, Australia.
jonathan.borwein@newcastle.edu.au

SCOTT T. CHAPMAN received a bachelors degree in mathematics and political science from Wake Forest University, a masters degree in mathematics from the University of North Carolina at Chapel Hill, and a Ph.D. in mathematics from the University of North Texas. Since 2008, he has held the titles of professor and *Scholar in Residence* at Sam Houston State University in Huntsville, Texas. In 2012, he became editor of the *American Mathematical Monthly*, and his term runs through 2016.
Sam Houston State University, Department of Mathematics and Statistics, Box 2206, Huntsville, TX 77341-2206, USA.
scott.chapman@shsu.edu

A. APPENDIX: I. NIVEN - THE TRANSCENDENCE OF π [14].

Among the proofs of the transcendence of e, which are in general variations and simplifications of the original proof of Hermite, perhaps the simplest is that of A. Hurwitz.[17] His solution of the problem contains an ingenious device, which we now employ to prove the transcendence of π.

We assume that π is an algebraic number, and show that this leads to a contradiction. Since the product of two algebraic numbers is an algebraic number, the quantity $i\pi$ is a root of an algebraic equation with integral coefficients

$$\theta_1(x) = 0, \tag{30}$$

whose roots are $\alpha_1 = i\pi, \alpha_2, \alpha_3, \ldots, \alpha_n$. Using Euler's relation $e^{i\pi} + 1 = 0$, we have

$$(e^{\alpha_1} + 1)(e^{\alpha_2} + 1) \cdots (e^{\alpha_n} + 1) = 0. \tag{31}$$

---

[17]A. Hurwitz, Beweis der Transendenz der Zahl e, Mathematische Annalen, vol. 43, 1893, pp. 220-221 (also in his Mathematische Werke, vol. 2, pp. 134-135).

We now construct an algebraic equation with integral coefficients whose roots are the exponents in the expansion of (31). First consider the exponents

$$\alpha_1 + \alpha_2, \ \alpha_1 + \alpha_3, \ \alpha_2 + \alpha_3, \ \ldots, \ \alpha_{n-1} + \alpha_n. \tag{32}$$

By equation (30), the elementary symmetric functions of $\alpha_1, \alpha_2, \ldots, \alpha_n$ are rational numbers. Hence the elementary symmetric functions of the quantities (32) are rational numbers. It follows that the quantities (32) are roots of

$$\theta_2(x) = 0, \tag{33}$$

an algebraic equation with integral coefficients. Similarily, the sums of the α's taken three at a time are the $_nC_3$ roots of

$$\theta_3(x) = 0. \tag{34}$$

Proceeding in the same way, we obtain

$$\theta_4(x) = 0, \ \theta_5(x) = 0, \ldots, \theta_n(x) = 0, \tag{35}$$

algebraic equations with integral coefficients, whose roots are the sums of the α's taken $4, 5, \ldots, n$ at a time respectively. The product equation

$$\theta_1(x)\theta_2(x) \cdots \theta_n(x) = 0, \tag{36}$$

has roots that are precisely the exponents in the expansion of (31).

The deletion of zero roots (if any) from equation (36) gives

$$\theta(x) = cx^r + c_1 x^{r-1} + \cdots + c_r = 0, \tag{37}$$

whose roots $\beta_1, \beta_2, \ldots, \beta_r$ are the non-vanishing exponents in the expansion of (31), and whose coefficients are integers. Hence (31) may be written in the form

$$e^{\beta_1} + e^{\beta_2} + \cdots + e^{\beta_r} + k = 0, \tag{38}$$

where k is a positive integer.

We define

$$f(x) = \frac{c^s x^{p-1} \{\theta(x)\}^p}{(p-1)!}, \tag{39}$$

where $s = rp - 1$, and p is a prime to be specified. Also, we define

$$F(x) = f(x) + f^{(1)}(x) + f^{(2)}(x) + \cdots + f^{(s+p+1)}(x), \tag{40}$$

noting, with thanks to Hurwitz, that the derivative of $e^{-x}F(x)$ is $-e^{-x}f(x)$. Hence we may write

$$e^{-x}F(x) - e^0 F(0) = \int_0^x -e^{-\xi} f(\xi)d\xi.$$

The substitution $\xi = \tau x$ produces

$$F(x) - e^x F(0) = -x \int_0^1 e^{(1-\tau)x} f(\tau x) d\tau.$$

Let x range over the values $\beta_1, \beta_2, \ldots, \beta_r$ and add the resulting equations. Using (38) we obtain

$$\sum_{j=1}^r F(\beta_j) + kF(0) = -\sum_{j=1}^r \beta_j \int_0^1 e^{(1-\tau)\beta_j} f(\tau \beta_j) d\tau. \tag{41}$$

This result gives the contradiction we desire. For we shall choose the prime p to make the left side a non-zero integer, and make the right side as small as we please.

By (39), we have

$$\sum_{j=1}^r f^{(t)} = 0, \quad \text{for} \quad 0 \le t < p.$$

Also by (39) the polynomial obtained by multiplying $f(x)$ by $(p-1)!$ has integral coefficients. Since the product of p consecutive positive integers is divisible by $p!$, the pth and higher derivatives of $(p-1)! f(x)$ are polynomials in x with integral coefficients divisible by $p!$. Hence the pth and higher derivatives of $f(x)$ are polynomials with integral coefficients, each of which is divisible by p. That each of these coefficients is also divisible by c^s is obvious from the definition (39). Thus we have shown that, for $t \ge p$, the quantity $f^{(t)}(\beta_j)$ is a polynomial in β_j of degree at most s, each of whose coefficients is divisible by pc^s. By (37), a symmetric function of $\beta_1, \beta_2, \ldots, \beta_r$ with integral coefficients and of degree at most s is an integer, *provided that* each coefficient is divisible by c^s (by the fundamental theorem on symmetric functions). Hence

$$\sum_{j=1}^r f^{(t)}(\beta_j) = pk_t, \quad (t = p, p+1, \ldots, p+s)$$

where the k_t are integers. It follows that

$$\sum_{j=1}^r F(\beta_j) = p \sum_{t=p}^{n+s} k_t.$$

In order to complete the proof that the left side of (41) is a non-zero integer, we now show that $kF(0)$ is an integer that is prime to p. From (39) it is clear that

$$f^{(t)}(0) = 0, \quad (t = 0, 1, \ldots, p-2)$$

$$f^{(p-1)}(0) = c^s c_r^p,$$

$$f^{(t)}(0) = pK_t, \quad (t = p, p+1, \ldots, p+s)$$

where the K_t are integers. If p is chosen greater than each of k, c, c_r (possible since the number of primes is infinite), the desired result follows from (40).

Finally, the right side of (41) equals

$$-\sum_{j=1}^{r} \frac{1}{c} \int_0^1 \frac{\left\{c^r \beta_j \theta(\tau \beta_j)\right\}^p}{(p-1)!} e^{(1-r)\beta_j} d\tau.$$

This is a finite sum, each term of which may be made as small as we wish by choosing p very large, because

$$\lim_{p \to \infty} \frac{\left\{c^r \beta_j \theta(\tau \beta_j)\right\}^p}{(p-1)!} = 0. \qquad \blacksquare$$

I Prefer Pi: Addenda

Jonathan Borwein and Scott Chapman

Abstract. In the rush to prepare our March 2015 article on Pi [**1**], several infelicities escaped our eye. Herein we repair the damage.

1. MATHEMATICAL CORRIGENDA. Lord Brouckner's continued fraction given in [**1**, (4)], and mentioned in the text above, should have been

$$\frac{4}{\pi} = 1 + \frac{1^2}{2} + \frac{3^2}{2} + \frac{5^2}{2} + \frac{7^2}{2} + \frac{9^2}{2} + \frac{11^2}{2} + \frac{13^2}{2} + \cdots,$$

and the corresponding identity in [**1**, (12)] should have been

$$\pi = 3 + \frac{1^2}{6} + \frac{3^2}{6} + \frac{5^2}{6} + \frac{7^2}{6} + \frac{9^2}{6} + \frac{11^2}{6} + \frac{13^2}{6} + \cdots.$$

2. BIBLIOGRAPHIC CORRIGENDA.

- References [33] and [74] omit the names of co-authors. The references should read as follows.

  33. R. Guy and Y. Matiyasevich, A new formula for π, *Amer. Math. Monthly* **93** (1986) 631–635.

  74. J. Sondow and H. Yi, New Wallis- and Catalan-type infinite products for p_1, e, and $\sqrt{(2 + \sqrt{(2)})}$, *Amer. Math. Monthly* **117** (2010) 912–917.

- The following paper, with 33 Google ciatations, was omitted from the regular bibliography and the special bibliography on p. 196. We offer our apologies to Professor Osler.

  T. J. Osler, The united Vieta's and Wallis's products for pi, *Amer. Math. Monthly* **106** (1999) 774–776.

ACKNOWLEDGMENT. The authors wish to thank the readers who drew our attention to these lacunae.

REFERENCE

1. J. M. Borwein, S. T. Chapman, I Prefer Pi: A Brief History and Anthology of Articles in the American Mathematical Monthly, *Amer. Math. Monthly* **122** (2015) 195–216.

http://dx.doi.org/10.4169/amer.math.monthly.122.8.800
MSC: Primary 01A99, Secondary 11Z05

Index

"0123456789" †, 331

A

Abel, H., 66, 89
"absolutely abnormal"†, 346
absolutely normal, 286, 343
absolutely strongly normal, 352
absolute normality, 434
Adamchik, V., 223
Adegoke, K., 334
Agarwal, R.P., 415
AGM, 87–89, 93, 95
Airy, 431
Akihiko Shibata, 161
Alexander, J., 417
Alexander Pope, 389
algebraic functions, 84
algebraic irrationals, 338
algebraic number, 106
algorithm, 15, 17, 18, 83, 95, 105, 262, 264,
     266–268, 449, 465, 485
algorithm, numerically stable, 16
al-Khwarizmi, 375
Almkvist, G., 311, 322
alternating series, 205
analytic functions, 91
analytic limit, 93, 94
Anaxagoras of Clazomanae, 369
Antiphon, 170
Apéry, R., 235
Apostol, A., 236
"April Fool's day"†, 472
Archimedes, 81, 82, 112, 329, 429, 439
Archimedes method, 107
Archimedes of Syracuse, 169, 370, 449, 479
arctangent formulas, 105
arctangent series, 112
Aristotle, 370
arithmetic-geometric mean, 3, 4, 11, 14–17,
     23, 31, 35, 84, 99, 106, 128, 129, 138,
     153, 154, 331, 460
arithmetic-geometric mean inequality, 84
arithmetic progression, 213
Aryabhatiya, 330
Aryabhatta, 330, 367
assembly language, 113
astronomy, 367
Aurefeuillian factors, 229
average normalized distance, 350

B

Babylonians, 83
Backhouse, N., 295, 298
Bailey, D.H., 155–157, 173, 307, 311, 313,
     323, 331, 335, 343, 353, 354, 409, 412,
     432, 487
Bakhshali manuscript, 330
Baruah, N.D., 317
Basel problem, 386, 389, 452, 484
Bauer, G., 321
"BBP" formula for π, 270, 329, 332, 333,
     334, 338, 354, 412, 433, 491
Beckmann, P., 82, 112
Bellard, F., 267, 413, 469
Belshaw, A., 352
Benford's principle, 291
Berndt, B.C., 167, 205, 306, 316, 317
Bernoulli, J., 29, 58
Bernoulli numbers, 200, 205, 314, 360,
     390, 452
Bernoulli polynomials, 484
Bessel functions, 338
Beukers, F., 235, 236
Bhargava, S., 306
Bhaskaracharya, 374
Bickford, N., 357, 487
binary algorithm for exponentiation,
     264, 332
binary exponent, 11
binary exponentiation, 466
binary exponentiation algorithm, 228
binary fraction, 11
binary representation, 11
binary scheme for exponentiation, 226
binary splitting, 12, 83, 95
BlueGene/P system, 337
Bonvein's quartic convergent formula, 153
Boole summation formula, 201, 202
Borchardt, C.W., 12
Borchardt's algorithm, 82
Borel, E., 105
Borel-Cantelli lemma, 336
Borodin, A., 117
Borwein, D., 333
Borwein, J.M., 106, 111, 112, 214,
     237, 295–297, 307, 312, 323, 333,
     343, 408
Borwein, P.B., 111, 214, 307, 312, 331,
     352, 408, 412, 432

---

* Italic for person, bold for definition, and dagger for quote.

© Springer International Publishing Switzerland 2016
D.H. Bailey, J.M. Borwein, *Pi: The Next Generation,*
DOI 10.1007/978-3-319-32377-0

Printed in the United States
By Bookmasters